D1829211

Sustainable Horticulture in Semiarid Dry Lands

Shrikant Hiwale

Sustainable Horticulture in Semiarid Dry Lands

Springer

Shrikant Hiwale
Fruit Crops
Central Horticultural Experiment Station
Vejalpur, Gujarat, India

ISBN 978-81-322-2243-9 ISBN 978-81-322-2244-6 (eBook)
DOI 10.1007/978-81-322-2244-6

Library of Congress Control Number: 2014960269

Springer New Delhi Heidelberg New York Dordrecht London
© Springer India 2015
This work is subject to copyright. All rights are reserved by the Publisher, whether the whole or part of the material is concerned, specifically the rights of translation, reprinting, reuse of illustrations, recitation, broadcasting, reproduction on microfilms or in any other physical way, and transmission or information storage and retrieval, electronic adaptation, computer software, or by similar or dissimilar methodology now known or hereafter developed.
The use of general descriptive names, registered names, trademarks, service marks, etc. in this publication does not imply, even in the absence of a specific statement, that such names are exempt from the relevant protective laws and regulations and therefore free for general use.
The publisher, the authors and the editors are safe to assume that the advice and information in this book are believed to be true and accurate at the date of publication. Neither the publisher nor the authors or the editors give a warranty, express or implied, with respect to the material contained herein or for any errors or omissions that may have been made.

Printed on acid-free paper

Springer (India) Pvt. Ltd. is part of Springer Science+Business Media (www.springer.com)

Foreword

Semiarid and arid zone areas account for almost 70 % of the total cropped area of the country. In these areas, physical constraints like low and erratic rainfall, high temperature, high wind velocity, low fertility, poor soil structure, and salinity of soil and ground water are the deterrents to assured crop production. In the absence of aggregation of any type, the soils are highly erodable, structureless and very coarse in texture with low water holding capacity. The ever-increasing population has put tremendous pressure on agricultural lands with the result that even the marginal and sub-marginal lands are put under cultivation. The net result is desertification and degradation of land, water and vegetation resources. Intensification of production of annual crops in rain-fed areas, if carried out in an unbalanced manner, as usually is the case, accelerates run-off and soil erosion and other degradation processes causing damage to the ecosystem. Resource conservation showed that the effect of intensive agricultural practices, increasing population pressure, climatic changes, environmental pollution, loss of biodiversity, soil erosion, salinisation and water depletion is all threatening the sustainability of agriculture. In view of the mounting demand for food and need for linking of enhancing food production with nutritional security, conservation of natural resources, enhancing farmer's income, employment generation and diversification of agriculture is necessary. Diversification in terms of fruits and vegetables, medicinal plants, flowers and agro forestry etc. contributed to the holistic approach of development.

Horticulture, particularly fruit trees, can play a major role in solving both the problem of nutrition and sustainability as fruits are a rich source of vitamins and minerals and apart from this are also known for their antioxidant properties. The fruit trees which are mostly deciduous types add lot of leaf litter to the soil which ultimately helps in improving the soil health. Apart from this, fruit trees are known to reduce soil erosion and reduce run-off thereby contributing to sustainability. They also provide fodder to the cattle and fuel wood to humans thus reducing the dependence of the farmer on trees saving them from cutting. The deep root system helps in extracting water from deeper layers thus helping in plant survival. The tree species also play a major role in purifying the environment as they are the known carbon sequesters. Fruit tree cultivation is assuming importance as a paying proposition even under rain-fed conditions with hardy fruit tree species helping in uplifting the farmer's economy. There is no scope to increase the land surface, and all the increase in productivity therefore has to be from the available land.

Apart from this, reducing post-harvest losses and value addition with technologies that require minimum infrastructure and can be practiced on farm will help to raise the income of the farmer leading to employment generation along with food and nutritional security.

Thus the book on sustainable horticulture under semiarid rain-fed conditions by S.S. Hiwale can help in increasing production per unit area making full use of the soil and water to the maximum under harsh climatic conditions of semiarid areas. It also results in improving the sustainability along with improving the availability of fruit which are vital for human health along with increasing the income of the small and marginal farmers.

Indian Council of Agricultural Research N.K. Krishna Kumar
New Delhi, India

Preface

Intensive agricultural policies, increasing population pressure, climatic changes, environmental pollution, loss of biodiversity, soil erosion, salinization and water depletion are all threatening the sustainability of agriculture. In view of the mounting demand for food and need for linking of enhancing food production with nutritional security, conservation of natural resources, enhancing farmer's income, employment generation and diversification of agriculture is necessary. Diversification in terms of fruits and vegetables, medicinal plants, flowers, agroforestry etc. contributed to the holistic approach of development.

Fruit crop are known since ages for their nutritive and medicinal value. The fruits are a rich source of vitamins and are known for their antioxidant value. Apart from this they also contain substantial quantity of minerals vital for human body.

Identification of superior varieties with high yield increased productivity even under semiarid rain-fed conditions on marginal degraded lands. The tree can beat the vagaries of monsoon, drought conditions and high temperature. There are a number of value added products that can be prepared and sold in the local as well as in overseas market. The demand for such products is on increase and around 20 t of aonla candy was produced last year in Gujarat itself. Thus, there is great potential for aonla cultivation in semiarid rain-fed areas.

The purpose of the book is to disseminate the technology developed in these underutilized fruit crops which are grown since ages which are of seedling origin. Due to seedling origin there is lot of variability in these fruit crops which needs to be exploited to identify superior types which can be cultivated on marginal lands under rain-fed conditions as the prime land is already used for feeding the teaming millions. I am sure the book will be of great use to researchers, horticulturists and growers.

Vejalpur, India — Shrikant Hiwale

Contents

Part II Crop Specific Production Technologies for Semiarid Rain Fed Areas

About the Author

Dr. Shrikant Hiwale holds a brilliant academic career. He started as a Scientist at the IIHR Bangalore, where he later became a Principal Scientist in Horticulture. At present he is serving as a Principal Scientist at CHES in Vejalpur. He has worked on various semiarid fruits, examining aspects like germplasm conservation, evaluation and characterization; standardization of agro techniques for successful cultivation of these fruits under rainfed conditions. He has also worked on fruit-tree-based cropping system for semiarid rainfed areas and successfully developed cropping system models. He worked as a Lead Center Principal Investigator for the NATP Project "Develop Sustainable Agri-Horti-Pastoral System on Marginal Lands". He is also Principal Investigator for the Project "Revolving Fund Scheme on Propagation Commercial Fruits", which is intended to supply high-quality planting material. He is at present working as a Principal Investigator on the DST project "Enhancement of livelihood of tribal farm households of Panchmahals district in Gujarat through agricultural diversification". He has also completed seven institutional projects and is now working on "value addition in semiarid fruits". He has to his credit three new varieties, one each in Ber, Aonla and Pomegranate, and wrote a book entitled *The Pomegranate*. Apart from this he has guided four M.Sc. Agri (Horti) students and one Ph.D. student. He has presented and published eleven lead papers at international/national symposium/seminars, seven research bulletins and five extension bulletins on various aspects of the cultivation of semiarid fruits, and has written ten book chapters on various aspects like breeding and agro techniques.

Abbreviations

@	At the rate
Anon.	Anonymous
B.C. ratio	Benefit cost ratio
C.D.	Critical difference
cm	Centimeter
cm^2	Square centimeter
Cv.	Cultivar
e	Coefficient of variation
e.g.	Example gratia (for example)
et al.	Et allii, and others
etc.	Etcetera, so on
Fig.	Figure
g.	Gram
GA	Gibberelic acid
Ha.	Hectare
Hr.	Hour
i.e.	That is (id est.)
Kg	Kilogram
km	Kilometer
m	Meter
Max.	Maximum
Mg	Milligram
Min.	Minimum
MT.	Metric ton
N	Nitrogen
No.	Number
NPK	Nitrogen, phosphorus and potash
NS	Non-significant
pH	Potential of hydrogen ion
ppm	Parts per million
pn	Photosynthesis
%	Percent
Q	Quintal
q/ha	Quintal per hectare
:	Ratio
Rs.	Rupee

Temp	Temperature
TSS	Total soluble solid
viz.	Videlicet (namely)
Wt.	Weight

1 Introduction

The semiarid tropics (SAT) extends over 2.1 billion ha and supports a large population. Ninety percent of the area and 99 % of the population are located in developing countries. Currently crop yield is low throughout the non-irrigated area of the semiarid tropics. Substantial increase in productivity is needed in the future to sustain population increase. India contains about 56 % of the SAT population on about 19 % of semiarid tropics. The area is predominantly marked by vertisols which are known to possess the best potential for increasing manifold. Though India has made significant stride in food production in the recent past, major emphasis was given on irrigated and resource-rich situations. However, in rainfed areas, the marginal farmers are still practicing the traditional way of farming and struggling below the poverty line. Therefore, this trend needs to be discouraged. Virtually, a sizable area (171 m ha) in our country is categorized as degraded lands. Such lands which are subjected to problems of water and wind erosion, salinity, alkalinity, acidity, water logging, gullied and ravine lands, shifting cultivation, etc. are either lying vacant or underutilized. Although these lands have their own biophysical farming constraints, they can be utilized efficiently on sustained basis for fulfilling the basic needs of food, fruit, fuel, fodder, fiber, fertilizer, etc. through horticulture-based farming systems by adopting practically feasible, economically viable, and eco-friendly technologies developed in the recent past.

In India tremendous pressure has been put on natural resources, owing to not only degradation of principal resources like land, water, and vegetation but also reduced per capita availability of land. At these places, the extent of land degradation has crossed such critical limits that it is difficult to bring it back to its original state. The unscientific cultivation of annual crops in rainfed areas, as usually is the case, accelerates runoff, soil loss, as well as other degradation processes causing fragile ecosystem. In India about 53.34 million tons of top soil (16.4 t/ha) containing 5.4 million tons of plant nutrients is lost annually due to agriculture and associated activities (Dhruva Narayan and Ram Babu 1983). The prime agricultural lands continue to be diverted to nonagricultural uses like industrialization and urbanization with the result that marginal rainfed lands are brought under cultivation with least attention on resource conservation. Out of 143 million hectares of arable land in India, nearly 74 % is rainfed, which contributes about 42 % of the total food production. However, soil moisture stress due to vagaries of weather, moderate to severe soil erosion, and poor soil fertility are the major limiting factors of low productivity in rainfed areas.

Trees are known for their conservative as well as reproductive nature. They provide not only fruits but fuel and fodder also. Apart from this, they are known to withstand vagaries of monsoon better than agricultural crops. High cost of earth structure construction and the fact that they are

S. Hiwale, *Sustainable Horticulture in Semiarid Dry Lands*,
DOI 10.1007/978-81-322-2244-6_1, © Springer India 2015

not well maintained so there is a greater need for biological means of conservation. Comparatively the establishment and maintenance of the agri-horti production system are moderate (Hiwale 2004). The conservation of basic resources on which production depends is the major contribution of agroforestry to sustainable land use (Young 1988). Very less work has been done to study the role of trees in reducing degradation under rainfed conditions with a farming system approach.

Fruit cultivation has so far been associated with fertile soil with assured irrigation, but presently the irrigated fertile land is already being used for meeting the food requirement of the ever-increasing population and industries. The marginal and degraded lands in arid and semi-arid regions of the country with practically no irrigation facilities are to be used for increasing fruit production with simultaneous need to meet the basic demand for food, fodder, and fuel. Hence, no time is to be lost in developing highly productive self-sustainable farming systems fully adapted to the precarious ecology of arid regions.

This seeks to achieve more productive, sustainable, and diversified output from the land than is possible with cropping system. The approach aims at optimizing the use of land, water, and vegetation in an integrated way and thus helps to alleviate drought and moderate floods, prevent soil erosion, improve water availability, and increase production of fruits, vegetables, fuels, and fodder on a sustained basis. Considering the above facts, horticulture-based composite farming systems were evaluated from 1985 to date to establish a viable model for rehabilitation and utilization of semiarid rainfed areas of western India.

References

Dhruv Narayan VV, Rambabu (1983) Estimation of soil erosion in India. Indian J Irrig Drain Engg 109(4):419–434

Hiwale SS (2004) Technical bulletin on "Develop sustainable Agri-silvi-Horti Production system for Marginal lands under arid conditions". Central Horticultural Experiment Station, Vejalpur, 389340

Young A (1988) Agroforestry in the control of soil erosion by water. Agroforestry Abstracts (172/3), pp 39–48

Part I

Sustainable Horticulture

2 Problems of Horticulture in Semiarid rain fed areas

Abstract

India has 56 % of the semiarid tropics population on 10 % of the semiarid tropics area. The major soils of semiarid tropics are vertisol which possess the best potential for increasing crop yield severalfold. The semiarid tropic is one of the poorest economic regions. There is increasing pressure on the land with the risk of soil degradation under existing practices. There is a high risk of soil loss with the use of traditional systems. The climate of semiarid region is marked by aridity index (0.2 P/ETP 0.50). FAO has based their classification on the growing period and indicated that the growing period of 96 days which is within the 75–119-day range is classified as semiarid. The rainfall is significantly less than the potential evapotranspiration throughout the year. Successful crop production depends on stored soil water in addition to rainfall. The area is marked by low and erratic rainfall, low soil organic matter, erosion, low inherent fertility, and frequent droughts which limit the scope for high horticultural productivity.

2.1 Introduction

The semiarid tropics extend to 2.1 billion ha and support a large population. The crop yields are low as most of the area is nonirrigated. Increased production of all crops will be needed in the future to sustain population increase. India has 56 % of the semiarid tropics population on 10 % of the semiarid tropics area. The major soils of semiarid tropics are vertisol which possess the best potential for increasing crop yield severalfold. Today's concern is not only to increase productivity but also to sustain it for a longer period. Vertisol and associated soil occupy 27 million ha of the area (Murthy et al. 1982). The environment of the area is drought prone and only one crop can be grown during the year. The semiarid tropic is one of the poorest economic regions. There is increasing pressure on the land with the risk of soil degradation under existing practices. There is a high risk of soil loss with the use of traditional systems. At Solapur, a land with a slope of 1–2 %, more than 50 cm of soil has been lost over the past 100 years (ICRIPDA, Hyderabad), resulting in reduced water storage capacity and lower production capacity of soil.

S. Hiwale, *Sustainable Horticulture in Semiarid Dry Lands*,
DOI 10.1007/978-81-322-2244-6_2, © Springer India 2015

The climate of semiarid region is marked by aridity index (0.2 P/ETP 0.50). FAO has based their classification on the growing period and indicated that the growing period of 96 days which is within the 75–119-day range is classified as semiarid. The rainfall is significantly less than the potential evapotranspiration throughout the year. Successful crop production depends on stored soil water in addition to rainfall. The area is marked by low and erratic rainfall, low soil organic matter, erosion, low inherent fertility, and frequent droughts which limit the scope for high horticultural productivity.

In such a scenario, horticulture particularly fruit trees can play a major role in increasing the production potential of these areas; these conditions greatly favor the production of high-quality fruits in a number of fruit crops such as date palm, ber, pomegranate, aonla, sapota, mango, guava, and many underutilized fruits and vegetables such as cucurbitaceous crops, spice, and some medicinal plants. The existing low productivity level can be increased by adopting suitable soil and water management technologies and inputs. It is now realized that there is limited scope for quantum jump in fruit and vegetable production in traditional areas. As the prime land is already occupied, the only scope is to increase area under arid and semiarid climate. This will not only help in increasing the production of fruits and vegetables but also help in the amelioration of environmental conditions in the area as trees are known as carbon sinks acting as environmental protectors. The recent awareness regarding the potential of these ecologically fragile lands for the production of quality horticultural produce has not only opened scope for providing economic sustenance for the people of the region but also for bringing new areas to increase the production.

2.2 Semiarid Region

The arid zone fruits are very poor in fertility. The soils of the northwestern arid region described as "desert soils" and "gray brown soils" of order arid soils are light textured. The solum of the soils is moderately calcareous (0.29 % $CaCo_3$), and below this solum at the depth of 40–120 cm, a sharply differentiated zone rich in alkaline earth carbonates (5–45 % $CaCo_3$) is present in the form of hard crystal-like concentration, which may be many-meter thick. Most of the arid areas (64.6 %) are duny where soils often contain 3.2–4 % clay and 1. 4–1.8 % silt. The brown light soils occupy 1.7 % area, which has loamy fine sand to fine sandy loam on the surface and a heavier subsoil underlain with calcium carbonate concentration. Besides this about 5.9 % area is covered by soils having hardpan, 5.6 % under hills and pediments, 6.8 % alluvial duny, and 1.6 % seirozome extending from the soils of Haryana and Punjab. In Gujarat also, gray brown soils are widespread besides a large area having deltaic alluvium with small area in Kutch having deep black and medium black soils. In the peninsular India, a considerable part of arid region has red sandy soils and some parts having mixed black soil. Burford and Virmani (1983) reported that the semiarid tropic extends over 2.1 billion ha., and supports a large population. Vertisols and associated soils occure extensively in India (72 million ha.), (Murthy et al. 1982).

The soils are poor in organic matter 0.03–0.1 % in stabilized dunes. Soil are rich in total potassium (8,250–18,980 ppm) and boron (1.9–12.2 ppm) but are low in nitrogen (150–185 ppm), phosphorus (285 ppm), and micronutrients such as copper (10–11 ppm), zinc (2.07 ppm), and iron (20 ppm) The soils often have high salinity.

The groundwater resources are not only limited owing to poor surface and subsurface drainage but are also highly saline. The depth of water ranges from 10 m to as high as 140 m. The other irrigation water resources are seasonal rivers and rivulets, surface wells, and some runoff water storage devices, e.g., tanks and khadins. Thus, the water resources in arid region are limited and can irrigate hardly 4 % of the area.

The mean annual rainfall in the Indian arid zone is very low and varies from 100 mm in northwestern sector of Jaisalmer to 450 mm in the eastern boundary of Rajasthan. In Gujarat, it varies from 300 to 500 mm and in Haryana and Punjab from 200 to 400 mm. In peninsular India, the rainfall varied from 520 mm in Bellary (Karnataka) to 748 mm in Cuddapah (Andhra

Table 2.1 Moisture balance scenario in semiarid region

States	Rainfall (mm)	PE (mm)	Moisture index
Rajasthan	100–450	2,063–1,503	−70–91
Gujarat	30–500	2,144–1,700	−79–82
Punjab/ Haryana	200–400	1,656–1,362	−59–73
Peninsular India	520–750	1,859–1,738	−60–71

Pradesh). Most of the precipitation in western arid region occurs during July to September in about 19–21 days. The moisture balance scenario of the region is given in Table 2.1.

2.3 Present Scenario

Although, India is the second largest producer of fruits in the world, the availability of fruits and vegetables to the people in arid region is very low. Considering the present trend of population growth to meet the minimum requirement of the population, there is a need to increase production of fruits to the tune of 6 million tons. Although the current scenario indicates increasing trend, the pace is slow.

In semiarid region, the area under fruit cultivation increased from 50,000 ha in 1984–1985 to about 66,000 ha in 1993–1994. Considerable area has come under fruits like ber, pomegranate, and aonla. Ber has spread from northern states to western and southern states from mere 12,000 ha in 1978 to nearly 70,000 ha in 1994–1995 with a production of 0.87 million tons. Similarly the area under pomegranate has also increased to 35,000 ha.

A good deal of germplasm of semiarid fruits is available in different parts of the country in the form of land races and primitive cultivars; many of them are mainly propagated by seed over the years generating a lot of variability, e.g., ber, bael, aonla, jamun, karonda, phalsa, jungle jalebi, pomegranate, custard apple, fig, etc. The fruits like ber, aonla, and phalsa are indigenous, whereas dates, pomegranate, and custard apple are exotic but owing to seed propagation over the years created lot of variability in several pockets of the country which needs to be exploited through intensive research.

Agriculture in India is valued at 252 billion US$ constituting 14 % of GDP. In the present era of globalization, several changes in Indian economy are taking place. Population growth, pollution of aquatic and terrestrial systems, food shortages, and industrial and infrastructural activities, singly or in combination, are going to impact Indian agriculture. To meet the challenge of food shortage, there is a need to narrow the gap between the potential crop yield and actual yield. The present world population of six billion is expected to reach to eight billion by 2025. Most of the population increase will be seen in developing countries. Projections made by FAO in world agriculture toward 2015–2030 have clearly indicated that nations with low income and with high dependence on agriculture in which food production determines the progress of food security and sustainability will remain under threat from the rural poor. With the increasing population, a great majority will depend upon agriculture for their livelihood which will put pressure on the already dwindling natural resources.

Agriculture intensification in a developing country like India with shrinking land base (per capita arable land in India will reduce to 0.08 ha for feeding the projected population 1.4 billion in 2025). The success of green revolution in transforming India from "sheep and mouth" like situ-

		Production "000" tons		Requirement "000" tons		
States	Population million 1996	1983–1984	1994–1995	Current	2001	2020
Rajasthan	17.59	22.4	26.4	546	617	963
Gujarat	4.96	81.6	105.6	256	174	278
Punjab	4.84	34.0	54.0	150	170	272
Haryana	3.56	41.0	48.8	111	125	200
Peninsular region	5.70	400.0	490.0	156	200	320
Total	36.65	679.0	724.8	1,117	1,286	2,033

ation of the mid sixties to over flowing granaries in 90s is the result. However, it has given rise to the second-generation problem like stagnation of productivity per unit area, declining factor productivity of inputs, and deterioration of the soil water and environmental quality resulting in widening socioeconomic disparities within the society. Unsustainable use of water has given rise to the problems like waterlogging, salinization, and declining water table in wells and tube wells in irrigation areas. There is a need to narrow the gap between the potential crop production and actual yield with rationalized use of natural resources (Table 2.2).

It is difficult to quantify these effects in isolation as they are interdependent and complex interaction. The advantage of cropping system with trees as one of the components cannot be quantified simply in terms of productivity because there are some benefits which help in the improvement of the environment also, e.g., given erosion control and increase organic matter.

Overall productivity (biomass of cropping system) is generally greater than that of the annual system which is due to more growth resources, e.g., given water and light or due to improved soil fertility transforming the system into sustainable landscape.

Scale	Direct impact	Indirect effect
Farming system	Food production	Soil organic matter
	Erosion control	Soil retention
	Nutrient cycling	Soil nutrients
	Microclimate regulation	Shady low ___
	Water cycling	
Water shade	Erosion and sediment control	Soil retention
	Water cycling	Water content
General	Gas regulation	Reduction of harms
	Carbon sequestration	Biomass production
	Climate regulation	Reduce environmental degradation
	Overcoming drought	
	Overcoming moisture shortage at critical crop growth stages	
	Rural poverty elevation	Improved living standards and increase incomes

Table 2.2 Effects of tree crops intervention

	Direct effect	Indirect effect
Increased productivity	Positive	Positive (Ong et al. 1991)
Improved soil fertility	Positive	Positive (Kang et al. 1990)
Nutrient cycling	Positive	Positive (Szott et al. 1991)
Soil conservation	Positive	Positive (Lal 1989b; Wallace et al. 1991)
Microclimate improvement	Positive	Positive (Monteite et al. 1991)
Competition	Negative	Negative (Ong et al. 1991)
Allopathic	0	Negative (Tian and Keng 1994)
Weed control	0	Positive (Rizvi 1991)
Sustainability and stability	0	Positive (Ong et al. 1991)
Pest and diseases	0	Negative and positive (Zhao 1991)

2.4 Soil Improvement

The potential of microsite enrichment by trees is an important aspect of the cropping system. Long-term process of capture of precipitates or a nutrient-rich litter, and the recent alley cropping system approach with nitrogen-fixing fast-growing trees in humid tropics, has substantially improved soil fertility in 2–3 years (Kang et al. 1990). The tree component produces large quantity of biomass and can be used to add leaf litter to the soil as they are deciduous in nature in the semiarid tropics (Hiwale 2004). Micro pores are quite common in soils under Agroforestry resulted in enhanced faunal activity (Lavelle et al. 1992).

Contour hedgerow has been highly effective in controlling soil erosion (Lal 1991). The woody hedgerow of trees provide semipermeable barrier to the surface movement of water, while mulch from trees reduces the impact of raindrops on the soil and minimizes splash and sheet erosion (Young 1989). Mulching also acts as means of reducing soil evaporation and improving microclimate condition. Physical properties of soil in terms of infiltration rate and bulk density and soil moisture holding capacity. Microclimate improved through shading.

Productivity improved when shading is about 37 % (Yu et al. 1991).

2.5 Competition

Tree and crops when grown in close proximity, below ground competition is not apparent. It is complicated due to proliferation of roots into nearby plots or due to effect of shading specially with tall trees (Huxley et al. 1989; Rao et al. 1991) One of the measures of the consequences of competition of crop yield can be done through land equivalent ratio LER = yield of intercrop + tree crop yield/sole crop yield. When LER is one, there is no advantage to intercropping; when LER is less than one more, land is required to produce given yield by each component. If LER values are exceeding one, then the mixtures are productive than sole crops (Ofori and Stern 1987). Tree components are dominant and account for higher LER. There is competition between species for the finite resources (light, water, nutrition, CO_2). The maximum attainable biomass for individual species depends primarily on the availability of light water and nutrients. To increase productivity, further crops must either capture more of these resources or use them more efficiently. In semiarid areas, water is the limiting factor or nutritional availability rather than light. Trees usually put forth growth on the onset of monsoon and also able to capture more resources due to extensive root system. It is possible to improve productivity by using species mixture if component species capture more of the available resources or use them more efficiently compared to sole crop.

2.6 Principles of Resource Capture

Capture of limited resources (light, water, and nutrition) depends on the number, surface area distribution, and effectiveness of individual element within canopy or root system of the species mixture involved. The conversion coefficient expressed in terms of total intercepted shortwave radiation is generally around 1.0 g m^2 for C_3 spp and 2.0 g m^2 for C_4 spp. Mixed cropping systems often have considerable scope for improving seasonal water use in the semiarid tropics. Intercropping increases the total harvestable produce by using available water more efficiently. Improved rainfall use efficiency through increased transpiration, i.e., dry matter production per unit area of water transpired.

2.6.1 Direct Effects

The traditional cropping system in the semiarid tropics often uses less than half the rainfall input since there is substantial losses of water via soil evaporation, runoff, and drainage. Direct soil evaporation can account for 30–60 % of the annual rainfall (Cooper et al. 1983; Wallace et al. 1991) in the Middle East and West Africa. In India the best cropping system only uses 40 % of the annual rainfall while the rest is lost as runoff (26 %) and deep percolation (33 %) (Ong et al. 1991). The drainage losses can be reduced by the presence of trees which are able to take the water from deeper layers due to extensive root system even after cessation of rainfall (Huda and Ong 1989). Reduction in runoff losses can be achieved in agroforestry system particularly on sloping land (Young 1989).

2.6.2 Indirect Effects

Improved Soil Condition and Microclimate

The large tree canopy modifies microclimate improving water use efficiency.

1. Shading the ground results in reduction of soil evaporation.
2. Reduction in vapor pressure deficit as a result of transpiration from the canopy (Monteith 1988) proved that transpiration efficiency is inversely proportional to the vapor pressure deficit of the air, i.e., plant uses more water to fix given amount of carbon when the air is dry.

3. Shading reduces soil and air temperature and gives beneficial effect to the crop growth. The exposure of large area of the soil in semiarid tropics may produce hot environment. Trees may bring the system closer to optimum temperature.

Most annual cropping systems use 30–33 % of the total rainfall because much of these is lost by soil evaporation from sowing until the canopy closes, as a runoff, or is left as residual moisture at final harvest. There is evidence that intercrops may make better use of available soil moisture than the shorter duration sole crops. Cropping systems may be the more effective in increasing the proportion of rainfall used for transpiration and hence overall productivity.

Light interception increases with increasing age of the plant and total light interception of intercropping is higher than the sole crops which shows spatial complementarity between the species. Morris and Garrity (1993) reported that there is a linear relationship resulting in apparent advantage of 30–40 %. Complementarity higher yield can be demonstrated by comparison with sole stand.

2.7 Mechanism Responsible for Increased Yield

Mixed cropping system offers considerable scope for improving seasonal water use in the semiarid tropics. Intercropping results in better use of available soil moisture than short-duration sole crop. Incorporation of trees in the system helps in more effective use of rainfall for transpiration and hence productivity. The tree component has an access to the below ground resources as they have well-established root system which is also true in case of light capture because of the age and size has considerable advantage in harnessing resources from a larger area both below ground as well as above ground, enhancing the soil physical and chemical properties under the canopy (Kessler 1992; Belsky et al. 1993). The organic matter content, total NPK, and Ca concentration were significantly higher and bulk density is significantly lower under tree canopies. Total transpiration is determined indirectly from the balance of all the components: precipitation, interception, runoff, soil surface evaporation, deep drainage, dry matter production, transpiration, and water use ratio (WUR) in line-planted and dispersed trees.

2.8 Water Balance of a Mixed Tree Crop System

Inclusion of trees or shrubs improves rainfall use efficiency either directly, by more rain being used for transpiration, or indirectly by improved transpiration efficiency, i.e., more dry matter production per unit of water transpired. Successful intercropping is supposed to be those which make better use of resources.

The gross precipitation is first intercepted by the trees and crop canopies giving rise to interception loss from the trees. The presence of plant canopy modifies the rainfall so that the input to the ground beneath the tree is different from beneath the crop. This input of water to the ground infiltrates at a different rate below the trees, producing different rates of surface runoff.

Water evaporates directly from the soil surface. The water content of soil zones beneath the trees and the crop is different due to surface input and transpiration rate particularly when the soil is saturated for a significant amount of time.

The presence of trees modifies the rainfall pattern, amount, and energy of the rainfall input to the ground. It reduces by 15 % to that of open soil surface. The rainfall increased on the windward side and reduced on leeward side. It also modifies the amount and spatial distribution of the rainfall. The kinetic energy of the rainfall is also modified by the tree canopy, by changing the size and velocity of raindrops. This effect is important in relation to soil erosion; during rainstorms, the size of raindrop is big which splits to smaller size when it comes in contact with the tree resulting in reduced kinetic energy. On the other hand, low-intensity storms have raindrops which are small in size; tree canopies can amalgamate drops so that drops from the canopy can be greater than that of the incident rainfall. Winchester and Smith (1978) reported that the most widely used USLE system for estimating the soil loss (Universal Soil Loss Equation) the standard plot size is 1.8×22 m. Larger leaf forms larger drop than small leaf. So tree selected to prevent soil erosion should have smaller leaves. Management of tree shaped by pruning is another option to reduce erosion. Calder and Rosier (1976) developed the design for large plastic sheet net rainfall gauges for measurement of rainfall.

2.8.1 Rainfall Interception

When precipitation falls, some of it is held for sometime on the plant canopy; if the storm is very light or if rainfall persists in excess of evaporation, water builds upon the canopy to the point where it will begin to drop off the leaves and run off from plant stems or fall directly on the soil as drops, a process called interception. The Rutter model which has been shown to work well generally with high leaf area index (Calder 1976; Gash and Mortan 1978).

Gash (1979) model requires less data which is based on the knowledge or the availability of the vegetation to store water in the canopy and average rate of rainfall and evaporation from wet canopy. These models do not work well in space canopies. Gash et al. (1995) modified the equation and got better results than the original model. The main rainfall rate used in the model was 2.3 mm/ha which was calculated from 642 h rainfall data from Machecos in 1992–1993. The main evaporation rate during the rainfall was assumed to be 0.2 mm/ha. It is predicted that the annual interception loss is likely to be between 3 and 10 % of rainfall.

2.8.2 Infiltration and Runoff

When rainfall reaches the soil surface, some of it infiltrate into the soil. If the rainfall is greater than the infiltration rate, the excess water starts to collect at the surface; when there is an excess of surface storage, runoff occurs. Infiltration therefore is the dynamic process that changes during the rainstorm depending on soil characteristic, slope of land, and rainfall intensity. Hoogmoeid et al. (1991) using rainfall simulator on cultivated millet fields in Niger reported that the initial infiltration rate exceeds 100 mm/ha but it drops in the final infiltration rate to 30 mm/ha. The difference between infiltration and the rainfall rate gives the runoff rate. Undisturbed soil has much lower infiltration rate than soil which has tillage. Tilling the soil has also altered the surface storage, there is the delay of 18 min in surface runoff in recently tilled soil. On slopes of 7 %, the presence of *Leucaena* and *Gliricidia* hedgerows reduced runoff by 64 and 80 % compared to sole annual crop on plow-tilled land (Lal 1989a), thus indicating a role of land preparation on infiltration and runoff.

Soil characteristics that affect infiltration are surface crust, surface storage, saturated hydraulic conductivity, and presence or absence of plant residues. Surface crustation had very marked effect on infiltration. Even in sandy soil, crust forms and generates significant runoff. Roose and Bertrand (1971) and Hoogmoeid and Strossnidger (1984) reported runoff between 16 and 25 % of rainfall in sandy soil with slope less than 3 %. Vegetation cover generally increases in infiltration rate and reduces runoff. In Senegal runoff decreased from 456 mm in bare soil to 264 mm in cultivated soil and to 200 mm in fallow land containing mixture of shrub (Lal 1991). Vegetation affects the surface infiltration by two ways by canopy modification and may alter soil particle detachment and crust formation. The second effect is via reduction in surface crusting and improved soil hydraulic conductivity as a result of incorporation of plant residues in the soil (Keipe and Rao 1994). Mulching is widely used in the tropics for conserving soil and reducing soil erosion (Stigter 1984). An additional advantage of mulching is the increase activity of soil fauna, further improving the soil structure (Lavelle et al. 1992). Planting of tree hedgerows on the contour of slopping land can have effect of forming natural terrace as water and soil are collected on the slope side of hedgerows (Lal 1989a, b; Young 1989; Keipe and Rao 1994).

The measurement of infiltration can be done by double-ring infiltrometers (Lal 1989b). One dimensional flow in the inner ring, while the rate of fall of the water in the outer ring gives the infiltration rate (FAO 1979). Rate of infiltration is affected by changes in soil surface brought by rainfall. Runoff can be measured by collecting the water from a given area which may range in size of plot less than 1 m^2 to the large catchments. The ULSE (Universal Soil Loss Equation), Winchester and Smith in 1978, standardized plot area 1.8 m×22 m (0.04 ha) and are usually large enough to give information on combined effect of surface dynamic of land flow process.

2.8.3 Soil Water and Drainage

The amount of water that enters the soil is the consequence of the combined effect of rainfall, infiltration, and runoff. Once within the soil matrix, water can be reevaporated from the surface taken up by plant roots and under conditions of excess water can flow laterally and leave the bottom of soil profile. Surface soil water content influences infiltration, runoff, and evaporation. The water content in deeper layers affects water movement and drainage. Soil water is also key parameter in determining plant growth and nutrition uptake and transpiration. In cropping systems, differences in soil water content beneath the trees and the cropped area can arise due to differences in infiltration and runoff under trees and crops. Drainage pattern may also be altered due to greater abstraction by tree roots and utilization out of cropping season by trees. McGowan (1974) used time series of soil water content to infer drainage. There is a point in the soil profile in which water moves upward and then moves downward which is known as zero flux plane (ZFP). Once this is identified, water content above the ZFP can be ascribed to evaporation and water content changes below the ZFP to drainage. In tropics, agricultural crops which do not cover the ground for sometime or all the season direct soil evaporation can be significant component in the semiarid region of Middle East and West Africa. Soil evaporation can be up to 30–60 % of rainfall (Cooper et al. 1983; Wallace et al. 1991).

To date alley cropping or hedgerow cropping was found to be the most widely used system wherein intercrops are grown by adopting pruning or loping and adding the leaf litter to the soil below. Hariah et al. (1992) inferred that reduced stem height after pruning reduced the chances of survival and maintenance of apical dominance. However, there is an effect of shade to which some species are highly susceptible; apart from this, above-ground effect of canopy competition for water and nutrition needs to be taken care of.

Moreover, below-ground competition affects the growth of intercrop through root growth and absorption of water and nutrients. Tree density definably has the effect on the growth of intercrops; therefore, low density will help in reducing the competition effect as well as shading and root interaction. The crop selection is important as the crop phonology affects the competition for resources. If it is different, then the requirement will also be different. This will result in temporal complementarity. Spatial complementarity has a greater scope in agroforestry than in intercropping because the trees have deeper root system than the associated crops. Dhany et al. (1990) reported that most of the fast growing trees are likely to compete with field crops. With the use of sap flow technology, it is indicated that 80 % of transported water may be drawn from below-ground (Howard et al. 1995).

With the adoption of fruit tree-based system of suitable soil and water management practices, it is possible to increase the production potential of rainfed areas and also ensure the sustainability of the soil by improving soil health in semiarid areas. The system also helps provide fodder and fuel to the farming families apart from providing vital minerals and vitamins and increasing the meager income of the farmers of the region.

References

Belsky AJ, Mwonga SM, Duxbury JM (1993) Effect of widely spaced trees and live stock grazing on understory environments in tropical savannas. Agrofor Syst 24:1–20

Burford JR, Virmani SM (1983) Improved farming systems for vertisols in semiarid tropics. Annual report, Farming systems research ICRISAT, Hydrabad, pp 17–21

Calder IR (1976) The measurement of water losses from the forested area using a natural lysimeter. J Hydrol 30:311–325

Calder IR, Rosier PTW (1976) The design of large plastic sheet net rainfall gauges. J Hydrol. 30:403–405.

Cooper PJM, Ketinge JDH, Hughes G (1983) Crop evapotranspiration – a technique for calculation of its components by field measurements Field. Crop Res 7:299–312

Dhany SK, Narayan P, Singh RK (1990) Studies on root distribution of five multipurpose tree species in Doon valley, India. Agrofor Syst 12:147–161

FAO (1979) Soil survey investigation for irrigation. FAO, Rome

Gash JHC (1979) An analytical model of rainfall interpretation by forests. Q J R Metrol Soc 105:43–55

Gash JHC, Mortan AJ (1978) An application of the Rutter model to the estimation of interception loss from Thetford forest. J Hydrol 38:49–58

Gash JHC, Lloyd CR, Lachaud G (1995) Estimating sparse forest rainfall interception loss with an analytical model. J Hydrol 170:79–86

Hariah K, van Noordwick M, Santoso B, Syekhfani MS (1992) Biomass production and root distribution of eight tree and their potential for hedgerow intercropping on an ultisol in Lampung. Agrivita 15:54–68

Hiwale SS (2004) Technical bulletin on "Develop sustainable Agri-Horti production system under rainfed conditions on marginal lands, pp 1–60

Hoogmoeid WB, Strossnidger L (1984) Crust formation on sandy soils in the sahel. I. Rainfall infiltration. Soil Tillage Res 44:5–23

Hoogmoeid WB, Klaij MC, Brouwer (1991) Infiltration, run off and drainage in the Sudano-Sahelian zone. In: Sivakumar MVK, Wallace JS, Renard C, Giroux, C (eds) Soil and water balance in the Sudano-Sahelian zone. Proceeding international workshop Niamey Niger, Feb 1991. IAHS Pub. Edits

Howard SB, Ong CK, Rao MR, Mathura M, Black CR (1995) Partitioning of light and water in Leucaena/maize agroforestry systems. In: Sinoquet H, Cruz P (eds) The ecophysiology of tropical intercropping. INRA Editions, Paris, pp 123–135

Huda AKS, Ong CK (1989) Crop simulation model and some implications for agro forestry systems. In: Proceedings of ICRAF/WMO/UNEP workshop on application meteorology to agro forestry system s planning and management, ICRAF, Nairobi, pp 115–124

Huxley PA, Akunda EE, Piney A, Darnhofer T, Gatama D (1989) Tree crop interface investigations, preliminary results with *Cassia siamea* and maize. In: Metrology and agroforestry. ICRAF, Nairobi, pp 361–370

Kang BT, Reynolds L, Atta-krah AN (1990) Alley farming. Adv Agron 43:315–359

Keipe P, Rao MR (1994) Management of agro forestry for the conservation and utilization of water and water resources. Outlook Agric 23:17–25

Kessler JJ (1992) The influence of karaite (*Vitellaria paradoxa*) and ne're' (*Parkia biglobosa*) trees on sorghum production in Burkina Faso. Agrofor Syst 17:97–118

Lal R (1989a) Agro forestry systems and soil surface management of tropical alfisols II. Water runoff, soil erosion and nutrient loss. Agrofor Syst 8:97–111

Lal R (1989b) Agro forestry systems and soil surface management of tropical alfisols V. Water infiltrability, transitivity and soil water sorptivity. Agrofor Syst 8:217–238

Lal R (1991) Current research on crop water balance and implications for the future. In: Proceedings of international workshop on soil and water balance in the Sudano- Sahelian zone, Niamey, Niger, February 1991

Lavelle P, Spain AV, Blanchart E, Martin A, Martin PS (1992) The impact of soil fauna on the properties of soils in the humid tropics. In: Myths and science of soils of the tropics. Soil Science Society of America (Special publication 29), Maidson, pp 157–185

McGowan M (1974) Depths of water extraction by roots. In: Isotopes and radiation techniques in soil physics and irrigation studies. International Atomic Energy Agency, Vienna, pp 435–445

Monteith JL (1988) Steps in crop climatology. In: Unger PW, Sneed TV, Jordan WR, Jenson R (eds) Challenges in dry land agriculture global perspective. Proceeding of international conference on dry land farming, Texas Agricultural Experiment Station, Texas, USA, August 1988, pp 275–282

Monteite JL, Ong CK, Corlett JE (1991) Micronutriets interactions in agroforestry systems. For Ecol Manage 45:31–44

Morris RA, Garrity DR (1993) Resource capture and utilization in intercropping, water. Field Crop Res 34:303–317

Murthy RS, Bhattacharjee TC, Lande RJ, Pofali RM (1982) Distribution, characterization and classification of vertisol. In: 2th international congress of soil science, New Delhi, 8–16 February 1982, pp 3–22

Ofori F, Stern WR (1987) Cereal legume intercropping systems. Adv Agron 41:41–90

Ong CK, Odongo JCW, Marshall F, Black CR (1991) Water use by tree crops, five hypothesis. Agrofor Today 5(4):2–5

Rao MR, Sharma MM, Ong CK (1991) A tree crop interface design and its use for evaluating the potential of hedgerow intercropping. Agrofor Syst 13:143–158

Rizvi SJH (ed) (1991) Allelopathy basic and applied aspects. Chapman and Hall, London

Roose JC, Bertrand R (1971) Contribution a' l'eutde de la Methode des bandes d' arrert pour contre perosion lutte contre l'erosion hydrique en Afrique de pouest. Agronomic Tropicale 20:1059–1064

Stigter CJ (1984) Mulching as a traditional method of microclimate management. Arch Meteorol Geophys Bioclimatol Ser B 35:147–154

Szott IT, Fernandes ECM, Sanchez P (1991) Soil plant interactions in agroforestry systems. For Ecol Manage 45:127–132

Tian G, Kang BT (1994) Evaluation of phytotoxic effects of Glyricidia sepium (Jacq) Walp. prunigs on maize and cowpea seedlings. Agrofor Syst 26:249–254

Wallace JS, Batchlor CH, Daabeesing DN, Teeluck M, Soopramnian GC (1991) A comparison of light interception and water us of plant and first ratoon sugar cane intercropped with maize. Agric For Metrol 57:85–105

Winchester WH, Smith DD (1978) Predicting rainfall erosion losses-a guide in conservation planning. USDA Agricultural Handbook no. 537. USDA, Washington DC

Young A (1989) Agro forestry for soil conservation. CAB International, Wallingford

Yu S, Wang S, Wei P, Zhu Z, Lu X, Fang Y (1991) A study of paulownia/tea intercropping system – microclimate modification and economic benefits. Agroforestry systems in China. CAF-DRC, Singapore, pp 150–161

Zhao L (1991) Biological control of insect pests and plant diseases in agroforestry systems. In: Avery ME, Cannel MGR, Ong CK (eds) Biophysical research in Asian agroforestry. Winrock International, Arlington, pp 73–90

3 Scope of Horticulture

Abstract

Dry land horticulture has a special significance, especially when there is a renewed emphasis on increasing the productivity of marginal lands. At present, the productivity of these lands is very low as most of these lands are located in either semiarid or arid areas. Most of these lands are not fit for crop farming, and crop farming on such lands is also very risky and is not remunerative. The cultivation of fruit crops in this region with conventional methods is not going to be a paying preposition. It is only by adopting farming system approach the cultivation of fruit crops can be successful in these regions. Ecoregional approach will have to aim at enhancing agricultural productivity and production on a sustainable basis, to meet the ever growing needs of farm family and livestock for food, fuel, and fodder. This would imply an up scaling of research activities within the ecoregions. Research and management of natural resources and a balance in the development and utilization of biodiversity would be important. The research should aim at improving the productivity of scarce resources while protecting the quality of soil and water and at the same time safeguarding biodiversity for posterity. Thus there is tremendous scope for horticulture in semiarid areas.

The scope of the semiarid horticulture depends on the adaptability of the trees to the adverse climatic conditions prevalent in the area. There are many underutilized fruit trees found growing in the fruit area. Future scope of expansion depends on the availability of inputs, market, and industrial support. Many of these fruits are known for their vitamins and mineral composition, and their ability to withstand droughts is unquestionable that is why probably these species are surviving in these areas over the decades. To bridge the gap between demand and supply, there is a need to increase the area under these fruits so that the basic minimum requirement of the public could be fulfilled. Results of the research show that fruit production can be doubled by adopting suitable crop species, variety, and improved orchard management.

S. Hiwale, *Sustainable Horticulture in Semiarid Dry Lands*,
DOI 10.1007/978-81-322-2244-6_3, © Springer India 2015

At present semiarid zones are marked by monocropping which is leading to low yields and pest buildup resulting sometimes in total failure of crops. This situation results in unstable income giving rise to poverty and debt trap. With the availability of land decreasing at a faster rate and use of prime land for raising food and cash crops like cotton, there is scope to increase the area on marginal degraded land; here no successful agronomical crops can be raised. There are certain fruit trees which once established can survive under the semiarid rainfed areas. It is proved beyond doubt at CHES Vejalpur that more than 30 fruit tree species can be grown successfully. Fruit crops like ber, aonla, sapota, mango, pomegranate, custard apple, wood apple, bael, jamun, karonda, chironji, tamarind, guava, and fig which are considered as underutilized can be highly profitable even under rainfed conditions in this region. The sick saline soils where no agronomical crops can be grown can be put to use by growing these fruits. The fruit trees are suitable for cultivation in rainfed areas which complete their reproductive phase well before the onset of stress period. Despite best efforts to bring large area under irrigation, only 40 % of area could be brought under permanent irrigation.

Adoption of cropping systems like agri-horti, horti-silvi, and horti-pasture can not only improve sustainability of degraded lands by addition of large biomass to the soil resulting in improved soil health. Trees also help in reducing runoff and soil loss. They also help in improving the environment by sequestering CO_2 from the atmosphere.

Malnutrition in these areas is a major problem particularly in pregnant woman and children. Fruits like aonla (Indian goose berry) and ber (Indian jujube) are a rich source of vitamin C. Ber contains more protein than even apple. Fruits like date palm, wood apple, and custard apple are rich in minerals which are vital for the maintenance of body.

Already there is revolution in these fruits, and with the identification suitable technologies and superior varieties, the area under these crops has increased tremendously. The semiarid tropics (SAT) extend over 2.1 billion ha and support a large population – 750 million in 1980. Ninety percent of this area and 99 % of this population are located in developing countries. India contains about 56 % of the SAT population on about 10 % of the SAT area. Although these fruits are highly perishable in nature, the marketing of which is a major problem, e.g., custard apple ripens in 2–3 days of harvesting and if not consumed it gets spoiled. Similarly the early variety of ber Gola fruits harvested in the morning become unmarketable in the very next day. Also with the glut in the market, the prices of these fruits drop drastically making it uneconomical for the farmers to sustain production; the result is farmers cut the trees. To avoid the situation, there is a need to extend shelf life of these fruits and to develop post-harvest value addition technologies which are simple and adaptable at the farm level. This will not only result in developing small-scale industry but will also provide employment to the rural poor throughout the year resulting in increased income of both farmers and workers. Apart from this, trees are a very good source of fuelwood; a 1 ha orchard of ber which needs to be pruned every year provides fuelwood sufficient to meet the requirement of a standard family of four throughout the year. In the year of drought, tree leaves are a good source of fodder for cattle. Thus, there is great scope to put the marginal degraded lands to cultivation of these fruits.

3.1 Importance and Scope

- Provides three basic necessities, i.e., food, fodder, and fuel.
- Drought alleviation through assured production.
- Economically viable.
- Labor intensive.
- No long gestation period.

- Improvement of soil health and thus sustainability.
- Reducing pressure on the existing ecosystem.
- Soil degradation and soil loss reduced.
- Fuelwood availability will reduce burning of cow dung cakes.
- Farmwomen need not waste their time in collection of fuelwood because of on-farm availability of fuel.
- Reduced malnutrition through increased fruit availability.
- Improvement in the socioeconomic life of farming families.

3.2 Criteria for Selection of Different Components

3.2.1 Perennial Component

1. Easy to establish.
2. Good cropping capacity.
3. Palatable foliage.
4. Fast-growing and high-biomass production.
5. Tolerant to biotic and abiotic stresses.
6. Ability to withstand browsing/grazing.
7. Reproductive phases coincide with the period of moisture availability.

3.2.2 Annual Component

1. Tolerant to biotic and abiotic stresses
2. Preferably leguminous in nature
3. Dwarf and spreading type
4. Short duration
5. Better soil binding capacity, thereby resisting erosion
6. Tolerant to partial shade
7. High yielding and quality

3.3 Recently Released Varieties

Sr. No.	Crops	Varieties
1.	Ber	Goma kirti, Gola, Seb, Umran
2.	Pomegranate	Bhagwa, Arka, Ganesh, Mridula, Ruby, Jyoti
3.	Aonla	Goma Aishwarya, NA-7, Chakaiya
4.	Date palm	Halawy, Medjool, Barhee, Khalas, Khuneji
5.	Custard apple	Balanagar, Arka sahan, APKCA-1
6.	Guava	Sardar, Allahabad Safeda, Arka Mridula
7.	Mango	Alphonso, Kesar, Dashehari, Totapuri, Langara, Amrapali, and Mallika
8.	Beal	NB-5, NB-9
9.	Sapota	Kalipatti, Cricket ball, PKM-1, PKM-2, PKM-3, DSH-1, DSH-2
10.	Fig	Poona fig, Black Ischia, Dianna, Conardia, Dinkar
11.	Tamarind	Pratisthan, Urigan
12.	Phalsa	No distinct varieties are available
13.	Wood apple	CHES Selection-1, CHES Selection-2
14.	Jamun	Konkan seedless, Lucknow seedless
15.	Chironji	No distinct varieties are available
16.	Mahua	No distinct varieties are available
17.	Cashew	Vengurla-1, Vengurla-2, Vengurla-3, Vengurla-7, Vengurla-8 Vridhachalam M 25/1, Bapatla Selection No. 129
18.	Manila tamarind	No distinct varieties are available

3.4 Challenges/Options for the Future

Lessons/constraints	Challenges/options
Full exploitation of the potential of arid areas for the production of a large number of horticultural crops was not possible since technologies specific to these areas were not available	There is need to enlarge the program of work to incorporate the fruit, vegetable, medicinal, and ornamental crops which can provide nutrition security to the people along with economic and ecological sustainability to the region
The available biodiversity, which faces danger of erosion, could not be properly conserved in the absence of repositories	Well-defined germplasm repositories and gene banks should be established at appropriate locations
Crop failures and productivity losses are common as a result of extremely high aridity leading to drought conditions	A major program should be directed to genetic improvement for drought and heat tolerance besides the study of physiological adoption mechanisms to these conditions
The fruit crops which are already established in the region face some productivity constraints	There is need to lay emphasis on integrated approach for management incorporating biocontrol agents
Absence of standard varieties and/or multiplication techniques for some indigenous/established fruit crops come in the way of their systematic production, e.g., *Prosopis cineraria*, *Capparis deciduas*, date palm, etc.	Development of fast multiplication techniques is required utilizing micropropagation and tissue culture approaches

3.5 Opportunities

3.5.1 Exploitation of Indigenous Variability

A reserve of considerable variability of a number of hardy plant species which yield edible fruits and products of horticultural value provides opportunity for their exploitation by genetic improvement and systematic plantation. Besides, there is an ample scope for introduction of useful plants from isoclimatic regions of the world.

3.5.1.1 New Introductions

Several fruit yielding species common in the arid region have not yet been commercially exploited mainly owing to the absence of their standard cultivars/types. Similarly, there is a scope for production of plant species introduced from the isoclimatic regions of the world, e.g., cactus pear, quandong, carob, African dove plum, argan tree, marula nut, nance, oyster nut, ye-eb nut, etc.

3.5.2 Genetic Improvement

Genetic improvement in the traditional fruits such as ber, pomegranate, and aonla and vegetables such as chilli, tomato, and cucurbits to induce resistance against abiotic stresses with particular reference to drought and heat should further boost their production in the region. Some of the traditional horticultural plants suffer from productivity constraints owing to the susceptibility to specific disease and pest problems, e.g. powdery mildew in ber and cucurbits. Thus, by genetic improvement in the traditional as well as presently wild but potential species and introduction of suitable exotic species, the major causal factors for the slow pace of spread of area under horticultural crops and for their extremely low productivity in the arid region can be mitigated.

3.5.3 Production for Export

Since arid zone agroclimate offers great potential for production of high-quality produce, there is great opportunity for production for export. The potential of quality production of citrus fruits such as sweet oranges and kinnow, ber, and pomegranate, vegetables such as tomato and chillies, and spice crops such as cumin and isabgol is well known.

3.5.3.1 Value-Added Products

There is great scope for developing high-value products from the indigenous produce such as *Capparis* and *Cucumis* and from the commercial horticultural crops.

3.5.4 Runoff Farming

Although soil conditions in arid region are poor, soil characteristics such as coarse texture and crust formation on soil surface, respectively, cut water losses through capillary movement and induce good runoff water yield even under low rainfall areas.

3.5.4.1 Cropping Systems

The high solar radiation resource in the arid areas provides opportunity for its harvesting by optimization of cropping system models incorporating multiplayer receptor crops.

3.6 Production and Productivity of Fruit Trees

The productivity of fruit trees was much higher than the traditional cropping system in the semi-arid areas under rainfed conditions. With the standardization of vegetative propagation techniques, the long gestation period has been reduced. The fruit trees like ber, custard apple, and pomegranate start bearing in the second year itself; however, trees like mango and sapota start commercial bearing in the fourth year of plantation. Among all the fruit trees, the highest productivity per hectare was recorded in ber (176.40 q/ha) due to accommodation of higher number of plants per hectare (400 plants) followed by guava (98.80 q/ha). The productivity was least in pomegranate and custard apple (17.6 and 20.2 q/ha, respectively). Though productivity was lowest in pomegranate, its net return was the second highest amounting to Rs. 65,400 owing to accommodation of higher number of plants and higher market rate compared to other fruits. The highest net return was recorded in mango (Rs. 75,000) in the 9th year. The highest B.C. ratio was recorded in mango (17.66) followed by pomegranate (14.08) which was three times higher than the traditional farming system. Economic evaluation of agroforestry system with sapota+trees+field crops species gave the highest income; it was maximum from field crop+sapota+teak (46 %) than field crop+sapota. Benefit-cost ratio was higher in sapota+teak (Mutanal et al. 2001) (Tables 3.1 and 3.2).

Table 3.1 Yield q/ha

Crop/year	2nd	3rd	4th	5th	6th	7th	8th	9th
Ber	18.8	50.80	83.00	107.60	126.00	134.8	166.00	176.40
Custard apple	6.92	17.84	21.08	32.32	45.20	61.60	71.60	80.80
Pomegranate	12.72	29.48	41.00	46.00	54.89	59.20	61.60	70.40
Guava	16.80	26.00	46.80	59.60	71.20	80.80	91.20	98.80
Aonla	7.9	25.8	39.3	61.7	75.4	83.3	79.4	94.5
Sapota	–	20.6	27.4	38.2	47.9	56.4	64.3	73.8
Mango	–	7.3	16.7	27.8	35.4	48.7	63.8	79.5
Maize+pigeon pea*	15.96	3.57						

*Traditional method

Table 3.2 Net return (Rs/ha)

Crop/year	2nd	3rd	4th	5th	6th	7th	8th	9th
Ber	2,640	1,990	21,400	28,530	33,800	35,940	45,050	47,920
Custard apple	960	6,170	7,540	12,910	18,850	26,800	31,550	35,900
Pomegranate	9,720	26,230	37,750	42,750	51,640	54,750	56,850	65,400
Guava	5,900	10,250	20,400	26,300	31,850	36,400	41,350	44,900
Aonla	1,450	10,150	20,400	16,550	27,150	33,700	37,400	42,750
Sapota	–	7,800	10,450	15,600	20,200	24,200	26,950	32,400
Mango	–	4,800	13,700	24,500	31,650	44,700	56,550	75,000
*Maize+pigeon pea								10,050

*Traditional method

Reference

Mutanal SM, Nadagoudar BS, Patil SJ (2001) Economic evaluation of an Agroforestry system in hill zone of Karnataka. Ind J Agric Sci 71(3):163–165

Part II

Crop Specific Production Technologies for Semiarid Rain Fed Areas

Ber (*Zizyphus mauritiana* Lamk.)

4

Abstract

Ber (*Ziziphus mauritiana* Lamk.) is one of the ancient and important fruit crops of India as it can stand drought and tolerate salinity to a considerable extent. It is often called the poor man's fruit. In India out of the 143 million ha of cultivated land, about 57 % is rainfed. Still this area contributes about ½ the food basket of the country. Horticulture can play an important role in rejuvenating the falling incomes of the farmers in these areas. Of late commercial importance of ber has been realized and ber cultivation has received a great impetus in India. The ber fruits are available right from October to May in different regions. It is an ideal fruit tree for growing in arid and semiarid zones of India. Exponential increase in human and livestock population in India has put tremendous pressure on natural resources, owing to not only degradation of principal resources like land, water, and vegetation but also reduced per capita availability of land. The ber is a multipurpose crop as it provides 3F's, i.e., fruit, fodder, and fuel. Ber fruits are generally consumed fresh, but dried ber, candy, preservers, beverages, etc. can also be prepared from it. Besides being palatable and delicious, the recent research has disclosed its high nutritive value. It is one of the richest sources of vitamin C next only to amla and guava but better than citrus and apple.

4.1 Introduction

In India out of the 143 million ha of cultivated land, about 57 % is rainfed. Still this area contributes about ½ the food basket of the country. Even with all the efforts, the percentage area under irrigation cannot be increased beyond 25–26 %; crop yields in dry lands are not only low but are also prone to risk and complete failure during seasons of erratic rainfall. Horticulture can play an important role in rejuvenating the falling incomes of the farmers in these areas. To bridge the gap of demand and supply, the fruit production will have to be increased from present level of 23.8 to 85 million tons; on the other hand, National Commission of Agriculture proposed that the area under fruits should be raised to 4 million ha by 2020 AD. Hence, any expansion in area or to increase the production to meet the

S. Hiwale, *Sustainable Horticulture in Semiarid Dry Lands*,
DOI 10.1007/978-81-322-2244-6_4, © Springer India 2015

ever-increasingly demand can be met by adopting dry land horticulture on marginal lands.

Ber (*Ziziphus mauritiana* Lamk.) is one of the ancient and important fruit crops of India as it can stand drought and tolerate salinity to a considerable extent. It is often called the poor man's fruit. However, grafted varieties of ber are no longer the poor man's fruit as they fetch even higher prices than several other fruits. Of late commercial importance of ber has been realized and ber cultivation has received a great impetus in India. The ber fruits are available right from October till May during the year. It is an ideal fruit tree for growing in arid and semiarid zones of India.

Ber being xerophyte in nature and its ability to stand drought makes it a prime fruit suitable for the semiarid rainfed conditions. The important characters are deep and extensive taproot system capable to penetrate through hard pan and draw water from deeper layers of soil. Ber plant goes into dormancy during hot summer months by shedding leaves and growth cessation thus reducing water loss from the plant. Hard and pubescent scales develop on the buds resulting in reduction of dehydration of buds. The fruiting is completed during the period of maximum moisture availability.

Exponential increase in human and livestock population in India has put tremendous pressure on natural resources, owing to not only degradation of principal resources like land, water, and vegetation but also reduced per capita availability of land. At places, the extent of land degradation has crossed such critical limits that it is difficult to bring back to its original state. The unscientific cultivation of annual crops in rainfed areas, as usually is the case, accelerates runoff, soils loss, as well as other degradation processes causing fragile ecosystem. In India about 53.34 million tons of top soil (16.4 t/ha) containing 5.4 million tons of plant nutrients is lost annually due to agriculture and associated activities. The prime agricultural lands continue to be diverted to nonagricultural uses like industrialization and urbanization with the result that marginal rainfed lands are brought under cultivation with least attention on resource conservation. Out of 143 million hectares arable land in India, nearly 74 % is rainfed, which contributes about 42 % of total food production. However, soil moisture stress due to vagaries of weather, moderate to severe soil erosion, and poor soil fertility are the major limiting factors of low productivity in rainfed areas.

Though India has made significant stride in food production in the recent past, major emphasis was given on irrigated and resource-rich situations. However, in rainfed areas the marginal farmers are still practicing traditional way of farming and struggling below the poverty line. Therefore, this trend needs to be discouraged. Virtually, a sizable area (171 m ha) in our country is categorized as degraded lands. Such lands which are subjected to problems of water and wind erosion, salinity, alkalinity, acidity, water logging, gullied and ravine lands, shifting cultivation, etc. are either laying vacant or underutilized. Although, these lands have their own biophysical farming constraints but can be utilized efficiently on sustained basis for fulfilling the basic needs of food, fruit, fuel, fodder, fiber, fertilizer etc. through horticulture-based farming systems by adopting practically feasible, economically viable and eco-friendly technologies developed in recent past. Considering above facts, horticulture-based composite farming systems were evaluated from 1985 to 2000 to establish a viable model for rehabilitation and utilization of semiarid rainfed areas at Vejalpur, Godhra, Gujarat.

Fruit trees play a vital role in crop diversification. Their inclusion in the system reduces the risk of crop failures. Apart from this, the fruit trees are known to increase the income of the farmer substantially over traditional monocropping. Fruit cultivation has so far been associated with fertile soil with assured irrigation, but presently the irrigated fertile land is already being used for meeting the food requirement of ever-increasing population and industries. The marginal and degraded lands in arid and semiarid region of the country with practically no irrigation facilities are to be used for increasing fruit production with simultaneous need to meet the basic demand for food, fodder, and fuel. Hence, no time be lost in developing highly productive self-sustainable farming systems fully adapted to the precarious ecology of arid regions.

In the broadest sense, the term agri-horti-silvi-pastoral system encompasses any and all techniques that attempt to establish or maintain forests, horticultural crops, forage trees, and pasture grasses on the same piece of land. The science and practice of agri-horti-silvi-pastoral system aim at systematically developing land use system and practices where the positive interaction between trees and crops is maximized. This seeks to achieve more productive, sustainable, and diversified output from the land than is possible with conventional monocropping system. In this system, the fruit and vegetable crops provide seasonal revenue, while forest trees are managed at 5–10 years rotation giving extra returns of timber, fuelwood, and fodder. The approach aims at optimizing the use of land, water, and vegetation in an integrated way and thus helps to alleviate drought, moderate floods, prevent soil erosion, improve water availability, and increase production of fruits, vegetables, fuels, and fodder on sustained basis.

4.2 Nutritive Value

The ber is a multipurpose crop as it provides 3F's, i.e., fruit, fodder, and fuel. Ber fruits are generally consumed fresh, but dried ber, candy, preservers, beverages, etc can also be prepared from it. Besides being palatable and delicious, the recent research has disclosed its high nutritive value. It is one of the richest sources of vitamin C next only to amla and guava but better than citrus and apple. Ber fruits are also rich in vitamin A and contain B carotene at 81 mg per 100 g of fruit. Ber fruits are also rich in minerals like 0.03 % calcium, 0.036 % phosphorus, and 1.14 % iron. Its protein content is 0.94 % amino acids, viz., asparagine, aspartic acid, arginine, glutamic acid, glycin, serine, thiamine, X-alkaline, valine, methionine, leucine and isoleucine are also available from the fruit (Gupta 1984) (Table 4.1).

Ber leaves provide nutrition's fodder for the animals particularly goat and sheep. The leaves contain 5.56 % DCP and 49.7 % TDN and are rich in crude protein 12.9–16.9 and mineral matter.

Table 4.1 Nutritive value of important ber cultivars

Cultivars	TSS Brix	Protein %	Vitamin C mg/100 g pulp	Reducing sugar %	Total sugar %
Gola	19.8	5.25	302.18	4.75	15.62
Seb	14.2	5.00	189.59	2.79	12.09
Jogia	17.0	3.50	263.00	4.41	16.41
Kaithali	14.8	4.32	286.34	2.74	13.65

4.3 Climate

The semiarid climate is characterized dry low and highly variable precipitation with high evaporative demand. Therefore, moisture stress is the characteristic feature of semiarid zones. The average annual rainfall of the area is 700 mm, potential evapotranspiration 1,728 mm, and moisture index of 58.8. There is a need to improve cropping pattern in this area.

The ber is found growing under varying conditions right from Haryana to Tamil Nadu; however, it is sensitive to freezing temperature but can withstand extremely hot conditions going into dormancy. New growth, flowering, and fruiting are dependent on length of summer and period during the monsoon season.

In northern India, growth starts from late June to early July with advent of monsoon followed by flowering and fruiting in October and fruit maturity during February to April. In south India, there is not much fluctuation in the climate; because of this, the growth is nonstop resulting in flowering during May to July and maturity of fruits by October or by November (Randhawa and Kohli 1976). The ber prefers dryness for ideal performance; high atmospheric humidity is disadvantageous.

In semiarid western India and particularly in Gujarat, the average rainfall is 700 mm and the maximum temperature varies between 30 and 42 °C and minimum temperature 10–25 °C. The atmospheric humidity during fruit development period varies from 50 to 65 % providing the weather to be dry resulting in early fruit maturity from December, which is completed by February end.

4.4 Soil

Because of the quick development of taproot system within a short period, ber adopts itself to a wide variety of soils. Its trees are found growing on saline soils and marginal lands which are otherwise unfit for growing other fruit crops or even crops. It has in fact withstood and done better than the most of the fruit crops.

The ber however responds to better soil conditions. Ber is tolerant to salinity; it can tolerate EC of 14 mmhos/cm and to high PH of 9.2–10.5 (Singh et al. 1987). The soil of the experiment station was found to be saline with very poor fertility status. The pH value was 7.61 at 0 30 cm, with N content of 0.074 %, P content of 17.84 kg/ha and K content of 166.60 kg/ha.

The soil analysis carried out after 6 years indicated significant improvement in all the parameters. The pH value was 6.30 with organic content of 1.00 %. N content of soil increased to 0.084 %; similarly P content was 18.72 kg/ha and k content to 239.90 kg/ha indicating improvement in soil health (Tables 4.2 and 4.3).

Initial field view of experimental site (Photo)

Table 4.2 Soil analysis (before planting)

Details	Total N %	Av. P_2O_5 (kg/ha)	Av. K_2O (kg/ha)	EC (1:25)	PH
0–30 cm	0.074	17.98	166.60	0.2	7.61
30–60 cm	0.043	14.27	166.60	0.15	8.20

Table 4.3 Soil analysis after 7 years

Soil depth (cm)	O.C. (%)	EC (mmhos/cm)	pH	CEC NH_4 OA	N (%)	P (kg/ha)	K (kg/ha)
0–15	1.00	0.21	6.30	40.20	0.084	18.72	239.90
15–30	0.80	0.27	6.41	53.70	0.075	14.39	233.34
30–45	0.60	0.30	6.70	58.30	0.058	12.63	266.56

Hiwale (2002)

Pits preparation

Soil type at the site

4.5 Commercial Cultivar

There has been wide cultivars diversity in ber in India. Ber or jujube belongs to the genus of Ziziphus of the family Rhamnaceae; the genus consists of 135 species. However, only a few cultivars are commercially important which lack good productivity along with ability to withstand transport, storage, and resistance to pest and diseases. Earliness is a desirable trait in cultivars meant for arid and semiarid dry lands.

A wide range of variability exists in ber in India. Most of the present day, cultivars have been evolved through selection. Major emphasis has been laid on clonal selection and an early maturing clone (Pareek 1983). Ber has polyploidy and self-incompatibility in its cultivars, which are serious bottlenecks in its successful breeding programs (Pareek 1983). Most of the cultivated cultivars have been reported to be reciprocally cross incompatible.

For successful cultivation, choice of suitable variety is of paramount importance as there is variation in the climatic components which affects the performance of variety. The evaluation of varieties carried out at central horticultural experimental station for the last 4 years revealed that varieties Gola, Dandan, and Rashmi being early varieties giving good yield and quality fruits as well as Umran, Seb, and Aliganj being late variety good yielder and producing good quality fruits are recommended for cultivation under semiarid rainfed conditions.

4.5.1 Gola

It is an early maturing cultivar. The fruits are glossy with yellowish brown color at ripening. The fruits are round very attractive and its white semisoft and juicily flesh is very delicious because of the sugar-acid blend. It is best cultivars for rainfed cultivation; average weight of fruit is about 20 g with a TSS of 18.42 % and a high acidity of 0.64 and a pulp-stone ratio of 13.08. The yield per plant was about 28.4 kg in the 4th year. The variety cannot stand long transportation and hence should be planted near to big cities. The fruit ripens from the 3rd week of December to end of January.

4.5.2 Dandan

This is also an early variety. The fruit is large and oval in shape and has attractive golden yellow color, which turns into brown. The fruit is sweet with a TSS of 19.61 %. The yield per plant is 22.06 kg in the 4th year.

4.5.3 Rashmi

Though the fruit are smaller in size and weight about 8.81 g. The variety is early and the fruits are having good aroma. The yield/plant is 14 kg/plant in the 4th year.

4.5.4 Umran

This is a late variety, good yielder giving 16.0 kg yield per plant. The fruits are golden yellow in color. The fruit is large 19.91 g and oval in shape with golden yellow in color when it matures. TSS is 20.08 % and pulp-stone ratio is 19.67. The variety having firm flesh and hence can be transported to distant market. The fruits start maturing from the 1st week of January till the middle of February.

4.5.5 Seb

The fruit resembles an apple in shape and weighs about 18.42. It is a late variety with thick skin and can be transported to distant market. TSS is about 20 %. The fruits turn yellow when they are mature. The average yield/plant under rainfed condition is 20 kg. The fruits mature late in January.

4.5.6 Aliganj

Fruits are in shape with 17.39 g weight. The fruits are juicy with a TSS of 22.57 % and pulp-stone ratio is 16.29. The yield per plant is 14.8 kg.

4.6 Ber Cv."Goma Kirti" (*Ziziphus mauritiana* Lamk.) Salient Characters

1. The clonal selection was identified from the large germplasm collection made all over the country.

   Over the country at Central Horticultural experiment station Godhra.
2. Selection was multiplied by in situ budding and grown under rainfed conditions.
3. The selection started bearing in the first year of its budding.
4. It consistently flowered and fruited for the past 5 years.
5. The selection flowered 2 weeks earlier than the local check and the fruits matured 3 weeks earlier than the check.
6. The selection was found to be high, yielding 25.8–34.3 % increase.
7. The selection though early was found to have superior keeping quality. Compared to early varieties, which fruits have shelf life of 3 days, the selection could be stored up to 6 days under ambient conditions.
8. The physiological loss in weight was 27 % in early varieties whereas it was 17 % in the selection on 3rd day of storage in Cv. Goma kirti (Hiwale 2005).

BER- GOMA KIRTI IN BEARING

BER- GOMA KIRTI IN BEARING

GOMA KIRTI FRUITS

4.7 Propagation

In the recent year, ber has gained commercial importance and hence demand for plant material has increased manifolds with the advent of technology; raising ber nursery has also become a profitable preposition by raising rootstock plants in 200-gauge polythene bags of 25 cm length and 12 cm diameter. The seedlings are ready for budding in 150–200 days and they are ready for transplanting in another 40 days of budding. A net profit of Rs. 15,000 could be obtained (Table 4.4).

Best time for budding is observed to be from the first week of June till the onset of monsoon. During the rainy season, fungus development is fast due to high percent humidity. Patch budding was found to be the best method for budding ber plants either in the nursery or in field conditions. Ber species was the best rootstock which is *Ziziphus rotundifolia* found growing wild in India and is known for their extensive root system and hardiness. However, germination of seed is a problem because of stony endocarp. It is better to break the endocarp; even after breaking the endocarp, the germination percentage was found to be very low 46.66 %; hence, to increase germination percentage, the seeds were treated with various growth regulators revealed that treating extracted kernels with kinetin 100 ppm gave higher germination percentage (83.33 %) (Hiwale and Raturi 1996a, b). Seeds should be collected fully ripe fruits in the month of December and are sown in February to March in polythene bags or field. Other methods like stooling and air layering have also been tried by some workers (Singh et al. 1973; Misra and Jauhari 1970). In ringed shoots, 100 % success was obtained by the application of 8,000–12,000 ppm IBA; 100 % success was obtained in stooling by the use of mixture of IBA and NAA at 5,000, 7,500, and 10,000 ppm with 100 ppm boron (Table 4.5).

Table 4.4 Effect of time if budding on percent bud take in ber

	Percentage success		
Date	1988	1989	Mean
10/6	81.0	61.0	71.0
20/6	48.0	46.0	47.0
30/6	40.0	21.0	30.5
10/7	12.0	10.0	11.0

Ber in situ budding (root stock)

Budded plant with unwantedsprouts

Removal of extra growth

Successful in situ budded plant

Table 4.5 Effect of plant growth regulators on seed germination and seedling growth of ber rootstock

		Seedling growth		
Treatments	Germination %	Height (cm)	Diameter (cm)	Vigor index
GA 50 ppm	63.33	30.17	4.12	1,606.66
GA100 ppm	56.66	30.02	4.00	1,940.00
GA 200 ppm	70.00	29.67	3.97	2,386.66
Kinetin 50 ppm	70.00	30.27	4.12	1,816.66
Kinetin 100 ppm	83.33	30.82	4.27	3,466.66
Kinetin 200 ppm	73/33	30.00	4.03	2,280.00
Control	46.66	29.75	3.77	1,090.00
CD 5 %	13.00	NS	NS	413.21

4.8 Planting

Ber orchards can be raised by 2 months, i.e., by in situ budding and budding plants in nursery and transplanting in field. In situ seeds are sown in well-prepared pits at appropriate distance on the onset of monsoon. The plants are ready for budding in June to July in the coming year. This practice is highly suitable for semiarid conditions as percentage success in budding is very high; growth of the plant is very fast and the plant bears fruits in just 7 months after budding 5 even under purely rainfed conditions without any hand watering.

The nursery budded plants though require less period for transplanting take some time to settle, and also as the taproot system of the plant is disturbed, they need to be watered at least for the 1 year to get established. The comparative growth of the plant is slow and it bears no fruits in the first year of transplanting (Table 4.6).

The best time of planting ber is in the month of August to September as the plants are ready for transplanting during the end of July only (Hiwale 2004b). Varied recommendations are there for planting distance; it can be as much as 5 m × 5 m to 10 m × 10 m even under rainfed conditions depending on the vigor of the variety and its ability to bear in partial shady conditions. It is

Table 4.6 Performance of budding methods

	In situ budded		Nursery budded	
	1st-year-old	2nd-year-old	1st-year-old	2nd-year-old
% success in budding	85 %	–	45 %	–
Av. plant height (m)	0.25	1.78	0.085	0.56
Stock diameter (mm)	31.0	73.0	9.0	47.0
Scion diameter (mm)	30.0	68.0	8.0	43.0
Yield/plant kg	2.1	7.9	–	3.4

advisable to adopt planting at 5 m×5 m spacing under rainfed conditions. The pits of 1 m^3 size are dug during summer and are filled up before onset of monsoon with mixture of upper half soil, sand, and FYM. 100 g 10 % BHC is added to the mixture to prevent termite attack.

4.9 Flowering and Fruiting

The ber plants when pruned after the 25th of April to the 5th of May sprouted after about 30–40 days. The growth pattern of ber shoot was sylleptic type, i.e., on the main shoot; secondary shoots emerge after emergence of 5–9 leaves on primary shoot. Tertiary shoot emerges at the same interval as that of the secondary followed by the quaternary. Shoot growth was recorded at 15 days interval revealed that the growth is faster in the initial period from the first week of July to the last week of September and then it was stabilized. Slow growth however continued till the end of November. There was a variation in the emergence of secondary and tertiary shoots; it starts in the first week of July, adding 3–4 secondary's at 15 days interval (Tables 4.7, 4.8, and 4.9).

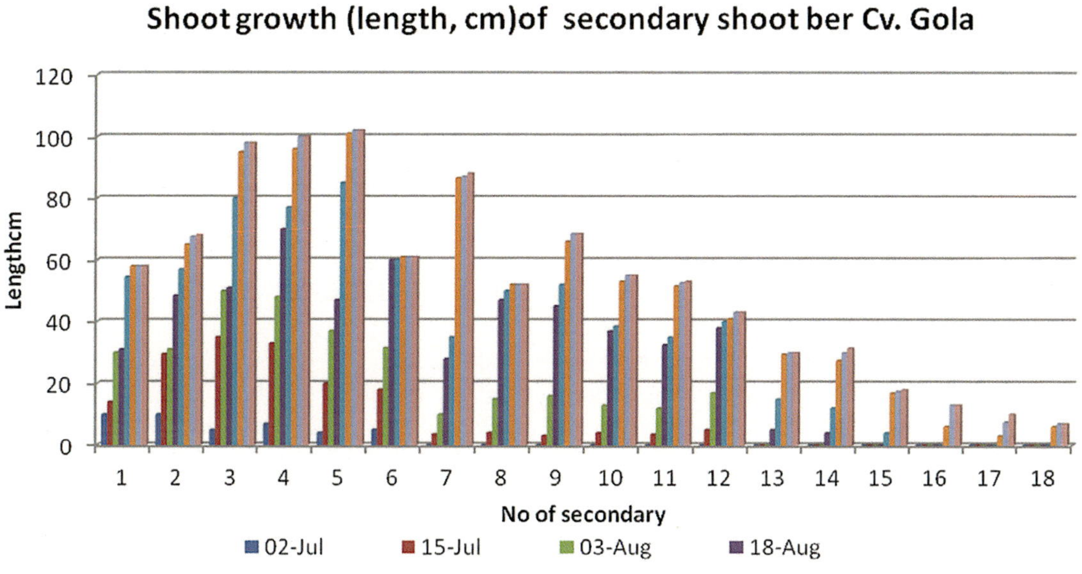

Table 4.7 Percentage flowering and mean number of leaves/shoot

Variety/shoot	Primary	Secondary	Tertiary	Quaternary
Gola	2.82 (68)	18.21 (430)	50.13 (615)	28.81 (280)
Umran	3.59 (64)	23.55 (332)	67.86 (523)	4.99 (124)
Seb	1.96 (71)	21.77 (409)	60.63 (715)	15.62 (88)
Mundia	3.20 (63)	22.10 (459)	65.77(547)	8.91 (110)

Figures in bracket indicate mean number of leaves/shoot

Table 4.8 Percent set and retention in ber (shoot wise)

Variety/shoot	Primary	Secondary	Tertiary	Quaternary
Gola	1.7 (4.9)	19.2 (20.4)	62.6 (54.6)	16.3 (19.3)
Umran	1.5 (3.5)	26.4 (33.9)	66.5 (55.4)	8.9 (7.0)
Seb	3.8 (2.9)	29.9 (33.6)	57.8 (61.2)	8.4 (2.2)
Mundia	0.87 (3.1)	22.60 (35.14)	69.63 (52.16)	8.8 (5.9)

Figures in parenthesis indicate percentage retention per shoot

Table 4.9 Percent yield (shoot wise)

Variety/shoot	Primary	Secondary	Tertiary	Quaternary
Gola	4.96	20.49	54.65	19.87
Umran	3.52	33.97	55.44	7.05
Seb	2.91	33.67	61.12	2.28
Mundia	3.6	32.30	58.53	5.26

There is a direct relationship between number of leaves and percent flowering. The trend is that a maximum number of leaves are on tertiary and percent set also higher in tertiary shoot followed by secondary and then quaternary.

Flower initiation normally starts in August to September. Flowering stops in the end of September except in some like Banarasi, Maharwali, Kaithali, Ilaichi, Thornless, and Kali. The fruit set starts in the third week of August and it is in peak during the middle of September and is over by the end of October, whereas in Uttar Pradesh, flowering starts in the second week of September which continued up to November. The flowers are borne in the axils of the leaves as main as well as the lateral shoot and the inflorescences are an axillary cyme (Teotia and Chauhan 1963). The fruit set begins from the 1st week of August and continues till the 1st week of November. Fruit set is reduced drastically if there are continuous rains during flower-opening period due to lack of pollinizers and washing away of the pollens due to continuous rains during flowering. There is profuse flowering on sunny days and bee's activity is maximum during flowering period. There were significant differences in respect to fruit set and fruit retention. Maximum fruit set was observed in variety Z-G-3 and Katha followed by Seb and Banarasi.

4.9.1 Fruit Set

Though there is profuse flowering but due to shedding of flowers and dropping of fruits at various stages is so heavy that certain varieties have been eliminated from commercial plantation also due to the phenomena of self and cross incompatibility and pollen sterility. Fruit set is also low percentage; fruit set in 22 ber varieties was found to be ranging from 3.17 to 8.75 %, lowest in Aliganj and highest in Z-G-3. However, fruit retention was ranging from 22.94 to 60.86 highest in Aliganj and lowest in Dandan, Maharwali, Tikdi, and Banarasi Pewandi. Attempts to increase fruit set by exogenous applications of plant growth regulators have been made. Both GA and 2, 4, 5-T nearly doubled the fruit set to 15.5, 13.6, and 14 % in "Dandan," Kaithali, and Umran (Dhillon and Singh 1968).

4.9.2 Fruit Drop

Under semiarid conditions, fruit drop in ber starts from the middle of October and continues till the middle of December. Teotia and Chauhan (1963) have found ovule disintegration to be the major cause of fruit drop in initial stages. The pattern and extent of drop varies with the varieties of their maturity period and the extent of initial set. The maximum number of fruits drop in the initial stages immediately after fruit set followed by pinhead drop. In cultivar Umran, maximum fruit drop was observed followed by Gola and Seb.

Exogenous application of growth regulators has not been very effective in controlling fruit drop. Dhillon and Singh (1968) found 2,4-D at

10–20 ppm to be the most effectively in reducing fruit drop, whereas GA 25 ppm and 2,4,5-T at 5–10 ppm (*Ziziphus mauritiana* Lamk.) showed little effect.

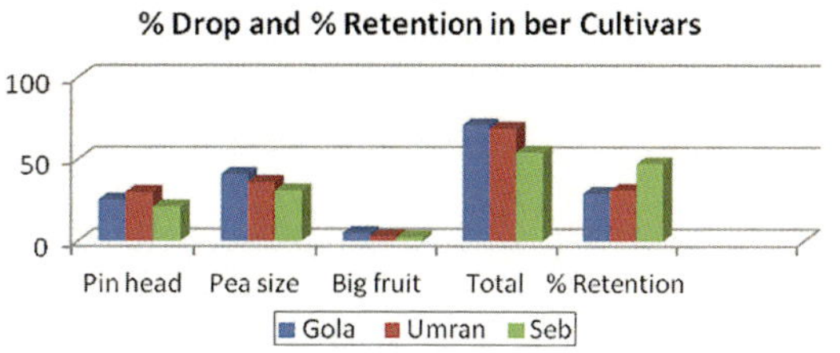

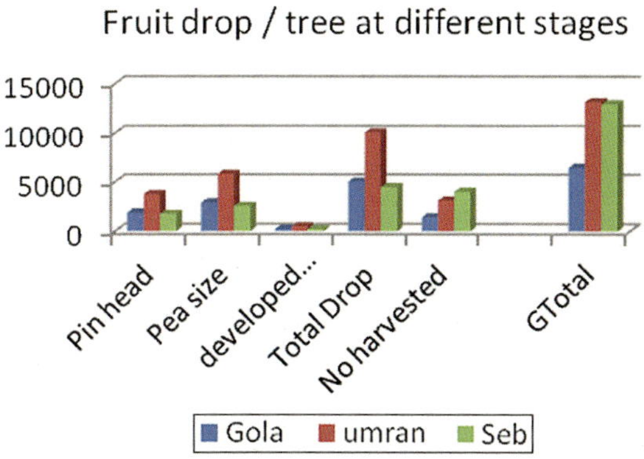

4.9.3 Fruit Development

Fruit development in ber follows double sigmoid curve in most of the cultivars. In Gola, Umran, and Seb, the growth of ber fruits under semiarid dry land condition was faster in the first 30 days, which reduced from 45 to 60 days. Again it was maximum in the second phase from 60 days onwards. The stone weight stabilized with increasing fruit age in all the three varieties.

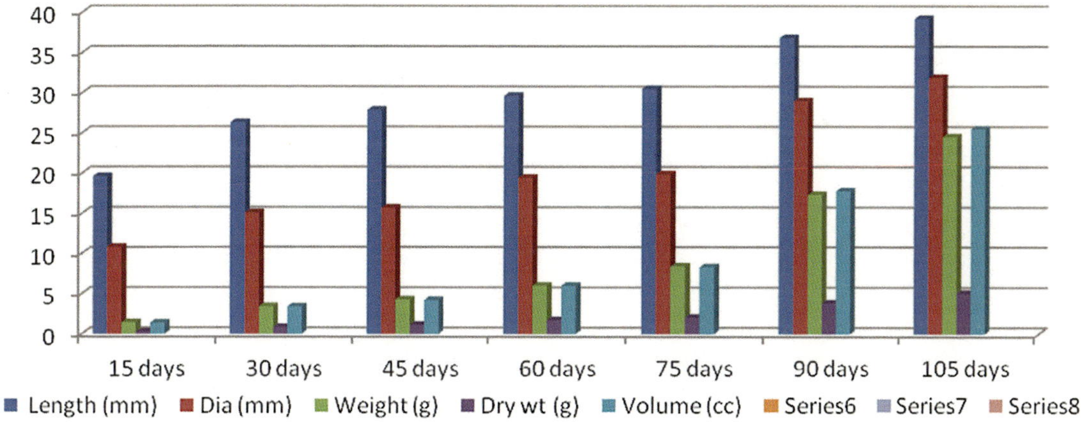

4.10 Manuring

Ber orchards need to be manured regularly to obtain sustained productivity. Under irrigated conditions, higher dose of fertilizers can be applied, whereas under rainfed conditions, the dose can be reduced drastically. In Gujarat under irrigated condition, a dose of 400 gN, 350 g P_2O_5, and 450 g K_2O per plant was found to be optimum (Gopani 1976); FYM, P_2O_5, K_2O, and two third of N is applied in July; the remaining one third N is added in September.

The work on nutrient management carried out at Godhra with the concept of boundary line indicating that there is a peak in yield increases, and then with additional dose, there is a drop in production. So application of single nutrient along with higher number of replications under rainfed conditions was tried; the dose varied from 50 to

Table 4.10 Estimation of leaf area in jujube

Cultivar	Actual leaf area (cm^2)	Apparent leaf area (cm^2)	Leaf area by factor method (cm^2)	Leaf area by regression method (cm^2)
Umran	38.272	49.824	38.269	38.272
Gola	39.840	51.720	39.839	39.846
Seb	33.490	44.140	33.489	33.490
Mundia	36.570	45.550	36.567	36.567

600 g with 10 treatments and 15 replications for N, P, and K. The results indicated that the dose of 30 kg FYM, 200 g N, 100 g P_2O_5, and 100 g K_2O applied in the month of July is the best. The fertilizer dose is reduced to almost 50 % with single application due to rainfed conditions as there are no rains after September.

4.11 Leaf Nutrient Guide

The leaf was found to be the best index tissue to indicate the nutritional concentration of essential minerals in trees and is one of the important factors in determining plant behavior. To develop need-based fertilizer, schedule for ber leaf analysis was done. Concentration of N, P, and K decreased with increase in age, whereas Ca and Mg showed marginal increase. Sulfur, Zn, and Cu recorded decrease in their content with increasing leaf age, whereas Fe and Mn have not followed any specific pattern. The data revealed that the 5th and 6th leaf from secondary shoots represent recently matured leaf, which can be collected as an index tissue for nutritional diagnosis (Bhargava et al. 1990).

4.11.1 Leaf Area Determination by Nondestructive Method

In ber leaf area/fruit ratio determines the size and quality of fruit. The optimum leaf area required for the development of fruit can be calculated by nondestructive method. Maximum correlation was recorded in the product of maximum length and maximum breadth with actual leaf area. The regression equation based on these parameters was worked out. The functional relationship between length × breadth and leaf area values was best described by the models (Hiwale 1991) (Table 4.10).

1. $Y = (-)1.483 + 0798x$ (Umran).
2. $Y = (-)2.72 + 0.823x$ (Gola).
3. $Y = (-)0.875 + 0.778x$ (Seb).
4. $Y = (-)0.008 + 0.803x$ (Mundia).

4.12 Canopy Management

Ber is deciduous in nature dropping its leaves during the summer months and flowers and set fruits on new growth; therefore, there is a need to prune the plants every year during the dormant period. An experiment was conducted at CHES; Godhra on 6- and 7-year-old ber plants (*Ziziphus mauritiana* Lamk.) budded on *Ziziphus rotundifolia* rootstock planted at 5 × 5 m spacing to standardize the time and severity of pruning in Cv. Gola. There were five pruning treatments which were applied at three different times starting from the end of April with an interval of 10 days. The study indicated that pruning ber plants in the first week of May is the best to obtain maximum number of sprouts, shoot length and shoot diameter. Another experiment on severity of pruning carried out at the station indicated that the growth parameters, viz., shoot length, shoot diameter, and number of sprouts per plant were highly influenced by the various pruning treatments. Shoot length was maximum when severity was higher. Similarly number of sprouts per plant from pruned branches decreased with increasing severity. Pruning helps in higher fruit set and retention per shoot compared to unpruned plants. Quality character was also improved by pruning. Higher TSS was recorded in pruned plants. Pruning plants to 60 cm of previous season's

Table 4.11 Effect of intensity of pruning on growth, yield, and quality of ber Cv. Gola

	Pruning intensity						
Intensity of pruning	30 (cm)	60 (cm)	90 (cm)	120 (cm)	No pruning	SEm	CD 5 %
Shoot length m	3.16	2.95	2.91	2.74	1.55	0.11	0.32
Shoot dia. mm	38.38	35.34	35.71	29.74	18.03	1.42	4.37
Number of sprouts/shoot	45.25	49.50	48.25	83.25	130.77	9.14	28.63
Fruit set (%)	2.71	3.27	2.94	2.83	1.67	NS	NS
Fruit retention (%)	30.82	36.75	32.33	28.69	21.09	1.86	5.71
Yield (t/ha)	23.25	28.50	26.08	25.05	17.50	1.27	3.91
Fruit weight (g)	19.52	20.90	19.27	18.25	15.05	0.23	0.73
Fruit length (mm)	29.41	31.26	29.75	28.73	25.40	NS	NS
Fruit dia.	30.81	31.98	31.28	29.91	27.71	NS	NS
Pulp-stone ratio	12.67	12.30	13.45	12.80	11.95	NS	NS
TSS ° Brix	25.22	25.72	24.92	26.10	24.98	0.33	1.00
Acidity %	0.34	0.32	0.33	0.35	0.29	NS	NS

Raturi and Hiwale (1993)

growth in the first week of May was found to be the best for obtaining higher yield and better quality of fruits under semiarid conditions, (Hiwale and Raturi 1993b) (Table 4.11).

4.12.1 Canopy Management Through Shoot Pinching on Growth Yield and Quality of Ber Cv. Gola

Canopy management in Indian jujube is one of the important management practices. Selective pruning is a desirable practice as the fruits are born in the axils of the leaves of newly emerged shoots developed during the current year. Extent to which the Indian jujube trees are to be pruned to maintain balance between growth and fruiting varies with climatic conditions prevalent in the area. The results of the study revealed that the growth pattern of the shoot is sylleptic type, i.e., on the main shoot there are secondary shoot, on secondary there are tertiary shoots, and on tertiary there are quaternary shoots. The flowering and fruiting pattern in Indian jujube was studied, and it was noticed that between 6th and 12th secondary, maximum fruits are set though the vegetative growth continues to 18th–22nd secondary. The large number of secondary, tertiary, and quaternary shoots resulted in compaction and causing difficulties in intercultural operation and fruit harvesting. To control the growth of the shoot and to have a manageable canopy, pinching the shoot after 6th, 9th, and 12th secondary was attempted. The mean main shoot length was found to be least in shoots pinched to 6th secondary followed by 9th and 12th secondary. However, main shoot diameter was maximum in shoots pinched to 12th secondary (29.16 mm) compared to control (24.42 mm). There was a significant effect of shoot pinching on number of fruit set per shoot. Maximum fruit set was recorded in plants, where main shoot was pinched after 12th secondary (1355 no./shoot) as compared to control (655 no/shoot). Percent fruit retention was also highest in the same treatment (7.77 %) and was least in control (3.54 %). This resulted in higher yield per plant when shoots were pinched to 12th secondary (19.24 kg/plant) compared to control (8.38 kg/plant). Thus, shoot pinching to 12th secondary in the first week of September was found to regulate the canopy as well as increase the production potential of Indian jujube significantly (Table 4.12).

4.13 Fuelwood Production

The system also resulted in the production of sufficient quantity of fuelwood. Leucaena served both the purpose of fuel and fodder. Besides forest trees, ber trees also provided fuelwood from

Table 4.12 Ber shoot pinching

Treatment	Shoot dia. (mm)	Fruit set/ shoot (no.)	Fruit retention/ shoot (no.)	Percent retention	Yield kg/plant
6th secondary	25.72	1,148.2	67.20	5.40	14.18
9th secondary	26.18	1,359.8	67.40	5.79	16.88
12th secondary	29.16	1,355.0	106.0	7.77	19.24
Control	24.42	655.6	55.20	3.54	8.38
CD %	2.91	317.57	32.76	1.88	2.81

Hiwale (2011)

Table 4.13 Fuelwood produced (t/ha) and cow dung saved

Crop/year	III rd year	IV th year	V th year	Mean	Cow dung saved tons	Nutrient saved N tons	P tons	K tons
Ber[a]	34.00	54.00	72.00	53.30	30.24	0.106	0.045	0.060
Leucaena	106.56	146.50	169.92	140.99	150.08	0.325	0.225	0.300
Eucalyptus	83.20	113.60	59.36	120.53	98.56	0.344	0.147	0.197
Bamboo	37.54	46.24	59.36	47.71	31.75	0.115	0.051	0.072

[a]Pruned wood

the annual pruning. Among different plant types, maximum production (140.99 q/ha) of fuelwood was recorded from Leucaena followed by Eucalyptus (120.53 q/ha) and ber (53.30 q/ha) and minimum in bamboo (19.79 q/ha), which started yielding from the 5th year onward while all other plant species started fuelwood production from the 3rd year itself. Such type of systems will reduce the pressure on forest for fuelwood (Raturi and Hiwale 1992).

According to an estimate, 18 % of the time spent on domestic work is spent on collection of fuelwood for cooking the food. The time spent in collecting fuelwood can be effectively used for other domestic works and farming activities if alternative arrangement has been done. Apart from this, precious cow dung can be saved from burning and diverted in the agricultural field. It is estimated that fuelwood obtained from 1 ha. Ber orchard results in saving of 30.2 t cow dung which can help in addition of 0.106 t nitrogen, 0.045 t phosphorus, and 0.06 t potash in the form of farmyard manure (Table 4.13).

Ber fuel wood production

Table 4.14 Amount of pruned wood removed during pruning (fresh wt. kg. per plant)

Year/variety	I st	2nd	III rd	IV th	V th	VI th
Gola	0.453	2.50	7.42	7.55	17.8	27.8
Seb	0.566	2.75	6.67	8.38	22.4	38.2
Umran	0.938	3.25	13.66	16.12	24.6	55.0
Mean	0.652	2.83	9.25	10.68	21.60	40.33

Raturi and Hiwale (1992)

Table 4.15 Root distribution pattern in ber Cv. Gola (on fresh weight basis g)

	Radial distance cm						
Soil depth cm	0–30	30–60	60–90	90–120	120–150	150–180	Total
0–30	983	684	339	218	146	66	2,436
	(622)	(431)	(213)	(132)	(86)	(39)	(1,523)
30–60	488	287	249	252	56	–	1,332
	(313)	(196)	(178)	(163)	(33)		(883)
60–90	3,820	963	930	830	598	238	7,359
	(2,295)	(577)	(542)	(483)	(347)	(133)	(4,377)
90–120	757	299	223	94	30	–	1,403
	(458)	(189)	(133)	(52)	(18)		(850)
Total	6,048	2,233	1,741	1,394	830	304	12,550
	(3,688)	(1,393)	(1,066)	(830)	(484)	(172)	(7,633)
CD 5 %	Depth – 61.1; radial distance – 58.5						

Figures in parentheses dry weight in g

The wood is hard and heavy, having a specific gravity of 0.93. It is an excellent fuel tree and makes good charcoal with a heat content of almost 4,900 K cal/kg. The trees coppice well and grow vigorously from stumps and root suckers. The other users of wood are for timber which is hard, strong, fine grained, and reddish color commonly used for agricultural implements, tent, pegs, golf club, living fence, tannin, silk, lac, and fodder (Table 4.14).

4.14 Root Distribution in Ber

Information about the spread of roots helps in fertilizer and water application in fruit trees. The root distribution system of 7-year-old budded ber plants of Cv. Gola was studied by excavation method at CHES, Godhra, under semiarid rainfed conditions. The radial spread was up to 180 cm from the tree trunk. Maximum root activity was recorded in 60–90 cm radial distance (7,359 g on fresh weight basis). As the radial distance increased, the percent root decreased. Depth wise the spread of root was up to 120 cm. Maximum root activity was observed at 60–90 cm depth (3,820 g on fresh weight basis). The roots were classified as thin (less than 10 mm diameter), medium (10–30 mm diameter), and thick (more than 30 mm diameter). Of the total roots excavated, medium-size roots were the highest (53.31 %) followed by thick roots (34.35 %) and then thin roots (9.95 %) (Tables 4.15 and 4.16).

4.15 Ber-Based Cropping Systems

Orcharding of fruit trees on fertile land and under irrigated conditions is popular in Gujarat, but least information is available on utilization of degraded shallow lands by horticulture under rainfed conditions. Moreover, the cropping under such type of land was also considered as uneconomical and only coarse millets were grown during *kharif* season.

Table 4.16 Root distribution according to size of root in g

	Radial distance (cm)							
Type of root	0–30	30–60	60–90	90–120	120–150	150–180	Total	% of roots
Thin ≤ 10 mm dia.	588	360	316	90	124	71	1,549	9.95
Medium ≥ 10–30 mm dia.	2,780	1,072	1,082	937	614	205	6,690	53.31
Thick ≥ 30 mm dia.	2,680	801	343	367	92	28	4,311	34.35
Total	6,048	2,233	1,741	1,394	830	304	12,550	–

To augment the income of the farmers in the initial years, ber orchards can be intercropped with various kharif vegetables as there is plenty of interspaces available; kharif vegetables like cowpea (*Vigna unguiculata*), okra (*Abelmoschus esculentus*), cluster bean (*Cyamopsis tetragonoloba*), brinjal (*Solanum melongena*), chilli (*Capsicum annum*), and leguminous fodder species *Stylosanthes hamata* and *Cenchrus ciliaris* were intercropped in a young ber orchards. There was no deleterious effect of intercropping of the growth and yield of ber crop (Hiwale and Raturi 1993a). The ber trees assumed full size covering the interspaces in the fifth year of planting. In the fifth year due to shedding, yield of intercrop was drastically reduced compared to the earlier years. Hence young ber orchard can be intercropped at least up to 5 years. Seed-sown crops like cowpea, cluster bean, and okra performed best compared to transplanted crops of brinjal and chilli. Thus, okra is less susceptible to climate fluctuations compared to cluster bean and cowpea yield of which was reduced during heavy rainfall years.

The 3-year average data revealed that sole plantation of ber has given 79.13 q/ha fruit yield, and there was no much variation in yield of ber when intercropped with either vegetables or grasses. Among intercropping of vegetables, brinjal ranked first (63.04 q/ha) followed by okra (54.91 q/ha) and cowpea (41.18 q/ha). The yield of grasses was reduced markedly in the 5th year of planting, and overall performance of grass yield with ber was poor. The annual return of ber + okra ranked first (Rs. 49,708.33 q/ha) followed by ber + cowpea (Rs. 45,722.66 q/ha) and ber + brinjal (Rs. 41,479.00 q/ha). The traditional systems of farming again rank at lowest and about 12 times less return than best combination of ber + okra. Thereafter, the yield of intercrops decreased due to shading effect of ber tree; however, there was no variation in the ber plant yield, which continued to increase (Hiwale 2004a, b) (Table 4.17).

Intercropping in ber orchard

Table 4.17 Economics of ber-based agroforestry systems

Crop combinations	Yield q/ha Main crop	Inter crop	Gross output (Rs./ha)	Gross input (Rs./ha)	Net Return (Rs./ha)
3rd year					
Ber	42.5	–	21,250.00	3,000.00	18,250.00
Ber + cowpea	41.7	55.86	37,608.00	6,000.00	31,608.00
Ber + okra	42.3	61.45	39,585.00	6,000.00	33,585.00
Ber + Dolichos	42.5	29.48	30,094.00	6,000.00	24,094.00
Ber + cluster bean	41.9	35.55	31,615.00	6,000.00	25,615.00
Ber + Cenchrus	43.1	72.32	25,166.00	4,000.00	21,165.00
Ber + Stylosanthes	41.8	87.84	25,292.00	4,000.00	21,292.00
Ber + chilli	42.8	28.80	30,040.00	6,000.00	24,040.00
Ber + brinjal	43.2	79.50	31,537.50	6,000.00	25,537.50
Maize + pigeon pea[a]	6.5	2.20	6,500.00	1,500.00	5,000.00
4th year					
Ber	86.4	–	43,200.00	3,000.00	40,200.00
Ber + cowpea	85.7	50.35	57,955.00	6,000.00	51,955.00
Ber + okra	85.2	55.20	59,160.00	6,000.00	53,160.00
Ber + Dolichos	86.4	28.74	51,822.00	6,000.00	45,822.00
Ber + cluster bean	86.2	34.35	53,405.00	6,000.00	47,405.00
Ber + Cenchrus	85.5	75.40	46,520.00	4,000.00	42,520.00
Ber + Stylosanthes	85.4	92.50	47,325.00	4,000.00	43,325.00
Ber + chilli	85.8	26.25	50,778.00	6,000.00	44,778.00
Ber + brinjal	86.0	74.99	52,373.75	6,000.00	46,373.75
Maize + pigeon pea[a]	5.20	2.40	6,200.00	1,500.00	4,700.00
5th year					
Ber	108.5	–	57,250.00	3,000.00	51,250.00
Ber + cowpea	108.8	17.35	59,605.00	6,000.00	53,605.00
Ber + okra	107.9	48.10	68,380.00	6,000.00	62,380.00
Ber + Dolichos	108.1	13.73	58,169.00	6,000.00	52,169.00
Ber + cluster bean	110.2	14.35	59,405.00	6,000.00	53,405.00
Ber + Cenchrus	109.0	27.54	55,877.00	4,000.00	51,877.00
Ber + Stylosanthes	108.3	22.80	55,288.00	4,000.00	51,288.00
Ber + chilli	108.6	12.27	57,981.00	6,000.00	51,981.00
Ber + brinjal	108.4	34.63	58,528.75	6,000.00	52,528.00
Maize + pigeon pea[a]	5.70	2.00	5,850.00	1,500.00	4,350.00
Mean of 3 years					
Ber	79.13	–	39,566.66	3,000.00	36,566.66
Ber + cowpea	78.73	41.18	51,722.66	6,000.00	45,722.66
Ber + okra	78.46	54.91	55,708.33	6,000.00	49,708.33
Ber + Dolichos	79.00	23.98	46,695.00	6,000.00	40,695.00
Ber + cluster bean	79.43	28.08	48,141.66	6,000.00	42,141.66
Ber + Cenchrus	79.20	58.42	42,521.00	4,000.00	38,521.00
Ber + Stylosanthes	78.50	67.71	42,605.00	4,000.00	38,635.00
Ber + chilli	79.06	22.44	46,266.33	6,000.00	40,266.33
Ber + brinjal	79.20	63.04	47,480.00	6,000.00	41,479.58
Maize + pigeon pea[a]	5.80	2.20	6,183.33	1,500.00	4,683.33

[a]Traditional system

4.16 Rejuvenation

The trees start bearing fruits in the 2nd year of planting. In ber it is a practice to prune the tees every year from the third year onward by keeping 60–90 cm growth of previous season wood. After 12–15 years, the plants' structure becomes compact due to mingling of branches with one another with formation of compact canopy.

This condition results in shedding of branches and low light interception. Also some of the old branches start drying and needs to be replaced. At this stage, it is necessary to rejuvenate ber trees by heading back the plant to the main framework. This should be done in the month of April to May when the trees go into dormancy. Three to four scaffold branches of 30–60 cm length are allowed on all sides of the plant. On the onset of monsoon, new sprouts start emerging which are then thinned allowing the healthy ones on all the four sides of the tree. This can be done manually by undertaking thinning of shoots. This results in balanced canopy development. The trees do not go into juvenile phase and start flowering and fruiting in the same season. However, the yield per plant is reduced to 20–25 % of the previous yield per plant. After 1 year, the plant's canopy looks balanced and the yield potential of the plant increases. The heading back also results in the production of large quantity of wood which can be sold in the market or can be used to prepare small agricultural equipments or can be burnt as fuelwood.

10 year ber plant regular pruning

Rejuvenated Ber plant

Ber can be cultivated on saline soils under semiarid rainfed conditions. *Ziziphus rotundifolia* is the best rootstock. In situ patch budding is found the best method to raise ber orchard under semiarid rainfed conditions. Additional income per unit area can be obtained from ber orchard raised at 5×5 m spacing with commercial varieties like Goma kirti, Gola, and Umran, up to 5 years of age. Due to shading effect, the intercrop cannot be successfully cultivated in the ber orchard. About 12 times higher income can be obtained from ber-based cropping system as compared to the traditional farming system of maize + pigeon pea.

There was improvement in soil health, i.e., organic carbon content, reduction in pH value, and addition of NPK to the soil. Apart from this, the pruned wood obtained from ber trees, which needs to be pruned every year, is useful as fuelwood and meets the yearly requirements of the standard family.

Thus, ber found to be multipurpose tree as it provides fruit, fodder, and fuel to the farming family apart from providing higher income through orchard intercropping in the initial years per unit area of land.

References

Bhargava BS, Raturi GB, Hiwale SS (1990) Leaf sampling technique in ber (*Zizyphus mauritiana* Lamk.) for nutritional diagnosis. Singap J Prim Ind 18(2):85–95

Dhillon BS, Singh K (1968) Effect of some plant regulators on fruit set and fruit drop in *Ziziphus* jujuba Linn. J Res Punjab Agric Univ Ludhiana 5:392–494

Gopani GD (1976) Arthik dristi se bor ki kheti. Gujarat Agricultural University, Junagadh Gujarat (In Guajarati)

Gupta AK (1984) Nutritive components of arid fruits crops and vegetables. In: Das HC (ed) Proceedings of summer institute on recent advances in Arid Horticulture, pp 1–11

Hiwale SS (1991) Determination of leaf area in jujbe (*ZiZIphus mauritiana*) using linear parameters. Indian J Agric Sci 61(5):335–336

Hiwale SS (2002) Root distribution studies in Ber (*Ziziphus amuritiana. Lamk.*). Ann Arid Zone 41(1):89–91

Hiwale SS (2004a) Develop sustainable agri–horti production systems on marginal lands. Technical Bulletin, pp 1–57

Hiwale SS (2004b) Cultivating Ber in the dry lands of Western India. (*Zizyphus mauritiana* Lamk). Indian Farm 53(12):20–23

Hiwale SS (2005) Goma kirti: anew ber variety. Indian Hortic (50) 1:24

Hiwale SS (2011) Canopy management through shoot pinching in ber (*Zizyphus mauritiana* Lamk.). Abstract of paper in proceedings of 2nd international Jujube symposium (In China)

Hiwale SS, Raturi GB (1993a) Horti-silvi-pastoral system for increased productivity of marginal and degraded lands under rainfed conditions. Adv Hortic For 3:114–117

Hiwale SS, Raturi GB (1993b) Influence of severity of pruning on growth, yield and quality of Ber (*Zizyphus mauritiana* Lamk.). Progress Hortic 25(3/4):161–163

Hiwale SS, Raturi GB (1996a) Propagation of Ber by in situ budding. Indian Hortic 41(3):24–25

Hiwale SS, Raturi GB (1996b) Effect of GA3 and Kinetin on seed germination and seedling growth in ber. In: Singh SP (ed) Advances in horticulture and forestry. Scientific Publisher, Jodhpur, pp 51–53

Misra AK, Jauhari OS (1970) Root induction in layers and stem cuttings of *Morus alba* and (*Zizyphus mauritiana* Lamk.) with special reference to plant growth regulators. Indian J Hortic 27(3/4):141–146

Pareek OP (1983) The Ber. ICAR, New Delhi

Randhawa GS, Kohli RR (1976) Ber cultivation un south India. Indian Hortic 21(2):25–27

Raturi GB, Hiwale SS (1992) Increasing productivity of degraded lands in dry land areas. Orissa J Hortic (20):8–12

Raturi GB, Hiwale SS (1993) Horti silvi pastoral systems for increased productivity of marginal and degraded lands under rainfed conditions. In: Singh SP (ed) Advances in horticulture and forestry. Scientific Publisher, Jodhpur

Singh P, Bajwa MS, Singh R (1973) Propagation of ber (*Zizyphus mauritiana* Lamk.) I. Effect of ringing and IBA on the stooling. Plant Sci 5:137–139

Singh SP, Singh IB, Singh VN (1987) A note on effect of gibberellic acid on fruit set, fruit retention and quality of jujube (*Zizyphus mauritiana* Lamk.) Cv. Banarasi Karaka. Progress Hortic 19(1–2):61–64

Teotia SS, Chauhan RS (1963) Flowering, pollination, fruit set fruit drop studies in ber (*Zizyphus mauritiana* Lamk.) I. Floral biology. Punjab Hortic J 3:60–70

Pomegranate (*Punica granatum* L.) 5

Abstract

Pomegranate (*Punica granatum* L.) is one of the important semiarid fruits cultivated commercially in over 1.25 Lakh hectare areas in Western India. Its cultivation is possible even on marginal degraded lands that have previously been found unsuitable for growing crops. Apart from this, its ability to withstand salinity in soil and water has, to some extent, led this crop to emerge as a hardy fruit crop. It is commercially grown for its sweet-acidic fruits, which provide a cool refreshing juice and are valued for their medicinal properties. Its popularity is also due to the ornamental nature of the plant, particularly the bright red flowers that appear throughout the year. Its antioxidant properties are well known. The juice and seed contain large quantities of tannin and agolic acid, which are essential in the cure of several diseases. The fruit can be successfully grown under even purely rain-fed conditions in semiarid areas. In recent years, pomegranate cultivation has become an economically viable proposition. Large acreage is covered with new varieties such as Bhagwa, Arakta, and Mridula. Furthermore, some anardana types have also developed Goma khatta.

5.1 Introduction

Pomegranate is emerging as one of the important fruit crops owing to its hardiness and ability to withstand adverse soil and climatic conditions. Today, India contributes 40–45 % of a global production of 10 Lakh tons. Currently, the export of pomegranates is around 10,000 t, which is just 5 % of the total export. With the increase in population, the domestic demand for fruit has also increased substantially. To date, of a total area of 1,25,000 ha. under pomegranate, the majority is found in the state of Maharashtra. The state has 85,000 ha. (68 % area) under the crop, of which only 51,000 ha is under production. This is followed by Karnataka at 12,727 ha. And Gujarat at 3,787 ha as at 2003–2004 (Anonymous 2005). Earlier orchards were raised mainly with the Ganesh variety, which was the leading variety. However, it is fast being replaced with new cultivars such as Bhagawa and Arakta, the skin and arils of which are an attractive color and, hence, are suitable for export.

The pomegranate is an ancient and favorite table fruit of tropical countries. It is commercially grown for its sweet-acidic fruits, which

S. Hiwale, *Sustainable Horticulture in Semiarid Dry Lands*,
DOI 10.1007/978-81-322-2244-6_5, © Springer India 2015

provide a cool refreshing juice and are valued for their medicinal properties. The fruit rind, juice, leaf, and roots are used in the preparation of various ayurvedic medicines. Its popularity is also due to the ornamental nature of the plant, especially the bright red flowers that appear throughout the year. The fruit juice is refreshing and contains both glucose and fructose. The juice and seed contain large quantities of tannin and agolic acid, which are essential in the cure of several diseases. The fruit can be successfully grown under even purely rain-fed conditions. In recent years, pomegranate cultivation has become an economically viable proposition.

Pomegranate (*Punica granatum* L.) belongs to the natural order Punicaceae; Punica perhaps is the only known genus of this family, which includes large shrubs or small trees with two species: *Punica protopunica* Balf. is found wild in Socotra Island, and the other, *Punica granatum*, is cultivated in sub-tropical parts of the world. *Punica granatum* has been classified into two sub-species: chlorocarpa and porphyrocarpa, each having two varieties. These sub-species have been established on the basis of the color of the ovary, a stable feature that is retained even when they are reproduced by seed. Sub-species chlorocarpa is mainly found in the *Transcaucasus*, whereas the second sub-species, porphyrocarpa, is mainly distributed in central Asia.

5.2 Uses

The pomegranate is a favorite table fruit in tropical countries. It is greatly liked for its cool refreshing juice and is also valued for its medicinal properties. The juice is considered useful for patients with leprosy, the bark and rind of the fruits are commonly used in dysentery and diarrhea, and the rind is also used as a dye for cloth. The fresh juice of pomegranate retains its flavor and keeps when bottled after sweetening with sugar. A kind of wine is also prepared from pomegranate juice, and is considered superior to grape wine.

5.3 Composition

Water	78.2 %	Potassium/100 g	133 mg
Protein	1.6 %	Iron/100 g	3 mg
Fat	0.1 %	Thiamine/100 g	0.6 mg
Fibers	5.1 %	Riboflavin/100 g	0.1 mg
Calcium/100 g	10 mg	Nicotinic acid/100 g	0.3 mg
Magnesium/100 g	12 mg	Vitamin C/100 g	14 mg
Phosphorus/100 g	70 mg	Oxalic acid/100 g	14 mg

Pomegranate is grown on large acreages in Maharashtra, followed by Karnataka (Table 5.1). However, productivity is highest in Tamil Nadu,

Table 5.1 Statewise area, production, and productivity of pomegranate

	Area (000' ha)			Production (000' Mt)			Productivity (Mt/ha)		
	2008–2009			2009–2010			2010–2011		
State	Area	Production	Pdy.	Area	Production	Pdy.	Area	Production	Pdy.
Maharashtra	82.0	550.0	6.7	98.9	555.5	5.6	82.0	492.0	6.0
Karnataka	14.3	138.1	9.7	13.2	138.5	10.5	13.6	142.6	10.5
Gujarat	4.0	39.3	9.8	4.4	45.6	10.4	5.8	60.3	10.4
Andhra Pradesh	6.5	64.7	10.0	5.6	56.4	10.0	2.8	27.8	10.0
Tamil Nadu	0.4	10.0	25.0	0.4	17.5	39.4	0.5	12.7	27.6
Rajasthan	0.6	3.5	5.5	0.8	4.8	6.4	0.8	5.5	6.6
Others	1.4	1.6	1.1	1.66	2.1	1.2	1.8	2.2	1.2
Total	109.2	807.2	7.4	125.0	820.3	6.6	107.3	743.1	6.9

although the area is the smallest. Growth in Maharashtra reduced over the years 2009–2010, probably due to die back and oily spot diseases. The same trend can be recorded in Andhra Pradesh. Thus, further increases in area under pomegranate must be planned carefully and with disease-free planting material.

5.4 Botany

Bearing Habit

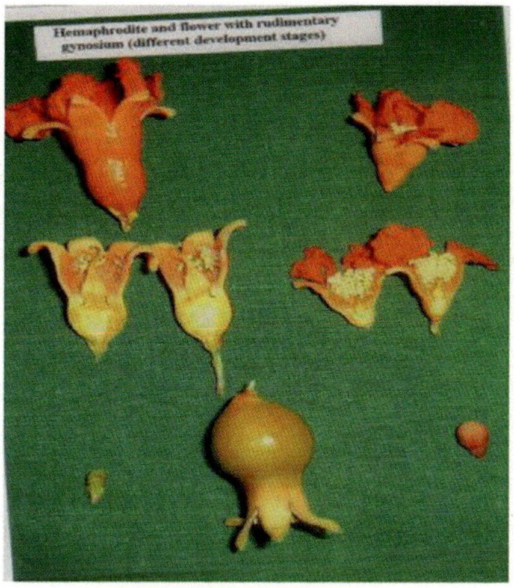

Types of flowers

5.5 Varietal Improvement

5.5.1 Germplasm Evaluation and Breeding

Several types of pomegranates with varying shapes, sizes, and colors are cultivated in India. Fruits are round, obvate in shape, and vary in diameter from 8 to 12 cm. The rind may be thick or thin, and the color ranges from pale yellow to crimson. The seed pulp in superior types is thick, fleshy, and very juicy; but it is thin in inferior types. The seed coat varies in hardness; some of the softer-seeded types are known as seedless (Bedana). A lack of lignifications of the testa is the main cause of so-called seedlessness in the pomegranate. Bedana and Kandhari are considered the best among the numerous types grown.

The varieties Ganesh, Mridula, Arakta, G-123, P-23, P-26, Mastani, and Bhagwa are popular in

Maharashtra. Bassein Seedless, Jyothi, Ganesh, Mridula, Arakta, and Ruby are popular in Karnataka; Dholka in Gujarat; Jalore Seedless, Jodhpur Red, and Jodhpur White in Rajasthan; and Kabul Red, Vellodu, Yercaud-1, and CO-1 in Tamil Nadu.

Although we know a great deal about the pomegranate and its origin, there are a number of seedling varieties in the pomegranate areas. It is better to select a variety with known qualities. Most horticulturists divide pomegranate varieties into three categories: sweet, sweet tart, and sour. However, we prefer to include the seed hardiness in this and divide them into six groups, as follows.

1. Soft-seeded sweet
2. Soft-seeded sweet tart
3. Early variety (mostly sweet)
4. Normal (harder) seeded sweet tart
5. Normal (harder) seeded sweet
6. Sour (nearly always normal seeded)

5.5.2 Breeding Criteria

Pomegranate is genetically heterozygous material, thus there is lot of variability existent in the nature. Therefore, there is scope for crossing two selected genotypes, raising of hybrid populations, and identification of superior types.

1. Cracking and splitting of fruits is a severe menace and occurs with variations in atmosphere and humidity.
2. Resistance to fruit borer and stem borer is desired.
3. To develop varieties with high sugar (16 %) and less acid (1.5 %) content.
4. All varieties bear some sterile flowers. This appreciably reduces yield. The selection of varieties bearing a high percentage of fertile flowers should be achieved.
5. To develop better varieties for Western India by hybridization between the locally adapted Alandi and Dholka types. Alandi types have medium-sized fruits, deep red to deep pink flesh, hard seeds, and quite sweet juice. However, they are slightly more acidic than Dholka types. Dholka types bear larger fruits with greenish white rind, whitish flesh, soft seeds, and sweet juice.
6. To develop better varieties for Southern India by hybridization between prolific bearing Paper Shell or Spanish Ruby, and the soft-seeded Musket Red and the large-sized-fruit-bearing Vellodu types.
7. Breeding varieties with dark red arils and skin.
8. Breeding Anardana-type varieties with dark red bold arils, high acidity, bigger fruit size, and a high yield.

5.5.3 Breeding Methods

5.5.3.1 Selfing

Bag the bud a day prior to anthesis. Tag and label it.

5.5.3.2 Crossing

Select an appropriate young bud that will open the next morning. Insert forceps through petals or remove one petal and take out the stamens. Take care that no anther is left inside the flower, and the ovary is not injured. This can be done easily, as stamens emerge from within the upper half or more of the hypanthium (Calyx tube) and are distant from the ovary. Bag and tag. Collect the desired anthers in yellow stage and incise them with a needle and transfer the pollen with the help of a brush onto the receptive stigma of the emasculated bud the next morning. Bag and label.

5.5.3.3 Hybridization

Hybridization is conducted in the morning before 11.00 am. The steps involved are as follows: emasculation (removing of anthers) using forceps; pollination with the pollen of the desired male parent; covering with butter paper bag and labeling. Fruits require about 120–130 days for maturity. Crossed fruits are harvested. The seeds are extracted and sown in plastic trays; they germinate within 10 days of sowing. The seedlings are transplanted to polythene bags and subsequently to field for further evaluation, Bhandari (1979).

5.5.4 Genetic Evaluation of Germplasm

The world's largest collection, comprising 1,117 entries, is maintained at Turkmenistan (CIS). They exhibit wide variability for fruit size and quality, skin color, resistance to abiotic and biotic stresses; however, most are temperate in nature and do not grow well in the Indian tropics. There is scope to use this material for breeding superior types.

Several varieties have adapted to both the tropical and the sub-temperate climate. They are either evergreen or semi-deciduous or deciduous. The flowering habit, fruiting, and flower physiology alters depending on the habitat. In the tropical climate with the mild winter of south India, growth and flowering are continuous processes; in the sub-tropical climate of north India, the trees remain dormant during the cold winter and flower in the following spring. In temperate climates, flowering occurs in summer.

In India, cultivar 'Ganesh' was one of the leading varieties. It is a seedling selection from 'Alandi' with a large fruit (400–450 g), sweet taste (16–17 °Bx TSS) and soft seeds; however, the aril color is not attractive (pink or light pink). Later on, two new varieties with Ganesh as a base and with red arils were released: Arkta from Mahatma Phule Krishi Vidyapeeth (MPKV) Rahuri and Ruby from the Indian Institute of Horticultural Research (IIHR Bangalore). In both varieties, the red aril color was imparted by the Russian varieties Gulsha Red and Gulsha Rose Pink. Ruby is a multiple hybrid resembling more of Ganesh, while Arkta is an F2 selection. Recently, a large area in Maharashtra has been planted with a new variety (Bhagwa), which is a selection made by a farmer. It has a deep red aril and an attractive skin color, but the variety takes longer to mature (160–170 days). Several other seedling selections are grown on a limited scale across the country: Madugiri, G-137, Muscat, Jyothi, Bassein seedless, Dholka, and Jalore Seedless in the states of Karnataka, Tamil Nadu, Haryana, Gujarat, and Rajasthan. In Himachal Pradesh, a sour type known as Daru is found abundantly in the wild and is mostly used for anardana preparation. A new F1 hybrid developed at IIHR Bangalore and named 'Amlidana' has been released. Most varieties have one or another superior trait that is susceptible to bacterial nodal blight, a disease that has threatened the cultivation of the pomegranate. The quality parameters are strongly influenced by soil and climatic conditions. New varieties with desirable traits such as bold arils with red color and soft seeds and resistance to blight, drought, cracking, and aril browning are needed.

Evaluation of pomegranate germplasm under semiarid rain-fed conditions indicated that seeded varieties bear fruits with higher fruit weight, whereas their aril seed ratio was the lowest and 100 seed-weight the highest. Seasonal variation affected all the physico-chemical characteristics studied. In hasta bahar, the quality of the fruit was the best compared with mrig bahar fruits. All characters showed increasing trends except acidity, which declined in hasta bahar. The seeds of hasta bahar were harder than mrig bahar fruits, but the juice color was attractive in hasta bahar. Fruit weight ranged from 200 to 95 g in mrig bahar but from 261 to 141 g in hasta bahar, fruit length varied from 75.5 to 50.0 mm and 75.5 to 66.5 mm in mrig and hasta bahar, respectively. Fruit diameter varied from 73.5 to 50.05 mm and 78 to 70.2 mm in both mrig and hasta bahar, respectively. Aril seed ratio varied from 1.42 to 0.63 in mrig and 3.02 to 0.69 in hasta bahar. Juice percent varied from 58.3 % to 77.3 % in mrig bahar and 78.25 % to 62.29 % in hasta bahar. Acidity of the fruit ranged from 1.45 % to 0.35 % in mrig and 0.90 % to 0.28 % in hasta bahar. Purohit (1985) studied the seed softness of commercial varieties and reported that the thickness of the testa, weight of whole seed, density of the seed, and weight of testa as % of seed weight can be used to determine the seed softness of the pomegranate. His scale indicated that variety in thickness of seed testa <0.5 mm, the density of whole seed <0.6 g/ml, the density of seed testa <0.4 g/ml, weight of the seed testa <50 % of the weight of whole seed. Bedana, Bassein seedless, Ganesh, and Dholka were found to be soft seeded. Although Bedana has soft seeds, it cannot be grown commercially owing to its deciduous nature and sparse flowering.

Evaluation of 24 pomegranate germplasm genotypes carried out by Meena et al. (2003) indicated that all the characters except total soluble solids (TSS) and total sugar had less pronounced differences between the value of phenotypic and genotypic coefficient of variation. Genotypic coefficients of variation were higher for seed content, weight of 100 aril, weight of 100 seeds, acidity, and average fruit weight. The results indicated that a higher magnitude of variation for the above traits offered better opportunities for improvement through selection. These results are in concordance with those of Chaturvedi et al. (1980) and Pandey and Bist (1998). Higher estimates of heritability were obtained for all the characters except TSS, ascorbic acid and total sugar content, which may be due to environmental effects on these traits. Acidity and weight of 100 seeds had high heritability in spite of low genetic advance, which may be due to introduction of replication in the system (Burton and De Van 1953), while low genetic variability and low genetic advance was noted in total soluble solid. High heritability estimates, coupled with high-expected genetic advance as percent of mean, were observed for mean seed content, weight of 100 seeds, acidity content, and average fruit weight, indicating that high heritability coupled with high genetic advance might be due to additive genes effect on these traits. These characteristics can be considered as reliable selection indices. High to moderate heritability, along with low genetic advance were recorded for total sugar and ascorbic acid content indicating that these characteristics might be controlled by non-additive gene action, and improvement through selection for these characteristics would be rather ineffective. Panse and Sukhatme (1978) noted that, if a characteristic is governed by additive gene action, heritability and genetic advance would both be high. High estimates of heritability, along with high genetic advance, provide good scope for further improvement in advance generation if these characteristics are subjected to mass progeny or family selection (Table 5.2).

Table 5.2 Germplasm evaluation in pomegranate

Variety	Season	Fruit wt. (g)	Fruit length (mm)	Fruit dia. (mm)	Aril / seed ratio	Juice %	TSS °Bx	Acidity %	Juice color	100 seed wt. (g)
P-26	Mrig	153.32	69.1	65.5	1.41	75.21	14.0	0.351	Light pink	2.14
	Hasta	233.00	74.5	75.5	2.19	73.15	15.4	0.31		2.57
Kabul	Mrig	105.15	62.0	59.5	0.88	71.28	13.4	0.427	Light pink	1.63
	Hasta	194.00	74.5	73.5	1.30	75.43	16.4	0.28	Red	2.12
Jalore seedless	Mrig	095.73	54.75	50.5	0.93	61.70	12.4	0.78	Cream	1.87
	Hasta	176.50	67.5	70.5	1.35	76.14	15.2	0.28	Pink	2.25
Musket	Mrig	198.20	75.5	73.5	1.42	74.06	14.4	0.46	Cream	1.55
	Hasta	227.00	71.5	76.2	1.57	77.17	16.4	0.28	Pink	2.49
Jyoti	Mrig	192.14	71.4	73.0	1.41	73.69	14.4	0.35	Light pink	1.29
	Hasta	215.00	73.5	75.5	1.64	78.25	18.0	0.35	Pink	1.85
Ramna-aram	Mrig	200.68	71.0	70.5	0.63	58.32	9.4	0.63	Light pink	3.76
	Hasta	261.00	75.5	78.0	0.69	64.92	13.4	0.49	Red	4.77
Gulsha Rose	Mrig	127.62	56.0	64.5	0.70	77.30	11.4	1.45	Red	2.98
	Hasta	188.00	66.5	75.0	0.90	73.75	13.6	0.90	Dark red	3.56
Appuli	Mrig	117.38	50.05	53.5	0.71	67.05	13.4	0.63	Light pink	2.89
	Hasta	216.00	67.5	74.0	1.34	62.29	16.2	0.49	Pink	3.66
Ganesh	Mrig	134.00	134.0	70.1	1.89	77.0	16.6	0.44	Light pink	1.508
	Hasta	141.00	141.0	70.2	2.70	77.0	17.6	0.34	Pink	1.96
P-16	Mrig	110.48	61.5	57.5	1.26	67.17	13.3	0.462	Cream	1.96
	Hasta	141.83	70.9	70.24	3.03	78.00	16.6	0.369	Light pink	2.57

Hiwale (2001)

Anardana germplasm lines collected at the center were evaluated under rain-fed conditions; all lines flowered and fruited in all the bahar. Some lines had desirable characteristics such as dark red arils, soft seeds, high acidity, bold arils, high sugar acid ratio, etc. Line H in particular had high acidity, large fruit size, soft seeds, and the aril color was light pink. Compared with other crosses, the least growth was observed in line H and hence it is suitable for high-density farming. Based on growth data, the cultivar is found to be dwarf and, at 5×4 m spacing, 500 plants can be accommodated in a hectare. It requires 150–160 days to mature.

There is no long gestation period. The plants start bearing fruit in the third year of planting. There was 15–20 % cracking of the fruits. Suitable for processing as Anardana, mixing with Juice and Squash with other sweet varieties. High acidity, bigger fruit size, and high TSS compared with different crosses make it suitable for processing and export of anardana. Yield per plant was 11.08 kg under the rain-fed conditions of Gujarat. The hybrid also yielded Anardana to the tune of 1.55 kg/plant (Hiwale et al. 2009) (Table 5.3).

5.6 Hybridization

The commercial cultivars Ganesh and Musket, though prolific bearers and soft seeded, lacked the dark red color of the Russian types, which were, however, sour in taste. The hybridization program therefore was initiated at MPKV, Rahuri (Anonymous 1979). The F1 progeny was screened (Kale 1986) but none of the hybrids could be used directly because of defects observed in them. The F2 progeny developed from open-pollinated F1 resulted in the identification of a superior type that combined all the desirable attributes and was named Mridula (Keskar et al. 1994). Hybridization work carried out at IIHR Bangalore resulted in the identification of multiple cross-hybrid 'Arka Ruby', which was released at the institute level but further multiplication was stopped due to the incidence of bacterial blight (Table 5.4).

Hybridization work done at IIHR Bangalore has resulted in the identification of a hybrid Arka Ruby, which was a multiple-cross.

Table 5.3 Physico -chemical characters in anardana-type pomegranate germplasm

Acc. No.	Fruit set/plant	Fruit wt (g)	Fruit length (mm)	Fruit Dia. (mm)	Skin wt (g)	Total aril wt (g)	Juice %	TSS °Bx	100 aril wt (g)	Aril length (mm)	Aril dia. (mm)	Juice color
A.	85.0	77.0	68.3	58.9	53.3	95.3	35.2	17.1	95.3	9.76	6.5	L. pink
B.	59.0	67.0	54.0	48.4	22.0	43.6	39.4	11.1	42.3	9.46	5.16	pink
C.	17.7	104.0	64.5	61.3	37.0	80.6	37.0	17.0	83.6	9.76	5.03	D. pink
E.	48.7	86.3	50.5	53.3	33.3	53.0	36.6	11.5	53.0	9.46	9.93	L. pink
F.	26.3	77.6	54.7	50.0	26.3	45.0	55.4	15.8	44.3	8.03	4.6	D. pink
H.	40.0	137.3	57.3	63.2	35.0	96.6	46.7	16.3	53.0	10.7	5.9	Pink
I.	18.0	88.6	49.3	55.4	25.3	62.0	40.9	16.6	96.6	9.76	8.9	Pink
J.	28.7	95.6	54.7	54.7	20.0	71.6	52.6	14.6	62.0	11.0	8.96	L. pink
K.	29.7	71.6	51.7	48.3	21.6	44.6	77.2	16.6	71.0	9.4	5.23	Pink
C D 5 %	14.9	27.6	NS	9.1	12.2	36.9	–	1.40	44.6	0.57	0.53	–

Table 5.4 Physico chemical character of some of the hybrids identified

Cross	Fruit wt. (g)	Fruit size (cm)	Aril colour	TSS (%)	Acidity (%)	Softness of seed
Ganesh × Shirin Anar	414	9.2×8.9	Pink	16.3	0.36	Soft
Ganesh × Gulsha rose pink	244	7.5×7.7	Dark red	17.9	0.47	Soft
Ganesh × Gulsha red	221	7.6×7.4	Red	19.1	0.55	Soft
Arka ruby	249	7.2×7.8	Dark red	17.2	0.64	Soft

Ganesh x Kabul ↓ F1 x Yercaud ↓ F1 x ↓ F2

Ganesh x Gulsha Rose Pink ↓ ↓ ↓ F1 x ↓ F2

F2 ----------------------- X ----------------------- F2

↓

F1 (Arka Ruby)

Table 5.5 Characters of hybrids developed at CHES Godhra

Character	Ganesh × Jalore seedless	Ganesh × Sindhuri	Ganesh × (Ganesh × Nana)	R1P2
Male flowers	30	21	10	45
Hermaphrodite flowers	41	20	5	10
No. of fruit set	35	31	43	25
No. of days to maturity	115	120	120	117
Fruit wt (g)	249.66	216	182.5	129
Length (mm)	78.1	66.5	62.06	65
Dia. (mm)	93.0	71.5	68.33	62
Skin wt. (g)	59.66	52.0	41.25	46
Aril wt. (g)	187.00	165.33	138	93
100 aril wt (g)	30.33	26.0	26	19
Aril length (mm)	10.5	10.0	10.75	9.25
Aril dia.(mm)	8.5	7.87	6.5	4.30
Aril color	Pinkish white		Dark red	Red
Aril	Bold	Bold	Medium	Small
TSS °Bx	14.83	15.1	15.1	14
Acidity	0.35	0.47	0.50	0.36

Breeding work initiated at CHES Vejalpur using various cross combinations along with chemical mutagen colchicine @ 0.5 % resulted in the identification of good F1 hybrids. However, mutation resulted in smaller-size fruits, which is not a desirable characteristic (Table 5.5).

5.6.1 Hybrid-1

Hybrid-1 is a hybrid between Ganesh × Line I of anardana-type fruit. The fruit size is superior than the male parent and comparable to the female parent. The hybrid has a dark red aril. The arils are of medium size (10.75 × 6.5 mm) with a fruit-to-aril ratio of 1.32 and fruit-to-skin ratio of 4.42. Percent fruit set was 67.64; fruit retention was 60.29 %. TSS were 15.1 °Bx, with an acidity of 0.48 %. The yield potential was 8.21 kg per plant under semiarid rain-fed conditions.

5.6.2 Hybrid-2

Hybrid-2 is a cross between Ganesh × Sindhuri. The hybrid was found to have bold arils (10.0 × 7.87 mm) and bigger fruit (216 g.). However, the arils are light pink in color, with a fruit-to-aril ratio of 1.31 and fruit-to-skin ratio of 4.15. Maturity occurs in just 120–125 days in the hasta bahar crop under semiarid rain-fed conditions. The hybrid is found to be superior to its parents in respect of fruit set (70.85 %), fruit retention (59.95 %), and ultimate yield per plant (8.86 kg) in 4-year-old plants.

5.6.3 Hybrid-3

Hybrid-3 is a cross between Ganesh × Jalore seedless. It is superior to both parents in fruit and aril size. The mean fruit size for the 3-year data was 249.66 g and aril size was 10.5 × 8.5 mm (bold aril), with a fruit-to-aril ratio of 1.33 and fruit-to-skin ratio of 4.18. Maturity occurs in just 110–115 days in the hasta bahar crop under semi-arid rain-fed conditions. The hybrid is found to be superior to its parents in respect of fruit set (71.69 %), fruit retention (51.88 %), and ultimate yield per plant (13.73 kg) in 4-year-old plants.

Pomegranate hybrid-1 – (Ganesh x R10 P10)

Pomegranate Hybrid -2 (F1-Ganesh x Sindhuri)

Pomegranate Hybrid -3 (F1- Ganesh x Jalore seedless)

'Selfing' is conducted by bagging the bud a day prior to anthesis. The bud should then be tagged and labeled. Select an appropriate young bud that will open the next morning. Insert forceps through petals or remove one petal and take out the stamens. Take care that no anther is left inside the flower and that the ovary is not injured. This can be done easily, as stamens emerge from within the upper half or more of the hypanthium (Calyx tube) and are distant from the ovary. Bag and tag. Crossing is carried out the next morning: collect the desired anthers in yellow stage and incise them with a needle and transfer the pollen with the help of a brush onto the receptive stigma of the emasculated bud. Bag and label. Hybridization is carried out in the morning before 11.00 am. The steps involved are emasculation (removing of anthers) using forceps. Pollinate with the pollen of the desired male parent. Cover with a butter paper bag and label it. Fruits require about 120–130 days for maturity. Crossed fruits are harvested, and the seeds extracted; they are then sown in plastic trays. Seeds germinate within 10 days of sowing. The seedlings are transplanted to polythene bags and subsequently to field for further evaluation, Bhandari (1979).

5.7 Breeding for Bacterial Blight Resistance in Pomegranate

Kalpitiya was identified as a source of resistance to bacterial blight in the pomegranate. A total of 60 crosses were made with Nana, Daru, Yellow, Double flower, and Ganesh. Advanced breeding lines with favorable traits have been raised.

5.8 Evaluation of Hybrids in Pomegranate

A total of 482 seedlings of Daru and Nana where subjected to challenge inoculation with bacterial blight culture. Screening of 52 hybrids between Nana, Yellow, and Double flower revealed that all had hard seeds. In the second round, Kalpitiya × Ruby (30 hybrids) did not show a positive reaction to challenge with bacteria. A total of 11 plants were planted to test field resistance to bacterial blight of pomegranate (BBP) and the remaining 14 plants were sent to NRCP, Solapur. A total of 50 seedling of a cross between Bhagwa × (Ganesh × Daru) × Daru and Bhagwa × (Ganesh

× Daru) × Nana showing resistance to Bacterial bblight of pomegranate were planted in the field. Screening results revealed that 315 of 476 plants expressed Bacterial bblight of pomegranate symptoms. Further screening of 161 plants not showing Bacterial bblight of pomegranate symptoms is underway.

Rahuri center:

The breeding program developing varieties that were resistant to bacterial blight was terminated as both male parents (i.e. Daru and Nana) were found to be susceptible. The genetic improvement of pomegranate through breeding initiated and crossing of Bhagwa × Kabul yellow, Mridula × Kabul yellow, Phule Arkta × Kabul yellow was undertaken. A total of 192 F1 hybrids of Musket × Mridula, Phule Arkta, and Bhagwa were planted in the field.

- Future Lines of Work

Breeding to resistance/tolerance of biotic and abiotic stresses, particularly bacterial blight, Fusarium wilt, fruit rot, and fruit borer. Identification of genes for increasing storage life through and its incorporation through different breeding methods.

5.9 Commercial Varieties

5.9.1 Ganesh

Ganesh is a selection from the cultivar Alandi. The growth habit of the tree is spreading type with evergreen nature. The leaf is entire and elliptic in shape. The tree flowers throughout the year, with three main flushes. Fruit is round, smooth, and pink with a reddish tinge. The maximum weight per fruit was 255 g in mrig bahar and 211 g in ambe bahar. The length of the fruit was found to vary: 7.84 cm in mrig bahar and 7.41 cm in ambe bahar. The breadth also showed a similar trend: 7.96 cm in mrig bahar and 7.23 cm in ambe bahar. Arils are sweet and light pink. The juice percent varied from 52.76 % in mrig bahar and 51.54 % in ambe bahar. The TSS of the juice was observed to be 16.6 °Bx in mrig bahar and 17.4 °Bx in ambe bahar. The acidity was 0.44 % in mrig bahar and 0.34 % in ambe bahar. Seeds were found to be the softest in ambe bahar (1.15 kg/ cm^2) compared with mrig bahar seeds (1.59 kg cm^2 pressure). The variety is susceptible to fruit borer (18.55 %), leaf spot (PDI-15.72), and fruit spot (PDI-22.86).

5.9.2 Jalore Seedless

The growth habit of the Jalore seedless tree is spreading type with evergreen nature. The leaf is entire and elliptic in shape. The tree flowers throughout the year, with three main flushes. Fruit is round, smooth, and pink with a reddish tinge. The maximum weight per fruit was 175 g in mrig bahar and 154 g in ambe bahar. Yield per plant on a 6-year-old plant was 9.5 kg. The length of the fruit varied from 7.24 cm in mrig bahar and 6.84 cm in ambe bahar. The breadth also showed a similar trend: 6.67 cm in mrig bahar and 6.35 cm in ambe bahar. Arils are sweet and light pink. The juice percent varied from 51.72 % in mrig bahar to 50.25 % in ambe bahar. The TSS of the juice was 13.2 °Bx in mrig bahar and 15.2 °Bx in ambe bahar. Acidity was 0.59 % in mrig bahar and 0.46 % in ambe bahar. Seeds were soft, with the softest in ambe bahar (2.24 kg/cm^2 pressure) compared with mrig bahar seeds (2.42 kg/ cm^2 pressure). The variety is susceptible to fruit borer (29.53 %), leaf spot (PDI-22.66), and fruit spot (PDI-29.66).

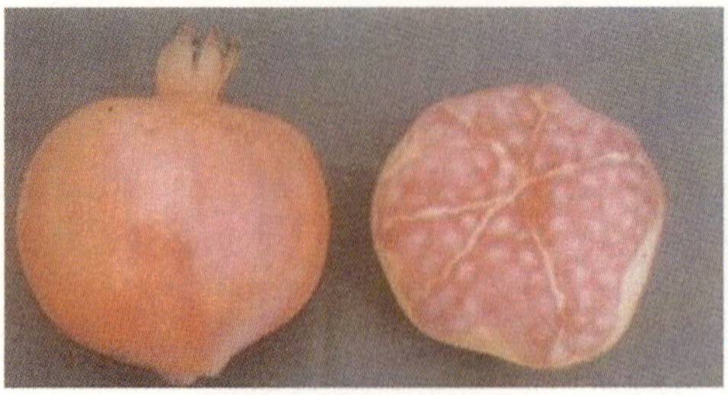

Jalore Seedless

Dholka

5.9.3 Dholka

The Dholka variety is grown in the Dholka area of Gujarat. The growth habit of the tree is spreading type with evergreen nature. The leaf is entire and elliptic in shape. The tree flowers throughout the year, with three main flushes. Fruit is round, smooth, and light green with a reddish tinge. Maximum weight per fruit was 298 g in mrig bahar and 211 g in ambe bahar. The length of the fruit varied from 8.35 cm in mrig bahar to 7.88 cm in ambe bahar. The breadth also showed a similar trend: 8.26 cm in mrig bahar and 7.92 cm in ambe bahar. Arils are sweet and light pink. The juice percent varied from 51.95 % in mrig bahar to 52.53 % in ambe bahar. The TSS of the juice was 16.2 °Bx in mrig bahar and 16.6 °Bx in ambe bahar. The acidity was 0.51 % in mrig bahar and 0.47 % in ambe bahar. Seeds were softest in ambe bahar (1.31 kg/cm^2 pressure) compared with mrig bahar seeds (1.25 kg cm^2 pressure). The variety is susceptible to fruit borer (25.44 %), leaf spot (PDI-18.00), and fruit spot (PDI-25.40).

5.9.4 P-13

The growth habit of the tree is spreading type with evergreen nature. The leaf is entire and elliptic in shape. The tree flowers throughout the year, with three main flushes. Fruit is round, smooth, and yellow with a reddish tinge. Maximum weight per fruit was 233 g in mrig bahar and 217 g in ambe bahar. The length of the fruit varied from 8.21 cm in mrig bahar to 7.84 cm in ambe bahar. The breadth also showed a similar trend: 7.51 cm in mrig bahar and 7.12 cm in ambe bahar. Arils are sweet and light pink. The juice percent varied from 49.52 % in mrig bahar to 53.5 % in ambe bahar. The TSS of the juice was 15.4 °Bx in mrig bahar and 14 °Bx in ambe bahar. Seeds were softest in ambe bahar fruit (1.64 kg/cm^2 pressure) compared with mrig bahar (1.82 kg cm^2 pressure). The variety is susceptible to fruit borer, leaf spot, and fruit spot.

5.9.5 P-16

The growth habit of the P-16 tree is spreading type with evergreen nature. The leaf is entire and elliptic in shape. The tree flowers throughout the year, with three main flushes. Fruit is round, smooth, and pink with a reddish tinge. Maximum weight per fruit was 245 g in mrig bahar and 201 g in ambe bahar. The length of the fruit varied from 8.11 cm in mrig bahar to 7.52 cm in ambe bahar. The breadth also showed a similar trend: 7.89 cm in mrig bahar and 7.35 cm in ambe bahar. Arils are sweet and light pink. The juice varied from 48.92 % in mrig bahar to 52.69 % in ambe bahar. The TSS of the juice was 16.2 °Bx in mrig bahar and 15.8 °Bx in ambe bahar. The acidity was 0.47 % in mrig bahar and 0.38 % in ambe bahar. Seeds were softest in ambe bahar (1.63 kg/cm^2 pressure) compared with mrig bahar seeds (2.14 kg cm^2 pressure). The variety is susceptible to fruit borer (26.97 %), leaf spot (PD-I9.50), and fruit spot (PDI-23.33).

5.9.6 P-23

P-23 is a selection from Muscat. The growth habit of the tree is spreading type with evergreen nature. The leaf is entire and elliptic in shape. The tree flowers throughout the year, with three main flushes. The fruit is round, smooth, and pink with a reddish tinge. Maximum weight per fruit is 260.5 g in mrig bahar and 231 g in ambe bahar. The length of the fruit varied from 8.07 cm in mrig bahar to 7.73 cm in ambe bahar. The breadth also showed a similar trend: 7.94 cm in mrig bahar and 7.62 cm in ambe bahar. Arils are sweet and light pink. The juice percent varied from 49.65 % in mrig bahar to 51.65 % in ambe bahar. The TSS of the juice was 17.2 °Bx in mrig bahar and 16.4 °Bx in ambe bahar. The acidity was 0.49 % in mrig bahar and 0.58 % in ambe bahar. Seeds were softest in ambe bahar (1.92 kg/cm^2 pressure) compared with mrig bahar seeds (2.06 kg cm^2 pressure). The variety is susceptible to fruit borer (34.82 %), leaf spot (PDI-15.5), and fruit spot (PDI-24.00).

5.9.7 P-26

P-26 is also a selection from Muscat. The growth habit of the tree is spreading type with evergreen nature. The leaf is entire and elliptic in shape. The tree flowers throughout the year, with three main flushes. Fruit is round, smooth, and yellow with a reddish tinge. Maximum weight per fruit was 268 g in mrig bahar and 229 g in ambe bahar. The length of the fruit varied from 8.08 cm in mrig bahar to 7.82 cm in ambe bahar. The breadth also showed a similar trend: 7.89 cm in mrig bahar and 7.51 cm in ambe bahar. Arils are sweet and light pink. The juice percent varied from 51.45 % in mrig bahar to 52.69 % in ambe bahar. The TSS of the juice was 14.62 °Bx in mrig bahar and 15.2 °Bx in ambe bahar. The acidity was 0.43 % in mrig bahar and 0.36 % in ambe bahar. Seeds were softest in ambe bahar (1.46 kg/cm^2 pressure) compared with mrig bahar seeds (2.2 kg cm^2 pressure). The variety is susceptible to fruit borer (29.65 %), leaf spot (PDI-11.00), and fruit spot (PDI-20.33).

5.9.8 G-137

The growth habit of the G-137 tree is spreading type with evergreen nature. The leaf is entire and elliptic in shape. The tree flowers throughout the year, with three main flushes. Fruit is round, smooth, and pink with a reddish tinge. Maximum weight per fruit was 270 g in mrig bahar and 232 g in ambe bahar. The length of the fruit varied from 8.41 cm in mrig bahar to 7.83 cm in ambe bahar. The breadth also showed a similar trend: 7.96 cm in mrig bahar and 7.23 cm in ambe bahar. Arils are sweet and light pink. The juice percent varied from 55.23 % in mrig bahar to 54.87 % in ambe bahar. The TSS of the juice was 17.0 °Bx in mrig bahar and 17.4 °Bx in ambe bahar. The acidity was 0.49 % in mrig bahar and 0.42 % in ambe bahar. Seeds were softest in ambe bahar (1.04 kg/cm^2 pressure) compared with mrig bahar seeds (1.27 kg cm^2 pressure). The variety is susceptible to fruit borer (20.42 %), leaf spot (PDI-18.45), and fruit spot (PDI-31.74).

5.9.9 Musket

The growth habit of the musket tree is spreading type with evergreen nature. The leaf is entire and elliptic in shape. The tree flowers throughout the year, with three main flushes. Fruit is round, smooth, and pink with a reddish tinge. Maximum weight per fruit was 310.27 g in mrig bahar and 219.14 g in ambe bahar. The length of the fruit varied from 8.34 cm in mrig bahar to 7.79 cm in ambe bahar. The breadth also showed a similar trend: 8.24 cm in mrig bahar and 7.21 cm in ambe bahar. Arils are sweet and light pink in color. The juice percent varied from 50.33 % in mrig bahar to 49.74 % in ambe bahar. The TSS of the juice was 14.4 °Bx in mrig bahar and 14.8 °Bx in ambe bahar. The acidity was 0.46 % in mrig bahar and 0.39 % in ambe bahar. Seeds were softest in ambe bahar (1.64 kg/cm^2 pressure) compared with mrig bahar seeds (1.82 kg cm^2 pressure). The variety is susceptible to fruit borer, leaf spot, and fruit spot.

5.9.10 Kandhari

The growth habit of the Kandhari tree is spreading type with evergreen nature. The leaf is entire and elliptic in shape. The tree flowers throughout the year, with three main flushes. Fruit is round, smooth, and pink with a reddish tinge. Maximum weight per fruit was 317.63 g in mrig bahar and 190.40 g in ambe bahar. The length of the fruit varied from 8.98 in mrig bahar to 7.24 cm in ambe bahar. The breadth also showed a similar trend: 8.44 cm in mrig bahar and 7.11 cm in ambe bahar. Arils are sweet and light pink in color. The juice percent varied from 52.33 % in mrig bahar to 50.15 % in ambe bahar. The TSS of the juice was 14.8 °Bx in mrig bahar and 14.8 °Bx in ambe bahar. The acidity was 0.41 % in mrig bahar and 0.40 % in ambe bahar. Seeds were softest in ambe bahar (1.26 kg/cm^2 pressure) compared with mrig bahar seeds (1.58 kg cm^2 pressure). The variety is susceptible to fruit borer, leaf spot, and fruit spot.

Kandhari

Kabul

5.9.11 Kabul

The growth habit of the Kabul tree is spreading type with evergreen nature. The leaf is entire and elliptic in shape. The tree flowers throughout the year, with three main flushes. Fruit is round, smooth, and yellow with a reddish tinge. Maximum weight per fruit was 298.18 g in mrig bahar and 193.67 g in ambe bahar. The length of the fruit varied from 8.17 cm in mrig bahar to 6.56 cm in ambe bahar. The breadth also showed a similar trend: 8.3 cm in mrig bahar and 7.03 cm in ambe bahar. Arils are sweet and light pink. The juice percent varied from 48.09 % in mrig bahar to 53.20 % in ambe bahar. The TSS of the juice was 16.2 °Bx in mrig bahar and 17.4 °Bx in ambe bahar. The acidity was 0.55 % in ambe bahar and 0.47 % in mrig bahar. Seeds were softest in ambe bahar (1.66 kg/cm^2 pressure) compared with mrig bahar seeds (1.78 kg cm^2 pressure). The variety is susceptible to fruit borer, leaf spot, and fruit spot.

5.9.12 Mridula

Mridula is a hybrid developed at MPKV Rahuri from the open-pollinated population of F1 from the cross Ganesh × Gul-E-Shah Red. The growth habit of the tree is spreading type with evergreen nature. The leaf is entire and elliptic in shape. The tree flowers throughout the year, with three main flushes. Fruit is round, smooth, and pink with a reddish tinge. Maximum weight per fruit was 252 g in mrig bahar and 244 g in ambe bahar. The length of the fruit varied from 8.94 cm in mrig bahar to 7.21 cm in ambe bahar. The breadth also showed a similar trend: 7.52 cm in mrig bahar and 7.19 cm in ambe bahar. Arils are sweet and light pink in color. The juice percent varied from 51.79 % in mrig bahar to 50.15 % in ambe bahar. The TSS of the juice was 14.92 °Bx in mrig bahar and 14.8 °Bx in ambe bahar. The acidity was on par in both the seasons, at 0.45 %. Seeds were softest in ambe bahar (1.09 kg/cm^2 pressure) compared with mrig bahar seeds (1.28 kg cm^2 pressure). The variety is susceptible to fruit borer, leaf spot, and fruit spot.

Mridula

Bhagwa

5.9.13 Bhagwa

The Bhagwa variety is known by various names: Shendri, Asthagandh, Mastani, Jai Maharashtra, and red Diana in various parts of Maharashtra. The variety is, to date, the leading variety in Maharashtra state. It is a heavy yielder and possesses fruit characteristics desirable for the domestic as well as the export market, such as an attractive fruit color, dark red arils, and no cracking. The fruits mature late at 180–190 days after fruit set. An average yield is 30–40 kg fruit per tree. The fruit has attractive thick, glossy, saffron-colored skin, and bold dark red arils. The variety is less susceptible to fruit cracking, fruit spot, and thrips. The market price is 2–3 times higher than that gained for other varieties.

5.9.14 Phule Arakta

The Phule Arakta variety is released by MPKV. It is a heavy yielder and suitable for export due to the attractive skin and aril color. Fruits are ready for harvest within 120–130 days.

5.9.15 Alandi

Alandi is a local variety found growing in Maharashtra. A spreading type variety, it produces medium-sized fruits with hard seeds, red arils, and a sweet taste.

5.9.16 Jyoti

Jyoti is a highly promising type previously known as GKVK-1. It is a selection from basin seedless and produces fruits with soft seeds and red aril.

5.9.17 Arka Ruby

Arka Ruby is a multiple hybrid between Ganesh × Gulsha rose pink. It is released from the IIHR Bangalore. It was developed for its aril color, which is dark red in winter. The seeds are softer. The fruit is small with deep red arils. The variety is a medium yielder.

5.9.18 Amlidana

Amlidana is an F1 hybrid (Ganesh×Nana). It grows well in the tropical climate and has good-quality fruit attributes. Amlidana is superior to the sour Daru variety, whose trees grow naturally in temperate regions of north India. Its fruits provide more acidic (16–18 %) anardana; the fruit yield per tree is higher, and short-statured trees are suitable for high-density planting, resulting in an increased fruit yield per unit area. It is therefore recommended for commercial cultivation (Jalicop et al. 2002).

5.9.19 Goma Khatta

Goma Khatta is an F1 hybrid between Ganesh × Nana cultivar identified for high acidity, high yield and yield of fruits and anardana. Subsequently releasd at Central Horticultural Experiment Station Godhra, Gujarat. The cultivar is suitable for preparation of anardana. Flowering takes in two flushes. A higher number of fruit set per plant than nine other crosses, resulting in higher fruit retention and, in turn, a higher yield. It also has high acid content, an ideal characteristic for making anardana (Table 5.6).

Goma Khatta

Table 5.6 Performance of Bhagwa and Mridula at CHES Vejalpur

	Bhagwa	Bhagwa	Mridula	Mridula	
Character	Mrig	Hasta	Mrig	Hasta	Ganesh
Male flowers	185.6	112	98.6	75	124.2
Hermaphrodite fl.	76.3	46	37.3	31	140
Total flowers	261.9	158	135.9	106	264.2
No. of fruit set	31.6	38.5	26.0	42.8	82.2
No. of fruit retained	27	22.6	21.3	13.6	65.0
% set	36.75	29.13	55.28	29.24	52.99
% retention	85.44	77.58	81.92	31.77	79.07
No. days to maturity	180	200	140	135	130
Fruit wt g	137	180	171	215	195
Length (mm)	64.1	65.4	60.6	69.5	68.7
Dia. (mm)	64.7	66.8	68,8	71.3	70.4
Skin wt. (g)	61	113	68	84.6	58
Aril wt. (g)	77	112	107	139.4	132
Aril color	Dark red	Dark red	Dark red	Dark red	Light pink
TSS °Bx	9.5	14	11.0	13	16
Acidity	0.58	0.47	0.33	0.38	0.32

5.10 Plant Propagation and Nursery Management

The pomegranate is emerging as an important fruit crop for semiarid and arid areas of India. With the identification of superior varieties such as Ganesh, Mridula, Arka Ruby, and Bhagwa, the area under pomegranate is rapidly increasing. Therefore, the demand for elite planting material is also rapidly increasing. Pomegranate has traditionally been propagated vegetatively, either by cutting or air layering. However, modern orchards raised from tissue culture are also very common. Nursery as a business proposition can be as profitable as raising a crop.

Different methods of pomegranate propagation

5.10.1 Seed

Raising of seedlings from seeds either in situ or in a nursery

In situ raising of seedlings is necessary because it helps in the development of strong root systems, creating sturdy plants that can better withstand the vagaries of the monsoon and have a longer life than notified plants. However, the gestation period is higher.

Another method of raising seedlings is by sowing seeds in raised beds at 30 × 30 cm spacing and, once the seedlings are 5–10 cm tall, transplanting them to already filled polythene bags with a mixture of soil, Farm yard manure (FYM), and sand in the ratio of 1:1:1; the use of vermicompost in place of FYM is advocated. The soil should also be drenched with copper fungicides at 3 g/L water to prevent infection with soil-born fungi. Before sowing, the seeds are treated with bacterial culture or vesicular arbuscular mahycorhize (VAM). The method is easy and less expensive, with a high percent of success with a long orchard life. It is currently mostly used for the evaluation of breeding materials and germplasm material obtained from overseas. The disadvantages are a long gestation period, genetic variability in the material, a long dormancy period and very low viability in some seeds. The seeds can be treated with bacterial culture of azotobacter at 30 g/kg seed or with VAM culture to improve germination and rooting.

5.10.2 Cutting

The commercial propagation of pomegranate through cuttings is not as successful as air layering due to unsatisfactory rooting. However, it is still successfully used in humid areas. Although root augmentation through growth regulators like auxin is a dependable option, it is costly.

Semi-hard woodcuttings of young healthy plants from the matured stem with light white streaks on them are more successful. Cuttings approximately 20–25 cm long are removed from the plants and their leaves are immediately removed to control loss of water from the cutting. The cuttings are then wrapped in moist sphagnum moss or newspaper if they are to be transported long distances. However, the percent rooting in these cutting is found to be very low at 20 %. Treating these cuttings with 4-indole-3-butyric acid (IBA) at 2,000 ppm for 15–20 min was found to be useful in increasing percent success to 60 %. The best time for taking a cutting is at the onset of the monsoon. Raised beds at 30 × 30 cm spacing are prepared, and the cutting receives a slanting cut at the bottom; a one-third portion is inserted into the soil to facilitate rooting. Once the seedlings are 5–10 cm tall, they are transplanted to already filled polythene bags with a mixture of soil, FYM, and sand at a ratio of 1:1:1. Vermicompost can be used in place of FYM as it contains a good amount of nutrients. The soil should also be drenched with copper

Table 5.7 Percent rooting of cutting in cultivar Ganesh

Treatment	Cutting rooted (%)	Survival (%)	Auxine (ưg/ml)
IBA 200 ppm	53.30	66.82	
Trichoderma harzianum	73.30	76.77	51
Azospirillum brasilense	23.30	74.68	32
Azospirillum lipoferum	36.20	60.07	21
Control	33.33	34.07	–
CD at 5 %	17.57	–	–

fungicides at 3 g/L water to prevent infection by soil-born fungi. Before sowing, the cuttings are treated with bacterial culture or VAM to obtain more roots. Using beneficial microbes is a novel approach to reducing the cost of chemicals and enhancing rooting/survival of the cuttings Liberman (1993). *Agrobacterium rhizogenes* and *Trichoderma harzianum* are known to produce growth regulators that have a beneficial effect on plant growth (Riker et al. 1930). The unique root-initiating ability of these microorganisms may lead to early setting and better rooting, resulting in improved vigor of freshly transplanted material. Percent rooting of cuttings was maximized in cultivar Ganesh in *Trichoderma harzianum-treated* rooting of cuttings. Maximum indole-3-acetic acid (IAA) (51 ưg/ml) was also noted in the same treatment Patil et al. (2004) (Table 5.7).

5.10.3 Air Layering

Air layering is one of the most successful methods for large-scale propagation of pomegranate. However, it is time specific, i.e. can be most successful when performed after one or two showers. The success percent is as high as 80 % even under rain-fed conditions. The matured shoots of healthy plants are selected, and a 1.5–2 cm portion of shoot cambium is removed; the cut portion is then covered with sphagnum moss dipped in water or in a solution prepared as below:

Potassium nitrate 0.28 g/L	Calcium nitrate 0.8 g/L
Potassium hypophosphate 0.2 g/L	Magnesium sulphate 0.2 g/L
Ferric tartrate 1 ml of 0.5 % solution	

Vermiwash can also be used as it also has been found to have hormonal and anti-fungal properties. Sphagnum moss can withhold about five times its weight in water and hence can provide a suitable atmosphere for rooting. The moss is then placed on 15 × 15 cm pieces of polythene sheeting, wrapped around the cut portion and tied with sutli at both ends. The air layers root in about 30 days, when they can be removed from the mother plant; their leaves are removed to prevent loss of water and to allow for easy transportation. After removing the polythene sheet, but keeping the moss intact, the layers are planted in polythene bags pre-filled with the soil mixture. Root length and weight was higher in air layers than in cuttings under semiarid rain-fed conditions. The maximum root length was 27 cm with air layering as compared with 12.45 cm in rooted cuttings. Fresh root weight was 1.45 g as compared with 0.513 g in cuttings, indicating greater success in the nursery after detachment of air layers from the plant as compared with cuttings.

Pomegranate air layer

5.10.4 Tissue Culture

Micro propagation is an improved technique that can be used for large-scale propagation of elite plant material under controlled conditions. This technology is increasing in India, and it is becoming an industry with the opening of trade for the export of plant material. This has the potential to be a profitable industry. In pomegranate, protocols have also been standardized for tissue-culturing of the shrub via a shoot tip culture or cotyledon callusing, as follows. Elite material was collected and sterilized with sodium hypochlorite 0.5 % for 5 min, followed by treatment with 0.1 % $HgCl_2$. This was followed by successive rinsing with distilled water and placing the sterilized materials on Murashige and Skoog (MS) media. Murukute et al. (2002) worked on profuse callusing and differentiation of callus into whole plants in pomegranate cultivar Ganesh using leaf segments and cotyledon explants. Cotyledon explants were found to be free from lethal leaching of phenolic compound on MS basal medium fortified with 6-benzyl amino purine (BAP) and naphthalene acetic acid (NAA). Better performance was obtained with a BAP + NAA combination; cotyledon explants responded better than leaf segments. Profuse callusing was obtained in a MS medium with BAP (1.0 mg L^{-1} + NAA 0.5 mg L^{-1}), and good rooting was obtained in a one-half MS + IBA 1.0 mg L^{-1} combination. Direct regeneration of shoots, roots, and whole plant with callusing intervention has also been investigated by Naik et al. (1999). Use of the anther wall (Morgud et al.) and leaf segment (Omula et al.) resulted in the production of only a few seedlings. In somatic embryo genesis (Narayana and Neelambica), there was no scope for screening of variants.

5.10.5 Green House Propagation

High-quality reliable green houses with controlled atmospheric conditions are required to achieve the desired success. Temperature control is most commonly manipulated via mechanical ventilation, along with fog cooling and adequate shading systems. In pomegranate, mist propagation increases success in cuttings to 80 %; however, the cost of establishing a mist chamber is around Rs. 1.5 Lakh.

5.11 Cultural Management

5.11.1 Soil and Climate

Commercial pomegranate orchards are confined mainly to the Mediterranean and semiarid regions of the world. It is a hardy fruit that can be grown successfully, even in low fertile soils. Although high-quality pomegranate can only be grown in locations with a cool winter and a hot dry summer, the trees grow under a wide range of climatic conditions—from the plains to an elevation of about 1,829 m. The tree can withstand frost but is injured by temperatures below 11.1 °C.

The tree is deciduous in areas with low winter temperatures but is evergreen or partially deciduous under tropical and sub-tropical conditions. The tree cannot produce sweet fruits unless temperatures are high for a sufficiently long period. The quality of fruit is adversely affected in humid climates. It is a hardy plant and can withstand a considerable amount of drought but does better when irrigated regularly. It can also flourish well on land that is too wet for many other crops.

The pomegranate can be grown on diverse soil types, including those that are considered unsuitable for most other fruit trees. It can grow quite well in both foothills and at high elevations. It can tolerate soils that are limy and slightly alkaline. The deep loamy or alluvial soils are ideal for its cultivation, though it can be grown in medium or light black soils of minimum depth of 60 cm.

5.11.2 Preparation of Land and Planting

Land selected for cultivation of the pomegranate is ploughed deeply with a furrow turning plough, followed by 3–4 fine tillage, and layout is according to a square or hexagonal system. Planting is in previously dug pits of 1 m^3. The pits should be filled with two to three baskets of FYM, 250 g of super phosphate, along with red earth and a small quantity of sand mixed thoroughly before the start of the rainy season. Planting should be carried out during monsoon for better and more rapid establishment of the plants. The planting should be spaced 3–6 m apart. In Maharashtra, a spacing of 3–5 m is generally followed, while in south India, a spacing of 2 m is common. However, the trees should be given a spacing of 5–6 m in poor, shallow soils of the Deccan and 6–7 m in the rich alluvial soils of Gujarat and North India.

- *Spacing*: In pomegranate, earlier recommendations were for spacing of 5 × 5 m; however, the mechanization of cultivation and adaptation of drip irrigation has seen the spacing adopted by farmers move to 14 × 14 ft, 12 × 12 ft, 15 × 10 ft, 10 × 10 ft, or 12 × 10 ft.

5.11.3 Bahar/Resting Treatment to Regulate Fruiting

Bahar treatments such as root pruning, root exposure, withholding of water, defoliation of plants by hand or by chemical means, etc. are practiced to induce moisture stress, so the plants drop their leaves and growth can be controlled. The main object of these treatments is to regulate the crop by forcing the tree to rest and profusely flower and fruit during any one of three bahars. The treatment also helps obtain uniform and good-quality fruits with maximum production and a lower incidence of pests and diseases. Normally, only two bahars are practiced in Maharashtra and the Dholka area of Gujarat. The trees shed their leaves, and recommended doses of fertilizers are applied depending on the selection of the bahar; however, a basal dose of FYM, phosphorus, and potash with a one-half dose of nitrogen are applied at the onset of monsoon, and fertilizers are applied.

Allowing a particular bahar or flowering depends on the market rate of fruits, the arrival of produce to the market, water availability to the plants, type of soil, insect pest threat, and weather conditions during the crop growth period. Under tropical south Indian climatic conditions, the pomegranate flowers continuously throughout the year. In central and western India, there are three distinct seasons for flowering: ambe bahar (January–February), mrig bahar (June–July), and hasta bahar (October). Ambe bahar is most commonly practiced by the growers of Maharashtra because of high yield, consequent to profuse flowering as compared with other bahars. However, assured irrigation is a must to utilise this bahar. In mrig bahar there will also be profuse flowering. In Karnataka, flowering is observed in June–October and March. Under north Indian conditions, the trees remain dormant during winter and flower only during spring. During monsoon, a flush of flowers are observed from April to June. Under Delhi conditions, depending on cultivars grown, flowering may be once or twice a year. In Himachal Pradesh, which is a temperate climate, flowering is seen during the middle of April.

For mrig bahar, plant growth must be suppressed during December to April by withholding water. This practice is generally followed in south India. By doing so, the trees shed their leaves in March and remain dormant until May. The land is then ploughed, followed by the application of manures and fertilizers. Plants are irrigated until the rainy season, and the fruits are harvested during October–November. In northwest India, irrigation is withheld from December to May, then regular digging is taken up followed by the application of manures and fertilizers during June along with normal irrigation until the rains start. Plants start growing by June, and fruits are harvested during October–December. Flower and fruit thinning is practiced to encourage good-size fruits. To induce flowering and fruit drop, applications of 2,000 ppm ethephon and 500–3,000 ppm alar have been tried. Fruit-let thinning has also helped in growing good-size fruit. At Rahuri, the fruit size in Muscat pomegranate during ambe bahar progressively increased when the number of fruits per tree was decreased from 70 to 20; the gross income was highest when only 50 fruits were retained per tree. Flowering varied widely during mrig and hasta bahar (Table 5.8).

Table 5.8 Effect of bahar on flowering in pomegranate cultivar Ganesh

	Mrig				Hasta			
	1986	1987	1988	Mean	1986	1987	1988	Mean
No. of hermaphrodite flower	219.5 (28.88)	236.57 (12.29)	301.7 (28.32)	**252.59 (20.21)**	249.6 (22.13)	304.8 (34.10)	284.5 (28.44)	**279.63 (27.76)**
No. of staminate flowers	468.57 (61.63)	1,054.71 (54 82)	568.6 (53.37)	**697.29 (55.80)**	788.8 (69.95)	420.4 (47.03)	531.2 (53.11)	**580.13 (57.59)**
No. of unopened flowers	72 (9.47)	617.14 (32.08)	195 (18.30)	**294.71 (23.58)**	89.2 (7.91)	168.6 (18.86)	184.5 (18.44)	**147.43 (14.64)**
No. of flowers dropped	540.57 (71.12)	1,671.85 (86.91)	763.6 (69.14)	**992.01 (79.38)**	878.0 (77.86)	489.0 (54.71)	715.7 (71.55)	**694.23 (68.92)**
Total flowers	760.07	1,923.57	1,065.3	**1,249.6**	1,127.6	893.80	1,000.20	**1,007.2**

Figures in parentheses are percent values

The highest total number of flowers was produced in mrig bahar (1,249.6). Total flower drop was highest (28 %), and maximum hermaphrodite flower production (27.76 %) and reduced flower drops (68.92 %) was observed in hasta bahar, which resulted in higher fruit set and ultimate yield (Raturi and Hiwale 1991).

5.11.4 Fruit Development

In mrig bahar, fruit development in cultivar Ganesh indicated that, after fruit set on the onset of monsoon, it took around 120 days for the fruit to reach maturity. Fruit length and diameter initially increased at a faster rate (up to 60 days) and remained constant. Fruit weight and volume increased constantly until maturity. A similar trend was also observed in hasta bahar crops.

5.12 Irrigation/Drip/Fertigation

Pomegranate is considered to be drought-hardy fruit crop. However, supplemental irrigation is necessary to realize a higher fruit yield and better quality. Irrigation is also used to regulate cropping in pomegranate; the total water applied therefore depends on the crop for a desired bahar. In Gujarat and Maharashtra, where ambe bahar is taken, regular irrigation is required during March to July and irrigation is withdrawn after the fruit harvest until January. For mrig bahar, crop irrigation is not required from July to September. However, in prolonged breaks in the monsoon, one or two supplementary irrigations are required at an interval of 10–15 days during October to December. In hasta bahar (September flowering), the prolonged rainy season helps reduce the requirements for irrigation. Three to four irrigations at monthly intervals after cessation of rainfall are necessary to obtain good fruit size and higher yield (Table 5.9).

Table 5.9 Crop coefficient for different crops

Name/age of tree	1st year	2nd year	3rd year	4th year	5th year
Pomegranate	0.4	0.4	0.6–0.65	0.6–0.65	0.60–0.65
Ber	0.4	0.4	0.6–0.65	0.6–0.65	0.60–0.65
Fig	0.4	0.4	0.6–0.65	0.6–0.65	0.60–0.65
Aonla	0.4	0.4	0.5	0.6	0.65–0.70
K. Lime	0.4	0.4	0.5	0.6	0.70–0.75
Guava	0.4	0.4	0.5	0.6	0.70–0.75

Formula for calculation of quantity of water to be applied per plant per day

$$A \times B \times C \times D \times E$$

A = Evapotranspiration/day × Pan Coefficient (0.7–0.8)
B = Crop Coefficient (From table)
C = Crop Canopy (At 12 noon)
D = Spacing
E = Area wetted = Crop canopy/Spacing (Table 5.10)

Table 5.10 Growth parameters of pomegranate Ganesh

Year	Pl. height (m)	Stem dia. (mm)	Spread (m) NS	Spread (m) EW	No. of fruit set/plant
3rd	1.4	14.1	1.35	1.4	45
4th	1.76	17.9	1.58	1.7	75
5th	2.36	36.4	2.35	2.2	120
6th	3.03	48.4	2.8	2.7	80

5.12.1 Quantity of Water to Be Applied Per Plant Per Day in Pomegranate Cultivar Ganesh

In case of flood irrigation during winter, irrigation should be conducted at an interval of 12–15 days, and during summer at an interval of 5–7 days depending on the bahar. However, with the modern limited water availability, drip irrigation/fertigation is becoming popular. According to one estimate, this method leads to a 98 % increase in production, with water savings to the tune of 45 %. The total area brought under drip in Maharashtra is around 27,217.39 ha (10.26 %).

5.12.2 Mulching and Drip Irrigation in Pomegranate Cultivar Jyothi

Drip irrigation and mulching maximizes water use efficiency of pomegranate. Total flower and hermaphrodite flower numbers were significantly higher at 0.70 E.pan and 200 gauge polyethylene mulch. The ratio of hermaphrodite to staminate flowers was maximum (2.54) in the interaction of three emitters per plant at 0.70 E.pan irrigation with 400 gauge polyethylene. Fruit set (56.39 and 51.94) and fruit per plant (55.36 and 45.89) were significantly higher with 0.70 E.pan replenishment with 400 gauge black polyethylene mulch, respectively Jagannath et al. (1999).

5.13 Fertigation in Pomegranate

Application of fertilizers through drip is gaining importance because of the following reasons:

- Reductions in the application costs of fertilizers, along with water.
- Balanced application of fertilizers based on the needs of the plant.
- The fertilizers are applied near the active root zone of the plant, which helps absorption by the roots.
- High fertilizer use efficiency can be achieved.
- Savings in the total quantity of fertilizer applied.
- Fertilizers are applied based on the stage of the crop.

5.13.1 New Growth, Flowering

For uniform and profuse flowering, phosphorus should be applied 3-4 times more than nitrogen, quantity of potash applied is lowered. In the beginning, to achieve uniform growth 20:20:20 fertilizer should be applied. Therefore, after shoot growth until fruit set, the ratio of fertilizer applied should be either 12:61:0 or 13:40:13.

5.13.2 Fruit Set and Development

To retain a higher number of leaves on the plant to ensure that maximum photosynthetic activity takes place during fruit development, the ratio of NPK (nitrogen, phosphorus, potassium) should be 2:1:3 (e.g. 16:8:24 or 13:5:26). Demand for calcium is increased at this stage; therefore, calcium nitrate should be applied via drip. This will help reduce fruit cracking and improve fruit size (Tables 5.11 and 5.12).

At MPKV Rahuri, fertilizers were applied via conventional methods and via drip. The study revealed that the yield obtained (11.88 t/ha) with the 100 % NPK recommended dose of solid soluble fertilizers via drip was significantly superior over the 100 % recommended dose of conventional fertilizer and was on par with 70 % nitrogen, 80 % phosphorus and potassium, and 70 % NPK. Thus, the maximum fertilizer saving (N, P, and K) was 30 % each compared with conventional fertilizer. The yield-contributing parameters were significantly influenced by different

Table 5.11 Application of liquid fertilizers

Liquid fertilizer	Total fertilizer required (kg/acre)	Weekly dose applied (kg/week/acre)	Time of application
20:20:20 + urea	10 + 10	10 + 10	First irrigation
20:20:20 + 12:61:0 + urea	15 + 50 + 22	3 + 10 + 4.40	6–25 days (5 weeks)
13:40: 13 + 16:8:24	50 + 25	10 + 5 (5 weeks)	36–70 days (weeks)
13.00:45	50	12.50 (4 weeks)	71–98 days (4 weeks)
00:00:50	50	10	91–133 days (5 weeks)
06:12:36	25	12.50	After 134 days

levels of solid soluble fertilizer over conventional applications. Fertigation levels did not significantly influence the juice quality and organoleptic evaluation for color and taste Firke and Kumbhar (2002) (Table 5.13).

Prasad and Bankar (2003) reported that fruits become ready for harvesting 4–5 months after fruit setting. They are harvested when their rind color turns yellow and pink to red. Fruits give a 'crunch' sound when pressed. Thus, a yield of 60–100 fruits per tree may be obtained. The right time for fruit picking is from the first week of December to the end of February. On average, 25–30 kg of fruit per plant may be harvested under arid conditions.

Table 5.12 Yield and yield-contributing parameters of pomegranate

Treatment	Fruit no. /tree	Fruit wt. (g)	Fruit size (cm)	100 aril wt. (g)	Yield (t/ha)	Water use efficiency (kg/ha/mm)
Conventional fertilizers						
T1 (100 % RDCF + SL)	73.45	191.0	7.00	31.27	8.71	7.71
T2 (100 % RDCF + DI)	76.50	200.0	7.46	33.60	9.54	12.08
T3 (100 % RDCF, N drip)	80.25	209.0	7.72	36.77	10.36	13.12
Solid soluble fertilizers						
T4 (100 % NPK + DI)	84.80	226.0	8.22	42.95	11.88	15.05
T5 (70 % N, 80 % P & K + DI)	82.60	220.0	8.10	40.20	11.27	14.28
T6 (70 % NPK + DI)	81.50	212.0	7.85	37.87	10.81	13.70
T7 (50 % N, 80 % P & K + DI)	75.15	145.0	7.23	33.40	9.14	11.58
T8 (50 % N, 70 % P & K + DI)	75.25	191.0	7.81	32.75	8.90	11.27
C. D. at 5 %	6.21	0.02	0.57	4.20	1.12	–

Table 5.13 Effect of drip irrigation on yield and quality of fruits

Treatment (L/h)	Juice (%)	Cracking (%)	Fruit weight (g)	TSS °Bx	Yield (kg/plant)
4	46.3	25.8	204.9	18.0	18.8
8	53.6	18.9	319.4	18.4	28.2
12	53.0	18.6	314.7	17.4	30.1
Control	44.4	30.1	160.2	19.3	17.7

5.13.3 Fruit Ripening

At the fruit ripening stage, the requirement for potash is higher than that for nitrogen and phosphorus, therefore 0:0:50 + 18 % sulphur is applied. However, light soils deficient in nitrogen are fertilized with 13:0:45 grade fertilizer. After the initial harvest of big fruit, a second dose of fertilizer is applied to increase development in under-developed fruits through 6:12:36 grade fertilizer.

5.14 Manures and Fertilizers

Although pomegranate grows well in soils of low fertility, production can be increased with the application of manures and fertilizers. Initial soil analysis is desirable for proper scheduling of fertilizer. The nutrition recommendation depends on fertility of the soil and also the age of the plant. For young plants (2–3 months old), application of 200 g Neem cake along with 3–4 kg FYM per plant is recommended. After 3 months, each plant may be given 250 g di ammonium phosphate (DAP), along with 250 g Neem cake and 5 kg FYM. Again, after 9 months, application of 500 g DAP, 250 g potassium sulphate, 1 kg Neem cake, and 10 kg of FYM per plant is recommended.

The first harvest is desirable when the plant is around 12–16 months old. There are three distinct periods of fruiting for the pomegranate, called 'bahars'. Before each bahar, the plant needs 1–1.5 months of rest. During this time, the nitrogen level in the soil should be low. Once the fruits are harvested, the plants must be supplied with a first dose of fertilizers along with a fungicide and insecticide spray. The recommended dose of fertilizer per plant is 20 kg of FYM, 1 kg (7:10:5), 1 kg DAP, 1 kg Neem cake, 100 g magnesium sulphate, and 10 kg of borax. The plant is given 1–1.5 months of rest. After this period, apply 250 g ammonium sulphate along with 250 g of 19:19:19, 500 g 7:10:5, 500 g Neem cake, and 500 g muriate of potash per plant. When the plants are 4–5 years old, the fertilizer dose should be increased 1.5 times. In general, a basal dose of about 25–35 cartloads of FYM per hectare, besides the recommended dose of NPK, must be applied during January, June, and September in equal proportions. From 4 to 5 years onwards, nitrogen should be applied in two split doses: the first during irrigation after bahar treatment, and the second after 3 weeks. The full dose of phosphorus and potassium should be applied as a single dose. The requirement for chemical fertilizers also varies from place to place. In Gujarat, recommendations are for 375 g nitrogen, 375 g phosphorus, and 375 g potassium per tree per year. Under Udaipur conditions, recommendations are that a 4-year-old tree needs 720 g nitrogen, 200 g phosphorus, and 220 g potassium per year. Under Yercaud conditions, recommendations are that young plants are started with 50 g nitrogen, 200 g phosphorus, 100 g potassium, and that, as the plant becomes old, the dose should be increased to 125 g nitrogen, 1,000 g phosphorus, and 500 g potassium. Under Jodhpur conditions, 700 g nitrogen per plant is ideal for 10-year-old Jalor seedless plants.

Under Rahuri conditions, plants that are 1.5–2 years old need 250 g nitrogen, 125 g phosphorus, and 125 g potassium; plants aged 2.5–3 years need 500 g nitrogen, 125 g phosphorus, and 125 g potassium; those aged 4.5–5 years need 500 g nitrogen, 250 g phosphorus, and 250 g potassium; and plants aged 6–7 years need 625 g nitrogen, 250 g phosphorus, and 250 g potassium (Tables 5.14, 5.15, and 5.16).

At IIHR Bangalore, the application of 500 g nitrogen + 250 g phosphorus + 125 g potassium per plant per year to the Bassein seedless variety of pomegranate gave the highest yield: 8.2 kg of fruit from a 3-year-old plant (first crop). Application of 500 g nitrogen, 250 g phosphorus, and 250 g potassium per plant per year to cultivar Ganesh gave the highest yield of 78.67 kg (50 fruits per plant) during the fourth year of cropping.

Based on the survey, it is recommended that combined foliar application of 0.3 % zinc sulphate and 0.1 % boron at pre-bloom, bloom, and post-bloom stages should be conducted to augment the deficiency of these nutrients. Soil application can also be carried out by incorporating boron (10 ppm) and zinc (50 ppm) in compost 3

Table 5.14 Critical levels of pomegranate leaf composition

Nutrient	Concentration (%)	Sr. no.	Nutrient	Concentration (%)
Nitrogen	2.5	6	Sulphur	0.18
Phosphorus	0.2	7	Iron (ppm)	95
Potash	1.47	8	Manganese (ppm)	45
Calcium	1.30	9	Zinc (ppm)	22
Magnesium	0.70	10	Boron (ppm)	30

Table 5.15 Pomegranate orchard survey in Maharashtra soil characteristic (mean six orchards)

Place	P^H	E.C. (dsm^1)	Org. carbon	K (ppm)	Zn (ppm)	Mn (ppm)	Cu (ppm)	Fe (ppm)	B (ppm)
Pune	8.05	0.81	0.53	174	0.51	8.83	0.25	6.42	0.47
Solapur	8.35	1.42	0.24	189	0.46	8.13	0.36	5.3	0.48

Sharma and Hiwale (2005)

Table 5.16 Leaf composition (mean of six orchards)

	Zn (ppm)	Mn (ppm)	Fe (ppm)	K (%)		B (ppm)
Place	Young leaf	Young leaf	Young leaf	Old leaf	Mg (ppm)	Young leaf
Pune	12.33	73.5	161.0	0.79	0.32	22.0
Solapur	12.00	44.66	99.33	0.59	0.31	27.0

Sharma and Hiwale (2005)

months before its application. Though the soils are rich in calcium, fruit cracking may be due to a combined deficiency of calcium and boron. Hence, two sprays of calcium chloride at 0.5 % and boric acid at 0.1 % is recommended at 15-day intervals to avoid cracking.

Commercial micronutrient formulations are used to increase fruit growth, yield, and quality. Two sprays of 0.33 % Macroliq proved to be the best, followed by 1 % Macroliq. Soil application of 12.5 g boron and 45 g zinc sulphate per tree per year gave fruits with higher sugar content. Application of three sprays of ferrous sulphate (0.4 %)+manganese sulphate (0.3 %)+boric acid (0.2 %) along with zinc sulphate (0.3 %) at monthly intervals before flowering, at full bloom and at fruit set stage increased tree growth, yield, and fruit quality in cultivar Ganesh.

5.14.1 Organic and Inorganic Fertilizers

An investigation carried out on pomegranate cultivar Ganesh regarding the suitability of bahar under rain-fed semiarid conditions revealed that hasta bahar was the best. Organic farming is increasing in importance in India, with rising threats to food security and natural resources. Indiscriminate use of chemical fertilizers, pesticides, and weedicides has adversely affected soil fertility, productivity, and produce quality. Soil and water pollution with heavy metals released by fertilizers and pesticides have resulted in health hazards. The organic farming system relies on crop rotation, crop residues, animal manures, and legumes in an intercropping system, and green manure organic cakes as byproducts of plants. A trial was conducted at Central Horticultural Experiment Station (CHES) Vejalpur, with the objective of reducing the use of chemical fertilizers and increasing the use of organic fertilizers without compromising the production potential of the pomegranate. In the study, a six-treatment combination of organic and inorganic fertilizers, along with the recommended doses for semiarid regions, was applied at the onset of monsoon in the 4th and 5th years of plant growth in cultivar Ganesh. Leaf sampling was conducted via the collection of the 8th pair of leaves from the growing tip as suggested by Bhargava and Dhandhar (1987).

Fruit retention and yield per plant were significantly influenced by various treatments. Maximum fruit set (96.50 fruits per plant), fruit retention (57 fruits per plant), and yield (10.75 kg per plant) were recorded in treatment application of nitrogen 50 % through FYM, 25 % through castor cake, and urea. Similar results were reported by Shinde (1977). Phadnis (1974) also recommended mixed doses of organic and inorganic fertilizers to obtain maximum fruiting and yield. Wavhal et al. (1985) also recommended the application of nitrogen along with *azotobacter* for obtaining higher production in pomegranate cultivar Ganesh.

Fruit weight and fruit length (but not fruit diameter) were significantly influenced by all the treatments, and TSS by various treatments. Fruit weight (188.75 g), fruit length (69.72 mm), fruit diameter (66.15 mm), and TSS (16.90 °Bx) were highest with the application of nitrogen 50 % through FYM and 25 % through castor cake and urea. Bankar et al. (1990) reported that increased doses of nitrogen did not affect fruit size and TSS in pomegranate.

Nutrient analysis of leaf samples collected (8th leaf pair from tip used as indicator tissue for all analysis) 1 month after fertilizer application showed significant influence in respect of nitrogen and potassium. No significant differences in respect of phosphorus content in various treatments were observed. Maximum leaf nitrogen content (3.15 %) was observed in full nitrogen application through urea. Maximum phosphorus (0.14 %) and potassium (0.74 ppm) content was observed in treatment application of nitrogen through FYM, 50 % and 25 % through castor cake and urea. Similar results are reported by Wavhal (1981). Oil cakes, though insoluble in water, are quick-acting organic manures; their nitrogen quickly becomes available to plants in about a week to 10 days (Guar et al. 1971) (Table 5.17).

5.14.2 Biofertilizers

Biofertilizers are preparations containing live or latent cells of efficient strains of nitrogen fixing, phosphate solubilizing, or cellulolytic microorganisms used for application to seed, soil, or compost with the objective of increasing the number of such organisms, in order to accelerate the process of nutrient availability to the plant. They are very efficient, even when applied in small quantities. Careless management of orchards and the ruthless use of chemicals have

Table 5.17 Effect of organic and inorganic source of nitrogen application on fruit and nutrient composition of pomegranate cultivar Ganesh

Treatment	Fruit wt. (g)	Fruit length (mm)	Fruit dia. (mm)	TSS °Bx	N (%)	P (%)	K (ppm)	Fruits retained/plant	Yield (kg)/plant
100 % FYM	179.0	65.7	62.5	16.2	2.74	0.13	0.69	46	8.23
100 % Castor cake	178.5	63.6	63.5	16.3	2.93	0.13	0.68	46	8.20
50 % FYM + 50 % castor cake	185.2	65.7	63.2	16.3	2.82	0.12	0.73	48	8.84
50 % FYM + 50 % castor cake	188.7	69.7	66.1	16.9	2.95	0.14	0.74	57	10.75
50 % FYM + 50 % urea	175.5	64.1	62.9	16.3	3.02	0.11	0.74	46	8.11
100 % urea	169.7	60.4	60.9	16.0	3.15	0.11	0.66	33	5.80
CD at 5 %	8.21	3.81	NS	NS	0.21	NS	0.07	12.05	2.19

Hiwale (2004)

Table 5.18 Effect of bio-fertilizers on yield and yield attributes of pomegranate cultivar Ganesh

Treatment	No. fruit set/pl.	No. fruit retained/pl.	Fruit yield kg/pl.	N (%)	P (%)	K (%)
PSB culture	123.50	73.50	8.37	2.88	0.156	0.68
Azospirillum	144.75	83.00	8.29	2.67	0.177	0.71
PSB + azospirillum	135.50	78.25	11.08	2.92	0.178	0.84
VAM culture	136.50	79.80	8.56	2.79	0.173	0.76
Control	72.50	51.00	7.16	2.52	0.121	0.70
CD at 5 %	34.77	14.18	0.872	NS	0.017	0.06

Hiwale (2004)

Table 5.19 Effect of bio-fertilizer application on quality of pomegranate cultivar Ganesh

Treatment	Fruit wt. (g)	Dia. (mm)	Length (mm)	Aril wt. (g)	Skin wt. (g)	Juice (%)	TSS
PSB culture	201.25	67.62	76.41	136.75	62.50	72.57	15.17
Azospirillum	173.75	67.27	74.75	106.25	53.25	73.89	14.65
PSB + azospirillum	213.00	70.12	78.87	137.37	66.62	74.62	16.05
VAM culture	188.75	98.02	72.75	117.00	58.12	73.29	15.86
Control	153.25	67.50	68.75	95.12	43.87	74.46	15.43
CD at 5 %	24.97	NS	NS	24.30	6.59	NS	0.86

Hiwale (2004)

already shown deleterious effects on pomegranate crop performance in Andhra Pradesh, Karnataka, and Maharashtra. Excessive fertilization and water application has led the orchards to succumb to pests and diseases, and soil health is also spoiled. Unless and until balanced management is practiced, it will not be possible to save these orchards. Organic fertilizers play a significant role in maintaining the sustainability of the soil and improving soil physico-chemical properties and micro-fauna activity. Therefore, an experiment was carried out with different biofertilizers to investigate their effect on sustainability and productivity of the pomegranate cultivar Ganesh.

Biofertilizers had no influence on the vegetative parameters of the pomegranate. Azospirillum at 100 g per plant resulted in a higher leaf nitrogen content as well as higher fruit set and retention; however, fruit weight and fruit size and yield was maximized with the combined application of phosphorus solubilizing bacteria (PSB) + azospirillum at 50 g per plant. Use of biofertilizers such as azotobacter has reduced the requirement for fertilizers. Application of 250 g azotobacter culture with 100 g nitrogen has provided the same effect as application of 300 g nitrogen per plant, i.e. saves 200 g nitrogen per plant. Nutrient analysis of leaf samples collected 1 month after fertilizer application showed a significant influence in terms of phosphorus and potassium, but no significant differences in respect of nitrogen content in various treatments (Tables 5.18 and 5.19).

Maximum leaf nitrogen content (2.92 %) was observed in azospirillum. Maximum phosphorus (0.178 %) and potassium (0.84 %) content were observed in treatment application of 50 g azospirillum + PSB culture, which may be due to solubilization of phosphorus by the bacteria, leading to higher uptake. Oil cakes, though insoluble in water, are quick-acting organic manures; their nitrogen is quickly available to plants in about a week to 10 days (Guar et al. 1971).

Maximum fruit set (144.75 fruits per plant) and fruit retention (83 fruits per plant) was recorded in azospirillum 100 g per plant and yield (11.08 kg per plant) in 50 g azospirillum + PSB culture (Table 5.8). Phadnis (1974) also recommended mixed doses of organic and inorganic fertilizers to obtain maximum fruiting and yield. Wavhal et al. (1985) also recommended the application of nitrogen along with azotobacter to obtain higher production in pomegranate cultivar Ganesh. Shinde (1977) reported similar results.

Fruit weight, aril weight, skin weight, and TSS were significantly influenced by all biofertilizers. Fruit weight (213 g), fruit length (78.87 mm), fruit diameter (70.12 mm), and TSS (16.05 °Bx) were highest with treatment application of 50 g azospirillum + PSB culture (Table 5.19), which may be due to the beneficial effect of phosphorus solubilizers with nitrogen fixers. Bankar and Prasad (1992) reported that increases in doses of nitrogen did not affect fruit size and TSS in the pomegranate. Ram and Rajput (2000) observed a good response to azotobacter in guava under Uttar Pradesh conditions. Day et al. (2005) reported a beneficial effect from phosphorus solubilizers with nitrogen fixers in guava L-49.

The application of biofertilizers influenced nitrogen content: maximum nitrogen content (2.92 %) was recorded with the combined application of PSB with azospirillum. Similarly, maximum phosphorus (0.178 %) and potassium content (0.84 %) was recorded in the same treatment. Leaf sampling was conducted by collecting the 8th pair of leaves from the growing tip as suggested by Bhargava and Dhandhar (1987). Plant material (20 g) was ground in a mixture grinder, and a 1 g sample was taken for nitrogen, phosphorus, and potassium analysis from the oven-dried leaf samples.

5.15 High-Density Orcharding

India is the second-largest producer of pomegranate (41.5million tons) in the world. However, the productivity of almost all fruits in India is very low (12 t/ha) compared with other fruit-growing countries of the world. This is a major cause of concern to our scientists, extension personnel, and orchardists. Many reasons may be attributed to the low productivity, and, to solve this problem, our scientists must develop high-yielding varieties/hybrids that are resistant to biotic and abiotic stresses and develop better agro-techniques for realizing the maximum potential of fruit cultivation in India. Although we have achieved this distinction, the per capita consumption of fruit in our country is still 80 g per day due to our huge population. This is far below the recommendation of 120 g per capita per day. There is little scope to convert the existing area to fruit cultivation because our first requirement is to have sufficient food. Therefore, the major area for expansion will be marginal degraded arid and semiarid land. Drought-hardy fruit plants that can withstand biotic and abiotic stresses and salinity will play a major role, as they can be grown on these lands successfully.

High-density orcharding results in early bearing, helping to minimize weed problems. However, the productive life of an orchard is reduced due to dense canopies that do not allow sunlight penetration, resulting in gradual decline. This problem can be overcome by removing plants in alternate rows. Other methods that can be used to overcome this problem include training and selective pruning. The productivity of almost all fruits in India is lower than in other fruit-growing countries of the world. To overcome this problem, pomegranate high-density orcharding in cultivar Ganesh has been standardized under Maharashtra conditions. To obtain higher production per unit area, a maximum number of plants were accommodated per hectare. High-density plantation at 5 × 2.2 m under semiarid conditions resulted in a 2.5 times higher yield than did the normal spacing of 5 × 5 m in a 6- to 7-year-old orchard.

	Yield q/ha.	
Spacing (m)	1981	1982
5 × 2	66.65	85.69
5 × 3	49.76	61.96
5 × 4	36.30	36.35
5 × 5	27.46	36.64
CD at 5 %	11.15	25.27

5.16 Canopy Management

Manipulation of tree vigor and the maximum utilization of sunlight results in increased productivity and quality. The basic principle of canopy management is maximum utilization of light. Avoiding the growth of a microclimate conducive to diseases and pests helps in cultivation operations. Furthermore, the inter space can be utilized for raising annuals as intercrops.

5.16.1 Training

The pomegranate plant is bushy in nature and throws a considerable number of shoots near ground level. Retaining all these shoots at the base increases the crowding of the tree frame and the incidence of shoot borers and diseases. Considering these factors, keeping three to four stems from the base was found to be satisfactory for maintaining good productivity as well as proper health of the plant.

Training is done only to give shape to the tree. At the time of planting, all the side shoots are removed. Young plants are allowed to grow into a bush with a number of main shoots arising at the ground level. Then, they are properly trained to form a single stem with a number of well-distributed scaffold limbs. The plant is topped at a height of 60–70 cm, and the side shoots (4 to 6 in number, well distributed along the main stem) are allowed to arise from about 40–50 cm from ground level. As and when the plant becomes too thick, very low branches are thinned out and removed. Trellises provide support to the branches to avoid the breaking of branches. Training is done systematically during the initial 3–4 years of plant growth. This helps ensure plants are uniform in the orchard. After 3–4 years of age, plant growth is tremendous. Therefore, training is necessary to control growth, avoid criss-crossing of twigs, and to allow air and light to the canopy. Sometimes non-bearing twigs and water shoots, along with dried and diseased twigs, must be pruned. If pruning is not carried out manually, it can be achieved by applying 1,000 ppm spray of Lihosin once, 2,3,5-triiodobenzoic acid (TIBA) 500 ppm twice, or maleic hydrazide 500 ppm twice. These sprays can reduce plant growth and does the job of pruning.

5.16.2 Pruning

Farmers in Maharashtra have adopted manual as well as chemical pruning to rest the plant and subsequently initiate bahar. Even though no regular pruning is recommended, it has recently been practiced by some farmers to improve fruit size. Pomegranate fruiting occurs on 1-year-old spur. Fruit that set on the terminal branches do not attain a good size. To avoid this, the terminal branches are pruned (15–20 cm).

Systematic work on this aspect was initiated at CHES Vejalpur, where manual pruning was carried out. Pruning in pomegranate was attempted to initiate new growth on 5-year-old plants. The plants were pruned in the second half of May. Two types of pruning were attempted: (1) heavy (removing 5–7 kg of fresh weight per plant; and (2) light (removing 2–3 kg of fresh wood per plant). Under rain-fed conditions, sprouting started 3–4 weeks after pruning. However, the subsequent flowering was delayed.

The mean fresh and dry weight of biomass removed in heavy pruning per plant was 5.67 and 2.92 kg per plant, respectively, and 2.90 and 1.65 kg per plant in light pruning. New shoot growth, as evidenced by an increase in shoot length, diameter, fresh and dry weight revealed that heavy pruning resulted in higher shoot growth than did light pruning and control (no pruning). Similar trends were also recorded in leaf growth (Tables 5.20 and 5.21).

Although fruit set was significantly higher in controls (no pruning), fruit retention was higher with heavy pruning, ultimately giving a higher yield per plant. There was an increase of about 7.36 q/ha (16.95 %) in yield, giving an additional income of Rs. 3,680/ha at Rs 500/q. Similarly, fruit weight and fruit diameter showed significant increases, which may have resulted in increased yield per plant.

In pruning of pomegranate cultivar Ganesh plants, subsequent growth started 1 month later. Leaves and shoots were sampled for the 4th and 5th month. The percent NPK content in shoot and leaf revealed that leaves had higher nitrogen but there was not much difference in terms of phos-

Table 5.20 Effect of pruning on new growth (fresh and dry weight of shoot)

	Shoot fresh wt. (g)		Shoot dry wt. (g)		Shoot length (cm)		Shoot dia. (mm)		Pruned wood/plant	
									Fresh	Dry wt.
Treat.	4/10	22/11	4/10	22/11	4/10	22/11	4/10	22/11	Wt. (kg)	(kg)
Heavy pruning	6.37	10.04	3.16	4.75	66.33	70.83	4.51	4.98	5.67	2.90
Light pruning	4.51	6.57	2.14	3.23	56.16	60.66	3.76	4.28	2.90	1.65
Control	1.58	2.49	0.70	1.22	26.83	30.5	2.42	2.95	–	–
CD at 5 %	0.51	0.73	0.29	0.47	8.50	9.04	0.26	0.23	–	–

Table 5.21 Effect of pruning on new growth and nutrient composition

	Leaf fresh wt. (g)		Leaf dry wt. (g)		N (%)		P (%)		K (%)	
Treatment	4/10	22/11	4/10	4/10	4/10	22/11	4/10	22/11	4/10	22/11
Heavy pruning	8.78	11.27	4.07	4.61	2.94	2.42	0.18	0.13	0.850	0.712
Light pruning	6.12	8.21	2.70	4.02	2.91	1.52	0.16	0.13	0.698	0.66
Control	2.39	4.87	1.16	0.98	2.58	1.81	0.14	0.18	0.782	0.84
CD at 5 %	1.02	0.68	0.53	0.72	–	–	–	–	–	–

Table 5.22 Effect of pruning on yield and yield parameters

Treatment	Fruit set no.	Fruit retention no./plant	% retention	Yield kg/plant	Yield q/ha	Fruit wt (g)	Fruit length (mm)	Fruit dia. (mm)
Heavy pruning	67.8	52.0	85.54	12.7	50.0	169.33	65.21	71.33
Light pruning	75.7	63.5	88.75	10.4	41.6	158.33	58.88	61.05
Control	114.5	54.3	68.39	10.86	43.4	147.00	57.98	63.05
CD at 5 %	20.3	6.8	06.16	–	–	14.45	NS	7.72

Table 5.23 Effect of pruning on NPK content of shoot

	N (%)		P (%)		K (%)	
Treatment	4/10	22/11	4/10	22/11	4/10	22/11
Heavy pruning	1.97	1.06	0.18	0.11	0.894	0.95
Light pruning	1.52	1.48	0.14	0.11	0.764	0.97
Control	1.81	1.52	0.18	0.13	0.834	0.77

phorus and potassium content. The NPK content in both leaf and shoot decreased just before initiation of flowering, probably due to diversion of nutrition, which may explain why the nitrogen and phosphorous was reduced but potassium content increased. Nitrogen content was higher in heavily pruned shoots in both leaf and in new growth, and it was least in control (Tables 5.22 and 5.23).

5.16.3 Fruit Thinning

Fruit thinning is a tool that horticulturists can use to reduce the burden of the plant and avoid the use of stored food material at one time. The sink source relationship plays an important role in managing the crop load. Fruit thinning helps in maintaining the sustainability of production over the years, particularly in fruit trees with a long life span. By reducing the number of fruit, the crop load can be evenly distributed all over the plant, thereby avoiding the breaking of branches and improving the exposure of fruits to sunlight, which helps improve fruit development.

Therefore, work was initiated at CHES Vejalpur under semiarid rain-fed conditions. Fruit weight showed maximum increase over control (79.92 %) when 25 fruits per plant were retained; this decreased to 69.08 % with 50 fruit, and 41.60 % with 75 fruits per plant. However, there was no change in percent juice content and acidity of fruits. TSS content was maximized when a lower number of fruits per tree were retained (17.08 °Bx), showing a decreasing trend as the number of fruits per plant increased. It was observed that keeping 75 fruits per plant produced the maximum yield per plant (11.41 kg) compared with control (10.24 kg). However, thinning to 25 fruits resulted in a yield reduction by 54.98, and thinning to 50 fruits resulted in a yield reduction to 12.56 %. Fruit thinning produced good size 'B' grade fruit, fetching higher market prices. Overall, keeping 50 fruits per plant was found to be an economically viable preposition for farmers as the economic returns were highest (Rs. 10,670/-) compared with control (Rs. 4,692/-), Hiwale (2009) (Table 5.24).

Rate of sale according to grade:	*Cost of cultivation* – Rs. 3,500/-,
B grade- Rs. 5/kg- 400/q,	*Cost of thinning*– 3 labor /ha @Rs. 50=Rs. 150/-
C grade- Rs. 3/kg- 300/q,	*Small fruits* – Rs. 2/kg- 200/q.

5.17 Rejuvenation of Pomegranate Orchards

Old orchards become non-productive due to the incidence of pests, diseases, and neglect. It is a general tendency of farmers in India to neglect such orchards. In pomegranates, when the survey was conducted, it was observed that 25 % of the old orchards became unproductive due to loss of vigor in the plant, a heavy incidence of pests and diseases and a deepening of the water table. There is still scope to review production from these orchards through intensive care. Rejuvenation can be achieved via mechanical measures, chemical measures, and by training and pruning of plants. Work carried out on non-selective pruning at CHES Vejalpur has shown encouraging results. A 15-year-old orchard that was showing signs of decline, mainly due to drying of old branches, resulting in sparse flowering and fruiting, was used for the study. To overcome the situation, non-selective pruning of the orchard was undertaken. The plants were headed back to 30, 60, 90, and 120 cm above ground level in the month of April; plants that were not pruned acted as controls. The cut ends were sprayed with chlorpyriphos 3 ml/L with bavistin 1 g/L to control the incidence of termites and various other pests and diseases. New sprouts emerged 20–25 days after pruning. The number of new sprouts per plant was reduced to 8–10 by allowing two to three healthy sprouts per branch to balance the framework of the plant. Upon onset of monsoon,

Table 5.24 Effect of thinning of fruit on yield and physico-chemical characteristics of pomegranate cultivar Ganesh

Thinning	Yield/ plant (kg)	Fruit wt. (g)	Fruit length (mm)	Fruit dia. (mm)	Aril skin ratio	Juice (%)	TSS °Bx	Acidity (%)	Yield (q/ha)	Gross income (Rs.)	Net return (Rs.)
25 fruit	4.61	197.74	66.26	65.86	1.66	74	17.08	1.82	18.44	7,336	3,726
50 fruit	8.95	185.82	63.22	61.89	1.77	73	16.16	1.84	35.80	14,320	10,670
75 fruit	11.41	155.62	61.74	60.02	1.87	73	16.0	1.79	45.64	13,692	10,042
Control	10.24	109.9	56.16	56.82	2.32	72	15.8	1.82	40.96	8,192	4,692
CD at 5 %	0.474	10.59	2.11	N.S.	0.01	NS	0.080	NS	–	–	–

Hiwale (2009)

the growth of the plant increased, and after 1 year the plant height was on a par with that of control plants. In due course, the plant growth slowed down due to diversification of food material to flowering and fruiting. Mean shoot diameter also increased at a faster pace after the initial setback. Plant spread was almost equal to that of the controls after 3 years of growth, i.e. non-significant differences were recorded in various treatments.

The fruit set in the initial year was reduced in the first year of pruning (19 per plant), which surpassed the controls in the second year (101 per plant) in plants pruned to 30 cm from ground level. They outperformed controls in the following years. Similar trends were recorded in terms of the number of fruit retained per plant. The maximum number of fruit was retained in plants pruned to 30 cm from ground level (Table 5.2). Yield (kg per plant) in the first year after pruning was reduced to 1.14 kg (81.18 % reduction) in plants pruned to 30 cm from ground level. However, the yield per plant surpassed that of controls in the second year, and the trend continued thereafter. The increase was to the tune of 16.96, 32.94, and 26.10 % in the 2nd, 3rd, and 4th years of pruning to 30 cm level (Tables 5.25, 5.26, and 5.27).

Pruning plants to 30 cm from ground level was the most effective method to rejuvenate the old pomegranate orchard, and can be recommended as standard practice in rejuvenating old orchards (Hiwale et al. 2006).

Table 5.25 Growth parameters of rejuvenated pomegranate cultivar Ganesh

					Plant spread			
	Plant height (m)		Stem dia. (mm)		NS (m)		EW (m)	
Treatment/year	1999	2001	1999	2001	1999	2001	1999	2001
T1 cutting 30 cm	2.68	2.42	21.23	30.94	2.39	2.55	2.43	2.73
T2 cutting 60 cm	2.61	2.66	23.88	32.67	2.61	2.51	2.66	2.62
T3 cutting 90 cm	2.68	3.34	17.18	34.03	3.46	2.72	3.46	2.72
T4 cutting 120 cm	2.7	3.42	17.56	32.81	3.34	2.29	3.48	2.32
T5 control	3.13	3.56	32.73	35.96	3.68	2.54	3.66	2.63
CD at 5 %	**2.93**	**0.35**	**2.62**	**7.13**	**0.26**	**NS**	**0.35**	**NS**

Hiwale et al. (2006)

Table 5.26 Fruit set and retained/plant

	No. fruit set/plant				No. of fruit retained/plant			
Treatment/year	1999	2000	2001	2002	1999	2000	2001	2002
T1 cutting 30 cm	19.0	101	89.2	53.00	7.6	52.4	60.8	51.8
T2 cutting 60 cm	50.8	92.2	88.8	140.75	16.2	44.4	52.0	74.4
T3 cutting 90 cm	63.4	70.8	82.2	42.75	20.0	35.4	38.4	34.6
T4 cutting 120 cm	72.8	48	64.8	48.25	22.0	29.0	35.0	46.4
T5 control	83.4	90.6	93.2	90.75	40.4	44.8	46.4	59.0
CD at 5 %	**5.97**	**21.05**	**27.33**	**47.62**	**2.74**	**4.58**	**5.43**	**11.16**

Hiwale et al. (2006)

Table 5.27 Percent retention and yield kg/plant

	% retention				Yield kg/plant			
Treatment/year	1999	2000	2001	2002	1999	2000	2001	2002
T1 cutting 30 cm	40.0	51.88	68.16	97.73	1.14	7.86	9.12	11.16
T2 cutting 60 cm	31.89	48.16	58.56	52.86	2.43	6.66	7.80	7.77
T3 cutting 90 cm	31.54	50.00	46.72	80.93	3.02	5.25	5.76	6.33
T4 cutting 120 cm	30.22	60.42	54.01	96.16	3.10	4.35	5.46	6.96
T5 control	48.44	49.45	49.78	65.01	6.06	6.72	6.86	8.85
CD at 5 %	–	–	–	–	**0.39**	**0.69**	**0.92**	**0.48**

Hiwale et al. (2006)

Pomegranate heading back

Pomegranate sprouting

Rejuvenated Pomegranate plant

5.18 Staggering of Crops

Staggering pomegranate plants is practiced with a view to regulating crop load on the plant, which helps in the availability of the fruit for a longer period, helping the farmer fetch remunerative prices in the market. It also helps avoid gluts in the market that lead to distress sales. The relationship between source and sink can be regulated, which helps improve fruit development. Pomegranates flower and fruit throughout the year if irrigation is provided. Taking of particular bahars leads to gluts in the market. This results in distress sales by farmers, as fruit crops are obviously perishable in nature. To avoid this, staggering the pomegranate crop was attempted to find the best time to allow flowering during various months of the year so that maximum production with the best-quality fruits can be obtained. The pomegranate was allowed to flower during different months of the year in 5-year-old plants of cultivar Ganesh by withholding irrigation and releasing it during the specified period to initiate flowering. The best crop was obtained when September flowering was allowed, because the fruit development coincided with maximum moisture availability and cool climate, leading to a lower incidence of insect pests and reducing cracking, thereby improving the fruit quality (Table 5.28).

Flowering in September was observed to produce the maximum yield per plant. Aril seed ratio increased and softness of seed decreased with this period. Time of flowering significantly influenced TSS and acidity: TSS increased with an increase in time of flowering, while acidity showed a decline. The best fruits were obtained with September flowering under rain-fed conditions with one or two supplementary irrigations. Similar observations were recorded by Sonwane and Desai (1989).

Reducing the fruit load per plant results in the production of fruits of superior size, which then fetch premium prices at market. Also, in the pomegranate, where the production of hermaphrodite flowers is higher and results in higher fruit

Table 5.28 Staggering in pomegranate

Month	Fruit length (mm)	Fruit Dia. (mm)	Fruit wt. (g)	Yield/plant (kg)	Aril seed ratio	100 seed wt (g)	TSS °Bx	Acidity (%)
August	67.4	78.10	134.20	9.83	1.89	1.54	16.18	0.44
September	70.99	70.24	151.68	11.83	2.70	1.89	17.64	0.35
October	70.92	70.70	141.88	9.98	3.03	1.96	16.72	0.37
November	70.75	71.05	141.66	2.88	3.28	2.36	18.12	0.34
CD at 5 %	2.11	N.S.	2.91	2.35	2.51	0.17	1.03	0.07

Anonymous (1999)

set, the thinning of fruits can be adopted as regular practice. Appropriate fruit load helps in proper distribution of food and maintenance of sustainability of production. Work at CHES Vejalpur on cultivar Ganesh in hasta bahar indicated that thinning fruits to 25 per plant resulted in larger-size fruits than all the other treatments applied but yield was significantly reduced. However, retaining 75 fruits per plant produced uniform medium-sized fruits with much reduction in yield per plant. The returns were higher than with other treatments.

5.19 Crop Regulation–Chemical Regulation

Fruit and leaf abscission have been reported with the application of alar at 500–3,000 ppm, which induced flower and fruit drop. However, chemical regulation is currently avoided, as most chemicals involved have been reported to be harmful and are either banned or are no longer manufactured.

The main purpose of crop regulation is to force a tree to rest and to induce profuse flowering during the regained season. For instance, the pomegranate has three flushes of flowering: mrig (June–July), hasta (September–October), and ambe (December–January) bahar. The methods used for imposing bahar can be broadly classed into cultural and chemical. The common cultural method is to withhold irrigation or use root exposure and root pruning, etc. In terms of chemical methods, most of the growth retardants and growth regulators can be used to impose the bahar treatment. The pomegranate naturally flowers in mrig bahar, which depends upon monsoon and is difficult to regulate. But withholding irrigation can regulate hasta and ambe bahar. Thus, imposition stress leads to leaf drop and cessation of growth before the bahar. Later, reducing the stress slowly will lead to profuse flowering and fruit set. At CHES Vejalpur, Raturi and Hiwale (1991) worked on pomegranate cultivar Ganesh under rain-fed conditions and reported that it is possible to raise a successful hasta bahar crop. Hand thinning was used to remove mrig bahar flowering. Comparative studies on mrig and hasta bahar indicated that the fruit set, retention, and yield per plant were highest in hasta bahar. The fruit of the bahar are available when there is no pomegranate glut in the market and therefore can fetch remunerative prices. The quality of fruit skin color and TSS was also superior, while the incidence of pests and diseases was lowest. Fruit borer infection in mrig bahar was recorded at up to 18.83 %, whereas in hasta bahar it was just 8.92 % (Table 5.29).

5.19.1 Improving Fruit Skin Color

When temperatures are high during the ripening period, the skin color of the Ganesh variety normally becomes dull and pale, and fruit lose market value. When there is excess nitrogen, irrigation, and magnesium in the soil, the skin color becomes light and dull. Therefore, when fruits are maturing and turning their color, the application of potassium helps improve the fruit color. Color of the aril and skin are due to

Table 5.29 Effect of cropping season on fruit, retention, and yield of pomegranate cultivar Ganesh

	% fruit set		% fruit retention		Yield/plant (kg)	
Cropping season (Bahar)	88–89	89–90	88–89	89–90	88–89	89–90
Mrig	3.90	3.99	39.97	47.63	3.18	5.16
Mrig + hasta	5.04	5.43	42.49	55.46	4.52	7.37
Hasta	8.12	9.26	54.06	65.81	5.82	10.25
CD at 5 %	1.26	0.79	9.43	5.55	1.01	2.18

Table 5.30 Root distribution pattern (fresh and dry weight g/plant)

	Depth of	0–50		50–100		100–150		Total	
Type of root	soil (cm)	Fresh wt.	Dry wt.	Fresh wt.	Dry wt.	Fresh wt.	Dry wt.	Fresh wt.	Dry wt.
Thin (>3 mm)	30	190	114	250	173	101	67	541	354
Medium (>7 mm)	30	282	162	185	122	90	50	557	334
Thick (>15 mm)	30	202	118	–	–	–	–	202	118
Thin (>3 mm)	60	272	213	216	153	138	100	526	466
Medium (>7 mm)	60	364	304	164	101	97	73	625	478
Thick (>15 mm)	60	287	228	105	40	–	–	392	268
Thin (>3 mm)	90	–	–	–	–	–	–	–	–
Medium (>7 mm)	90	–	–	–	–	–	–	–	–
Thick (>15 mm)	90	65	37	35	21	15	9	115	67
Grand total		1,662 (54.17)	1,176	955 (31.12)	610	441 (14.37)	299	3,068	2,085

Figures in parentheses are percent values

anthocyanin pigment (Du et al. 1975). Low temperatures stimulate anthocyanin synthesis and activity of the enzyme phenyl aniline ammonia lyase (PAL), which is directly related to anthocyanin synthesis (Akoi and Katayama 1970 and Hyodo 1971). A spray of potassium dihydro orthophosphate 1–2 g in 1 L of water, two sprays of 500 ppm lihosin, two sprays of 500 ppm of ethrel during the color-breaking stage, along with cytozyme100 ml +100 g potassium dihydro orthophosphate in 100 l of water, improve the skin color to reddish.

5.20 Root Distribution

Root distribution acts as a guide for the application of fertilizers as well as irrigation. Root distribution varies according to the type of soil and method of propagation. No such work had been carried out in India on the pomegranate. Therefore, studies were initiated to study root distribution patterns of air-layered 6-year-old pomegranate cultivar Ganesh. The active root zone was very shallow and was spread within the plant canopy. The root distribution pattern in air layers of pomegranate cultivar Ganesh under semiarid rain-fed conditions revealed that the root system is shallow in nature, as little root activity was recorded below a soil depth of 60 cm. Maximum root activity, on a fresh and dry weight basis, was observed at a radial distance of 0–30 cm from the tree trunk (54.17 %); at a distance of 30–60 cm it was 31.12 %; at a distance of 60–90 cm it was 2.12 %. However, in terms of soil depth, maximum root activity was observed at a soil depth of 30–60 cm, and 42.69 % activity was found at a soil depth of 0–30 cm. The least root activity (3.74 %) was observed at a soil depth of 60–90 cm (Hiwale et al. 2009) (Table 5.30).

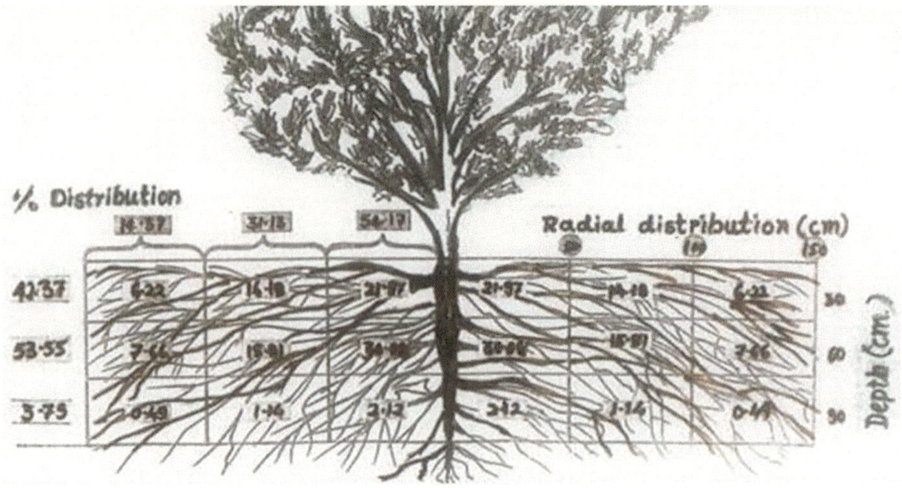

Spatial root distribution in pomegranate Cv. Ganesh

Ganesh

5.21 Intercropping

Pomegranate occupies the recommended spacing (6 m×6 m) after 8 years; however, the trees start bearing at 4 years and, until then, inter-cropping can be carried out with fodder such as berseem and lucerne, crops such as cowpea and green gram, and vegetables such as cucurbits, cabbage, cauliflower, beans, peas, tomato, carrot, onion, radish, potato, and brinjal. However, inter-crops should be regulated so that their cultural care does not interfere with the bahar regulation.

5.22 Post-harvest Technology

The pomegranate has a low respiration rate and a non-climacteric respiratory pattern. Freshly harvested pomegranate fruits of cultivar Ganesh were treated with wax coupled with carbendazim (0.1 %) and stored at room temperature (22–24 °C with 52–82 % relative humidity) as well as in cold storage. Results indicated that fruits could be stored for up to 75 days in cold storage as against 30 days at room temperature (Waskar et al. 1999). Chilling injury can be overcome by intermittent warming. Storing pomegranates at 2 °C with intermittent warming was optimum for minimizing chilling injury and maintaining the quality of the fruit up to 13 weeks (Artes et al. 2000). Controlled atmospheric storage of pomegranate fruits indicated that the color value of the juice was significantly increased in fruit stored at 10 % O_2, 5 % CO_2 (Artes et al. 1996). The pomegranate fruit reached its horticultural maturity when its soluble solid content reached 15 %. The maturity period of 105–140 days from fruit set can be used as a criterion for fruit maturity (Hayes 1953). However, Kulkarni et al. (2005) reported that harvesting of fruit on the 100th day from fruit set was crucial to prevent browning during storage.

5.23 Value Addition

With increases in area and production of fruit, steps need to be taken to develop industries to add value to a crop; if this does not occur, a glut in the market results and the crop becomes uneconomical, leading to either distress sales or removal of orchards. Pomegranate fruit is known for its value-added products such as juice, jelly, anardana, anar rub, rind power, etc., which can be easily prepared at the field level. Fruits are mainly used as fresh fruit or juice.

5.23.1 Fruit Juice

Pomegranate varieties with dark red arils are preferred. The juice can be extracted by two methods: (1) keeping the skin intact and removing the lower portion, the fruit is separated into four parts and the juice is extracted via a juice extractor; (2) the skin is removed, arils are separated, and juice is extracted. Juice is preserved with the addition of potassium metabysulphite and storing it in cold temperatures in steel or glass bottles/vessels. A temperature of 77 °C for 33 s is recommended for pasteurization of fruit juice. This process eliminates *Aspergillus niger* and lactic strains. Color stability was maintained at temperatures up to 80 °C for 30 min, after which a slight color reduction was observed. Pasteurized juice showed comparable quality to unpasteurized juices (Rage and Pai 1999). The pomegranates were frozen at –40 °C, either with or without sugar and stored at –18 °C for 9 months (Bilsil and Cevic 1997). Clarification of pomegranate juice via ultra filtration resulted in a better retention of ascorbic acid (2–5 % loss), high yield (97.8 %), and better quality than conventionally clarified juices (Iboyaima et al. 1993). Effective stabilization of color in juice stored at room temperature can be achieved with polyvinylpyrrolidone (PVP) and trypsin (Khrameeva et al. 1987).

5.23.2 Pomegranate Syrup

Pomegranate syrup is prepared by taking one-quarter fruit juice and raising the sugar to 60–64 %. The syrup can be stored under normal conditions for about 3–4 months, and is served by mixing water in appropriate proportions according to taste.

5.23.3 Pomegranate Nectar

Fresh pomegranate juice can be preserved for a long time by increasing the sugar content to 20–30 %.

5.23.4 Juice Concentrate

Juice is converted to powder form either by freeze drying or by drying on a conveyer belt and is vacuum packed in polythene bags.

5.23.5 Pomegranate Chutney

Wild pomegranate (Anardana type) found in Himachal Pradesh can be used to prepare chutney. The product was dried in a mechanical cabinet at 65 °C for 10–14 h. Samples were packed in polythene pouches and stored at 37 °C. The quantity of green chilies added increased the vitamin C content, total sugars, and PH, while protein levels and acidity decreased. Chutney had a shelf life of more than 6 months.

5.23.6 Carbonated Pomegranate Beverages

The pomegranate beverages were prepared with different levels of juice (5, 10, 15, and 20 %). Drinks were formulated to a pH of 3.14–3.16 and acidity of 0.24 % and stored for 80 days at ambient temperatures. The beverage with 10 % pomegranate juice and initial content of 11.78 % sucrose, 1.3 % reducing sugar, and 1.0 mg ascorbic acid was the best (Riaz and Elahi 1992).

5.23.7 Pomegranate Wine

Pomegranate wine is one of the important products with which it is possible to absorb the glut of this fruit in the market, simultaneously providing a healthy drink and creating job opportunities. Around 500 ml of juice can be obtained from 1 kg of pomegranate fruit, from which

400–450 ml of wine can be prepared through fermentation with the addition of yeast.

5.23.7.1 Methodology

Preparation of starter: A spoonful of standard wine yeast *Saccharomyces ellipsoids* no.101 maintained on yeast extract peptone dextrose (YEPD) agar was transferred to a test tube containing 5 ml of YEPD broth and incubated overnight at 25 °C. After 22 h, it was shaken well in an incubator at 120 rpm, which is used for inoculation after activating the yeast cell at 40 °C for 1.5 h.

5.23.8 Anardana

Wild pomegranate seeds and aril are sun dried and commercially marketed as 'anardana', which is used as a condiment (Sharma and Sharma 1990). Chemical characteristics such as high juice content and sugar/acid ratio, more intensive color, and less tannin content are good attributes for a better quality of anardana. To improve color stability by preventing oxidation of the anthocyanin, a mixture of glucose and citric acid is used as stabilizer at a ratio of 1.0:0.20:0.003 of juice to glucose to citric acid (Gafizov et al. 1990).

In Himachal Pradesh, making anardana is a household activity and it is mainly used as a substitute for tamarind and mango slices. It is also used in the garnishing of ice cream, fruit salad, chutney, and panmasala. There is huge demand for the product in North India. The culled fruits can be used for preparation of anardana. Anardana can be prepared from varieties with acidity of 7.8–15.4 %; the Russian types are especially suitable. The varieties under large-scale cultivation have acidity of just 0.3–0.5 %, which can be used with the addition of 5 % citric acid to the arils. The total expenditure for the preparation of 1 kg anardana is around Rs. 60.

Mahajan et al. (1992) processed wild pomegranate for the preparation of anardana with three thermal treatments: deep sand roasting of whole fruits for 5 min; dry roasting on a heater; and blanching in boiling water for 5 min. The first method was found to be the best. Three drying techniques were also tried: solar drying; conventional sun drying; and via a hot air dehydrator. Solar drying was the best. Hussein et al. (2004) analyzed pomegranate Whole fruit remove as seed powder samples, and seed powder samples as suggested standard grades for sensory evaluation.

References

Akoi SAk,k, Katayama O (1970) Phenyl nine ammonia lyase activities in Japanese chestnut, strawberry, apple fruit and bracken. J Food Sci Technol 17:507–511

Anonymous (1979) Research review subcommittee report, Department of Horticulture, MPKV, Rahuri

Anonymous (1999) Annual report IIHR, Bangalore

Anonymous (2005) Pomegranate Smaranika. Maharashtra Pomegranate Growers Association, Pune

Artes F, Gains Martin J, Martinez JA (1996) Controlled atmospheric storage of pomegranate. J Zeitschrift Fuer-Lensmittlel-untersuchung-und-Forschung 203(1):33

Artes F, Tudala JA, Villaescusar R (2000) Thermal PH treatments for improving pomegranate quality and shelf life. J Postharvest Biol Technol 18(3):245

Bankar GJ, Prasad RN (1992) Performance of important pomegranate cultivars in arid regions. Ann Arid Zone 31(3):181–3

Bankar GJ, Das HC, Prasad RN, Vasistha BB (1990) N nutrition in pomegranate under arid conditions. Ann Arid Zone 29:63–66

Bhandari MM (1979) Practical's in plant breeding. Oxford & IBH Publishers Co., New Delhi, pp 23, 52, 79

Bhargava BS, Dhandhar DG (1987) Leaf sampling techniques for pomegranate (*Punica granatum* L.). Prog Hortc 19(3–4)

Bilsil A, Cevic I (1997) Studies on stability of some pomegranate varieties for freezing preservation publication II-Kontrol-Laboratuvar-Mudurlugu-Canakkale No. 37, 12 pp

Burton GW, De Van W (1953) Estimating heritability in tall fescue (*Festuca aruninancea*). From replicated clonal material. Agron J 45:478–81

Chaturvedi KN, Rai JN, Gupta DB (1980) Variability in grape. Indian J Hortc 37:262–4

Day P, Rai MI, Kumar S, Vishalnath DV, Reddy NN (2005) Effect of bio fertilizer on physico chemical characteristics of guava (*Psidium guajava* L.) fruit. Indian J Agric Sci 75(2):95–96

Du CT, Wang PL, Francis FJ (1975) Anthocyanins of pomegranate. J Food Sci 40:417–418

Firke NN, Kumbhar DB (2002) Effect of different levels of N, P and K ferigation on yield and quality of pomegranate. J Maharashtra Agric Univ 27(2): 146–148

Gafizov GK, Semochkina LG, Gakharmonov MS (1990) Method for improving the color stability of pasteurized natural pomegranate juice during storage. USSR Patent SU 1 535 517

Guar AC, Neelkanthan S, Dargan KS (1971) Organic manure. ICAR Pub, New Delhi, pp 3–10

Hayes WB (1953) Fruit growing in India, 3rd rev. ed. Kitabisthan, Allahabad

Hiwale SS (2001) Evaluation of pomegranate germplasm. Annual report, CIAH Bikaner, 2001

Hiwale SS (2004) Effect of organic and inorganic fertilizers on growth, yield and quality of pomegranate cv. Ganesh. In: Paper presented in "National Seminar on Organic farming in horticulture for sustainable production, 29–30 Aug 2003

Hiwale SS (2009) Effect of fruit thinning on growth, yield and quality of pomegranate cv. Ganesh. Ann Arid Zone 48(2):183–184

Hiwale SS, Dhandhar DG, Bagle BG (2006) Rejuvenation of pomegranate through non-selective pruning. In: Paper presented in "National symposium on production, utilization and export of underutilized fruits with commercial potentialities, BCKV, Kalyani, West Bengal, 22–24 November 2006

Hiwale SS, More TA, Bagle BG (2009) Root distribution pattern in pomegranate cv. Ganesh. In: International conference on pomegranate and minor including Mediterranea fruits by ISHS at U.A.S., Dharvad, India, 23–27 June 2009. Zone 41(1):89–91.

Hussein ML, Devikar GY, Singh G, Jaiswal PK (2004) Quality evaluation of pomegranate seed and powder. Spice India 17:28–31

Hyodo H (1971) Phenylalinine ammonia lyase in strawberry fruits. Plant Cell Physiol 12:989–991

Iboyaima Singh N, Madaiah N, Nangundaswamy AM (1993) Preliminary studies on clarification of fruit juices by ultra filtration. Ind Food Packer 47(5):9

Jagannath S, Raju B, Krishna Manohar R, Narayan Gowda JV, Khan MM (1999) Yield and yield attributes of pomegranate cv. Jyothi as influenced by drip irrigation and black polyethylene mulch. In: Paper presented in national seminar on problems and prospects of micro irrigation – a critical appraisal, held at GKVK, Campus, Bangalore, pp 46–51

Jalicop SH, Tiwari RB, Sampath Kumar (2002) Amlidana: a new pomegranate hybrid. Indian Hortic 47:22–23

Kale RH (1986) M.Sc. (Agri) thesis. MPKV., Rahuri MS. India

Keskar BG, Karale AR, Dhawale BC, Kaulgud SN (1994) Variety release proposal. Department of Horticulture, Mahatma Phule Krishi Vidyapeeth Rahuri

Khrameeva NP, Kosalkina OT, Ruanet VV (1987) Spectral characteristics of pomegranate juice after treatment with stabilizers. J Izavenstiya-Vysshikh-Uchebnykh-Zavednil-Pishchevaya teknologeya 3:96.

Kulkarni AP, Aradhya SM, Divakar S (2005) Chemical changers and antioxidant activity in pomegranate arils during development. Food Chem 93:319

Liberman RG (1993) Effect of biocontrol agents on plant growth. Cambodia Proc Int Plant Propag Soc 43:249–252

Mahajan BVC, Chopra SK, Sharma RC (1992) Processing of wild pomegranate (*Punica granatum* L.) for anardana: effect of thermal treatments and drying modes on quality. J Food Sci Tech 29:327–328

Meena KK, Singh R, Singh SK (2003) Genetic variability and genetic advance to average fruit weight and its components traits in pomegranate (*Punica granatum*) genotypes. Indian J Agric Sci 73(11):630–2

Murukute A, Patil S, Patil BN, Mayakumari (2002) Micropropagation in pomegranate callus induction and differentiation. South Indian Hortic 50(1–3):49–55.

Naik SK, Patanaik S, Chand PK (1999) In vitro propagation of pomegranate (Punica granatum L.) cv.Ganesh through axillary shoot proliferation from nodal segment of mature trees. Scientific Hortic 79:175–183

Pandey G, Bist HS (1998) Variability, correlation and path analysis in pomegranate germplasm. Hortic J 11:7–12

Panse VG, Sukhatme PV (1978) Statistical methods for Agril. Workers, pp 381. ICAR, New Delhi

Patil PB, Patil CP. Swamy GSK, Kumar N (2004) Influence of micro propagation on rooting of pomegranate cuttings. In: Proceedings organic farming in Horticulture, published by Director Central Institute for Subtropical Horticulture, Rehmankhera, Lucknow, Uttar Pradesh, pp 209–211

Phadnis NA (1974) Pomegranate for desert and juice. Indian Hortic 19(3):9–13

Prasad RN, Bankar GJ (2003) Pomegranate cultivation. Indian J Hortic 48:26–29

Purohit AG (1985) Soft seediness of commercial pomegranate varieties. Indian J Agric Sci 55(5):367–368

Rage AR, Pai JS (1999) Development of thermal process for clarified juice of pomegranate (*Punica granatum*). J Food Sci Technol 36(3):261

Raturi GB, Hiwale SS (1991) Regulation of cropping in pomegranate under rainfed conditions. Indian J Hortic (Indian) 48(1):13–15

Ram RA, Rajput MS (2000) Role of biofertilizers and manures in production of guava (*Psidium guajava* L.) cv. Allahabad safeda. Haryana J Hortic Sci 29(3–4):193–194

Riaz RA, Elahi M (1992) Studies on preparation and storage stability of Carbonated pomegranate drink J. Sci Indian Res (Pakistan) 35(11):467

Riker AJ, Benefield UM, Wright WH, Ketit GW, Sagen HE (1930) Studies on infectious hairy root of nursery apple trees. J Agric Res 41:507–540

Sharma BD, Hiwale SS (2005) Survey report submitted to Pomegranate Growers association, Pune, Maharashtra

Sharma SD, Sharma VK (1990) Variation for chemical characters in some promising strains of wild pomegranate (*Punica granatum* L.). J Euphytica 49(2):131

Shinde DG (1977) Effect of N, P and K on growth, yield and quality of pomegranate (*Punica granatum* L.) M.Sc. (Agri) thesis, M.P.A.U.Rahuri

Sonwane PC, Desai UT (1989) Performance of staggered cropping in pomegranate. J Maharashtra Agric Univ 14(3):341–342

Waskar DP, Khedkar RM, Garnade VK (1999) Effect of post harvest treatments on shelf life and quality of pomegranate in evaporative cool chamber and ambient conditions. J Food Sci Technol (India) 36(2):111–114

Wavhal KN (1981) Varietal improvement of pomegranate. Maharashtra J Hortic 2:49–52

Waval KN, Rane DA, Kamble KB, Choudhary SM (1985) Nutritional studies in pomegranate. 3rd National workshop on arid zone fruits, MPKV. Rahuri 5–7 July 1985

6 Aonla (Indian gooseberry)

Abstract

Aonla (*Emblica officinalis* Gaertn.) or Indian gooseberry is an important fruit crop of Indian origin and is known since for its nutritive and medicinal value. The fruit is a rich source of ascorbic acid, a vital vitamin for its antioxidant value. The vitamin is not oxidized due to the presence of tannins and is least destroyed during storage. Dried fruits of aonla are useful in the treatment of dysentery, jaundice, dyspepsia, and cough.

Aonla is emerging as one of the important minor fruits of India. The area under this fruit has gone up to approximately 25,000 ha over the past decade. Identification of superior varieties with high yield increased productivity even under semiarid rainfed conditions on marginal degraded lands. The tree can beat the vagaries of monsoon, drought conditions, and high temperature. There are a number of value-added products that can be prepared and sold in Indian as well as in overseas market. The demand for such products is on an increase, and around 20 t of aonla candy was produced last year in Gujarat itself. Thus, there is great potential for aonla cultivation in semiarid rainfed areas.

6.1 Introduction

Aonla (*Emblica officinalis* Gaertn.) is also known as Indian gooseberry. It is an important crop of Indian origin which is grown throughout northwest Indian plains. It is very rich from the nutritional point of view, being next to Barbados cherry in terms of vitamin C content (500–750 mg/100 g pulp) which is not oxidized due to the presence of tannins and is even least destroyed during storage. Besides, it has excellent medicinal value. It can also be grown successfully in arid climate and in the soils of higher pH and poor fertility wastelands. Aonla is emerging as one of the important minor fruits of India. The area under this fruit has gone up to approximately 10,000 ha over the past decade. Identification of superior varieties with high yield increased productivity even under semiarid rainfed conditions on marginal degraded lands. The tree can beat the vagaries of monsoon, drought conditions, and high temperature. There are a number of value-added

S. Hiwale, *Sustainable Horticulture in Semiarid Dry Lands*,
DOI 10.1007/978-81-322-2244-6_6, © Springer India 2015

products that can be prepared and sold in Indian as well as in overseas market. The demand for such products is on an increase, and around 20 t of aonla candy was produced last year in Gujarat itself. Thus there is great potential for aonla cultivation in semiarid rainfed areas.

6.2 Uses and Nutritive Value

The aonla fruit has special significance in view of its high nutritive value and rich medicinal properties (Table 6.1). It is a rich source of vitamin C (ascorbic acid), and the bark, leaves, and fruits are used for tanning. Its fruits are also a rich source of pectin which is used for tanning and preparing jelly. Dried aonla fruit is useful for the treatment of dysentery, jaundice, dyspepsia, and cough. It is also a good treatment for eye problems and stomach ailments. Aonla fruit is also an important ingredient of chyawanprash which is an ancient well-known Ayurvedic medicine. Aonla is also used for manufacturing hair oil. Fruits are also utilized for the preparation of preserves and pickles.

6.3 Botany

Aonla (*Phyllanthus emblica* L.) belongs to the family Euphorbiaceae, to which also belongs another species, *Phyllanthus acidus* (*distichus*). In both these species, the leaves are small and arranged in two rows along small branches, some of which are deciduous. The tree is hardy and of medium height. However, the seedling tree can attain height up to 30 m. It is evergreen; however, leaf shedding takes place during February–March. The tree trunk is smooth and whitish which resembles guava. The bark peels off and each shoot has at least about 100 leaves. Leaves are simple and are clustered on their branchlet in such a way that the branchlet appears to be a compound leaf. On these branchlets, male and female flowers appear during the late spring season. Plants are monoecious. Flowers are borne on young shoots in the axil of leaves. Lower flowers are usually female. Flowers are small and have small pedicles. Pistillate flowers are borne singly and are green in color. Staminate flowers are borne in groups of three which are light yellow in color. Sepals vary from 5 to 6. Anthers open vertically. Fruits are borne on the same shoot which carries leaves. The fruit is berry and round in shape. It is smooth with firm flesh and has six lobes. The stone is hexagonal with six flattened seed ends, and the fruit is almost stalkless.

Table 6.1 The nutritive value of aonla fruit

Constituent	Value
Moisture	81.2 %
Protein	0.5 %
Fat	0.1 %
Carbohydrate	14.1 %
Calcium	0.05 %
Phosphorus	0.02 %
Iron	1.2 %
Vitamin C	600 mg/100 g

Source: Singh et al. (1963)

Aonla fruits are borne on the branches and the plant thus appears to bear flowers on its leaves, giving the name of the genus *Phyllanthus* which means literally "leaf flower." A third species of the genus, *P. fischeri*, found in the forests of south India, also bears fruits for human consumption. The genus *Phyllanthus* comprises about 350 species, mostly shrubs, some herbs, or trees.

Aonla cultivars have not been standardized and are identified mostly on the basis of size, color, or location where they grow. The popular cultivars are Banarasi, green tinged, red tinged, and white streaked. Some important varieties of aonla recommended for commercial cultivation are Banarasi, Deshi, Chakaiya, and Hatijhool. The Banarasi is the best variety for making preserves. The fruits are large, yellowish, and shiny, with good keeping quality. Chakaiya is a very hardy variety.

6.4 Cultivars

Banarasi, Chakaiya, and Francis (Hatijhool) are major commercial cultivars of aonla. The salient characteristics of these cultivars are described below.

Description of Aonla Cultivars
The aonla cultivars based on their evaluation at CHES, Godhra, can be divided into three groups:
- *Early season*
- Maturity from mid October to the end of October. Banarasi, Chakaiya, and Anand
- *Mid season*
- Maturity from the end of October to middle of November. Francis, NA-7, and Krishna
- *Late season*
- Maturity from the first to third week of November. Kanchan and Anand-1

6.4.1 Banarasi

It is an early bearer and bears good crop of quality fruits having a good keeping quality. The fruits are medium to large and roundish with six linear grooves extending from the base to apex and are light green in the early stage and whitish green on maturity. The skin is thin, translucent, adhered to the flesh, and slightly fibrous, and the stone is lightly embedded in the flesh, dull green, and hexagonal with six small elliptical seeds. Juice content is medium and is moderately astringent. The TSS is 13.0 %, acidity 2.34 %, total sugars 8.0 %, and seed/pulp ratio 1:21 (Teotia et al. 1968).

6.4.2 Chakaiya

It is a hardy cultivar with very heavy bearing habit. It is a mid-season cultivar having a very good keeping quality. Fruits are medium sized and flattened at the base and apex. The stalk is roundish and inserted. The skin is thin, green in color, and smooth. Segments are 6–8 in number, flesh soft, fibrous, adhered to the stone, and astringent, and the TSS is 9.0 %, acidity 2.17 %, total sugars 9.6 %, and seed/ pulp ratio 1:17 (Teotia et al., 1968). The stone is small and dull green with 6–8 furrows and ridges. It is a good pollinizer. It is a late and heavy bearer and has poor keeping quality. Fruits are large, oval roundish, and short stalked with six linear grooves. The skin is thin and light green. The flesh is firm, slightly fibrous, firmly adhered to the stone, and moderately astringent, having a TSS of 11.5 %, acidity 2.52 %, and total sugars 7.3 %. The stone is small, flattened, and dull green (Teotia et al. 1968).

6.4.3 NA7

It is a selection from open pollinated seedlings of Francis and bears profusely with 5.06 female flowers per branchlet. It has a precocious and prolific bearing. Fruits are medium to large (47.5 g). Oval with irregular basis and conical in shape, smooth yellowish green skin. The flesh is almost fibreless soft, has moderate keeping quality, and is free from necrosis and suitable for processing.

New strains of aonla have been identified at CHES, Godhra, through clonal selection which are high yielding and possess other good quality attributes.

6.4.4 Kanchan

It is a chance seedling of Chakaiya. The tree is tall and has a spreading growth habit. Fruit medium, flattened oblong, skin is smooth, light green, segment six and difficult to separate. The flesh is fibrous and hard. It has good keeping quality and is ideally suitable for preparation of pickles. It matures late and is free from fruit necrosis. The TSS is 9.5 %, acidity 2.11 %, and vitamin C 422.70 mg/100 g.

6.4.5 Krishna

This is a chance seedling of Banarasi from Pratapgarh District. The tree is semi-tall with spreading growth habit. Fruits large, skin smooth, whitish green to apricot yellow in color with red dpot on exposed surface. Segments are six in number and can be easily separated, and the flesh is pinkish green, less fibrous, and highly astringent. It is a shy bearer, matures earlier, and has no evidence of fruit necrosis. It has a moderate keeping quality. The TSS is 11.00 %, acidity 2.32 %, and vitamin C 549.20 mg/100 g.

6.5 Aonla Variability and Its Performance

The performance of eight varieties of aonla over the last 5 years revealed that varieties differ in growth and bearing habit. Trees of cultivars Anand-2, Anand-1, Banarasi, and Krishna were vigorous and spreading, whereas those of NA-7, Francis, and Chakaiya showed less vigor and an upright growth pattern. The cultivars clearly differ in the bearing number of female flower per branchlet, which largely affects their yield potential (Hiwale and Raturi 1999) (Tables 6.2 and 6.3).

Consequently cultivar NA-7 outyielded all the others, with a yield potential of 102 kg per plant, followed by Kanchan with a 88.4 kg yield per plant in the 10th year (Table 6.2). Similar results were reported by Singh et al. (1963) and Pareek and Nath (1996). The fruit maturity in cultivars Banarasi, Krishna, and NA-7 occurred during mid October to mid November, and these are classified as early cultivars. The mid-season cultivars Francis, Anand-1, Anand-2, and Kanchan mature during mid November to mid December. Cultivar Chakaiya was late to mature. Maximum fruit weight, length, and breadth were recorded in cultivar Banarasi. It was least in NA-7. Banarasi and Anand-2 have the highest TSS. Krishna has the maximum vitamin C content. The results clearly indicated the supremacy of aonla cultivar NA-7 in terms of production as well as income. Cultivar NA-7 was found to be early and fruit development coincides with maximum moisture availability and hence recommended for commercial cultivation.

6.6 Pollination Techniques and Pollinizers on Fruit Set and Fruit Quality in Aonla

NA-7 was found to be a promising cultivar, but there was a self-incompatibility problem in this cultivar and it needs pollinizers to improve fruit set (Alemullah and Santram 1990). Investigation carried out of study self-incompatibility of cultivars NA7 revealed that under bagging the lowest fruit set (2.35 %) and no fruit retention at maturity was recorded. Fruit set and fruit retention were significantly higher under open pollination (61.43 and 18.43 %, respectively). Among different cross combination, the highest fruit set and fruit retention were recorded in cv. NA-7 × NA-6 (58.42 and 10.41 %, respectively), closely followed by NA7 × Krishna (46.86 and 9.83 %, respectively). Poor fruit set under bagging clearly indicated the problem of self-incompatibility in NA-7. Most of the aonla varieties are found to be self-incompatible. Aonla cv. NA-6 and Krishna are found to be suitable pollinizers for NA-7 (Table 6.4).

6.7 Flowering Fruit Set and Fruit Drop Studies in Aonla

Aonla or Indian gooseberry can be successfully cultivated under semiarid rainfed conditions. In newly evolved cultivars, their performance in respect of flowering, fruit set, and fruit drop was considerably influenced by the rainfed climatic conditions. Cultivar NA-7 was found to be superior to other nine cultivars evaluated in respect to earliness, higher number of female flowers per branchlet, lowest male/female ratio, maximum fruit set, retention, and thus productivity and therefore is recommended for commercial cultivation under semiarid rainfed conditions of western India.

Table 6.2 Performance of aonla varieties

			Spread			Physicochemical character						
Cultivar	Height (M)	Dia. (cm)	NS	EW	Yield/plant (kg)	Fruit (wt. g)	Length (cm)	Breadth (cm)	TSS °Brix	Acidity (%)	Vit. C mg/100 g	Total sugar (%)
Chakaiya	5.18	30.6	3.8	4.5	78.5	31.8	3.4	4.1	10.5	1.65	527	4.0
Francis	4.33	30.9	5.1	5.3	71.1	32.0	3.4	3.7	12.0	1.70	516	4.5
Kanchan	4.98	35.3	5.4	5.1	88.4	34.2	3.6	4.2	10.0	1.45	504	4.0
Krishna	5.29	33.3	5.3	5.3	69.6	34.6	3.7	4.1	11.5	1.40	549	4.9
NA-7	4.52	29.9	3.8	3.9	102.3	30.6	3.4	3.7	11.5	1.60	500	4.6
Anand-1	5.51	35.7	5.7	5.3	45.0	32.8	3.5	3.6	12.0	1.82	512	5.0
Anand-2	5.67	36.9	5.4	5.6	59.5	37.2	3.6	4.3	12.5	1.75	528	4.9
Banarasi	5.43	37.2	5.6	5.7	75.2	41.5	3.8	4.3	12.5	1.50	484	4.6

Hiwale and Raturi (1999)

Table 6.3 Fruit set and fruit retention in aonla cultivars

Variety	Male flowers/cluster	Female flowers/cluster	Male to female ratio	Fruit set (%)	Fruit retention (%)
Banarasi	520.29	2.33	253.96:1	27.01	6.22
Krishna	597.04	2.00	355.49:1	29.69	8.13
Francis	570.58	4.84	159.74:1	46.03	9.19
NA-7	503.60	9.79	61.11:1	58.94	12.37
Chakaiya	358.73	7.86	64.05:1	52.55	8.44
Kanchan	329.64	7.77	66.28:1	50.79	12.28
Anand-1	724.37	3.21	225.66:1	13.75	3.42
Anand-2	694.42	4.56	152.27	14.53	3.75

Table 6.4 Effect of pollination techniques and pollinizers on fruit set and fruit quality in aonla

Crosses	Fruit set (%)	Fruit retention (%)	Fruit size (cm^2)	Fruit wt. (g)	TSS (%)	Ascorbic acid (mg/g)
'NA-7 (bagged)	2.35	3.72	15.71	47	11.6	
'NA-7 (open pollination)	61.43	18.54	14.5	45	12.0	590
'NA-7x Chakaiya	29.98	6.87	14.23	40	9.5	525
'NA-7x'Francis	31.24	2.42	15.01	43	8.0	490
'NA-7x Krishna	46.86	9.83	15.91	50	11.0	512
'NA-7x Banarasi	22.02	2.77	16.11	52	12.5	586
'NA-7x NA-6	58.42	10.42	15.00	41	10.5	568
'NA-7x NA-10	38.69	6.39	16.28	53	13.0	600
CD at 5 %	4.81	2.1	NS	1.03	2.3	NS

The problem of erratic flowering and low fruit set and heavy fruit drop is still persistent to the extent that some of the varieties have become commercially uneconomical. The present investigation was therefore undertaken to evaluate the performance of new cultivars for flowering, fruit set, and fruit drop under semiarid rainfed areas of western India.

Germplasm of nine cultivars of aonla, viz., Chakaiya, Francis, and Banarasi (the traditional ones), were compared with those of NA-7, Krishna, Kanchan, Balawant, Anand-1, and Anand-2 at the central horticultural experiment station, Vejalpur, Gujarat, under semiarid rainfed conditions.

Flowering and fruit set studies were carried out by counting the total number of branchlet per shoot, no of male/female flowers per determinate branchlet and on this basis male/female ratio was worked out. Similarly the number of fruit set per branchlet, number of fruits retained at important fruit development stages, and number of fruit dropped after the fruit set were counted on a whole tree basis by putting polythene sheet below the tree at periodical interval, and their number was counted. Percent retention was worked out for both the stages in July and September.

6.8 Flowering and Sex Ratio

Cultivars Chakaiya, NA-7, and Balawant were early to flower (in the last week of February), whereas flowering started in the middle of March in cultivars Krishna, Banarasi, Anand-1, Anand-2, Francis, and Hathizool. Flowering was late in Kanchan. Data on the total no. of determinate branchlet per shoot indicated significant differ-

Table 6.5 Flowering and sex ratio in aonla

Cultivar	No. of branchlet	No. of male flower/branchlet	No. of female flowers/branchlet	Male/female ratio
Chakaiya	236.50	325.67	4.66	141.74
Francis	460.17	270.33	2.50	33.52
NA-7	363.83	312.17	6.50	22.51
Banarasi	324.33	349.67	3.00	79.38
Anand-1	424.83	306.00	3.50	90.56
Anand-2	415.00	352.67	3.50	106.80
Kanchan	325.33	207.33	2.17	32.42
Krishna	370.83	208.33	3.33	32.08
Hathizool	401.17	316.67	2.33	109.58
Balawant	348.50	239.33	4.16	33.96
CD at 5 %	85.35	69.67	1.37	36.54

ences among the cultivars. It was least in Chakaiya (236.50) and was maximum in Francis (460.17), probably due to vigorous growth in this cultivar. Flower buds appeared on determinate shoot immediately on sprouting. In the initial stages, male flowers were first to appear and female flowers at a later date. Flower opening initiated before the branchlets were fully developed and male flowers were first to open. Bajpai (1968) and Karale et al. (1991) reported similar results. There were significant differences among the cultivars in respect to the number of male and female flowers per branchlet as well as sex ratio. Male flowers were highest in Anand-2 (352.67) and least in Kanchan (207.33). The maximum number of female flowers/branchlet was observed in NA-7 (6.5) and minimum in Hathizool (2.33), whereas the male/female ratio was highest in Chakaiya (141.74) and was least in NA-7 (22.51) (Table 6.5).

6.9 Fruit Set and Retention

Fruit set/shoot was recorded based on the no. of female flowers retained after flowering was completed in the first week of April. Significant differences were recorded among the cultivars. Fruit set was highest in NA-7 (989.33 no.) and was least in Anand-1 (337 no.) (Table 6.6). After this the fruits underwent dormancy for about one and half month, and further fruit development started on the onset of monsoon. Fruit retention was counted at a regular interval at the initiation of development, at the pea-size stage, and at the fully developed stage. Data showed significant differences among the cultivars. Fruit retention/shoot was maximum in NA-7 (658.33 and 409.66, respectively) in both July and September and was minimum in Anand-1 (174 and 85.66, respectively) (Table 6.6). Percent retention showed significant differences among different cultivars. NA-7 had highest percent retention in both July and September (66.58 and 41.40, respectively). Retention was least in Hathizool (43.75 % and 21.65 %, respectively) (Table 6.6).

6.10 Fruit Drop

Fruit drop studies were carried out on whole tree basis by laying polythene sheet under the tree canopy and counting the number of fruit dropped at a regular interval. Fruit drop was maximum after the first rainfall which may be due to the sudden change in the temperature from 40 °C to around 30 °C. The next wave of fruit drop was mainly due to the incidence of fruit borer and the third wave due to again the sudden rise in temperature in September causing sunburn and fruit drop. The intensity of fruit drop varied among the cultivars (Fig. 6.1). Fruit drop was recorded from the 30th of July to the 2nd of September. The maximum number of fruit dropped was in NA-7

Table 6.6 Fruit set and fruit retention/shoot in aonla

Cultivar	Fruit set (April) No./shoot	Fruit retained (July) No./shoot (%)		Fruit retained (Sept.) No./shoot (%)	
Chakaiya	668	310.33	46.74	221.66	32.62
Francis	713.66	392.33	54.90	224.66	31.30
NA-7	989.33	658.33	66.58	409.66	41.40
Banarasi	681	392.33	54.12	176.00	25.96
Anand-1	337	174.00	52.19	85.66	34.00
Anand-2	387	232.66	60.07	132.33	25.75
Kanchan	477.66	249.33	54.03	163.33	35.04
Krishna	627	395.33	62.83	194.33	31.18
Hathizool	510.33	220.33	43.75	107.66	21.65
Balawant	660.33	334.33	49.68	150.00	22.25
CD at 5 %	153	107.86	9.11	71.81	9.39

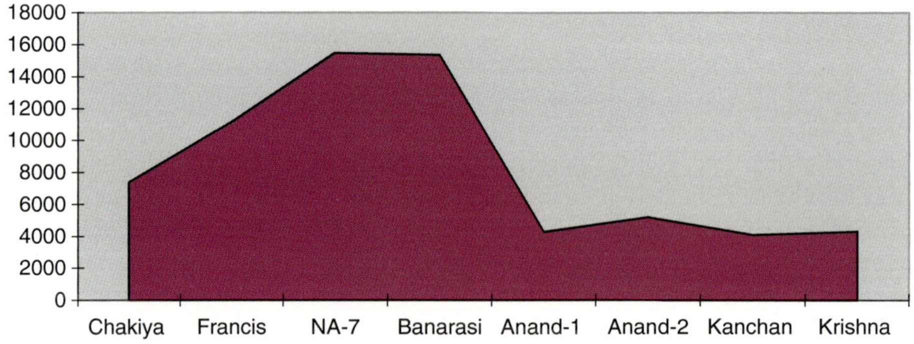

Fig. 6.1 Total fruit drop no./tree

owing to the heavy set, and it was least in Anand-2. Allemullah and Sant Ram (1990) and Bajpai (1968) also reported variation in fruit set and fruit drop in different cultivars. Cultivar NA-7 was found to be commercially superior due to the high fruit retention and therefore yield under semiarid rainfed conditions.

6.11 Aonla Clonal Selections and Its Performance

The objective of breeding was to evolve the aonla variety for earliness, high yield, and suitability to high-density orcharding under semiarid conditions. The breeding method followed was clonal selection. The clonal election was made from the collections done at CHES, Vejalpur, over the past 20 years. Selection was made from plus trees observed from the collection. The clonal selection identified was evaluated and then multiplied by in situ budding. Evaluation of the selection for the past 5 years revealed that vegetative growth parameters of all the plants selected were on par. The timely flowering resulted in early fruit development with change in weather conditions. There are a more number of female flowers per branchlet, a less number of empty clusters per shoot resulting in early fruit maturity and heavy fruit set, fruit retention, and therefore yield. As for the quality attributes, it has a low fiber content which makes it suitable for processing and export.

The selection matures 2 weeks earlier than the parent material. Observation on flowering and fruiting recorded indicated that selection-6 had the least number of empty branches (53.33) as compared to the control (165.33). The number of female flowers per branchlet was highest (16.66) in the selection and was least in control (6.66). The number of branchlet bearing female flowers was maximum in the selection (70.1 %) compared to the control (30.33 %). This resulted in higher fruit set and retention leading to higher yield per plant. The yield increase is to the tune of 22.9–54.15 % over the control under dry land conditions of Gujarat (Table 6.7). The results clearly indicated the superiority of selection-6 over other selections and control and was therefore identified and released as "Goma Aishwarya".

6.12 Cultivation

6.12.1 Soils

Sandy loam soils are the best suited for aonla cultivation. Soil should be at least 2 m deep; otherwise, the plant will not survive beyond 12 years in fruiting (Sant Ram 1983). Alkali soils having a pH of 8.5 suppressed root growth due to its adverse effect on the availability of micronutrients (Sant Ram and Rao 1976). Similarly, soils rich in lime are also not much suitable for growing aonla. A six-month-old seedling could be planted and subsequently budded in alkali soils having an ESP of 32 and ECe of 10 dSm^{-1} (Anonymous 1988).

6.12.2 Climate

Its cultivation is more successful in subtropical regions. It can be grown successfully up to the elevation of 1,800 m from sea level. It can also be grown in semiarid to arid regions.

Table 6.7 Performance of aonla selections

Sel. no.	No. of fruit set/shoot	No. of fruit retained/shoot	% retention	Total no. of branchlet/shoot	No. of empty branchlet/shoot	No. of branchlet with female flower/shoot	% of branchlet with female flower /shoot	No. of female flower/branchlet	Yield/ plant (kg)	% yield increase over control
Sel.-1	222.33	66.66	29.98	348.33	219.00	107.66	30.93	12.66	79.00	23.43
Sel.-2	375.00	97.83	26.09	220.00	115.66	104.3	47.40	8.66	78.66	22.90
Sel.-3	533.33	127.00	23.81	367.33	238.66	148.39	40.39	16.00	80.00	25.00
Sel.-4	365.00	85.33	23.37	187.67	188.66	32.33	17.22	14.66	71.00	10.93
Sel.-5	516.00	171.66	33.21	189.33	94.86	94.66	49.99	14.33	81.33	27.07
Sel.-6	762.33	284.33	37.27	182.00	53.33	129.00	70.87	15.66	98.66	44.15
Sel.-7	749.00	269.33	35.94	249.00	112.00	137.00	55.00	16.66	87.00	35.93
Control	248.33	62.33	17.89	237.33	165.33	72.00	30.33	6.66	64.00	–
CD at 5 %	280.53	107.33	14.86	108.51	65.29	NS	–	3.19	14.60	–

6.13 Propagation

Aonla can be propagated by both seed and vegetative methods. Seeds are extracted from the fruit after ripening and are dried in shade. The seeds which float in the water are described, and settled seeds on the bottom are taken for sowing. According to Bhujbal (1975), dipping of seeds in GA3 50 ppm for 24 h gave maximum seed germination. Seeds are sown in the nursery either during March or June–July. The seedling plants after attaining proper stem girth are budded during June–August. Shield, forkert, and patch are common methods of budding. Besides, inarching is also successful to the extent of 25–30 % (Singh 1952). Cutting and layering are not successful in aonla. Aonla cultivation is picking up fast and hence demand for plant material has increased manifolds with the advent of technology. Raising an aonla nursery has also become a profitable preposition by raising rootstock plants of Deshi aonla available in the forest areas in 200 gauge polythene bags of 25 cm length and 10 cm diameter. The seedlings are ready for budding after 5–6 months, and they are ready for transplanting in another 40 days of budding (Table 6.8).

Table 6.8 Performance of budding methods

	In situ budding	Nursery budding
Success in budding (%)	85.70	62.80
Average plant height (m)	01.33	85.00
Average stock diameter (mm)	31.39	09.00
Average scion diameter (mm)	25.91	08.00
No. of fruits per plant in the 3rd year	52.00	27.00
Yield per plant (kg) in the 3rd year	15.00	00.45

The best time for budding is observed to be from the first week of May till the rest of the monsoon. Patch budding was found to be the best method for budding. Deshi aonla is found growing wild in India and is known for their hardiness. Seed germination can be improved by wetting the seed in water containing 2 % salt and removing the floating seeds. The seeds which settle down are tied in to a gunny bag and then buried into FYM for 3–4 days. The seed germination is improved to 70 % by this method. Seeds should be collected from fully ripe fruits in the month of November and are sown in the month of February in polythene bags. For in situ budding, the seeds are sown in prepared pits on the onset of monsoon and are budded subsequently when they attain the diameter of 4–5 mm.

6.14 Establishment of Orchard

6.14.1 Plantation

Planting is done during the rainy season (July–August). However, it can also be done during March–April. Plantation can be done at a space of 8–11 m apart which may vary according to soil and climatic conditions and growth habit of a particular cultivar.

6.14.2 Pruning and Training

Aonla trees do not require special pruning and training. However, during the initial years, pruning should be done to have a balanced training of the plant. In addition, dried branches should be

In situ budding Aonla

removed periodically. As the branches carrying heavy crop load break off, the plants should be trained to develop a low-headed foundation frame of main branches on the trunk within 1 m from the ground. The frame should be developed by encouraging growth of 4–6 well-spaced branches with fairly wide angles. Bearing trees can be pruned after the termination of crop each year by removing dead, diseased, broken, or weak cross branches and suckers from the rootstock.

6.14.3 Irrigation

During the active growth period from March to June, trees are required to be irrigated at a weekly interval to secure higher fruit set and reduce fruit drop. During October–December, irrigation at 20 days' interval may help in better fruit development. Care must be taken to ensure that water does not stagnate in the orchard during the rainy season failing which trees may die.

6.14.4 Nutrition

Nutrition plays an important role in aonla cultivation. By way of proper nutrition, the orchard can be maintained in good fruiting for long, and 180–360 g nitrogen, 540–1,080 g phosphorus, and 180–360 g potash per tree per year have been recommended in cultivar Banarasi (Singh 1973). The application of fertilizers helped in increasing female flowers compared to male flowers, increased fruit size and TSS, and also reduced fruit drop. Phosphorus and potassium improve vitamin C content. Boron helps in flowering and also in improving TSS and vitamin C content. Foliar spray of 0.6 % borax solution thrice at 10–12-day interval during September–October has been recommended for controlling internal necrosis in cultivar Francis (Sant Ram and Rao 1976). 15 kg FYM per year of age to the maximum of 75 kg and 1 kg calcium ammonium nitrate and 2–5 kg super phosphate per tree in February and 1 kg calcium ammonium nitrate in July–August have been recommended, in Haryana.

6.15 Harvesting and Storage

Fruits become ready for harvesting by November–December and can be harvested up to February. The color of fruits starts turning to light yellow at maturity stage. Care must be taken at the time of picking so that fruits do not get injured due to dropping to the ground.

Under cold storage conditions, fruits can be stored for 7–8 weeks, keeping temperature at 32–35 °F and relative humidity of 85–90 %.

6.16 Yield

Deshi aonla trees start bearing fruits after 10 years of plantation, whereas budded or grafted trees start bearing after 3 years. From the full-grown tree of 10–15 years of age, approximately 100 kg fruit yield can be obtained per year per tree.

6.17 Economics

The yield stabilized in the 10th year with a production of 95 q aonla/ha. A net return of Rs. 42,750 and BC ratio of 1:10.98 were recorded (Hiwale 2014).

Annex

Rainfall (mm) – 2002, 2003, and 2004 – CHES, Godhra

Date (2002)	Rainfall (mm)	Date (2003)	Rainfall (mm)	Date (2004)	Rainfall (mm)
25/6	3	17/6	15	9/5	4.1
26/6	51	19/6	20	5/6	22
27/6	12	21/6	6.08	14/6	15
28/6	104	22/6	9.04	15/6	10.2
29/6	5	23/6	16.06	16/6	1.2
30/6	8	24/6	15	17/6	1
16/7	12	28/6	3	18/	1.8
17/7	11	3/7	35	6/7	14.3
22/7	3.2	8/7	10	7/7	1.8
26/7	6	9/7	33	17/7	16.4
5/8	4	10/7	3	20/7	.7
6/8	11.6	11/7	38.06	21/7	14
7/8	8	12/7	10.04	22/7	14.8
8/8	3.2	16/7	35	26/7	10
10/8	2	17/7	17	27/7	23.5
16/8	14	19/7	16	28/7	3
24/8	16	21/7	35.02	29/7	1
25/8	86	22/7	22	30/7	38
26/8	18	23/7	32	31/7	68
31/8	8	24/7	49	1/8	43
3/9	7	25/7	31	3/8	6.3
4/9	55	26/7	11	4/8	18
5/9	10	27/7	12.06	5/8	44
Total	*458*	28/7	97	6/8	46
		29/7	13	7/8	31
		2/8	34	8/8	74
		3/8	7.02	9/8	34
		4/8	9.08	10/8	19
		5/8	5.04	11/8	58
		24/8	161	12/8	20
		25/8	25	13/8	42
		26/8	17	14/8	102
		27/8	27	15/8	25
		28/8	5.5	16/8	8
		20/9	15	17/8	2.4
		22/9	23	19/8	20
		23/9	24	20/8	1
		Total	*940.5*	21/8	18
				23/8	9
				24/8	26.2
				25/8	0.2
				Total	*907.09*

References

Anonymous (1988) Annual report. All India Co-ordinated Research Project on Arid Zone Fruits. Technical document no. 26

Allemullah M, Sant Ram (1990) Causes of low fruit set and heavy fruit drop in Indian gooseberry (Emblica officinalis Gaertn). J Hortic 47(3):270–277

Bajpai PN (1968) Studies on flowering and fruit development in Aonla (*Emblica officinalis* Gaertn.). Hortic Adv VII:38–67

Bhujbal BG (1975) Improvement in seed propagation of aonla. Res J MPAU 6(1):73–75

Hiwale SS, Raturi GB (1999) Performance of aonla (*Emblica officinalis* Gaertn.) under rainfed condition. In: Proceeding of national seminar on sustainable horticulture production in tribal region, CHES Ranchi, 25–26 July 1999

Hiwale SS (2014) Bulletin on "Aonla based cropping system" no. 55. CHES, (CIAH) (ICAR), Vejalpur

Karale AR, Keskar BG, Dhawale BC, Shetye MP, Kale PN (1991) Flowering, sex expression and sex ratio in Banarasi aonla seddling trees. J Maharashtra Agric Univ 16(2):270–272

Pareek OP, Nath V (1996) Booklet on Coordinated fruit research in Indian Arid zone. A two decades profile, pp 45–54

Sant Ram (1983) Aonla. In: Pandey RM, Mishra KK (eds) Phal Utpadan. Directorate of Translation and Publication, GBPUAT, Pantnagar

Sant Ram, Rao R (1976) Naturally occurring cytokinin in aonla fruit. New Phytol 79:441–448

Sant Ram, Rao R (1978) Studies on naturally occurring gibberellins in aonla fruit. New Phytol 81(3):513–517

Sant Ram, Dwivedi TS, Bist LD (1976) Internal fruit necrosis in aonla. Prog Hortic 8(3):5–12

Singh LB (1952) Top working of aonla. Sci Cult 17:345–346

Singh DB (1973) Studies on the problems of fruitfulness in aonla. Ph.D. thesis, Agra University, Agra

Singh S, Krishnamurthi S, Katyal SL (1963) Fruit culture in India. I.C.A.R., New Delhi

Teotia SS, Singh DB, Singh RD, Singh RN (1968) Studies on some varieties of aonla of Eastern Utter Pradesh. Punjab Hortic J 8:241–244

Mango (*Mangifera indica* L.) 7

Abstract

Mango (*Mangifera indica* L.) is the number one fruit tree in respect to area, production, as well as income. It is known for its liking by the people, attractive color, vitamins, and mineral composition. The area under mango from 1978 to 1979 was 942,560 ha which was 42.6 % of the total area under fruits with a total production of 8.215 million tons. In 2010–2011, the area was 2,297,000 ha, a phenomenal 2.43 times increase, but there is a decrease in percentage of total area (36.0 %). The leading mango growing states in India are Uttar Pradesh, Andhra Pradesh, Bihar, Maharashtra, Orissa, and West Bengal. The varieties suitable for export are Alphonso, Kesar, and Dashehari which are grown on large scale. Fresh mango export was to the tune of 83,703 t worth Rs. 17,071 Lakhs in 2008–2009. Large quantity of mango pulp is also exported in 2008–2009; it was 173,013 t worth Rs. 75,298 Lakhs.

7.1 Introduction

Mango (*Mangifera indica* L.) is the number one fruit tree in respect to area, production, as well as income. It is known for its liking by the people, attractive color, vitamins, and mineral composition. The fruit appears to be linked with the cultural history of India. It is one of the ancient fruits of India. It is known in India since 2000 BC and has historical, religious, cultural, and economic importance. Mango is not only important for its fruits but also its timber, fuel, and fodder value; apart from this it provides shade to the cattle and human beings. In olden days, mango was mostly propagated by seeds, which helped in soil conservation. Mango cultivation got real boost during the Mogul period; many of the choice seedling trees were identified and named and propagated vegetatively. Mango is believed to have originated in the Indo-Burma region (De-candole 1904; Popeno 1932; Mukherjee 1951). Besides India, mango is cultivated in most of the countries of Southeast Asia, namely, the Philippines, Indonesia, Thailand, Burma, Malaysia, and Sri Lanka. Other important countries growing mango are Egypt, Southeast Africa, South Africa, Israel, tropical Australia, the USA (Hawaii and Florida), Mexico, Brazil, Cuba, and West Indies. In India, mango is distributed throughout the length and breadth of the country except in hilly regions above 915 m from mean sea level. The area under mango during 1978–1979

S. Hiwale, *Sustainable Horticulture in Semiarid Dry Lands*,
DOI 10.1007/978-81-322-2244-6_7, © Springer India 2015

was 942,560 ha or 42.6 % of the total area under fruits with a total production of 8.215 million tons; the leading mango growing states in India are Uttar Pradesh, Andhra Pradesh, Bihar, Orissa, and West Bengal. Commercial planting of the variety Alphonso is located in Maharashtra (Ratnagiri) and Gujarat (Bulsar).

7.2 Area and Production (Tables 7.1 and 7.2)

7.3 Botany

Mango belongs to the family Anacardiaceae; other well-known plants of the family are cashew and pistachio. There are other edible species of mango; they are *Mangifera odorata*, *Mangifera foetida*, and *Mangifera caesia* which are cultivated. The genus *Mangifera* consists of 41 species (Mukherjee 1949a) distributed throughout Malaysia from India and Sri Lanka in the West to

Table 7.1 All-India area, production, and productivity of mango

Year	Area (000' ha)	% of total area	Production (000' MT)	% of total production	Productivity (MT/ha)
1991–1992	1,077.6	37.5	8,715.6	30.4	8.1
2001–2002	1,575.8	39.3	10,020.2	23.3	6.4
2002–2003	1,623.4	42.9	12,733.2	28.2	7.8
2003–2004	1,906.7	40.8	11,490.0	25.2	6.0
2004–2005	1,970.4	39.7	11,829.7	24.0	6.0
2005–2006	2,080.7	39.1	12,663.1	22.9	6.1
2006–2007	2,154.0	38.8	13,734.0	23.1	6.4
2007–2008	2,201.0	37.6	13,997.0	21.3	6.4
2008–2009	2,309.0	37.8	12,750.0	18.6	5.5
2009–2010	2,312.3	36.5	15,026.7	21.0	6.5
2010–2011	2,297.0	36.0	15,188.0	20.3	6.6

Kumar et al. (2011)

Table 7.2 State-wise area, production, and productivity of mango

	Area (000' ha)			Production (000' MT)			Productivity (MT/ha)		
	2008–2009			2009–2010			2010–2011		
State	Area	Prod.	Pdy.	Area	Prod.	Pdy.	Area	Prod.	Pdy.
Utter Pradesh	271.2	3465.9	12.8	276.4	3588.0	13.0	267.2	3623.2	13.6
Andhra Pradesh	497.7	2522.0	5.1	480.4	4058.3	8.4	391.1	3363.4	8.6
Karnataka	141.3	1284.4	9.1	153.8	1694.0	11.0	161.6	1778.8	11.0
Bihar	144.1	1329.8	9.2	146.0	995.9	6.8	147.0	1334.9	9.1
Gujarat	115.7	299.8	2.6	121.5	856.7	7.0	130.1	911.3	7.0
Tamil Nadu	148.8	821.4	5.5	132.7	636.3	4.8	148.0	823.7	5.6
Orissa	164.3	449.7	2.7	177.6	577.5	3.3	190.1	642.0	3.4
West Bengal	86.0	548.9	6.4	88.1	578.0	6.6	89.5	620.2	6.9
Jharkhand	–	–	–	–	–	–	38.9	427.9	11.0
Kerala	76.7	445.4	5.8	63.8	373.2	5.9	62.2	380.9	6.1
Maharashtra	457.0	712.8	1.6	474.5	597.0	1.3	477.0	331.0	0.7
Others	206.2	869.5	4.2	197.42	1071.7	5.4	194.0	951.1	4.9
Total	2309.0	12749.8	5.5	2312.3	15026.7	6.5	2296.8	15188.4	6.6

Kumar et al. (2011)

the Philippines and New Guinea in the East; most of them are wild and economically unimportant. It has been observed that all the five species and 23 wild and horticultural varieties of mango have the same chromosome number $2n=40$. Their meiotic behavior is regular, showing regular pairing and disjunction into 20 bivalents. The chromosomes are small in size, varying from 0.4 to 2.0 μm, but they can be morphologically distinguished into eight types according to size, presence, and absence of primary and secondary constrictions and satellites. From cytological evidences, Mukherjee (1950) concluded that mango varieties originated through allopolyploids. Further differentiation of mango varieties took place through gene mutation and intervarietal hybridization, selected varieties being retained through vegetative propagation.

7.4 Flowering and Fruiting

Flowering in mango depends on the local climatic conditions; it starts as early as in November in Rayalaseema region of Andhra Pradesh and the south Konkan region on the west coast Gandhi (1955). North Indian mango flowers from February to March. Mango flowers earlier in the eastern region. There are certain varieties of mango, which flower twice or thrice such as Rumani, Bangalore, Neelum, and Alipasand (Naik and Rao 1943) when these varieties are grown in the Kanyakumari region. The duration of flowering in North India is about 20–25 days (Singh 1960).

7.4.1 Fruit Set

In mango the inflorescence is primarily terminal, but auxiliary and multiple panicles may arise from auxiliary buds sometimes. The panicle consists of a main axis bearing many branched secondary axes. The secondary axes may bear cymes of three flowers, or tertiary branches may rise on them, which bear cymes of three flowers, each flower born on branched pedicle; the flowers are either male or hermaphrodite. The total number of flowers in a panicle varies from 1,000 to 6,000 depending on the variety (Mukherjee 1953). The panicle also varies in length from few centimeters to 60 cm. Percentage of hermaphrodite flowers will depend on the time of emergence of the panicle and the variety. Singh (1954) reported that in north Indian conditions, the percentage of perfect flowers in Dashehari and Langara was 30.6 and 69.8. For south Indian mangoes, it varies from 16.41 in Neelum to 3.17 in Alampur Baneshan (Naik and Rao 1943). In Dashehari mango, late flushes of flowering had seven times more perfect flowers than in the early flush in a panicle (Singh et al. 1966). Fruit set is therefore high in mid and late flushes. Close relationship was observed between high temperature and higher number of perfect flowers. The percentage of perfect flowers in some of the south Indian varieties when grown in North India is very low (Singh et al. 1965). The percentage of perfect flower can be increased significantly by application of 200 ppm NAA resulting in higher fruit set. Mango is a cross-pollinated crop and pollination is essential for fruit set, which is primarily done by housefly (*Musca domestica*); pollen viability is reported to be 93 %; the stigma remains receptive up to 5 days.

7.4.2 Fruit Drop

In mango, fruit drop is a continuous process, yet fruit drop occurs at three periods. The first drop consists of young fruitlets from fruit set to 20 days after. The second drop takes place when the developing fruits are 28–35 days old after pollination and fertilization. In the third drop, fruits drop regularly. In the first and second periods of drop, high level of inhibitors and low level of promoters appear to be the major factors. The fruit drop in mango accounts for about 99 % (Mukherjee 1949a). In Bihar in commercial varieties, 13–28 % fruit set is obtained, out of which 0.1–0.25 fruits only mature.

- *Causes*
  1. Mango hopper, mealybug, powdery mildew, and anthracnose
  2. Deficiency of nutrients
  3. Degradation of embryo in self-pollinated flowers
- *Control measures*

Gill and Mukherjee (1967) reported that 2, 4 days was found to check fruit drop better than NAA and 2,4,5-T when used at a concentration of less than 20 ppm, when sprayed 6 weeks after fruit set. Urea at a rate of 2 % when sprayed on the variety Dashehari in the early stages of fruit development increased final fruit harvest. Spraying BA (1.5×10^3 µm) at anthesis followed by GA (7.2×10^2 µm)+NAA (3.1×10^2µm) application at the young fruit stage reduced fruit drop in mango. Single spray of 2,4-D(2,4 dichlorophenoxy acetic acid) or NAA each at 20 ppm or alar at 100 ppm at the pea stage of fruit resulted in higher fruit retention.

7.5 Self-Incompatibility

Singh et al. (1962) reported the existence of self-incompatibility in Dashehari mango. The varieties Langra, Chausa, and Bombay Green also showed self-unfruitfulness. Self-pollination resulted in negligible fruit set of 0.0–1.68 % compared to cross-pollination (6.4–23.4 %). Reddy and Ramayya (1976) observed that the use of Himauddin as the polinizer resulted in 50 % more fruit set and increase in fruit size compared to open-pollinated Rumani fruits. The variety Dashehari is cross-incompatible with Chausa and Safeda Malihabad and is cross-compatible with Langra, Ratual, and Bombay Green. Neelum is partially self-incompatible, whereas Mallika is self-incompatible.

7.6 Soil and Climatic Requirements

The mango grows well on any well-drained soil; however, in waterlogged areas, the trees may not die but will remain unhealthy and chlorotic; besides this flowering, fruiting, and vegetative growth will be adversely affected. Thus, the areas frequented by floods may not be suitable for mango growing. The mango being deep rooted needs soil profile of at least 2 m in depth. The mango is found growing equally well on alluvial as well as lateritic soils. The west coast of India has the later type of soil, which is well drained and suitable for growing mango. Red soil characterized by the presence of peroxide or iron and medium black soils of peninsular India are equally suitable for growing mango. However, deep black cotton soils have been considered to be unsuitable (Gandhi 1955). Soils with appreciable quantity of gravel or *kanker* ($CaCO_3$ in the profiles) may be considered good for mango cultivation if not appreciably alkaline. Alkalinity is injurious to young plants. The most suitable soil for mango is the one with medium texture and deep and is well drained having a pH of 5.5–7.5 and water table below 180 cm around the year. Although mango is a tropical fruit and grows well under semitropical conditions, the annual temperature at which mango thrives well is around 26.7 °C and optimum temperature is 23.9–26.7. Temperature below 1.1 °C adversely affects the mango crop. Special care therefore needs to be taken into account for growing mango under frost conditions. Establishment of mango orchard in frost pockets is not possible. Apart from this, temperature affects sex expression in mango. Mango grows well in rainfall ranging from 25 to 250 cm annually. However, with the rainfall more than 75 cm and above, it can be grown with little or no irrigation. One of the prerequisites for successfully growing mango is the absence of rain during the flowering time. Cloudy weather with resultant increased humidity in the atmosphere encourages greater incidence of pest and disease. The areas exposed to high-velocity wind during the fruiting period are not suitable for mango growing. Besides branch breaking and heavy fruit drop, extreme heat also results in low fruit set.

7.7 Varietal Situation

The first systemic record of the good and bad varieties available is from the account of this fruit tree in *Ain-i-Akbari*, 1590 AD. Maries (1901–1902) made the first attempt to describe mango varieties scientifically and collect many varieties from India. Woodhouse (1909) described the mango varieties of Bhagalpur (Bihar) and suggested a system based on fruit characteristic. Rolphs (1915) attached a special importance to fruit characteristic, particularly shape, while classifying the mango varieties grown in Florida. Sturock (1951) used an artificial key for the identification of mango varieties commonly grown in Florida, using fruit characteristic alone. Heartless (1913) was the first to emphasize the importance of floral characteristic in classifying mango varieties. Pepone (1932) for the first time classified the mango varieties in a natural way on the basis of fruit characteristic, color of panicle axis, laterals and pubescence on the panicle branches, and number of embryos in the seed. Mukherjee (1948, 1949b) classified and described 72 varieties of Bengal and divided all the varieties into three broad groups, taking fruit shape as the main criterion. Apart from fruit characteristic, he took into account the color of emerging leaves, panicle axis and laterals, size of the flowers, intensity of pubescence of panicle branches, nature of bracts, and length of inflorescence. Nail and Gangolly (1950) described 335 south Indian mangoes with a logical method, and Singh and Singh (1956) based their work on the classification of Uttar Pradesh mango on almost the same line. These characteristics include fruit shape and beak, venation on the stones, color of panicle, leaf apex, and folding of leaves (Table 7.3).

Table 7.3 Commercial mango varieties grown in different states

States	Varieties
Andhra Pradesh	Banganpalli, Subarnarekha, Neelum, and Totapuri
Bihar	Bombay Green, Chausa, Dashehari, Fazil, Gulabkhas, Kishenbhog, Himsagar, Zardalu, and Langara
Gujarat	Kesar, Alphonso, Rajapuri, Jamadar, Totapuri, Neelum, Dashehari, and Langara
Haryana	Chausa, Dashehari, Langara, and Fazil
H.P.	Chausa, Dashehari, and Langara
Karnataka	Alphonso, Totapuri, Banganpalli, Pairi, Neelum, and Mulgoa
Madhya Pradesh	Alphonso, Bombay Green, Dashehari, Fazil, Langara, and Neelum
Maharashtra	Alphonso, Kesar, and Pairi
Punjab	Chausa, Dashehari, and Malda
Rajasthan	Bombay Green, Chausa, Dashehari, and Langara
Tamil Nadu	Alphonso, Totapuri, Banganpalli, and Neelum
Uttar Pradesh	Bombay Green, Chausa, Dashehari, and Langara
West Bengal	Fazil, Gulabkhas, Himsagar, Kishenbhog, Langara, and Bombay Green

7.8 Commercial Varieties

1. *Alphonso*: It is one of the choicest varieties in India. It is mainly grown in the Ratnagiri area of Maharashtra. It is grown commercially to a small extent in Valsad district of Gujarat. Fruits are medium in size averaging about 250 g and have attractive blush toward the basal end. Pulp is soft, firm, and fiberless with excellent orange color. It has good sugar-acid blend. The keeping quality of the fruit is good. It is susceptible to spongy tissue. The variety is most suitable for exports.
2. *Bombay Green*: It is one of the earliest varieties of North India. It is a biennial bearer. The fruits are medium in size weighing

around 250 g. Fruits have strong and pleasant flavor. Fruits taste sweet.

3. *Banganapalli*: It is a widely cultivated early season variety of south India. It is the commercial variety of Andhra Pradesh. The fruits are large weighing on average 350–400 g. The pulp is fiberless, soft, and firm with sweet taste. Pulp is yellow in color. Fruits have good keeping quality.
4. *Dashehari*: One of the choicest varieties of north India, and it is a midseason variety. It is a biennial bearer. Fruits are small to medium in size, have pleasant flavor, and taste sweet; pulp is orange colored and fiberless. Its keeping quality is good.
5. *Langra*: It is one of the important commercial varieties of north India. It is a biennial bearer. It is a midseason variety. Fruit quality is good. Flesh is firm with lemon yellow color, scarcely fibrous. It has the characteristic turpentine flavor. Its keeping quality is medium.
6. *Himsagar*: It is one of the choicest varieties of West Bengal. It is a regular bearer. It matures early. Fruits are medium in size with excellent flavor. Taste is sweet. Flesh is firm and fiberless. Its keeping quality is good.
7. *Kesar*: This is a famous variety of Gujarat. It is an irregular bearing variety. Fruits are medium in size. Flesh is sweet, fiberless, and sweet in taste. It has excellent sugar-acid blend. Fruits ripen to attractive apricot-yellow color. It has good processing quality.
8. *Krishnabhog*: It is indigenous to West Bengal. It is a midseason variety. Fruits are medium to large in size. Fruit quality is good, and the flavor is pleasant with traces of turpentine. Flesh is firm with few fibers. Its keeping quality is good.
9. *Neelum*: It is a heaving yielding late season variety of south India with regular bearing habit. Fruits are medium in size with good quality and good flavor. Flesh is soft, yellow, and fiberless. Its keeping quality is good.
10. *Pairi*: It is native to Maharashtra and matures early. It is a heavy and regular bearer. Fruits are medium in size with good quality. It has good flavor with good sugar-acid blend. Flesh is soft, primuline yellow, and fiberless. Its keeping quality is poor.
11. *Rajapuri*: It is one of the commercial cultivars of Gujarat. It is a heavy and regular bearer. Fruits are large in size. This variety matures during the early to midseason. Flesh is firm, yellow, and fiberless. Keeping quality is medium.
12. *Totapuri*: This variety is grown widely in south India. It is a regular and heavy bearer. Fruits are medium to large with prominent sinus. Fruit quality is medium. It has pleasant flavor and tastes sweet. Flesh is cadmium yellow and fiberless.

Apart from these commercial varieties, there are other varieties, which are equally important due to their special quality attributes.

7.8.1 Less-Known Superior Varieties

Fazri Jafrani, Amankhurd Buland Baugh, Zamurrad, Sona Tol, Nisar Pasand, Aziz Pasand.

7.8.2 Polyembryonic Varieties

The phenomenon of polyembryony, characterized by the formation of more than one embryo in the seed, is known to occur in a number of mango varieties. The seedling arising from adventive embryos of nucellar origin is highly uniform. These can therefore be used as such for vegetative multiplication of a polyembryonic variety. If found suitable, they can be used as standard rootstocks for some of the monoembryonic varieties. Polyembryonic varieties are found on the west coast of southern India. They are Chandrakaran, Bappakai, Kurukaan, Olour, Goa Kasargod, and Nileshwar dwarf, Salem. Some of the polyembryonic varieties of the Philippines are Cambodia, Carabao, Corazon, Paho, Pahutan, Pico, Sonera, and Strawberry.

Dashehari mango

Kesar mango

7.8.3 Colored Varieties

Most of the varieties developed in Florida (USA) are characterized by brilliant red blush on the cheeks. It adds to the desirable characteristic of the variety and enhances consumer appeal. In India now emphasis has been given to this aspect, and many new hybrids have come up; they are Arka Anmol, Arka Neelkiran, Jawahar, and Sabri. Some well-known varieties having brilliant red blush on the cheeks are Siduria (Gulab khas Red), Surkha Calcutta, Zafran, Husnara, Janardan Pasand, Suvarnarekha, Lal Mulgoa, and Vanraj; Sensation from Florida is the brilliantly colored variety.

Mango Vanraj

7.9 Clonal Selection from cv. Kesar

To identify superior types from existing variability in Kesar, a survey was conducted in Talala region and other places in Gujarat. The five collections were made and were evaluated for physicochemical characteristics. The data indicated that there is variability in the variety. The fruit numbers was maximum in local collection (6.57) followed by Kesar block I (5.82) TSS was also maximum in local collection followed by Kesar block I which was on par with check. The bud wood of the identified collections was grafted in nursery at CHES, Vejalpur nursery. The plants were transplanted in the field in 2011. The plants put up good growth and stated bearing in 2013. A total of five collections were planted in the field and were evaluated for growth parameters. There were significant differences in respect to plant height, stock, and scion diameter of the collections. Plant height was maximum in collection CHES-1 (2.43 m). It was least in CHES-5 (1.47). The highest stock and scion diameter were recorded in line CHES-2 (68.43 and 62.53 mm). Data in respect to plant spread indicated that both north-south and east spread was maximum in line CHES-2 (1.70 and 1.66 m) closely followed by CHES-1. Some plants started flowering, and the initial data recorded indicated that the number of clusters per plant was highest in line CHES-4 (16.66) and minimum in line CHES-3 and CHES-4 (2.33). Fruit set per cluster was highest in line CHES-4 (31.00) and least in line CHES-5 (11.66). Similar results were obtained in respect to fruit set per plant which was maximum in the collection from the station (413 g) followed by collection from Umethi (365 g); it was least in check (152 g) almost 2–3 times the existing check. Pulp-stone ratio was highest in local collection 6.57 (Table 7.4).

Table 7.4 Comparative data of mango collections for variability from Kesar

Variety	Hadmatia	Umethi	Local	Kesar block III	Kesar block I	Local check
Fruit wt. (g)	330	365	257	413	347	152
Fruit length (mm)	105	124	94	157	153	93
Fruit dia. (mm)	77	73	75	82	78	61.7
TSS	16	17	21	16	18	18
Pulp-stone ratio	4.86	4.63	6.57	5.28	5.82	3.85
Seed wt. (g)			28	50	38	24.6
Pulp wt. (g)			184	264	225	94.80
Spl. characteristic	Traditional belt	Traditional belt	Heavy bearing	Good fruit size, heavy bearer	Good fruit size, medium bearer	Small fruit size

Kesar Hadmatia

Kesar Umethi

7.10 Growth and Fruit Set in Mango Kesar Collection at CHES, Vejalpur

Efforts were continued to collect the variability in mango cv. Kesar to identify superior types. A total of five collections were planted in the field and were evaluated for growth parameters. There were significant differences in respect to plant height, stock, and scion diameter of the collections. Plant height was maximum in collection CHES-1 (2.43 m). It was least in CHES-5 (1.47). The highest stock and scion diameter were recorded in line CHES-2 (68.43 and 62.53 mm). Data in respect to plant spread indicated that both north-south and east spread were maximum in line CHES-2 (1.70 and 1.66 m) closely followed by CHES-1. Some plants started flowering, and thc initial data recorded indicated that the number of clusters per plant was highest in line CHES-4 (16.66) and minimum in line CHES-3 and CHES-4 (2.33). Fruit set per cluster was highest in line CHES-4 (31.00) and least in line CHES-5 (11.66). Similar results were obtained in respect to fruit set per plant. Maximum yield per plant was obtained in CHES-4 (25.41 kg.) followed by CHES-2 (17.93 kg.). The data is of the first year of bearing after 3 years, and hence, no conclusion can be drawn (Table 7.5).

7.11 An Ideal Variety of Mango

1. It should be dwarf.
2. Fruits medium sized, golden apricot in color with red blush on the cheeks for export.
3. Regular bearer.
4. Tolerant to various fungal and bacterial diseases.
5. Should have stable pleasant flavor for processing.
6. The keeping quality should be high.

7.12 Propagation

Criteria for the selection of better types:

1. The parent plant must have been tested for its performance over the years.
2. It must be free from transmittable diseases and should be healthy.
3. Fruit shape, size. and quality must conform to the typical specification of the variety.

- *Methods*
- Although a number of methods are being used, inarching is practiced on commercial basis to propagate mango. Veneer grafting in northern and southern India and soft wood grafting in Gujarat and Maharashtra are emerging as alternative methods. However, success of these methods has varied considerably with the climatic conditions of the area.

Table 7.5 Growth and fruit set in mango Kesar collection at CHES, Vejalpur

Line	Plant height (m)	Stock dia. (mm)	Scion dia. (mm)	Plant spread (m)		Cluster no./plant	Fruit set/cluster	Fruit set/plant	Yield/plant (kg)
				NS	EW				
CHES-1	2.43	66.73	51.2	1.66	1.41	4.66	19.66	30	6.18
CHES-2	1.96	68.43	62.53	1.70	1.66	11.6	27.66	155	17.93
CHES-3	1.69	47.03	37.66	1.06	1.13	2.33	12.66	26	5.20
CHES-4	1.76	42.43	32.04	0.98	0.96	16.66	31.00	182.66	25.41
CHES-5	1.47	44.43	42.43	0.90	1.01	2.33	11.66	22	4.53
CD 5 %	0.57	9.26	9.02	0.46	0.46	7.28	11.44	27.56	–

7.12.1 Inarching

The method of inarching or approach grafting is quite cumbersome and time-consuming but is still the leading method of commercial propagation of mango plants. The method consists of uniting the selected shoot (scion) of a desired parent tree (mother plant) to the potted or transplanted seedlings which are most suitable when they attain a height of about 30–45 cm and thickness ranging 0.75–1.5 cm. These seedlings are either grown in pots or under the mother plant from which the grafts are to be prepared depending upon the availability of suitable branches. Generally, a 1-year-old twig of the scion tree about 60 cm in length and nearly of the same thickness as that of the stock is chosen for grafting. Young and non-bearing trees should not be selected as mother plants.

Inarching should be done during the growing period when the tree is in active sap flow or active growth period. A hot and very dry period, as well as heavy rainfall, during the inarching period is not suitable. The end of the monsoon in heavy rainfall areas and early monsoon in the light rainfall areas is the best period for inarching. In north India, July is the best month for inarching. In the more even climate of south India, the operation can be done any time between July and February.

A thin slice of bark and wood, about 5 cm in length, 7.5 mm in width, and 2 mm deep, is removed by means of a sharp grafting knife from the stem of the stock as well as from the scion branch. The dimensions can be proportionately increased or decreased according to the thickness of the stock and scion. The cuts thus made should be absolutely flat, clean, and even and smooth. The ends of these cuts should be round and not angular. The cut surfaces of both, i.e., stock and scion, are made to coincide, facing each other, so that there remains no hollow space between the two. These are then tightly tied by polythene strips of about 1.5 cm in width and preferably of 200 gauge thickness which has proved to be a good tying material. After about 1 month of operation, the scion below the graft union and stock above the graft union should be given light "V"-shaped cuts at weekly intervals, and the grafts can be finally detached while giving the fourth cut. In the last stage, the top of the stock above the graft union should also be removed completely.

7.12.2 Veneer Grafting

This method holds promise for mass-scale commercial propagation. The method is simple and can be adopted with greater success. The rootstocks as mentioned for inarching are suitable for this method also. For veneer grafting, a downward and inward cut 30–40 mm long is made in the smooth area of the stock at a height of about 20 cm. At the base of the cut, a small, shorter cut is done to intersect the first so as to remove the piece of wood and bark. The scion stick is given a long slanting cut on one side and a small short cut on the other so as to match the cuts of the stock. The graft union is then tied with polythene strip as recommended for inarching. After the scion takes and remains green for more than 10 days, the rootstock should be clipped in stages. The scion wood to be used for veneer grafting requires proper preparation. The desired shorts should be defoliated at least 1 week prior to grafting so that the dormant buds in the axils of the leaves become swollen. The best time for this method is the same for different regions as for inarching.

7.12.3 Budding

Although success of budding in mango was reported in the beginning of this century, budding still continues to remain in the experimental stage as far as commercial mango propagation is concerned.

7.12.4 Stone Grafting

The grafts prepared by this method require less time and expenditure as compared with inarching, veneer grafting, or budding. Success is usually as high as 75.80 percent. However, there is some mortality after final transplanting. For grafting, sprouted mango stones are used as rootstocks and semi-mature terminal shoots of 12–15 cm length, which have passed the purple coloration stage as scion. The grafting is done either by cleft or whip method. The grafts should be placed in a polythene house in nursery for 12–14 days where high humidity (more than 80 %) is maintained. During this period, proper union takes place and the scion sprouts. The grafts are then potted and kept in greenhouse or in a shady place for some time before final transfer to the field. The presence of stored food material in stones and high meristematic activity help in proper healing and subsequent growth of the scion.

7.12.5 Softwood Grafting

The technique of softwood grafting is similar to that of cleft or wedge grafting. In this case, grafting is done on newly emerged flush having bronze-colored leaves and stem. This method is useful in in situ grafting. The scion wood to be used should be defoliated 10 days prior to the grafting and have the same thickness as that of the terminal shoot. The graft should be secured firmly using 1.5 cm wide and 4.5 cm long, 200 gauge polythene strip. July and August are the best months for softwood grafting.

Work was carried out at CHES, Vejalpur, on the response of different mango (*Mangifera indica* L.) varieties to softwood grafting for large-scale commercial nursery propagation. With the increasing emphasis on bringing more and more area under horticultural crops to meet the minimum requirement of fruits and to earn precious export earnings, there is an increased demand for the elite plant material of important commercial mango cultivars and hybrids.

The selection of a propagation method depends on the time taken, percent success, survival of the grafted plants, favorable climatic conditions, the age and thickness of scion, and stock and cost involved, have been found to exert significant influence on the success of any method. A new technique named softwood grafting was found to be a quicker method with high success and of low cost. Performance of commercial cultivars to soft wood grafting method for large-scale multiplication under the semiarid agroclimatic conditions of Godhra, Gujarat (India), was studied.

The rootstock seedlings were raised by selecting local rootstock based on their performance and plant canopy. The seeds were collected from a single plant during the month of June and were sown in specially prepared nursery beds after treating them with copper fungicides (0.2 %). Similarly healthy scion material of the desired variety about 3 months old bearing shoots from previous flush was defoliated 1 week before initiating grafting.

Healthy rootstock plants were uprooted from the beds and were grafted by wedge grafting method and tied with polythene sheet and were kept under partial shade of green polythenehouse. From the perusal of the data, it is very clear that there were significant differences among different varieties in respect to percent graft take, percent nitrogen content, and percent carbohydrate content in defoliated scion sticks. Nitrogen content of activated scion sticks varied significantly between cultivars.

Percent nitrogen content of activated scion sticks was maximum in the variety Alphonso (0.73 %), and it was minimum in Dashehari (0.68 %), which was at par with Amrapali (0.69 %). Carbohydrate content was maximum in variety Dashehari (18.32 %) and was least in the variety Alphonso (15.27 %). C:N ratio was maximum in the variety Dashehari (27.07). Varieties which have low nitrogen content and higher carbohydrate content showed high C:N ratio and vice versa (Table 7.6). Dashehari has the least recorded mortality (3.25 %), and the varieties having high N and low C content with low C:N ratio had high graft mortality (23.26 %) (Raturi and Hiwale, 1999).

Soft wood grafting was found to be the most successful method in the large-scale propagation of mango. However, there were significant differences among different varieties. Initially, after 15 days of grafting, maximum graft take was recorded in the variety Kesar (87.18 %).

Table 7.6 Influence of nutritional content of scion shoots in mango varieties on graft take and survival

Name of variety	Percent success	Percent survival	Nitrogen (%)	Carbohydrate (5)	C:N ratio	% mortality
Kesar	87.18	70.63	0.72	16.27	22.75	16.55
Rajapuri	76.55	66.80	0.71	17.21	24.35	9.75
Alphonso	74.64	51.38	0.73	15.27	20.91	23.26
Amrapali	80.86	70.15	0.67	18.06	26.09	10.71
Mallika	84.62	66.60	0.71	16.56	23.22	18.42
Dashehari	81.92	78.67	0.68	18.32	27.07	3.25
S. Em. +/−	1.46	1.88	0.01	0.37	0.84	–
CD at 5 %	4.42	5.67	0.03	1.65	2.65	–
CV %	3.60	5.45	2.57	3.74	5.99	–

Subsequent observations recorded on percent survival of grafts after three months revealed that Dashehari had the maximum survival (78.67 %), whereas the variety Alphonso showed the least survival (51.38 %) (Raturi and Hiwale, 1999).

This could be possibly due to nutritional balance in the defoliated scion stick, higher nitrogen and low carbohydrate content. Carbohydrate-nitrogen balance which was least in Alphonso might have resulted from the higher mortality of successful graft (23.26 %). In the case of Dashehari, which had higher C:N ratio, the maximum graft survival was recorded (3.25) (Table 7.6) (Raturi and Hiwale, 1999). Prasad et al. (1990) attributed higher graft take to sufficient stored food material and climatic conditions during the month of grafting and Singh and Srivastava (1979) to mobilization of food material.

Thus, it is clear from the studies that carbohydrate-nitrogen balance in the activated shoots plays a vital role in the percent survival of the grafts. The results are in line with those of Singh and Srivastava (1991).

The number of successful plants was obtained after 15 days of grafting and percentage worked out. Similar readings were recorded in the coming month also. The results revealed that in the first month, maximum percent success was recorded in the variety Kesar (92.75 %) which was at par with Rajapuri (90.46 %). It was least in the variety Mallika (87.56 %). However, after another one month in the nursery, it was observed that the percent survival was higher in Dashehari (22.07 %) followed by Amrapali (21.56 %) and the variety Kesar had the least percent survival (10.19 %) in the initial grafting.

7.12.6 Air Layering

Air layering can be done successfully in mango using IBA or NAA 10,000 ppm in lanolin paste. Success up to 50 % has been obtained by using Seradix B as root promoter. The air layers can be used for permanent planting or for raising uniform rootstocks.

7.13 Production Technology

In mango, different high densities in plantation have been tried to boost yield per unit area. In north India, the cultivar Amrapali was planted with a spacing of 2.5×2.5 m accommodating 1,600 plants per hectare. Some of the new varieties have also been found to be amenable for high-density planting, as their growth habit is not very vigorous. However, in mango varieties, which behave, as dwarf under one agroclimatic condition do not behave as dwarfs under another agroclimatic condition. Pruning can also be adopted to check the vegetation growth. Apart from planting closely, application of growth regulators like paclobutrazol has also helped in reducing the vegetative growth. Soil drenching of paclobutrazol at 2.5–5.0 per tree depending on the age during August–September has induced flowering shoots during off years and also has induced dwarfing in most of the alternate bearing varieties. It is generally observed that the production of new shoots soon after the crop harvest appears to be essential for regular bearing.

7.13.1 Nutritional Requirement

73 g N (170 g urea), 18 g P_2O_5 (112 g single super phosphate), and 68 g K_2O (114 g of muriate of potash) per year of age from the first to the tenth year and thereafter 730 g N, 180 g P_2O_5, and 680 g K_2O per plant per year have to be applied in two split doses, one half in June–July and the other half in October. Foliar spray of 3 % urea is recommended before flowering. Application of organic manure and phosphoric fertilizer should be taken up after harvest, and fertilizers like ammonium sulfate can be given before flowering. Actual fertilizer recommendation depends on the soil and leaf analysis in a particular area.

7.13.2 Leaf Analysis

Work carried out has shown that 4–5-month-old leaves from the middle of the shoot should be collected for sampling. The percentage of NPK at 1–1.50, 0.08–0.25, and 0.40–0.09 has been found to be optimum, respectively. In mango, of the micronutrients, zinc deficiency is the predominant one. Three sprays of 0.3 % zinc sulfate in February, March, and May correct this disorder. Spraying of borax at 0.5 % after fruit set twice at monthly intervals and 0.5 % manganese sulfate after blossom have shown to correct boron and manganese deficiencies. With regard to magnesium and iron, it has been found that spraying 0.5 % magnesium sulfate twice in February and April and spraying 0.3 % ferrous sulfate with 0.1 % base have shown to correct these deficiencies.

7.13.3 High-Density Orcharding in Mango

Most of the commercial varieties of mango are irregular, and they are generally planted at a conventional planting distance of 10–15 m, accommodating about 70–100 plants/ha. Wider spacing (10–15 m) often results in an underutilization of the interspaces during the early stage of an orchard life. Thus, a mango orchardist has to suffer a loss in terms of the productivity and net return as compared to other fruit crops. Therefore, necessity was felt to make mango orcharding adequately remunerative by utilizing the high-density planting concept. Available information suggests that the concept of high-density orcharding in mango took practical shape after the development of a dwarf and regular mango hybrid, Amrapali at IARI, New Delhi (Majumder et al. 1982), and dehorning technique developed at the GBPUAT, Pantnagar (Ram and Sirohi 1989).

7.13.4 Amrapali in High-Density Planting Concept

Amrapali, being of short stature, regular, and prolific bearing variety, has been found to be the most suitable variety for the high-density planting system. Planting distance of 2.5 m × 2.5 m in a triangular method has been recommended for the Amrapali. Thus, a mango orchardist may accommodate 1,600 plants/ha compared to a merged 70–100 plants/ha under the traditional planting systems in other commercial varieties of mango (Majumder et al. 1982). A full bearing orchard (11th year) of the Amrapali may yield as high as 22 tons of fruits per hectare.

Some special hints, for growing the Amrapali under a dense planting system, have been recommended. One must establish in situ orchard by growing the seedlings of any mango variety and grafting Amrapali scions on the seedlings. In order to grow the plant bushy, "pinching off" of the terminal buds up to 2 years of growth is essential (Majumder and Sharma 1989). Furthermore, the Amrapali is a prolific bearer, and ultimately, fruits that reached maturity are very high. As a result of this, fruits are not of uniform size. Thus, fruit thinning soon after the fruit set is essential for the fruits to have a uniform size. Otherwise, it is likely that the orchardist may not get good price of the produce due to size variation. As regards management practices, fertilizer application should be made as per recommendations based on the age of the plant. However, a systematic study is necessary to find out the root activity and root distribution pattern of the Amrapali under the high-density planting concept, so that the judicious application of fertilizers may be made.

The experiences at IARI, New Delhi, have shown that the high-density Amrapali orchards usually show decline in the yield after the 12th year of the planting due to the overcrowding and intermingling of the tree canopies. Therefore, pruning has been recommended after the 12th year of planting. It should be done every year soon after the harvesting of the fruits. In this way, the productivity and vitality of such orchards may be maintained for many more years.

7.14 Use of the Dehorning Technique in Dashehari

At GBPUAT, Pantnagar, the commercial varieties like Dashehari may also be grown under the high-density concept by following the dehorning technique (Ram and Sirohi 1989). A planting distance of 2.5 m × 3 m (1,333 plants/ha) has been recommended for Dashehari. Plants are given normal doses of manures and fertilizers. Under this planting density, plants are allowed to grow normally up to 10 years. When there is intermingling of the tree canopies after the 10th year of planting, 50 % of the branches should be dehorned in the 11th year and another 25 % in the 12th year. Dehorning should be done soon after the harvesting. Such orchards may give as high as 18 tons of the fruits/ha at the 10th year of planting. By staggered dehorning practice, one may harvest Dashehari fruits regularly, which is otherwise recognized as an irregular bearer.

7.15 Intercropping

Intercropping in mango can be done with black gram + wheat and brinjal + onion in the case of vegetables. Intercropping with turmeric and radish as well as intercropping with papaya can be carried out without any adverse effect on productivity or development of mango.

7.15.1 Salinity

Soil salinity is causing major concern in over 7 m ha. This problem aggravated because of poor drainage, improper irrigation methods, insufficient water supply, and rise in ground water table. Mango is considered moderately tolerant (4–6 dSm^{-1}) in irrigation water. Mango was generally found to be more sensitive to sodium salt. It has been found that application of FYM at the rate of 25 t/ha helped the salt-stressed mango to recover from salt injury and salinity threshold; the value of Bappakkai rootstock increased from 5.11 to 6.09 dSm^{-1}.

7.16 Yield

The yield of mango varies greatly, depending on the variety and agroclimatic conditions prevalent in the region. Grafted mango trees start bearing in the fifth year of plantation; however, seedling trees take 8–10 years. In the beginning, the yield per plant is just 10–15 fruits (2–3 kg) per tree, which increases to 50–75 fruits (10–15 kg) in subsequent years to about 500 fruits (100 kg) in the tenth year. After 20 years, the yield is stabilized to 1,000–3,000 fruits (200–600 kg) in a year. The productive age of mango is 40–50 years.

7.17 Maladies

7.17.1 Alternate Bearing

Alternate bearing has been one of the major problems. Most of the commercial varieties of the south and north are alternate bearers. Studies carried out to correct the disorder by nutrient management have not yielded any favorable result. However, experimental evidence shows that the regular bearers put forth vegetative flushes immediately after fruiting. Paclobutrazol was found to be a promising chemical for flower initiation in mango. Paclobutrazol as soil drench at 5–10 g per tree of more than 20 years of age has helped in managing this problem.

7.17.2 Mango Malformation

It causes losses from 50 to 80 % in north India and is a complex malady. It is caused by *Fusarium moniliforme* var. subglutinans. In malformed panicles, individual flowers are greatly enlarged in a compact mass with very low percentage of bisexual flowers. Malformed shoots do not set fruit and result in loss of crops. Deblossoming alone coupled with a spray of 200 ppm NAA was found to lower the incidence of malformed panicles. Chemical deblossoming with cycloheximide and cultar application has been tried and found useful in reducing the incidence of malformation. It is also postulated that *Fusarium moniliforme* may be producing malformins, which cause malformation.

Mango malformation

7.17.3 Black Tip

It is reported to occur in the regions of Uttar Pradesh, Bengal, Bihar, and Punjab in orchards located near brick kiln. Fruits affected by black tip lose their marketability. Spraying borax (0.6 %) at 10–14 days intervals has been recommended apart from keeping the brick kiln away.

7.17.4 Clustering

A cluster of fruitlets at the tip of the panicle giving an appearance of clustered panicle called "Jhumka" has been observed to be associated with adverse climate condition during flowering particularly in the months of February to March. Affected panicles have 92 % of the fruits possessing aborted and shriveled embryos. Delaying of panicle emergence and development helps in avoiding the adverse effect of cold.

7.17.5 Spongy Tissue

Spongy tissue is a major and specific problem in Alphonso, wherein a pulp patch fails to ripen. Several reasons have been associated to this malady, which includes inactivation of ripening enzyme due to high temperature, convective heat, postharvest exposure to sunlight, etc. This problem has also been tackled from a genetics angle at various regions with the development of new hybrids. It was noticed that there is segregation for the occurrence of spongy tissue in the resultant progenies at Bangalore. Use of sod culture and mulching has been found to be useful in reducing the incidence. Postharvest exposure to low temperatures between 10 and 15 °C for a period of 10–18 h reduces this malady. The application of various chemicals including growth regulator had almost negligible effect.

7.17.6 Sunburn

Temperature started rising from the middle to the last week of April (Table 7.7). The maximum temperature during the day was above 40 °C

Table 7.7 Incidence of sunburn in different mango varieties under local conditions

	Variety		
Observations	Alphonso	Kesar	Rajapuri
Average no. of fruits examined (per plant)	66.0	93.00	57.00
No. of fruits affected	16.0	10.0	6.00
Percent incidence	24.00	10.75	10.52

which caused sunburn. Maximum sunburn was recorded in cv. Alphonso (24 %) and least in Rajapuri (10.52 %) which was on par with Kesar. The sunburned fruits developed watery spot near the neck which slowly spread to the pedicel attachment; in 2 days, the spot turned black and ultimately the fruit dropped. Maximum sunburn happens at 2–4 o'clock when the temperature and intensity are at its peak in the southwestern direction. Sunburn was caused by increased light intensity in the southwest direction, which has a maximum of 6.4 k lux in the afternoon and 6.3 k lux in the northeast direction in the morning hours up to 12 o'clock. The variety Alphonso was severely affected due to the bearing of fruits on the outside of the canopy, thus getting a maximum exposure to sunlight, whereas cv. Rajapuri and Kesar bear fruits inside the canopy and suffered less.

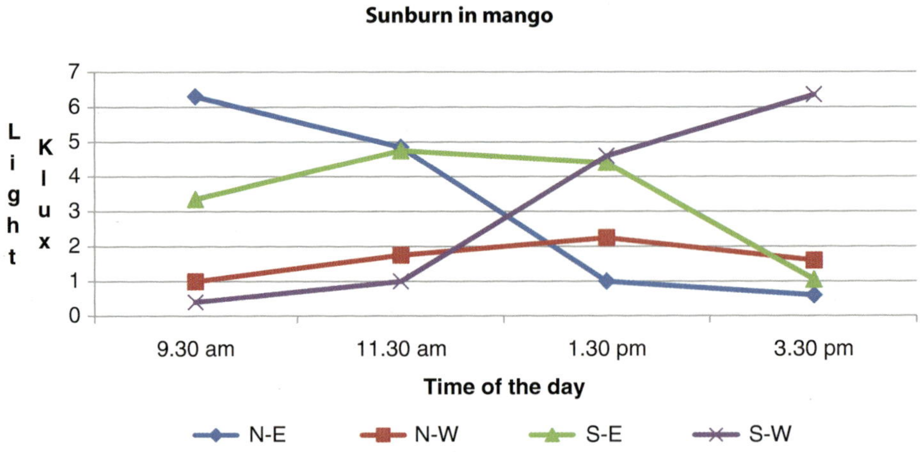

Sun burn in mango

Table 7.8 Export of mango from India (product: fresh mangoes)

	2008–2009		2009–2010		2010–2011	
Country	Qty	Value	Qty	Value	Qty	Value
UAE	24,570.9	7,818.3	25,608.2	10,383.0	25,725.0	10,066.9
Bangladesh	45,104.5	4,085.7	33,549.9	3,295.8	23,049.7	1,859.4
UK	2,527.4	1,403.2	2,958.7	1,746.9	2,723.5	1,453.8
Saudi Arabia	2,141.3	836.3	3,147.1	1,345.4	1,592.2	618.0
Kuwait	546.1	398.8	804.2	520.1	580.3	377.8
Bahrain	1,154.6	324.6	1,238.5	402.3	980.7	355.4
Nepal	4,765.0	375.0	4,058.2	378.6	1,991.3	209.6
Singapore	307.1	156.9	367.6	190.3	387.8	206.0
Qatar	265.4	117.5	659.0	512.8	375.0	199.1
USA	202.6	292.9	175.4	256.6	136.7	193.9
Others	2,118.4	1,262.1	1,894.0	1,022.2	1,678.7	752.2
Total	83,703.2	17,071.2	74,460.6	20,054.0	59,220.8	16,292.1

7.18 Export Potential

Although India is the largest producer of mango in the world, negligible portion of our produce is exported. The export trend of fresh mangoes indicates that there is great potential in Western Europe. At present, the varieties like "Alphonso" grown in Maharashtra and Gujarat and "Dashehari" in Uttar Pradesh are exported to countries, viz., Afghanistan, Bahrain, France, Kuwait, Malaysia, Nepal, Qatar, Singapore, and the UK. The export of fresh mango has increased from 1,200 t in 1969–1970 to 12,500 t in 1983–1984, valued at Rs. 110 million. Export of processed products amounts to 44,000 t in 1983–1984 which increases to 83,703.2 t in 2008–2009. However, in 2009–2010 and 2010–2011, the export declined. Bangladesh and UAE are the countries with the maximum mango export from India (Table 7.8). Mango nectar and pulp has great scope to European and American markets.

References

De-candole A (1904) Origin of cultivated plants. Kegan Paul, London

Gandhi SR (1955) The mango in India. Farm Bull Indian Coun Agric Res 6

Gill APS, Mukherjee SK (1967) Control of fruit drop in mango. Proc Int Symp Hortic 454

Heartless AG (1913) The flowering of the mango. Agric For India 8:90

Kumar B, Mistri NC, Singh B, Gandhi CP (2011) Indian horticulture data base, pp 93–94

Majumder PK, Sharma DK (1989) A new concept of orcharding in mango. Acta Hortic 231:335–338

Majumder PK, Sharma DK, Singh RN (1982) A study on high-density orcharding in mango (*Mangifera indica* L.) var. Amrapali. Punjab Hortic J 22:123–127

Mukherjee SK (1948) The varieties of mango (*Mangifera indica* L.) and their classification. Bull Bot Soc Bengal 2

Mukherjee SK (1949a) The mango and its relatives. Sci Cult 15:5

Mukherjee SK (1949b) A monograph on the genus *Mangifera* L. Lloydia 12:73

Mukherjee SK (1950) Cytological investigation of the mango (*Mangifera indica* L.) and the allied Indian species. Proc Nat Inst Sci India 16:287

Mukherjee SK (1951) The origin of mango. Indian For Genet 2:49

Mukherjee SK (1953) The mango- its botany, cultivation, uses and future improvements, especially as observed in India. Econ Bot 7:130

Naik KG, Gangolly SR (1950) A monograph on classification and nomenclature of South Indian mangoes. Suptd. Govt. Press, Madras.

Naik KG, Rao MM (1943) Studies on blossom biology and pollination in mangoes (*Mangifera indica* L.). Indian For Hortic 1:107

Popeno W (1932) Manual of tropical and subtropical fruits. Macmillan, New York

Prasad V, Suryanarayana V, Naran NL (1990) Studies on certain aspects of veneer grafting in mango

(*Mangifera indica* L.) Cv. Banganpally. South Indian Hortic 38(1):1–7

Ram M, Sirohi SC (1989) Feasibility of high-density orcharding in Dashehari mango. Acta Hort 191:207–212

Reddy and Ramayya (1976) Borax is the answer for mango black tip. Indian Fmg 15:13

Raturi GB, Hiwale SS (1999) Response of different mango (*Mangifera indica* L.) varieties to stone grafting technique for large-scale nursery propagation. In: Paper presented in sixth international organized by International Society at Mahidol University, Pattaya, Thailand, 11–15 April 1999

Rolphs PH (1915) Mangoes in Florida. University of Florida. Agric Expt Stud Bull 127

Singh RN (1954) Studies on floral biology and subsequent development of fruit in the mango (M. *indica* L.) varieties Dashehari and Langra. Indian For Hortic 11:1

Singh LB (1960) The mango: botany, cultivation and utilization. Leonard Hill, London

Singh LB and Singh RN (1956) Alternate bearing in mango. Punjab Hortic For 3:137

Singh NP, Srivastava RP (1979) Studies on the different aspects involved in veneer grafting in mango. Prog Hortic 11:67–74

Singh RN, Majumdar PK, Sharma DK (1962) Self-incompatibility in mango (*M. indica* L.) Var. Dashehari. Curr Sci 31:209

Singh RN, Majumder PK, Sharma DK (1965) Studies on the bearing behaviour of some South Indian varieties of mango (*Mangifera indica* L.) under North Indian conditions. Trop Agric Trinidad 42:171

Singh RN, Majumder PK, Sharma DK (1966) Sex expression in mango (*Mangifera indica* L.) with reference to prevailing temperature. Proc Am Soc Hortic Sci 89:228

Sturock D (1951) Frost protection for young mango trees. Fla Mango Forum Mango Stud 47

Woodhouse EJ (1909) The mangoes of Bhagalpur. Q Four Dept Agric Bengal 3:168

8 Sapota [*Manilkara achrus* (Mill) Forsberg]

Abstract

Sapota or sapodilla (*Manilkara achras* (Mill) Forsberg), commonly known as *Achras sapota* L., *Manilkara zapota* (L.) Van Royen, and *Sapota zapotilla*, belongs to the family Sapotaceae and is a gift of tropical America (Mexico). Sapota is a native of Mexico and Central America and now widely cultivated throughout the tropics. Due to wider adaptability and high economic returns, there has been a phenomenal increase in area (1,091.2 %) and production of sapota in India. Sapota was cultivated in the West Indies and long back in the Philippines in the early days and from there spread westward to Malaysia and other countries. It is used for making sherbet, halva (Singh et al. Fruit culture in India, ICAR. New Delhi, 1963), edible powder (6.2 % protein and 1.39 % acidity of dry powder) used for making milk shake and Indian sweets (burfi), and mixed jams, and they provide a valuable source of raw material for the manufacture of industrial glucose, pectin, and natural fruit jellies. They are also canned as slices. The sapota fruits are a good source of sugar which ranges between 12 and 14 %.

8.1 Introduction

Sapota or sapodilla (*Manilkara achras* (Mill) Forsberg), commonly known as *Achras sapota* L., *Manilkara zapota* (L.) Van Royen, and *Sapota zapotilla*, belongs to the family Sapotaceae and is a gift of tropical America (Mexico). Gholvad village in Maharashtra state is credited to have the first plantation of sapota in 1898. Thereafter, it has spread to different parts of southern and western India. The total area under sapota is estimated to be 23,824 ha with an annual production of 3.38 lakh tons. Average productivity is 14.19 t/ha with the maximum in Karnataka (17.74 t/ha) followed by Maharashtra (16.00 t/ha). The productivity is lowest in West Bengal (6.0 t/ha).

In Karnataka, the area under sapota is widely distributed with the maximum in Belgaum, Kalor, Bangalore, and Dharwad districts. It is grown in both irrigated and marginal levels and has high productivity under protected irrigation.

S. Hiwale, *Sustainable Horticulture in Semiarid Dry Lands*,
DOI 10.1007/978-81-322-2244-6_8, © Springer India 2015

Table 8.1 *Composition and uses* – ripe sapota fruit (per 100 g of edible portion)

Constituents	Amount (g)	Constituents	Amount (mg)
Moisture	73.7	Calcium	28
Carbohydrate	21.4	Phosphorus	27
Protein	0.7	Iron	2
Fat	1.1	Potassium	269
Mineral	0.5	Ascorbic acid	6
Fiber	2.8		

Due to wider adaptability and high economic returns, there has been a phenomenal increase in the area (1,091.2 %) and production of sapota in India. Sapota was cultivated in the West Indies and long back in the Philippines in the early days and from there spread westward to Malaysia and other countries (Purseglove 1968). Sapota cultivation was taken up for the first time in Maharashtra in 1898 in a village named Gholvad (Cheema et al. 1954).

It is cultivated in the states of Maharashtra, Gujarat, Andhra Pradesh, Karnataka, Tamil Nadu, and Kerala and on small scale in Punjab, Haryana, Assam, Orissa, Arunachal Pradesh, West Bengal, and Madhya Pradesh (Cheema et al. 1954; Purseglove 1968; Singh 1969). It is cultivated all over the world in countries like the USA, Sri Lanka, Indonesia, Burma, Venezuela, Guatemala, the Philippines, and Caribbean Island (Table 8.1).

8.2 Uses

It is used for making sherbet, halva (Singh et al. 1963), edible powder (6.2 % protein and 1.39 % acidity on dry powder used for making milk shake and Indian sweets (burfi), and mixed jams, and they provide a valuable source of raw material for the manufacture of industrial glucose, pectin, and natural fruit jellies. They are also canned as slices. The sapota fruits are a good source of sugar which ranges between 12 and 14 %.

Table 8.2 Area and production

Year	Area (000 ha)	Production (000 MT)	Productivity (MT/ha)
1991–1992	27.2	396.2	14.8
2001–2002	52.0	593.5	11.4
2002–2003	84.2	913.1	10.6
2003–2004	120.6	921.3	7.6
2004–2005	134.1	1076.5	8.0
2005–2006	139.7	1114.0	8.0
2006–2007	149	1216.0	8.2
2007–2008	152.0	1258.0	8.3
2008–2009	156.0	1308.0	8.4
2009–2010	158.9	1346.8	8.5
2010–2011	160.0	1424.0	8.9

Kumar et al. (2011)

8.3 Origin

Sapota is a native of Mexico and Central America and now widely cultivated throughout the tropics.

8.4 Area and Production

There is an increase of more than 2–3 % area under the crop, on a year-on-year basis. Similarly, the productivity also increased by almost 2 %, on a year-on-year basis. The yield per hectare however showed a decreasing trend from the earlier 14 tons to present 8.5 tons. The state-wise area under sapota is highest in Maharashtra (65,400 ha) followed by Karnataka with the least area in Orissa and West Bengal, 3.3 and 3.8, respectively. The productivity was maximum in Tamil Nadu (25 t/ha) during 2008–2009 (Tables 8.2 and 8.3).

8.5 Soil and Climate

Soil: It can be grown on a wide range of soils. Drainage is the most important factor. The most ideal soils are deep alluvium, sandy looms, red

Table 8.3 State-wise area production and productivity of sapota

State	2008–2009			2009–2010			2010–2011		
	Area	Production	Productivity	Area	Production	Productivity	Area	Production	Productivity
Karnataka	27.5	297.8	10.8	29.3	360.0	12.3	30.8	377.8	12.3
Maharashtra	65.4	298.7	04.6	69.1	298.0	04.3	70.0	322.0	04.6
Gujarat	26.7	255.4	09.6	27.4	272.6	10.0	28.8	288.0	10.0
Tamil Nadu	07.4	184.0	25.0	06.9	173.5	25.0	09.0	242.3	27.0
Andhra	20.6	205.9	10.0	17.1	171.4	10.0	12.2	122.1	10.0
West Bengal	03.8	043.0	11.2	04.0	043.4	10.9	04.0	043.6	10.9
Orissa	03.3	015.0	04.6	03.3	016.6	05.0	03.4	017.0	05.1
Others	01.6	008.1	05.4	01.7	011.1	06.6	01.9	011.3	06.0
Total	156.1	1307.8	08.4	158.9	1346.6	08.5	160.0	1421.1	08.9

Kumar et al. (2011)
Area, 000 ha; production, 000 MT; productivity, MT

laterites, and medium black soils. Sapota is sometimes planted in dry riverbeds with alluvial soil. In south India, particularly in the Western Ghats of Karnataka, sapota is grown successfully on gravelly laterite soils (Cheema et al. 1954; Singh et al. 1963; Anonymous 1975).

Climate: Sapota is a tropical fruit crop which can be grown from sea level up to 100 m. It prefers a warm and moist weather and grows well in both dry and humid areas and coastal climate. In south India, it is grown on the hills up to 1,000 m. At higher altitudes, the fruit quality and tree health suffer. Areas with an annual rainfall of 125–250 cm are highly suitable. Rain or cloudy weather during any part of the year does not do any harm to fruit set. The temperature should be between 11 and 34 °C.

8.6 Pollination and Fruit Set

Sapota is cross-pollinated with a chromosome no. of $2n=26$. Sapota trees produce many numbers of flowers, but the number of those reaching maturity is small. Most of the flowers either fail to set fruit or there is shedding of fruits, problem of shedding, and low fertility in sapota has been reported by Hayes (1957) and Cheema et al. (1954). Ramamurthy and Ramlingam (1954) reported the failure of fruit set when the flowers were bagged with or without emasculation. It is reported that crops in sapota have often bcen unsatisfactory, but hand pollination with pollen from common varieties increased the fruit set from 10 % with normal pollination to 39.6 % with hand pollination. In the cv. Kalipatti, 22 % of natural fruit set was observed and the maximum fruit drop occurred immediately after fruit setting (Patil and Narwadkar 1974). The first drop starts during the course of fruit development. It was also observed that the flowers situated at the base of inflorescence opened earlier. Such early fruit set developed rapidly and the remaining, which set relatively late, often dropped down. Pollination studies with five cultivars revealed that self-pollination by bagging or by hand produced little or no fruit set, and open pollination and hand pollination with pollen from the other gave a much higher fruit set, indicating that sapota is crop pollinated.

8.7 Flowering

Sapota starts bearing small crops from the second or third year of planting, but economical yields can be obtained from the seventh year onward. Under tropical conditions, flowers are seen almost throughout the year. However, there are

Table 8.4 Pollen studies and fruit set in four varieties of sapota

Varieties	Average pollen size (μ)	Pollen stainability (%)	Fruit set self-pollination (%)	Fruit set open polli. (%)
Kalipatti	16.25	97.7	40	100
Cricket ball	16.25	97.7	20	85
Calcutta round	16.25	97.6	10	60
Local	14.75	96.3	30	95

Nalawadi et al. (1977)

Table 8.5 Stigma receptivity (%) in sapota

Varieties	On the day of opening of flower	A day of opening of flower	On the 3rd day of opening of flower
Kalipatti	45.0	25.0	1.0
Cricket ball	51.0	18.0	0.0
Calcutta round	50.0	22.0	2.0
Local	50.0	20.0	5.0

Nalawadi et al. (1977)

Table 8.6 Pollination fruit set in cv. Kalipatti

Types of pollination	No. of buds	Fruit set (no.)	Fruit set (%)
Natural pollination	50	11	22
Cross-pollination	50	3	6
Self-pollination	50	1	2
No pollination	50	–	–

Patil and Narwadkar (1974)

two main seasons of flowering and hence two harvesting seasons. Bhuva and Chundawat (1978) reported that the flowering and fruiting in sapota in south Gujarat is a continuous process with two peak seasons of crop in September–November and February–March. Studies on floral biology, pollination, and fruit set, the prerequisites of breeding are reported by Sambamoorthy and Ram lingam (1954), Sundararajan (1960), Madhav Rao and Mohamed Abdul Khader (1960, 1962), Patil and Narwadkar (1974), Farooqi and Rao (1976a, b), Nalawadi et al. (1977), and Bhuva and Chundawat (1978). Under rainfed conditions, the sapota flowered and fruited twice a year, i.e., during October–

Table 8.7 Percentage of fruit set in intervarietal pollination

Male parent/ female parent	Kalipatti	Cricket ball	Calcutta round	Oval
Kalipatti	–	50.0	0.0	77.7
Cricket ball	67.4	–	15.	72.0
Calcutta round	65.1	72.2	–	10.0
Oval	51.0	54.0	45.0	–

Farooqi & Rao (1976a, b)

Female parent	Good pollen parents
Kalipatti	Cricket ball, oval
Cricket ball	Kalipatti, oval
Calcutta round	Kalipatti, Cricket ball
Oval	Kalipatti, Cricket ball

November and June–July (Hiwale and Raturi 1991) (Tables 8.4, 8.5, 8.6, and 8.7).

8.8 Use of Growth Regulators

Self-unfruitfulness is seen in stray plants and pollination becomes difficult (Singh 1969). Use of growth regulators has been found to be beneficial. In an attempt to get higher fruit set and retention of the fruits, several growth substances, such as GA3, ethereal, cycocel, and planofix (a NAA preparation), and SADH, were tried on cv. Cricket ball (Das and Mahapatra 1973). These chemicals were applied in spray solutions before flowering and again at the pen stage. Of all the regulators, ADH at 100 ppm resulted in the highest fruit set and planofix at 300 ppm resulted in the highest fruit retention and largest fruit, followed by 10 ppm GA3. In similar studies made at the Gujarat Agricultural University, Navsari, India, of the two growth substances tried at 25–100 ppm as spray at flowering for 15 days, NAA produced a better fruit than GA3 (Rathod and Amin 1981).

8.9 Cultivars

There are many cultivars of sapota in India. Tree habit, nature of branching, foliage color, leaf shape, leaf tip, fruit shape, texture of the fruit skin, and color and pulp quality are certain distinguishing characteristics in grouping of the sapota varieties. A total of 35 cultivated varieties have been recognized from the different sapota growing parts of the country. Out of which, important varieties are the Kalipatti, Cricket ball, Pilipatti, CO-l, CO-3, PKM-1, Kirthabarthi, Singapore, Baramasi, etc. Nearly 99 % of sapota in Gujarat is under Kalipatti grafted on khirni rootstock. In Florida recently, a new chance seedling named "Tikal" has evolved which is superior to the present popular variety "Prolific." Noteworthy characteristics of this variety are earliness, regular bearer, and high yielder (Campbell et al. 1988).

Crop improvement through clonal selection has resulted in the release of two high yielding varieties, viz., Co2 and PKM 1. Through planned hybridization, two hybrids of sapota, namely, DHS-1 and DHS-2, have been released.

A good table sapota should have a few seeds with melting sweet pulp. Thick-skinned, hard-fleshed cultivars with sandy texture are considered inferior. The cultivars need to be standardized. The following are the different cultivars grown in India.

8.9.1 Kalipatti

It is a leading cultivar of Maharashtra Gujarat and North Karnataka. It has dark green, broad, and thick leaves and spreading branches. Fruits are oval shaped and less seeded with sweet, mellow flesh of excellent quality. Fragrance is mild. Each fruit has 1–4 seeds. Fruits appear singly. The main harvest is in winter.

8.9.2 Chhatri

It is similar to Kalipatti but the branches have a dropping nature similar to that of an umbrella. The branches appear in whorls on the plant in all directions and are horizontal. The leaves are light green in color. Fruits are similar to Kalipatti in appearance, but the fruit quality is not as good as in Kalipatti. It is a fairly good crop but not as heavy as Kalipatti. It is also an important cultivar in Maharashtra.

8.9.3 Vanjet

The tree is slow growing in nature. Knots are seen on the branches. Root emerges from these knots. It is a shy-bearing cultivar, but fruits are of good quality. This is found in Maharashtra.

8.9.4 Pala

It is a popular cultivar in Andhra Pradesh and Tamil Nadu. The fruits are small to medium in size and oval or egg shaped, with apex broadly pointed and very sweet. The bearing is heavy and fruits are borne in clusters. The fruit has thin skin and good flavor.

8.9.5 Kirthabarthi

It is a popular cultivar of Andhra Pradesh. The fruits are small to medium sized and oval or egg shaped. On the rind, 4–6 ridges are seen. The fruit skin is rough, medium thick, and buff colored; the pulp is sweet. Fruit apex is rounded. The fruits can be transported to distant markets.

8.9.6 Oval

The fruits are small to medium in size and oval or egg shaped. Pulp is coarse grained and less sweet. It is shy-bearing cultivar.

8.9.7 Calcutta Round

It is commercially grown in West Bengal, Karnataka, and other states. The foliage is light green in color. Fruits are large, but the flesh is gritty and of moderate quality. It is susceptible to leaf spot disease.

and least in Jumakhia due to small fruit size. Pulp seed ratio and TSS were maximum in cv. Kalipatti, 42.2 and 30.4 ° Brix (Table 8.8).

At CHES, Vejalpur, vegetative growth parameters recorded indicated that cv. PKM-2 was most vigorous in respect to all the growth parameters followed by PKM-1. The least growth was recorded in DSH-2. Data recorded indicated that the maximum no. of fruits per clusters were recorded in cv. PKM-1 (4.33), while it was least in DHS-2 and CO-3 (1.66). However, the maximum no. of fruit retained per shoot was in PKM-1 (36.00) and the least in DHS-1 (12.33). The maximum no. of fruits set per plant was recorded in cv. PKM-1 (1216.66) and the least in CO-3 (51.66). Yield per plant was highest in PKM-1 in the fifth year of planting (73.17 kg/plant) (Tables 8.9 and 8.10).

Hybrid PKM in Bearing

Cv. Kalipatti in bearing

Evaluation of germplasm of sapota from the 5th to 7th year under semiarid rainfed conditions indicated that all the varieties were growing successfully under semiarid rainfed condition. Though the maximum number of flower cluster per shoot (79.78) was recorded in Jumakhia, the number of flowers per cluster was very few (1.67). The highest number of flowers per cluster was recorded in Co-1 (8.83). As the number of flower cluster per shoot was highest in cv. Jumakhia which had the maximum number of fruit per shoot (169.0), fruit retention was highest in Pilipatti, but the fruit weight was low (49.6 g) resulting in lower yield (65.2 kg/plant). The highest yield per plant was recorded in cv. Kalipatti (77.8 kg). Fruit weight was highest in Co-2 (116.6 g) and least in Jumakhia (43.0 g). Skin weight per fruit was maximum in Co-2 (11.3 g)

8.10 Propagation

Both seed and vegetative propagation methods are used to multiply sapota. In the earlier days, seedling was used for planting, but they had some disadvantages, such as slow growth, very long pre-bearing period of 8–10 years, growing to a huge size or height, and showing too much of variation or not being true to type (Cheema et al. 1954; Sayeed 1963). Seeds are collected from elite trees and sown in pans or beds with light sand soil as germinating medium and watered at

Table 8.8 Yield and physicochemical characteristics of sapota cv. Kalipatti under rainfed conditions of Gujarat

Variety	No. of flower cluster/shoot	No. of flower/cluster	Fruit set/shoot	Fruit retained/shoot	Yield/plant	Fruit wt. (g)	Fruit length (mm)	Fruit dia. (mm)	Skin wt. (g)	Pulp seed ratio	TSS ° Brix
Kalipatti	69.77	7.61	57.66	38.99	77.8	83.0	51.2	50.7	6.7	42.2	30.4
Pilipatti	44.44	3.5	88.94	79.66	65.2	49.6	42.3	44.2	6.0	8.70	25.6
Cricket ball	31.98	8.55	33.98	23.89	37.8	86.0	53.7	53.7	8.3	21.3	30.0
Singapore	45.22	8.78	48.11	31.22	43.9	65.6	46.6	48.8	6.0	24.0	30.0
Jumakhia	79.78	1.67	169.0	47.89	38.1	43.0	43.5	40.6	5.5	7.13	26.2
Co-1	47.88	8.83	41.89	13.44	35.2	97.67	52.6	58.6	10.0	26.3	26.7
Co-2	50.44	8.77	47.66	29.44	25.6	116.6	62.1	59.5	11.3	33.1	25.0
CD5%	22.47	1.47	19.09	15.49	5.6	13.48	2.86	3.09	1.67	8.75	2.79

Hiwale (2002)

Table 8.9 Performance of sapota varieties and hybrids developed in India

Hybrid	Parentage	Fruit shape	TSS (°Brix)	Fruit weight (g)	Special characteristics
CO1	Cricket ball × oval	Oval	18	125	Upright growth
CO2	Clonal selection from Baramasi	Obovate to round	23	400	Upright growth, big size fruits
PKM-1	Clonal selection from Guthi	Oblong	24	88–110	Heavy bearer, fruit size small
PKM-3	Guthi × Cricket ball	Oval	24	100–110	Upright growth suitable for high density. Yield 14 t/ha
PKM-4	Clonal selection from PKM-1	Oblong	21	100–120	Yield 100.9 kg/tree, an increase of 138.29 %
DHS-1	Kalipatti × Cricket ball	Round	26	154	30–40 % increase in yield
DHS-2	Kalipatti × Cricket ball	Obovate	23	180	30–40 % increase in yield

Table 8.10 Growth and fruit retention in sapota collection

Hybrid	Plant height (m)	Stock dia. (mm)	Scion dia. (mm)	Plant spread (m) NS	EW	Fruit set/ cluster	Fruit retained/ shoot	Fruit set/ plant	Yield/ plant (kg)
PKM-1	2.83	108.96	88.33	3.48	3.64	4.33	36.00	1216.66	73.17
PKM-2	3.25	121.67	109.00	3.53	4.18	4.00	29.66	312.66	27.19
PKM-3	3.01	103.67	83.67	2.34	2.40	2.00	14.66	95.00	11.49
DHS-1	1.85	66.16	52.90	1.28	1.36	2.33	12.33	64.66	8.58
DHS-2	2.18	78.93	61.40	1.83	1.86	1.66	13.00	58.00	7.59
CO3	1.88	70.16	55.77	1.55	1.06	1.66	15.33	51.66	3.55
CD5%	0.84	24.39	22.49	0.917	0.95	1.36	4.54	45.02	–

regular intervals. GA3 promoted the germination percentage and increased the growth of seedlings. Soaking in 400 ppm GA3 resulted into large seedling (Pampanna and Sulikeri 1999).

8.10.1 Rootstocks

To get commercial success by grafting or budding, it is necessary to decide suitable rootstocks. In the literature, various workers reported the following species as rootstocks tried for sapota:

1.	Rayan/khirni or pala	*Manilkara hexandra or Mimusops hexandra*
2.	Adam's apple	*Manilkara hexandra or Mimusops kauki*
3.	Mee tree	*Madhuca longifolia or Bassia longifolia*
4.	Mahua	*Madhuca longifolia or Bassia latifolia*
5.	Miracle fruit	*Sideroxylon dulicifolium*
6.	Seedling sapota	*Manilkara achrus*

- Rootstock performance

Seedling sapota takes a long time to attain graftable size; hence, it is now in vogue. Earlier Adam's apple was reported to be a rootstock in Andhra Pradesh and Uttar Pradesh, while in Sri Lanka, mee tree and miracle fruit are used as rootstocks. But in India, it is noticed by several workers that there are defects like pronounced distortion of bud union, overgrowth of stock, and poor quality of fruit with the use of mee and miracular stocks. Venkatratam (1973) observed that trees grafted on rayan bear sweet fruits of good quality, while those on mahua are usually of inferior quality, which he suggested to be due to alkaloid saponin in mahua transmitted to sapota. The most popular, widely adaptable, and better-suited rootstock at present is rayan or khirni (Sayeed 1963; Gandhi 1956; Cheema et al. 1954; Bose 1985; Shanmugavelu 1987). It is necessary to identify other rootstocks and their vigorous strains, e.g., star apple (*Chrysophyllum cainito* and *Manilkara kauki*). In the Philippines, *Palaquium* sp. tree is used as rootstock. Trees on bakul

Table 8.11 Effect of time of inarching and time of detachment of graft from mother plant

Month/ detachment	5 months (%)	6 months (%)	6.5 months (%)	Mean (%)
Octo. 1978	50	55	55	53.3
Nov. 1978	70	75	55	66.6
Dec. 1978	75	85	65	75.0
Jan. 1979	80	85	95	86.6
Feb. 1979	90	95	85	90.0
Mar. 1979	80	75	75	76.6
Apr. 1979	70	80	70	73.3
May 1979	75	75	70	73.3
June 1979	65	70	65	66.6
July 1979	35	30	10	25.0
Aug. 1979	15	10	10	11.6
Sept. 1979	10	15	10	11.6
CD at 5 %	Detachment, NS; survival, 6.54			

Bhuva et al. (1990a, b, c)

(*Manilkara elangi*) were not successful in Tamil Nadu. In view of the slow growth and larger spacing, it is necessary to plan for dwarfing rootstock for high-density planting like wild dilly. Wild dilly (*Manilkara emarginata*) small trees (5cm), Similar to Sapota in appearance, slow in growth is used as dwarfing stock. As per Sayeed (1963), seedling rayan has been found to show variability in growth rate (Vigorous to slow); hence, it is essential to isolate vigorous stains; *Lucuma multiflora* reported from Karnataka needs testing (Table 8.11).

Seed and seed treatment: Seedlings were used for planting in earlier days. However, seedlings have slow growth and very long gestation period (8–10 year), do not behave true to type, and exhibit variation in fruit characteristics. However, in Sri Lanka, seeds were selected from outstanding plus trees of merit, they were sown in light sandy soil beds, and transplanting was done at 2 1/2 months in optimum. Cheema et al. (1954) also reported that seed propagation was in vogue in Florida. In India, khirni is the most popular rootstock for multiplication of sapota. The rootstock exhibits problem of low seed germination and slow growth. Also khirni seeds have short viability and short period of availability of fresh fruits (April–May). Seeds are obtained in April and May and variably last for about 25–30 days. Use of fresh seed and soaking in water for 12–18 h or wrapping the seeds in wet sphagnum moss after treatment with fungicide for 4–6 days resulted into faster and better germination.

Shanmugavelu (1987) reported seed treatment to sapota and observed that seed treatment with GA 50 ppm and IAA 50 ppm for 24 h gave 70 % germination. Treating the seeds of Khirni in hot water (40 °C) for 24 h resulted in 48 % germination. Soaking seeds of *Manilkara hexandra* for 24 h in dung slurry and soaking them in 1 % thiourea solution for 6 h resulted in highest germination percent of 78.7 and 74.3 % respectively. At KKV, Dapoli, rayan seeds incubated at 35 °C for 12 h gave 55 % germination. Also soaking seeds in 50 ppm ethephon for 24 h at room temperature gave 54.66 % germination than 18.7 % under control. GA 300 ppm sprays on rayan seedling increased its height up to 85.3 % (Anonymous 1982). At IIHR, Bangalore, germination of rayan seed was improved (84 %) by soaking in hot water for 24 h. The germination was faster (in 30 days) than control, when seeds were soaked in thiourea 0.5–1.0 % also resulted in highest germination by 28th day. Fruit size of cv. Cricket ball used as rootstock revealed that bigger sized fruit are produced seedlings with high vigor. Spraying khirni seedlings with GA 100 ppm resulted in increased seedling height, leaf number and girth. Raja et al. (2001) reported that soaking the khirni seeds in thiourea at 1 % conc. resulted in early germination (14 days) and enhanced germination (85 %).

8.10.2 Vegetative Propagation Methods

1. *Cutting*

It is not a common method. Shanmugavelu (1987) reports that it can strike roots if 2 mg solution of IBA is used. Ramsunder and Khader (1986) at Coimbtore tried softwood grafting, shoot etiolation (for 5 days) with ringing and then cutting taken after 45 days and treated with IBA. They observed that etiolated + ringed + IBA

2000 ppm treatment gave 39.4 % roots and 80 % survival.

2. *Budding*

Budding by forkert method in coastal region of India and Sri Lanka has been reported, and in South Florida, shield budding is practiced (Bose 1985; Shanmugavelu 1987). In Jawa, forkert method is used for grafting by choosing a 1-year-old budwood.

3. *Layering*

Layering was a commonly used method of propagation till the 1950s (Sayeed 1963). It is also called marcotting or Gooty making. Layering can be accomplished by various ways like ground layer, pot layer, or air layer.

4. *Ground Layering*

It is not a common method, but followed in Dahanu area of Maharashtra state. This is more convenient in trees having branches close to ground level. It is easier and simpler than air layer. However, the limitation for this method is the limited number of branches available to the ground (Sayeed 1963).

5. *Pot Layering*

It is similar to ground layering except that shallow pot with notches is used to insert the ringed/half cut branch. This method is more common for guava.

6. *Air Layering (Gooty)*

This is the most common method in Maharashtra state, Gujarat, Karnataka, Andhra Pradesh, and Tamil Nadu (Bose 1985). By this method, it is possible to get sizeable plants in a short time, but mortality is higher and the root system is shallow. Such trees are not suitable for areas having high winds. Many scientists have done a good deal of work on the season, size of shoots, and use of PGR. In the old method, earthen pot with string is kept to water the Gooty, and mud plaster is the rooting medium, while in the improved method, alkathene/polythene film (200–300 gauge) with wet sphagnum moss on vermiculite as rooting medium is included. Work on plant growth regulators (PGR) to air layer for better and quicker rooting has been earlier reported by Chinappa (1960), Chinappa and Kalolgi (1961), and Sulladamath and Kalolgi (1969) from Karnataka, Singh et al. (1962) from Uttar Pradesh, Alam and Faruqui (1974) from Bangladesh, and Shanmugavelu (1987) from Tamil Nadu. The conclusion that can be drawn from all these trials is that application of 20,000 ppm IBA in April to August or 10,000 ppm each IBA + NAA can produce maximum rooting (70–80 %) and early rooting. Studies on the effect of different preconditioning treatments on the rooting of air layers were reported by Shanmugavelu (1987). The results showed that etiolation + ringing 30 days before layering followed by IBA + NAA 10,000 ppm (both) helped in increasing root number and cumulative root length. Chatterjee et al. (1990) reported that air layering done on 15-year-old plants revealed that 91.7 % rooting and 90.9 % survival could be obtained by invigoration of shoot through etiolation and application of 10,000 ppm IBA to the air layers.

7. *Grafting*

In view of demerits of layering, grafting is becoming increasingly popular in sapota. This is more significant in certain varieties of sapota, which do not strike roots easily. Out of the different grafting methods, the most common is inarching and the new one is softwood grafting.

8. *Inarching*

It is a scion attached method of grafting and commercially practiced method over the last 40 years. Rootstocks are railed in pots and pots are hung or tied to scion tree on platform and grafting is accomplished. It is a laborious and troublesome method. Rootstock raising to the graftable stage takes a long period (2–2.5 years); hence, it needs replacement by softwood method. Malo (1970) used a modified inarching method for grafting sapota with 100 % success in Thailand. Bhuva et al. (1990a, b, c) reported that inarching should be done in the month of February for maximum success (90 %) followed by January (86.65 %). Date of detachment had no influence on percent success. Significant relationship was observed between mean graft survivals with minimum temp., RH, and rainfall. Maharana et al. (1994) used a modified technique for approach grafting by keeping khirni seedlings in shade and covering the inarched portion with wet sphagnum moss in cv. Cricket ball. Detachment from mother plants after 40–50 days was recommended.

9. *Softwood Grafting*

Amin (1978) has first reported softwood grafting on in situ grown rayan seedlings in Gujarat. Softwood method has been experimented at PKV, Akola, from 1982 to 1986 in detail for faster multiplication of sapota grafts under nursery condition. This included raising of stock seedlings, suitability of season, defoliation stage, grafting technique, care in preparing stock-scion success and survival percentage, etc. Kulwal et al. (1985), PKV, Akola, has recommended this method for commercial propagation. The merits of this method are high success in grafting (90 %) and survival (80 %) in July–August season; plants are ready for sale after 18–20 months against 2.5–3 years in inarched ones. Hence, 40–50 % time is reduced, thereby reducing nursery cost, being scion detached method, large-scale production of grafts possible in stipulated time with ease. Since grafts are in poly-bags, mortality in transport and planting is less (Pampanna et al. 1994). For obtaining the highest graft take, softwood grafting in cv. Kalipatti should be done on 15 May (63.33 % survival after 180 days). Pampanna and Sulikeri (1995) demonstrated that 3-month-old scion sticks grafted on *M. hexandra* rootstock gave higher graft take (78.7 %). Pampanna and Sulikeri (2001) reported that fresh scions (without storage) defoliated 10 days prior to grafting required the lowest no. of days for sprouting (29.67). Scions without defoliation and stored for 6 days required the maximum no. of days for sprouting in cv. Kallipatti.

10. *Other Grafting Methods*

Cleft grafting has been tried at Saharanpur on *M. kauki*; similarly, side grafting for top working has been carried out at Saharanpur and Kodur by wedge grafting. In western coast Konkan, sapota growers provide support of rayan to sapota tree air layered and planted in the field; this is done by bridge grafting of old trees to achieve reduction in water requirement of sapota and to save sapota trees damaged by diseases/rodents (Gandhi 1963; Sayeed 1963). Sandhu and Subhadrabandhu (1992) found that veneer grafting was better when a 6-month-old 20 cm long shoot was grafted on a 2-year-old rootstock at the height of 20 cm from the ground level under Punjab conditions.

8.10.2.1 Performance of Air Layer, Inarching, and Softwood Grafts

The results of comparative study of layers and grafts from Pune and Gandevi indicated that grafts on rayan proved best in vigor and bearing at Gandevi and the yield was 50% more than those of layers and twice (100 %) than those on sapota stock (Cheema et al. 1954; Sayeed 1963). Goswami (1954) reported an experiment from Anand, wherein cumulative (1947–1954) production of inarched trees was 3,159 lbs and of layers was 1,091 lbs. Trees on rayan stock were found to be healthy and strong even after four decades. Sham Singh et al. (1963) noticed that in Gujarat, grafts on khirni under low rainfall and deep loam soil have proved better in respect to yield than air layers and grafts on sapota seedlings. An experiment was laid out at PKV, Akola, in 1985 to compare performance of inarched and softwood grafted sapota on rayan stock under field conditions at 10 × 10 m spacing. The variety used was Kallipatti. The results of 6 years (1985–1991) of observations clearly indicated that softwood grafts attained the maximum height (296 cm) than inarched (244 cm). The softwood grafts recorded 14–39 % more height than inarched ones at 6 year of age. In March 1991, 80 % of softwood graft trees and 66 % of inarched trees were in bearing. This clearly points out the superiority of softwood grafts on khirni/rayan stock than inarched ones under assured rainfall, hot climate, and medium heavy soils (Akola) (Anonymous 1991). Patel et al. (1996) studied comparative performance of inarching and softwood grafted plants in field conditions after 10 years of planting. The results revealed that plant spread and height up to 12, 48, and 96 months was significantly higher in approach grafting. Yield was also higher in 10-year-old plant by 66 % in approach grafting.

8.10.3 Tissue Culture (Micropropagation)

Very little work is reported. They observed that mesocarp and endocarp produced callous only. Sachdeva and Mehta (1986) cultured various explants from sapota seedling germinated in vitro on 3 basal media supplemented with growth regulator. Nitsch medium (NM) + 15 % coconut water (CW) + 4 ppm NAA and 2 ppm kinetin were best for callus induction and growth from root, hypocotyls, stem, and shoot-tip explants. After 2 and 4 subcultures of 1-month duration each, most cells were diploid ($2n = 26$), while some were polyploid or aneuploid (Purohit et al. 1997). Micropropagation of sapota using cotyledon node segment was developed, schenk and Hildbrandt's medium supplemented with 2 mg/L BAP which induced 3 shots/node. An average shoot length of 2.17 cm in 42 days was recorded. Further subculturing by including GA3 1 mg/L was done which improved shoot multiplication as well as elongation. Half the strength of SH medium with pre-autoclaved IBA 200 mg/L for 30 min induced rooting in 60 % shoot. Irrigation with ¼ SH medium on soilrite helped in rooting and hardening. 90 % of plantlets were transferred successfully to pots. Out of 500 plantlets, 440 were successfully established in the soil.

8.11 Cultivation

8.11.1 Land Preparation

A location free from water stagnation should be selected. It is plowed two or three times and leveled. Undulated land should be divided into terraces depending on the topography of the land and leveling is done. If the soil is poor, it would be advisable to grow a green manure crop and plow it in so as to improve its physical and chemical conditions before planting operations are taken up. Pits of size 60 cm^3 or 100 cm^3 are prepared at a distance of 8–10 m, both ways depending upon the planting materials and the soil. In low rainfall areas and soils with low fertility, close spacing is followed, while in heavy rainfall tracts and fertile soils, a wide spacing is recommended. In a high-density plantation trial in sapota cv. Kalipatti, Patel et al. (1993) observed that spacing had significant effect on plant height and yield. The maximum height of plant, number of fruits, and fruit weight were obtained at a spacing of 5 m × 5 m, which was mainly due to increased population per unit area.

8.11.2 Planting

The best time of planting is during early monsoon. Graft/budded plants or layers are planted one each pit in the center and care should be taken to see that the roots are gently and firmly pressed and stakes are provided to avoid wind damage; the plants are then watered. At CHES, Vejalpur, the plants are planted at 10 × 10 m spacing under rainfed condition, the vegetative growth of the plants is less, and hence after 30 years of age the spacing seems to be ideal (Hiwale (2004)).

Training and Pruning

In sapota, a central stem is necessary, and air layers in most of the branches are put forth very near the ground level. In the beginning, the basal branches help in developing a thick central stem, and hence, care must be taken to maintain proper distribution of branches on all sides. But later, with an advance in age, the lowermost branches bend down to the ground and become unfruitful. Sapota, in general, has a well-balanced distribution of branches and the tree assumes a uniform shape. There is no necessity of pruning every year. All the growth that appear on the rootstock below the graft or bud joint must be removed. After 3 to 4 years of planting, the lowermost branches up to a height of 60–90 cm may be removed. Similarly, overshaded and crowed branches are also to be removed. In sapota, new growth and flowering occur simultaneously and it is a mixed type of bearing habit. Flowers and fruits appear in the leaf axils on the new growth and hence pruning of branches should not be done; however, as the plants grow old and the branches start intermingling with each other, the plants need to be pruned to remove the dead and diseased branches, allowing sunlight into the orchard to make it bear fruits.

8.11.3 Irrigation

Though sapota can tolerate drought conditions to some extent, it responds well to irrigation. Young plants are watered regularly during dry season and in winter at an interval of 6–12 days (Singh et al. 1963). In Maharashtra, young plants are given irrigation once in 8 days from October onward till monsoon starts. Protective irrigation is given during the first two seasons in Karnataka for better establishment of plants (Anonymous 1975). Insufficient irrigation results in dropping of a large number of flowers leading to a loss of up to 40 % in yield (Cheema et al. 1954). In the beginning, small basins are made and watered, but as the plant grows in size, the size of the basin is also widened. In coastal Gujarat, there is a practice of raising 0.5 m high mounds of earth around the trees to protect them from strong winds and to conserve moisture (Singh 1969). Bhuva et al. (1990a, b, c) suggested that for obtaining economic return from sapota cv. Kalipatti, 10–11 irrigations each at 80 mm depth are to be applied at 32 days interval in winter and 18 days interval in summer. However, care should be taken not to over-irrigate the plants as it may cause flower and fruit shedding and increase the number of misshapen fruits.

- *Drip Irrigation*

Based on the following formula, water requirement for sapota was calculated for 1–10-year-old plant. The data in Table 8.12 gives idea about the dripper to be used along with the growth of the plant. The formula for calculating water requirement /day is

Evapotranspiration × Crop K × Wetted area × Spacing = Water lit/day/plant

In sapota, during the year 1996–1997 of the total area of 34,833 ha, only 2,506 ha area was under micro-irrigation only. As water is becoming a scarce commodity over the years, there is a need to put a maximum area under micro-irrigation to save water. Drip irrigation is used for enhanced growth and early cropping. Irrigation at 0.50 IW/CPE ratio was optimum for cv. Kalipatti. The system resulted in saving water up to 40 % with 70 % higher income. The system is laid out with two drippers spaced 50 cm from the tree in the initial stage during the first 2 years and then four drippers about 1 m away from tree trunk till it attains the age of 5 years. A field trail was conducted with antitranspirants, viz., phenyl mercuric acetate, kaolin salicylic acid, power oil, and liquid paraffin, on rainfed sapota cv. Kalipatti. It was found that antitranspirant affected the plant height, plant girth, and relative water content (RWC). Minimum transpiration was recorded

Table 8.12 Quantity of water required L/day/plant

Month/age	2nd year	3rd year	4th year	5th year	6th year	7th year	8th year	9th year	10th year
Jan.	2.22	5.18	9.91	13.99	21.2	23.36	35.60	46.72	58.74
Feb.	2.69	6.3	12.05	17.02	25.78	28.40	43.30	56.81	71.43
Mar.	3.71	8.67	16.58	23.41	35.47	39.07	59.56	78. 15	98.26
Apr.	5.38	12.59	24.08	33.99	51.50	56.74	86.49	113.48	142.68
May	6.28	14.68	28.09	39.65	60.08	66.18	100.88	132.37	166.43
June	5.05	11.82	22.64	31.92	48.36	53.28	81.21	106.56	133.98
July	3.1	7.26	13.89	19.61	29.71	32.73	49.89	65.46	82.31
Aug.	2.77	6.49	12.42	17.53	26.57	29.27	44.61	58.54	73.60
Sept.	2.80	6.56	12.55	17.71	26.83	29.56	45.06	59.12	74.33
Octo.	2.40	5.62	10.74	15.16	22.97	25.30	38.57	50.62	63.63
Nov.	2.29	5.36	10.25	14.47	21.92	24.15	36.81	48.30	60.73
Dec.	2.01	4.72	9.03	12.74	19.30	21.26	32.42	42.53	53.48

with power oil (2 %) and maximum stomatal resistance was recorded with liquid paraffin (1.5 %), while salicylic acid (1,500 ppm) was found to be best when fruit yield was concerned (Tables 8.13 and 8.14).

The studies revealed that among the growth parameters, stock and scion diameter and plant spread were significantly influenced, whereas plant height was not affected by different treatments. Maximum stock and scion diameter (32 and 29 mm), respectively, were recorded in treatment irrigating at 0.4 IW/CPE ratio daily. Plant spread both in north-south and east-west direction (7.96 m) was highest in the same treatment. Fruit set (26 fruits/shoot) and yield per plant (87.44 kg) were maximum in the treatment irrigating at 0.4 IW/CPE ratio daily. Physicochemical analysis of the fruits indicated that fruit weight (79.66 g), length (50.76 mm), and diameter (47.53 mm) were highest in the same treatment. However, TSS was maximum in control (30.26 ° Brix) and acidity was the least, indicating more sweetness in less frequently watered plants compared to frequently watered plants (Table 8.15).

There were significant differences in respect to vegetative and yield parameters of sapota under fertigation: the highest plant height in treatment in T5-fertigation at 80 % WSF (4.36 m) and spread in T6-fertigation at 100 % WSF (46.50 m^2). The number of fruits per plant and yield per plant were highest in T5-fertigation at 80 % WSF (1,141.2 and 75.22 kg, respectively). Thus, fertigation of sapota with 80 % of water-soluble fertilizer is the best treatment for sapota as it resulted in saving 20 % of fertilizer due to increased nutrient use efficiency.

8.11.4 Weed Control

Weeds control nutrients and moisture and hence should be removed frequently when intercrops are grown; weeds may not be a problem during that period, and in such cases, the basins only are kept free from weeds.

Table 8.13 Effect of drip irrigation on growth parameters of sapota cv. Kalipatti

Treatment	Plant height (m)	Stock dia. (mm)	Scion dia. (mm)	Plant spread (m)	
				NS	EW
Irrigation at 0.2 IW/CPE ratio	4.9	32.00	29.00	7.96	7.96
Irrigation at 0.4 IW/CPE ratio	4.96	32.00	28.66	7.73	7.70
Irrigation at 0.6 IW/CPE ratio	4.83	31.33	27.66	7.83	7.80
Irrigation at 0.8 IW/CPE ratio	4.8	31.66	26.66	7.86	7.70
Irrigation at 1.0 IW/CPE ratio	4.73	30.66	28.00	7.73	7.63
Flood irrigation	4.9	31.00	28.66	7.70	7.56
Control	4.5	27.00	22.66	7.10	7.30
CD at 5 %	NS	3.75	2.97	0.36	0.34

Hiwale (2008)

Table 8.14 Effect of drip irrigation on yield and physicochemical parameters of sapota cv. Kalipatti

Treatment	No. of fruits/ shoot	Yield kg/plant	Fruit wt. (g)	Fruit length (mm)	Fruit dia. (mm)	TSS °Brix	Acidity (%)
Irrigation at 0.2 IW/CPE ratio	25.66	87.44	79.66	50.76	47.33	25.60	0.71
Irrigation at 0.4 IW/CPE ratio	26.33	84.7	77.0	50.30	47.23	26.26	0.72
Irrigation at 0.6 IW/CPE ratio	24.00	77.72	70.66	47.96	45.86	25.53	0.71
Irrigation at 0.8 IW/CPE ratio	25.33	81.68	70.00	47.53	45.33	26.60	0.72
Irrigation at 1.0 IW/CPE ratio	22.33	71.34	67.33	46.93	44.13	27.36	0.70
Flood irrigation	18.33	59.53	64.33	45.56	42.66	28.13	0.65
Control	17.00	47.85	59.00	39.56	36.70	30.26	0.62
CD at 5 %	4.16	17.88	5.45	1.94	1.36	1.36	0.08

Hiwale (2008)

Table 8.15 Fertigation of sapota through drip irrigation

Treatments	Plant height (m)	Spread (m^2)	Trunk girth (cm)	No. of fruits/plant	Fruit wt. (g)	Fruit (kg/plant)
T1-100%NF+rainfed	3.18	34.00	39.00	582.50	50.72	17.62
T2-100%NF+ basin	3.38	35.10	54.04	735.00	57.82	31.78
T3-100%NF+drip	3.62	39.20	57.36	795.00	59.45	31.34
T4-Fertigation at 60 % WSF	3.80	43.82	60.26	314.25	51.55	48.54
T5-Fertigation at 80 % WSF	4.36	46.50	65.80	1141.2	56.97	75.66
T6-Fertigation at 100 % WSF	4.32	45.20	62.76	990.50	64.20	68.82
CD at 5%	0.78	8.49	11.10	186.10	4.15	–

8.11.5 Manure and Fertilizers

For healthy growth and good quality fruits, manures and fertilizers should be applied in the required doses. When intercrops, such as banana, are grown in sapota, they get heavy fertilization indirectly since banana is heavily manured and fertilized. Besides this, the pseudo-stem is cut and incorporated into the soil (Singh 1969). In Andhra Pradesh, on an average each bearing tree is given 100 kg farmyard manure, 6 kg castor cake, and 2 kg superphosphate per year (Narasimham 1966). In Maharashtra, the sapota trees are given yearly application of manures and fertilizers and a 1-year-old tree gets 20 kg farmyard manure and 400 g castor cake, and this dosage is increased every year by 4 kg and 400 g, respectively. Thus, a ten-year-old tree gets 40 kg farmyard manure and 6 kg of castor cake. For bearing trees of 10 years and more, each tree is given 2 kg bone meal (Singh et al. 1963). Application of superphosphate to provide phosphorous is said to improve the size of fruits in old orchards.

8.11.6 Time and Method of Application

For best results, the manure and fertilizers should be applied in two split doses. In Maharashtra and Karnataka, they are first applied in the beginning of monsoon in dry zones and at the end of southwest monsoon in heavy rainfall areas. They are applied in a shallow circular trench, 15–22 cm deep, 15 cm wide, and 60–90 cm away from the trunk. Split application is followed in the trees that are bearing heavy crops. The fertilizers are incorporated into soil and irrigated.

8.11.7 Integrated Nutrient Management

The area under sapota cultivation is increasing fast; the estimated area under the crop is 23,824 ha. with a production of 3.38 lakh tons (Singh 1989). The productivity/ha has shown downward trend from 14 to 8 t/ha. Indiscriminate use of inorganic fertilizers has deleterious effects on the overall soil health resulting in the deterioration of plant health and ultimately its productivity. Organic fertilizers are in use since the time immemorial, and also the organic farming adopted by our ancestors has maintained the productivity of the soil over the years resulting in prosperous civilization. Sapota can be grown under semiarid rainfed conditions in semiarid areas (Raturi and Hiwale 1991). The concept of integrated nutrient management is helpful in improving the soil and plant sustainability. Under rainfed condition, application of high doses of chemical fertilizer can lead to the death of the plant, and therefore, an experiment was conducted on sapota cv. Kalipatti to reduce the dose of chemical fertilizers to half of the recommended dose and to meet the plant hunger supplement it with organic cakes.

Application of organic cakes with half the dose of chemical fertilizers had beneficial effect not only on soil but also improved nutrient composition of sapota cv. Kalipatti (Hiwale et al.

Table 8.16 Nutrient replenishment through organic fertilizers in cv. Kalipatti under rainfed conditions

Treatment	Leaf nutrients (%) N	P	K	Ca	Mg	S
Full castor cake	2.40	0.12	0.67	1.68	0.40	0.82
½ castor cake + ½ RDF	2.67	0.12	0.40	1.55	0.41	0.88
Full FYM	2.32	0.12	0.82	1.85	0.51	0.94
½ FYM + ½ RDF	2.30	0.12	0.58	1.53	0.39	0.91
Full G. nut cake	2.34	0.12	0.37	1,65	0.41	0.80
½G. nut cake + 1/2 RDF	2.61	0.11	0.48	1.63	0.41	0.79
RDF	2.61	0.12	0.51	1.25	0.35	0.68
C.D. 5 %	0.24	NS	0.08	0.20	0.03	0.10

Hiwale et al. (2010)
Castor cake,10 kg/plant; FYM,100 kg /plant; ground nut cake, 8 kg
RDF – Recommended dose of fertilizer 1,000 g N, 500 g P, and 500 g potash

Table 8.17 Effect of organic and chemical fertilizers on soil moisture content and leaf RLW

Treatment	RLW (%) October	March	Soil moisture (%) October	March
Full castor cake	105.17	92.82	10.57	6.57
½ castor cake + ½ RDF	106.47	93.10	9.75	6.40
Full FYM	104.77	91.53	9.82	6.53
½ FYM + ½ RDF	106.75	93.43	12.22	7.26
Full G. nut cake	107.40	93.58	9.35	7.13
½G. nut cake + 1/2 RDF	105.55	91.92	9.47	6.53
RDF	94.77	79.73	8.57	6.13
C.D. 5 %	0.38	4.19	0.78	0.41

RLW – Relative water content (Hiwale et al. 2010)

2010). Composition of leaf in respect to nitrogen and potassium was significantly improved by castor cake, ground nut cake, and FYM in combination with reduced dose of chemical fertilizers. Application of ½ dose of castor cake and ½ doses of fertilizers had maximum N content in leaves of cv. Kalipatti (2.67 %). Phosphorus content of the leaves was found nonsignificant and it ranged from 0.11 to 0.12 %. Potash was maximum in application of full dose of FYM (0.82 %) followed by castor cake (Table 8.16).

There was improvement in soil moisture content of soil as well as leaf when organic fertilizers like castor cake (10.57 %), FYM (9.82 %), and ground nut cake (9.35 %) singly or in combination of ½ dose FYM with ½ RDF (12.22 %) compared to recommended dose of fertilizers (8.57 %) was applied; when the plants are in bearing stage in October and March, similar trend was observed with the highest soil moisture in ½ FYM with ½ RDF (7.13 %) and the least in RDF (6.13 %). The relative leaf water content also showed similar results with maximum RLW by application of ground nut cake (107.40 %) and the least in RDF (94.70 %) in October. With similar trend in March, i.e., maximum RLW (93.58 %) in ground nut cake and least in RDF (8.57 %). Kale and Jagtap (1991) also reported increased RWC in sapota Kalipatti with application of 2 % power oil (Table 8.17).

There was improvement in the soil health also. The pH and EC of the soil was reduced from 7.44 to 6.52 in ½ castor +1/2 RDF and EC of 0.46 mmhos/cm. Organic carbon content of the soil increased from 0.71 to 1.11 % with application of castor cake and FYM (1.11 %) (Table 8.18).

The improvement in plant and soil health has improved the fruit set per shoot to 60.00 fruits in ½ dose of FYM ½ RDF. This also resulted in higher yield per plant of 92 kg per plant in the same treatment. Kale and Jagtap (1991) also revealed that increased dose of nitrogen resulted in increased productivity of sapota cv. Kalipatti. Similar results are reported by Patil et al.(1981). Bhatt et al. (1991) observed that application of 500 g castor cake significantly increased fruit production. Thus, it can be concluded that application of organic fertilizers along with chemical fertilizers improves soil and plant health and results in increased production.

8.11.8 High-Density Planting

Work carried out at fruit research station Gandevi of south Gujarat region on cv. Kalipatti grafted on khirni (*Mimusops hexandra*) rootstock indicated that as the spacing decreased, the number of fruits and yield per plant increased. However, after 10 years of plant age, there is crowding of plants and it adversely affected the fruit set and yield (Table 8.19).

Table 8.18 Effect of organic and chemical fertilizers on soil properties, fruit set, and yield

Treatment	pH	EC (Mmhos)	O.C. (%)	Fruit set/shoot	Yield/plant (kg)
Full castor cake	6.70	0.71	1.11	46.2	81.2
½ castor cake + ½ RDF	6.52	0.46	1.02	48.7	84.0
Full FYM	6.46	0.53	1.11	60.0	92.0
½ FYM + ½ RDF	6.54	0.79	1.08	54.5	65.0
Full G. nut cake	6.53	0.58	1.08	44.5	80.7
½G. nut cake + 1/2 RDF	6.61	0.64	1.09	44.5	78.5
RDF	6.42	0.63	0.83	50.7	86.0
Original soil	7.44	0.20	0.71	–	–
C.D. 5 %	0.38	0.16	0.12	9.8	5.6

Hiwale et al. (2010)

Table 8.19 High-density planting in sapota cv. Kalipatti (6-year-old plants)

Treatment	No. of trees/plot	Plant height (cm)	Plant girth (cm)	No. of fruits/plot	Wt. of fruits/plot (kg)
10 × 10m	2	302.5	27.91	259.6	17.5
10 × 5 m	4	343.3	27.70	758,7	44.7
5 × 5 m	8	367.3	27.58	1182.6	73.6
CD at 5 %		41.83	NS	326.7	22.6

Katrodia and Bhuva (1991)

8.11.9 Intercropping

Sapota has a long pre-bearing age, and till the trees cover the entire area with their leaf canopy, intercrops can be grown. This period may be as long as 10 years, and short duration fruits like banana, papaya, or vegetables can be profitably grown. Leguminous vegetables and pulses like cowpea, cluster bean, lima bean, peas and pigeon pea, and others benefit in several ways.

8.11.10 Fertilizer Schedule for Sapota

Farmyard manure is applied at 40 kg per year per plant

Age of the plants	N (in g)	P (in g)	K (in g)
1–3 years	50	20	75
4–6 years	100	40	150
7–10 years	200	80	300
11 years and more	400	260	450

Anonymous (1975)

8.11.11 Harvesting

Sapota starts bearing in the third year of planting, but economic yield can be obtained in the 5th year onward. The two main seasons of flowering are October–November and June–July with two harvesting season, i.e., April–May and November–December. Sapota takes 5 months from flowering to maturity of fruit. Sapota is a climacteric fruit and it improves in quality after harvesting, but immature fruits should not be harvested. The fruits to be harvested must be fully matured, and maturity can be judged by several external symptoms. Sulladmath and Narayana (1990) suggested the following symptoms of maturity:

1. Fruits at full maturity develop a dull orange or potato color, with a yellowish tinge.
2. A mature fruit when scratched lightly shows a yellow streak instead of a green streak, which is a sign of immature condition.
3. Brown scaly material disappears from the fruit surface as the fruits approach full maturity.
4. As the fruit matures, the milky latex content is reduced.

5. The dried spine like stigma at the tip of fruit falls or drops off easily when touched.

Kariyanna et al. (1993) recommended harvesting sapota fruit after 245 days of fruit set, and morphological parameters like size external color, scurf content, and color of the skin after scratching could be used in deciding on the optimum harvesting stage in the field. Several workers suggested fruit weight at maturity as one of the maturity indices, and the different varieties showed variations in fruit weight and maturity, which, however, differ in different agroclimatic conditions.

Variety	Fruit wt. (g)
Calcutta round	48.2
Cricket ball	140.0
Pala	30.8
Mohangootee	101.43
Kalipatti	98.23
Pilipatti	82.40
Jumakhia	57.47
Co1	122.22
Co2	87.41
Gavarayya	90.48
Guthi	59.62

The fully matured fruits are harvested with the stalk intact, individually by giving a twist, and collected without bruising. The fruits thus harvested are spread in a thin layer on bamboo mats under shade for an hour or two. To avoid brushing of fruits, they are better collected in gunny bags and lowered to the ground carefully. The peak harvest periods are January–February and May–June in the west coast of Maharashtra and March–May and September–October in Andhra Pradesh and Karnataka.

8.12 Yield

The yield depends upon several factors, such as the age of the tree, cultivars, agroclimatic conditions of the locality nutrition, and plant protection measures. In round and large-sized cultivars such as Cricket ball and Calcutta round, less number of fruit is obtained, while in others with oval or long fruits, more number of fruits is produced but their size is small. In very prolific bearers like Pala, the number of fruits is much more. On an average, a 3-year-old tree bears 800 fruits/year; a 10-year-old tree, 1000 fruits/year; an 11-year-old tree, 1,500 fruits/year; a 15-year-old tree, 2,000 fruits/year; and a 30-year-old tree, 2,500 to 3,000 fruits/year (Gandhi 1963; Narsimham 1966; Singh 1969; Gopalaswamiengar 1970; Anonymous 1975). Purseglove (1968) also reported similar yield from 30-year-old trees of sapota.

8.13 Economics

The plants started flowering and fruiting from fifth year onward. Two crops were harvested under semiarid rainfed condition in a year, one in March–April and the second crop in August–September. The summer season fruits were better in size and yield per plant compared to rainy season crops which may be due to moisture availability. The maximum benefit-cost ratio of 1:5.70 was recorded in the 8th year, and then, it decreased due to low rainfall in the preceding years (Table 8.20).

Table 8.20 Economics of sapota cultivation

Age	Yield q/ha	Cost of cultivation	Gross return (Rs.)	Net return (Rs.)	Benefit-cost ratio
5	9.68	1,460	3,872	2,412	1.65
6	14.9	1,595	5,960	4,365	2.73
7	25.21	1,745	10,084	8,339	4.77
8	36.7	2,190	14,680	12,490	5.70
9	43.5	3,540	17,400	13,680	3.86
10	51.3	3,975	20,520	16,545	4.16

Raturi and Hiwale (1991)

Table 8.21 Export: country-wise export of sapota from India

	2008–2009		2009–2010		2010–2011	
Country	Qty (mt)	Value (Rs. Lakh)	Qty (mt)	Value (Rs. Lakh)	Qty (mt)	Value (Rs. Lakh)
UAE	1,602.3	294.7	2,121.0	247.1	788.0	179.0
Bahrain	825.6	102.8	839.2	104.3	631.6	61.6
UK	429.5	91.7	240.1	034.7	78.2	22.6
Canada	527.0	32.6	346.1	029.2	264.4	11.9
Qatar	195.6	15.7	249.0	026.5	046.3	46.3
Oman	068.5	13.7	102.5	015.7	056.1	11.9
Saudi Arabia	150.2	22.8	108.0	015.8	049.4	11.5
USA	126.4	17.2	28.4	006.3	052.7	7.8
Singapore	002.1	00.6	24.8	004.3	016.3	32
South Africa	000.0	00.0	02.0	000.5	011.7	11.8
Others	185.0	15.9	136.8	020.0	044.6	11.6
Total	4,112.3	607.9	4,197.7	504.5	2,039.3	353.3

Kumar et al. (2011)

8.14 Export

The sapota fruits are exported mostly to Arabian countries like UAE, Bahrain, Qatar, Oman, and Saudi Arabia. The export is 4,112.3 mt during 2008–2009; however, it declined to 2,039.3 mt in 2010–2011 indicating that there is no stability in the export market (Table 8.21).

References

Alam T, Faruqui AHM (1974) Success of air layerage of Sapota (A.zapota). Bangladesh Hortic 2:45–48

Amin RS (1978) Softwood grafting – a new technology for hardwood plants. Curr Sci 47:468–69

Anonymous (1975) Cultivation practices for fruits. Department of Horticulture University of Agricultural Sciences/I. I.H.R., Bangalore, pp 31–34

Anonymous (1982) Annual report of research, K.K.V. Dapoli (M.S.) pp 50–51

Anonymous (1991) Annual report of research, University Department of Horticulture, PKV, Akola, pp 10–11

Bhatt RI, Katrodia JS, Chundawat BS, Padhiar BV (1991) Fertilizing sapota Cv. Kalipatti in rain fed orchards in sub humid regions of south Gujarat. In: National seminar on optimization of productivity and utilization of sapota held at Navsari on 8 October 1991

Bhuva HP, Chundawat BS (1978) Flowering and fruit growth in Sapota. (*Achras sapota* L.). Udyanika 2:54–56

Bhuva HP, Katrodia JS, Raja KRV (1990a) Indian J Hortic 47:167–169

Bhuva HP, Katrodia JS, Chundawat BS (1990b) Influence of environment on success of sapota propagation. Hortic J 3:6–9

Bhuva HP, Katrodia JS, Patel RG (1990c) Haryana J Hortic Sci 19:71–78

Bose TK (1985) Fruit of Indian-tropical & sub-tropical, vol 6. Naya Prokash, Calcutta, pp 446–461

Chatterjee BK, Ray DP, Chakladar BP (1990) Orissa J Hortic 18:70–74

Cheema GS, Bhat SS, Naik KC (1954) Commercial fruits of India with special reference to Western India. Macmillan & Co, Calcutta, pp 359–378

Chinappa KS (1960) Thesis submitted for M.Sc. degree to Karnataka University (unpublished)

Chinappa KS, Kalolgi SP (1961) Lalbagh J 6:29–32

Das RC, Mahapatra S (1973) Plant Sci 7:93–94

Farooqi AA, Rao MM (1976a) Studies on Metaxenia in sapota (Achras sapota). Mysore J Agric Sci 10:413–423

Farooqi AA, Rao MM (1976b) Studies on fruit set in some sapota varieties in relation to intra and inter-varietal pollination. Mysore J Agric Sci 10:28–34

Gandhi SR (1963) Chiku, farm bulletin, Ministry of Agriculture, Government of India

Gopalaswamiengar KS (1970) Complete gardening in India. Kalyani Press, Bangalore

Goswami KN (1954) Bansilal Amritlal Agric Mag 3(7):6–7

Hayes WB (1957) Fruit growing in India. Kitabistan, Allahabad

Hiwale SS (2002) Annual report. CIAH, Bikaner

Hiwale SS (2004) Lead lecture delivered on sapota cultivation. In: Proceeding of winter school on strategies for horticulture development under arid condition on 7–29 December 2004

Hiwale SS (2008) Annual report. CIAH, Bikaner, p 56

Hiwale SS, Raturi GB (1991) Studies on the performance of sapota (Manilkara achras Mill Forsberg.) Cv. Kalipatti under semiarid rainfed conditions. National seminar on Optimization of productivity and utilization of sapota at Gujarat Agril University, Navsari, p 5

Hiwale SS, Apparao VV, Dhandhar DG, Bagle BG (2010) Effect of replenishment of N through organic fertilizers in sapota cv. Kalipatti. Indian J Hortic 67(2):274–276

Kale PB, Jagtap KB (1991) Nutritional studies in chiku (*Manilkara achrus* Mill) Forsberg. A review of work carried out at MPKV, Rahuri National seminar on optimization of productivity and utilization of sapota held at Navsari on 8 October 1991

Kariyanna S, Bojappa KM, Reddy TV (1993) Golden jubilee symposium on horticultural research- changing Scenario, Bangalore, 24–28 May, Abstract no. 19.10, p 347

Katrodia JS, Bhuva HP (1991) Orchard management, Nutrition and irrigation in sapota. National seminar on Optimization of productivity and utilization of sapota at Gujarat Agriculture University, Navsari, pp 7–8

Kulwal LV, Tayde GS, Deshmukh PP (1985) Studies on softwood grafting Sapota. PKV Res J, Akola 9(2):33–36

Kumar B, Mistry NC, Singh B, Gandhi CP (2011) Horticulture data base publisher. National Horticulture Board, New Delhi, pp 122–127

Madhav Rao VN, Mohamed Abdul Khader JBM (1960) Estimation of pollen production in fruit crops. Madras Agric J 49:152–156

Madhav Rao VN, Mohamed Abdul Khader JBM (1962) Effect of gibberelic acid on the germination and tube growth of pollen in sapota (Achras zapota L.) var. 'Long Oval'. Madras Agric J 49:412–414

Maharana T, Sahu BC, Tripathi P (1994) A note on the effect of detachment time in modified, approach grafting in sapota cv. Cricket ball. Orissa J Hortic 22(1–2):87–90

Malo SE (1970) Propagation of mamey sapota. Proc Am Soc Hortic Sci 14:165–174

Nalawadi UG, Dasappa MA, Sulikeri GS (1977) Floral biology of some varieties of sapota (Achras sapota L.). Prog Hortic 9(1):27–32

Narasimham B (1966) Horticulture part I., Agricultural College, Bapatla, Andhra Pradesh, pp 200–201

Pampanna Y, Sulikeri GS, Hulmani HC (1994) Effect of season on the success of soft wood grafting in sapota Cv. Kalipatti. South Indian Hortic 42:5,303–308

Pampanna Y, Sulikeri GS (1995) Effect of age of scion on the success of soft wood grafting in cv. Kalipatti. Karnataka J Agric Sci 8(1):56–59

Pampanna Y, Sulikeri GS (1999) Seed Res 27:49–53

Pampanna Y, Sulikeri GS (2001) Effect of pre curing and storage of scion sticks on the success and growth of soft wood grafts in Sapota cv. Kallipatti. Karnataka J Agric Sci 14(4):1025–1029

Patel CB, Patel DB, Patel BN, Katrodia JS (1993) Golden jubilee symposium on horticultural research- changing scenario, Bangalore, 24–28 May, Abstract no. 8.28, p 176

Patel RM, Virdia RR, Patel HC, Chundawat BS, Barot DV (1996) Study on comparative performance of sapota (Manilkara achras Mill. Forsberg.) Propagated by different methods. J Appl Hortic 2(1–2):110–115

Patil AV, Wavhal KN, Kamble KB, Patil BS (1981) Effect of varying levels N, P and K on growth, yield and quality of Chiku (*Achras sapota* L.) Cv.Kalipatti. Paper presented in national seminar on Tropical and subtropical fruits, Bangalore, pp 58

Patil VK, Narwadkar PR (1974) Studies on flowering, pollination and fruit drop in chiku. Punjab Hortic J 14:39–42

Purohit SD, Singhvi A, Nagori R, Kishan PBK (1997) In vitro propagation of A. zapota, optimization of conditions for rooting, hardening and acclimatization. Proceedings of symposium. Plant tissue culture and Biotechnology, emerging trends at Hyderabad, 29–31 January 1997, pp 134–145

Purseglove JW (1968) Tropical crops, dicotyledons. Longman Group Ltd, London

Raja K, Palanyswamy V, Selvaraj P (2001) Enhancing the germination and seedling vigour of Khirni seeds (*Manilkara hexandra*) by pre sowing treatment. IUFRO joint symposium on tree seeds Technology College, Laguna, Philippines, 30 April–3 May 2001

Ramamurthy K, Ramalingam B (1954) A note on hybridization I sapota (*Achras zapota* L.). Indian J Hortic 8:57–60

Ramsunder V, Abdul Khader JBM (1986) Studies on the rooting the sapota cuttings. South Indian Hort 34:40–41

Rathod RP, Amin HD (1981) National symposium on tropical and subtropical fruit crops, Bangalore, p 115

Raturi GB, Hiwale SS (1991) Studies on performance of sapota Manilkara achrus Mill Forsberg. In: National seminar on optimization of productivity and utilization of sapota held at Navsari on 8 October 1991, pp 5

Sachdeva S, Mehta PN (1986) Phytomorphology 36:315–324

Sambamoorthy K, Ramalingam V (1954) A note on hybridization in Sapota (*Achras zapota* L.). Indian J Hort 9:57–60

Sandhu MK, Subhadrabandhu S (1992) Standardization of grafting technique in sapota (*A. zapota* L.). In: International symposium on tropical fruits Frontiers in tropical fruit research, Pattaya, Thailand 20–24 May 1991. Acta Hort 332:610–615

Sayeed IA (1963) Propagation of Sapota. In: Fruit nursery practices in India, Ed. Venkatraman, Directorate of Extension, New Delhi

Shanmugavelu KG (1987) Production technology of fruit crops. S.B.A Publication, Calcutta

Singh R (1969) Fruits. National Book Trust, New Delhi

Singh HP (1989) World bibliography on fruits. IIHR, Bangalore

Singh S, Krishnamurthi S, Katyal SL (1962) Fruit culture in India. ICAR, New Delhi

Singh S, Krishnamurthy S, Katyal SL (1963) Fruit culture in India. ICAR, New Delhi

Sulladamath UV, Kalolgi SD (1969) South Indian Hortic 17:10–17

Sulladmath UV, Narayana Reddy MA (1990) Sapota. In: Bose TK (ed) Fruits of India tropical and sub-tropical. Naya Prakashan, Calcutta, pp 565–591

Sundararajan S (1960) Studies on certain aspects of growth, flowering and fruiting in same varieties of sapota (Achras zapota L.). M.Sc. (Ag) Dissertation submitted to Madras University

Venkatratnam L (1973) Horticulture in central India. Directorate of Extension ICAR, New Delhi, p 112

Custard Apple (*Annona squamosa* L.)

9

Abstract

Custard apple (*Annona squamosa* L.) is a delicious and important minor fruit crop which is cultivated in tropical and subtropical climate. It comes under family Annonaceae and native of the West Indies, but it was cultivated since early times throughout Central America to Southern Mexico.

Custard apple is an important dry land fruit of India. It is popular by virtue of its spontaneous spread in forests, wastelands, rocky slopes, and other uncultivated places. Its plants come up unattended in parts of Andhra Pradesh, Assam, Bihar, Karnataka, Maharashtra, Madhya Pradesh, Orissa, Rajasthan, and Tamil Nadu as a scrub or hedge plant. Bullock's heart is more commonly found in south India than in north India. Cherimoya is mostly restricted to Assam and hills of south India. Atemoya and soursop are cultivated in some gardens as miscellaneous fruits. Atemoya, cherimoya, and ilama also provided excellent opportunities for a large-scale exploitation in India. Softwood grafting in the month of March gives best results in propagation of custard apple. Identification of superior varieties like Balanagar and hybrids like Arka Sahan and APK-Ca-1 has resulted in regular plantation on large scale. An eight-year-old plantation yielded 22–25 q fruits/ha under semiarid rain-fed condition.

9.1 Introduction

Custard apple is a small group of edible fruits of genus *Annona and* family Annonaceae are collectively known as annonaceous fruits. Genus *Annona* has 120 species, six of them having pomological significance. *Annona* fruits are formed by fusion of pistil and receptacle into a large flashy aggregate fruit. Annonaceous fruits have morphological affinity for each other, but each type is unique in its taste, flavor, pulp color, and texture.

The annonaceous fruits originated in tropical America and are widely distributed in tropics and subtropics. Among annonaceous fruits, custard apple is the most favorite in India. Its plants come up unattended in parts of Andhra Pradesh, Assam, Bihar, Karnataka, Maharashtra, Madhya Pradesh, Orissa, Rajasthan, and Tamil Nadu as a scrub or hedge plant. Of late, custard apple has gained

S. Hiwale, *Sustainable Horticulture in Semiarid Dry Lands*,
DOI 10.1007/978-81-322-2244-6_9, © Springer India 2015

Table 9.1 Area and production

State	Area (ha)	% of total	Production (mt)	% of total
Maharashtra	9,424	64.45	65,968	64.49
Gujarat	1,426	9.75	9,223	9.01
Madhya Pradesh	3,590	24.55	25,050	24.48
Rajasthan	180	1.23	2,050	2.00
Total	14,620	–	1,02,291	–

commercial significance, and exclusive orchards are emerging in Maharashtra, Andhra Pradesh, and Gujarat (Table 9.1).

9.2 Uses

- As dessert fruit
- In ice creams and other milk products
- As jam and jelly
- In Ayurvedic and Unani systems of medicine like in seeds as abortifacient and roots as strong purgative
- Seed oil (30 %) in soap and paint industry
- Seed cake (4 % N) as cattle feed and as manure
- As insecticide with Neem oil

9.3 Edible *Annona* and Their Fruit Characteristics

Other annonnas are cultivated on a limited scale. Bullock's heart is more commonly found in south India than in north India. It is usually associated with gardens and compounds and not commercial orchards. Cherimoya is mostly restricted to Assam and hills of south India. Atemoya and soursop are cultivated in some gardens as miscellaneous fruits. Atemoya, cherimoya, and ilama also provided excellent opportunities for a large-scale exploitation in India.

9.3.1 Custard Apple: Sweet Soup Sugar Apple (*Annona squamosa*)

Indian name Sitaphal, Sharifa, woody plant, semi-deciduous. Fruits 250–300 g; globular; green skin; sweet (20 % sugar); nonacidic; pulp creamy white; distinct segment, 60–80 seeds/fruit.

9.3.2 Bullock's Heart and West Indian Custard Apple (*Annona reticulata*)

Indian name Ramphal plant; semi-deciduous reaching 6–7 m height. fruit large (350,400 g); heart shaped; yellowish red; smooth rind with hexagonal markings, pulp pale, gritty, flavored, 12.5 % sugar, and few seeds (30–40).

9.3.3 Cherimola and Cherimoyar (*Annona cherimola*)

Local name Hanuman Phal, semi-deciduous, tree reaching to a height of 8 m. Fruits wt. 250–300 g, pale green when ripe, subglobose, pine-like aroma 18 % sugar, segment fused, 10–15 seeds/fruit.

9.3.4 Annona atemoya (*A. squamosa* × *A. cherimoya*)

Local name Lakshman Phal, semi-deciduous, large spreading tree, 5 m ht. Fruit wt. 500 g globular green; white smooth pulp is very juicy with excellent sugar acid blend large segments 10–15 seeds/fruit.

9.3.5 Soursop and Prickly Custard Apple (*Annona muricata*)

Ever green tree, 6–8 m in height, fruits 1.5–3 kg heart shaped; dark green; fleshy pines, pulp is

Table 9.2 Composition of fruits (g/100 g edible portion)

Constituents	Sweet soup	Cherimola	Atemoya	Soursop
Moisture	75.97	68.71	78.7	80.1
Protein	1.89	1.54	1.4	0.69
Fat	0.57	0.13	0.6	0.39
Carbohydrate	20.82	28.95	15.8	18.23
Fiber	1.41	–	2.5	0.95
Ash	0.75	0.67	0.5	0.58
Energy (kj)	360	460	310	247
Calcium	17.0	9.0	17.0	9.0
Magnesium	22.0	–	32.0	22.0
Phosphorus	54.0	24	–	29.0
Potassium	142	–	250	320.0
Sodium	2.0	–	4.0	22.0
Iron	0.3	0.25	0.3	0.82
Ascorbic acid	35.9	12.20	43.0	16.4
Thiamine	0.10	0.11	0.05	0.07
Riboflavin	0.06	0.11	0.08	0.12
Niacin	0.89	1.0	0.8	1.52

white; fibrous, juicy with mango like flavor; 11–14 % sugar.

9.3.6 Ilama: White *Annona* (*Annona diversifolia*)

Not popular in India. Slender tree; fruit resemble custard apple/cherimoya pulp; good quality and highly acceptable (Table 9.2).

9.4 Collection, Introduction, and Evaluation of Custard Apple Germplasm

The results revealed significant differences among eight cultivars of custard apple in respect of most of the vegetative as well as physicochemical characteristics of fruits under rain-fed conditions of Panchmahals (Table 9.3). As regards vegetative growth parameters, stem diameter and plant spread were significantly influenced; however, plant height was found to be nonsignificant but was maximum (4.16 m) in Island gem. However, stem diameter and plant spread was found to be significantly influenced. Maximum stem diameter (103.33 mm) was recorded in Island gem. Plant spread (N–S & E–W) was maximum in Washington (5.60 and 5.30 m). Observations on fruit set per tree recorded revealed that it was maximum in Atemoya × Balanagar (263.67). Fruit and pulp weight was highest in Island gem (195.50 and 99.25 g, respectively). Whereas, maximum skin weight was noted in Balanagar (75.75 g) and seed weight in Local Sitaphal (15.75 g). TSS was maximum in local Sitaphal (30.00 °Brix) and least in Pink Mammoth (16.50 °Brix). Yield per plant was highest in Atemoya × Balanagar (26.84 kg/plant) (Table 9.3).

Table 9.3 Evaluation of custard apple germplasm

Variety	Plant Ht (m)	Stem Dia. (mm)	Pl. Spread NS (m)	EW (m)	Fruit wt. (g)	Pulp wt. (g)	Skin wt (g)	Pulp skin ratio	Seed Wt (g)	TSS °Brix	Yield/ plant (kg)
Balanagar	2.78	65.1	3.12	3.25	176.25	61.0	75.7	0.81	15.0	20.2	9.74
Washington 98797	3.84	102.5	5.60	5.30	128.00	47.7	59.5	0.80	14.0	18.0	1.68
Seedless atemoya	3.39	73.4	4.0	4.05	173.50	85.7	58.5	1.46	13.2	20.0	6.29
Pink Mammoth	3.46	96.1	3.9	4.28	152.00	62.5	53.5	1.16	14.0	16.5	25.17
Island gem	4.16	103.3	5.27	4.66	195.50	99.2	68.5	1.44	6.50	23.0	2.64
Atemoya× Balanagar	3.21	88.3	4.05	3.75	188.75	77.2	66.2	1.16	11.5	20.0	26.84
Local Sitaphal	2.87	67.8	3.52	3.47	132.75	55.7	43.2	1.28	15.7	30.0	20.70
CD 5 %	NS	26.8	1.08	0.90	47.13	22.5	18.7	–	4.55	1.44	15.89

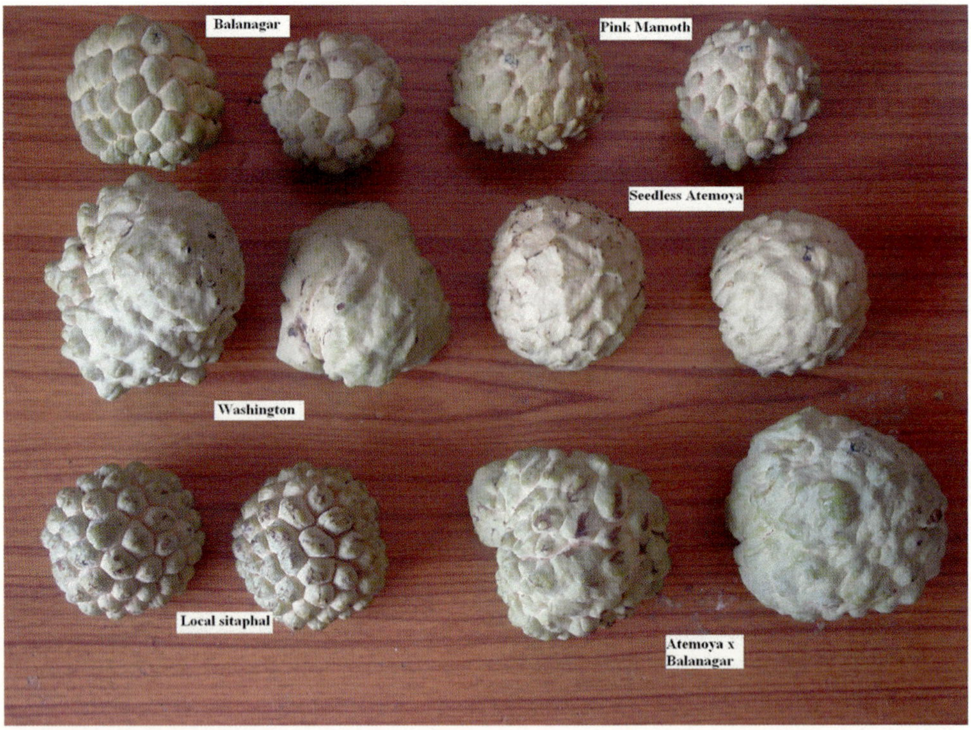

Fruit variability in custard apple

9.5 Cultivar Development

Promising cultivars such as Balanagar, Local Sitaphal, Red Sitaphal, and Mammoth were identified at fruit research station, Sangareddy. Karale et al. (1989) identified promising types at MPKV, Rahuri No. 5, 6, 8, 10, 15, 19, and 20 in Maharashtra. However, there is still scope to identify promising strains from the existing variability from available variability particularly in the states of Andhra Pradesh, Tamil Nadu, Rajasthan, and Madhya Pradesh. The bullock's heart or Ramphal is grown in most of the tropical countries and to some extent in India. The fruit is considered inferior to cherimoya or sweet soup but has fewer seeds. At Sangareddy two selections have been identified from *A. squamosa* one with heart shape and another with reticulata type, of which the latter was more productive with more solid fruit than that of sweet soup type.

The cherimoya or Lakshman Phal has performed well under moderate climate of south India and is found growing on the hill slopes of the Nilgiri, Nandi Hills, in Karnataka and Deccan Plateau.

9.6 Varietal Evaluation

1. *Arka Sahan* – A promising hybrid "Arka Sahan" (*A. atemoya*, Island gem × *A. squamosa*, Mammoth) has been developed at IIHR, Bangalore. The hybrid yields very sweet, fragrant, low seeded fruits having longer shelf life. Fruit weight 210.7 g, color greenish white. Rind surface smooth, rind thickness 0.5 cm, mesocarp color white, edible pulp 48.6 %, the number of seeds per 100 g fruit 8.9, TSS 30.8 °Brix, acidity 0.6 %, yield per tree 16.8 kg, with a shelf life 4 days. However, though there is profuse flowering, fruit set was negligible. The fruit set can be improved by hand pollination with the pollen of *Annona squamosa* only. Hand pollination with *A. squamosa* improved the set to 83 %, while natural set was just 1.2 % hand pollination with *A. atemoya*, *A. cherimoya*, and *A. reticulata* did not give satisfactory results (Jalikop and Sampatkumar 1998).
2. *AKP-CA-1* – The clonal selection developed by TNAU, Tirunelveli center. It is high yielding in rain-fed vertisols with a yield potential of 14.90 kg/tree, which is 30.7 % higher than Balanagar. Fruit weight was 207 g with 72 fruits/tree. TSS was found to be 24.50 Brix, acidity 0.2 %. Though the flowering is prolific, the fruit set is nil under semiarid rainfed conditions of CHES, Godhra.
3. *Balanagar* – The selection was identified at fruit research station, Sangareddy, Andhra Pradesh. The number of flowers/shoot was 552.33; fruit set was 24.62 %; the number of fruits/tree was 48, with a fruit weight of 137–264 g; seeds/fruit were 44 and seed weight was 5.7 g/fruit; pulp percent was found to be 44.9 %; TSS was reported to be 20.7 °Brix, acidity 0.20 %, reducing sugar 15.0 %, and total sugar 17.9 %; and yield/plant was 19.74 kg. The cultivar performed best at MPKV, Rahuri, IIHR Bangalore, and MPUAT, Udaipur.
4. *Washington P.I. 98797* – The number of flowers/shoot was observed to be 515 and fruit set 1.07 %, which was lowest among the varieties tested. Data on the number of fruits/trees recorded to be 21, fruit weight ranged from 146 to 161 g, seeds/fruit 48, seed weight 6.9 g/fruit, pulp percent. 38.1 %, TSS 20.80 Brix, acidity 0.28, reducing sugar 12.7 %, and total sugar 17.9 %. Yield/plant was 10.58 kg.
5. *Mammoth* – It is an introduction from West Indies, evaluation of the variety indicated that the mean number of flowers/shoot was 280, percent fruit set was 25.64, the number of fruits/tree was 29 with a fruit weight 154–195 g, seeds/fruit was 27, and seed weight was 8.7 g/fruit. Pulp percent was 44.8 %, TSS 20.0 °Brix, acidity 0.19 %, reducing sugar 15 %, and total sugar 17.9 %.Yield/plant was 25.17 kg.
6. *Barbados* – This is also a introduction from West Indies. The number of fruits/shoot was 35, fruit weight was 154 g, seeds/fruit was 60, and seed weight was 8.0 g/fruit. Pulp percent was 37.9 %, TSS was 21.3 °Brix, acidity was 0.27 %, reducing sugar was 15.0 %, and total sugar was 17.9 %.
7. *British Guiana* – This is also a introduction from West Indies. The number of fruits/tree was 35, fruit weight was 151 g, seed weight was 8.4 g/fruit, pulp percent was 37.9 %, and seeds/fruit was 60, TSS was 23.3 °Brix, acidity was 0.27 %, reducing sugar was 15.0 %, and total sugar was 17.9 %.
8. *Red Sitaphal* – Indian selection. Deep pink colored fruits and floral parts. The number of flowers/shoot was 243, percent fruit set was 3.97, the number of fruits/tree was 22, fruit

weight was 119–156 g, seeds/fruit was 50, seed weight was 5.2 g/fruit, pulp percent was 30.5 %, TSS was 22.3 °Brix, acidity was 0.24 %, reducing sugar was 13.7 %, and total sugar was 15.9 %. Yield/plant was 15.57 kg.

9. *Local Sitaphal* – The traditional Sitaphal growing wild in the forest on India. The number of flowers/shoot was 236.33, percent fruit set was 16.34, the number of fruits/tree was 31, fruit weight was 127–137 g, seeds/fruit was 48, seed weight was 9.2 g/fruit, pulp percent was 44.5 %, TSS was 28.8 °Brix, acidity was 0.26 %, reducing sugar was 12.7 %, and total sugar was 16.5 %. Yield/plant was 20.70 kg. The performance of the variety is best under semiarid rain-fed conditions. The only disadvantage is greater number of seeds/fruit.
10. *Island gem* – Data on the number of flowers/shoot was 256.66, percent fruit set was 1.21 %, and fruit weight was 195–235 g. Seeds/fruit was 31, seed weight was 6.5 g /fruit, pulp percent was 51.7 %, TSS was 26.6 °Brix, acidity was 0.26 %, reducing sugar was 11.8 %, and total sugar was 15.3 %. Yield/plant was 2.64 kg. Lowest yield/plant was recorded in this variety.
11. *Seedless atemoya* –Observations recorded on the number of flowers/shoot revealed 154.66, percent fruit set was 6.34 %, and fruit weight ranged from 186 to 215 g. Yield/plant was 8.29 kg.
12. *Atemoya x Balanagar*– One of the highest yielder due to high fruit set. The number of flowers/shoot was 44.00, percent fruit set was 49.23 %, fruit weight ranged from 216 to 255 g, seeds/fruit was 55 and seed weight 11.5 g /fruit, pulp percent was 44.5 %, TSS was 20.0 °Brix, acidity was 0.26 %, reducing sugar was 12.7 %, and total sugar was 16.5 %. Yield/plant was 26.84 kg.
13. *Annona 2* –The variety is developed by farmer in Solapur district of Maharashtra. The fruits have less seeds and fruit size ranges from 500 to 750 g. He claimed that, it produces 16–19 tones fruits per hectare, with revenue of Rs. 6–8 Lakhs.

Table 9.4 Comparison of selections developed by farmer (Kaspate, Solapur)

Variety	Productivity (tones/ha)	Income Rs./ha
Balanagar	10–12	1.25
NMK-1	13–15	7–8
Annona 2	16–19	6–8

9.6.1 Selections Developed by Farmer from Solapur

It is seen from the Table 9.4 that the selection is superior in respect of yield, quality, and income per ha. Selection Annona 2 has very few seeds. The average weight of fruit is 700–800 g under drip irrigation and pulp percent is 70–75 %.

9.7 Leaf and Leaf Area Determination

Studies on leaf area determination by nondestructive method carried out in seven cultivars of custard apple revealed that correlation coefficient (cv) of the actual leaf area with leaf length, maximum breadth, and product of leaf was highly significant for all the parameters. However, maximum values were obtained when actual leaf area was correlated with the product of leaf length and breadth. The values were 0.92, 0.91, 0.98, 0.86, 0.99, 0.99, and 0.77 for different cultivars, respectively. Based on which regression equation and factor values were calculated. A linear relationship was established by $Y=a+bx$ for regression method and $Y=Kx$ for factor method. The regression equation fitted and is given in Table 9.5 for calculation of leaf area in custard apple by nondestructive method (Tables 9.5, 9.6, and 9.7).

9.8 Climate and Soil

Most annonaceous fruits are acclimatized to tropical climate. Although custard apple withstands heat and drought conditions, high atmospheric humidity is necessary during flowering to improve fruit set. But continuous rains during

Table 9.5 Leaf length, breadth, and actual leaf area in *Annona* germplasm

Treatment	Length (cm)	Breadth (cm)	$L*B$ (cm^2)	Actual leaf area (cm^2)
Balanagar	7.33	3.23	23.90	29.15
Washington	11.39	5.64	64.67	55.67
Seedless atemoya	15.51	8.27	129.11	106.58
Pink Mammoth	14.56	6.07	88.89	69.82
Island gem	16.37	10.05	166.25	137.87
Atemoya * Balanagar	18.18	9.87	182.06	139.27
Local Sitaphal	9.14	3.86	35.42	32.01

Table 9.6 Leaf length, breadth, and actual leaf area in *Annona* germplasm

Treatment	Regression equation
Balanagar	$L \times B = (4.29 \times 1.04x)$
Washington	$L \times B = (10.09 \times 0.705x)$
Seedless atemoya	$L \times B = (-2661.54 \times 21.44x)$
Pink Mammoth	$L \times B = (14.91 \times 0.62x)$
Island gem	$L \times B = (9.23 \times 0.77x)$
Atemoya * Balanagar	$L \times B = (-3.49 \times 0.78x)$
Local Sitaphal	$L \times B = (5.52 \times 0.78x)$

Table 9.7 Correlation of leaf area in custard apple germplasm

Treatment	Length	Breadth	$L \times B$
Balanagar	0.62	0.77	0.93
Washington	0.72	0.78	0.91
Seedless atemoya	0.81	0.81	0.98
Pink Mammoth	0.65	0.74	0.86
Island gem	0.83	0.92	0.99
Atemoya * Balanagar	0.45	0.89	0.96
Local Sitaphal	0.48	0.65	0.77

fruit set are not desirable. An annual rainfall of 60–80 cm is optimum. It cannot stand frost or a long cold period. The trees remain dormant from December to February and shed its leaves. When the summer temperature rises above 103 °F, the tree sheds its flowers resulting in low fruit set.

Bullock's heart grows well in humid regions of south India and cannot withstand severe summer. It tolerates frost to some extent. Cherimoya prefers subtropical climate, but it can flourish on higher elevations (2,000 m) in tropics. While climatic requirements for atemoya are quite similar to those of custard apple, soursop in contrast is a fruit of the humid tropics.

Custard apple thrive naturally in rocky terrain with shallow, gravely, well-drained soils. However, they may grow well in arable, red, sandy shallow soil slightly acidic in reaction. Heavy soils are not suitable, especially in waterlogged areas. In Andhra Pradesh, annonnas come up chalka-red sandy or gravely soils. They can grow well even on calcareous soils containing lime as high 50 %.

Custard apple seedlings are found growing wild in India. Since custard apple is a cross-pollinated crop, wide variation in forms and sizes of fruit as well as color of the pulp is available. The natural variability available within the species is often exploited to identify superior genotypes, which are usually named after the place of collection or selection and fruit color. Depending on external fruit color, custard apple is distinguished into green, red, and yellow. But green ones are by far more common and popular than the other types. Balanagar, Barbados seedling, British Guiana, Kakarlapahad, Local Sitaphal, Mahaboobnagar, Saharanpur Local and, Washington are some of the varieties with green skin. Most of these varieties are not easily identifiable. Some of the traits that distinguish them are the fruit shape and size, the form of the areoles, and the number of seeds/fruit. But in fruits of a given tree, these attributes vary considerably as pollination and the environment largely influence them.

9.9 Ideotypes of Custard Apple

- Prolific bearing
- Low seed content
- Better keeping quality
- Large fruit
- Sweet (TSS more than 25 °Brix)
- Pleasant aroma
- Resistance to drought
- Resistance to salinity

9.10 Varieties

The varietal or genetic differences get masked confusing the varietal identification. Moreover, variety-specific pulp qualities are not clearly explained. However, some varieties can be recognized by the plant habit and foliage attributes. Two natural hybrids (mostly between custard apple and cherimoya), Israeli selection and Israeli hybrid, have been introduced. Fruits of both are less seeded. A hybrid Arka Sahan has slow ripening (6–7 days), better shelf life (2–3 days), less number of seeds (10/100 g fruit weight), and high Brix (31°). On average its fruit weigh was 210 g each. A 6-year-old plant yields 17 kg fruits (Table 9.7).

Exotic – Washington PI 98797, Washington PI 107005, British Guiana, Barbados seedling.

Exotic hybrids – Island gem, bullock's heart, Pink Mammoth

Selections – Balanagar, Mammoth, Red Sitaphal, Yellow Sitaphal

Indian hybrid – Arka Sahan

Arka Sahan – A promising hybrid "Arka Sahan" (*A. atemoya* × *A. squamosa*) has been developed at IIHR, Bangalore. The hybrid yields very sweet, fragrant, low seeded fruits having longer shelf life. Fruit weight (g) was 210.7, rind surface smooth, rind thickness (cm) 0.5, mesocarp color white, areoles per fruit 54, edible pulp (%) 48.6, number of seeds per 100 g fruit 8.9, TSS (°Brix) 30.8, acidity (%) 0.6, yield per tree (kg) 16.8, and shelf life (days) 4.

Bullocks Heart

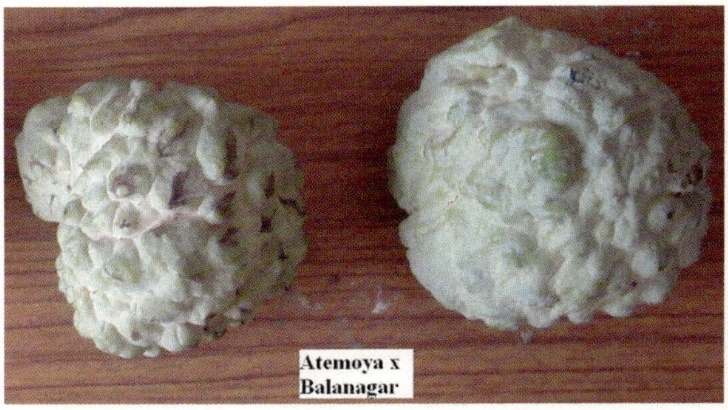

Island Gem

AKP-CA-1

9.11 Propagation and Rootstocks

Seeds traditionally propagate most of the *Annona*. The seed viability lasts for 3–4 years. However, fresh seeds germinate better when soaked for 24 h and resulted in highest germination of 71 % (Ratan et al. 1993). Hard seed coat can be softened either by soaking the seeds in water for 2–3 days of keeping them under running water for 50–70 h. Treating seeds with GA3 at 5,000 ppm increased germination (de Smet et al. 1999). Seeds are sown 2 cm deep either in nursery beds or in pots under partial shade. Regular watering is necessary to maintain good soil moisture. Seeds are slow to germinate and take 3 weeks. Nevertheless, it may extend to as long as 8–10 weeks. When seedlings are 10–12 cm tall, they are transferred to pots or plastic containing sand and peat or equal parts of garden soil, sand, and decomposed farmyard manure. The 30 cm tall seedlings become ready for transplanting.

Seed-originated plants are not true-to-type, lack precocity, and vigor, whereas grafting or budding helps largely to overcome to these drawbacks. *A squamosa, A. reticulata, A. cherimola, and A. atemoya* are grafted or budded on their own species and each other. *A. muricata* can be grafted on *A. reticulata* and *A. glabra*. However, *A. reticulata*, which promotes vigor and shows good graft congeniality, is commonly employed as a rootstock for most of the *Annona* (Khan and Rao 1952).

Table 9.8 Effect of method of propagation on percent success in custard apple cv. Balanagar

Method of grafting	Percent success
Softwood grafting	51.74
Patch budding	33.54
Cutting	13.62
Air layering	8.68
SEM	14.09
CD 5 %	58.57

Table 9.9 Effect of time of grafting/budding on percent success in custard apple cv. Balanagar

	Percent success	
Month	Softwood grafting	Budding
January	28.45	17.4
February	41.72	33.07
March	59.37	41.20
April	30.27	27.17
May	12.03	9.17
June	9.57	7.02
SEM	0.988	7.90
C.D. 5 %	2.96	23.83

Generally, 18 months old or 30 cm tall plants having pencil thickness are selected for grafting. Scion of well-matured wood from which the leaves have dropped at the end of the dormant phase is used to graft either by veneer or cleft technique. Shield or T-budding carried out in spring is also equally effective. Patch and chip bedding are other methods. Large buds, about 4 cm length, are collected from 1-year-old wood after the leaf drop gives good success. Since propagation by cutting and air layers give poor results, they are not widely practiced. Commercial production of plants through tissue culture is not yet successful. However, multiple shoot production from leaf explants of seedlings and root initiation from shoots are successful.

Propagation studies were undertaken to find out best method, viz., bud take percentage, softwood grafting success, rooting of cutting, and air layering in *Annona* (*Annona squamosa* L.) cv. Balanagar. It was recorded that softwood grafting was the most successful method for commercial propagation of *Annona*. Maximum graft take of 58.37 % was observed in cv. Balanagar fallowed by budding (33.54 %). Percent success in propagation was least in air layering (8.68 %). Softwood grafted plants showed maximum increase in the growth parameters compared to budded plants 6 months after grafting (Hiwale et al. 2009). Gholap et al. (2000) also reported maximum success in softwood grafting fallowed by budding (Table 9.8).

Another experiment on comparative performance of time (January to June) and method (softwood grafting and budding) of propagation revealed that softwood grafting in the month of March gave the maximum success of 59.37 %, compared to 41.20 % in budding in the same month, when leaf fall starts. Least success was recorded in the month of June (9.57 and 7.0 %) in both the methods (Table 9.9). Growth of the plant one month after softwood grafting/budding revealed maximum growth in softwood grafted plants (Table 9.10). Thus, vegetative propagation of *Annona* by softwood grafting method in the month of March was found to be the best (Hiwale et al. 2009).

Table 9.10 Effect of method of propagation on growth of custard apple cv. Balanagar

Treatment	Pl. height (cm)	Stock diameter (mm)	Scion diameter (mm)	Leaf area/ plant (cm^2)
Softwood grafting	356.17	4.91	4.80	311.18
Budding	246.17	4.13	3.58	197.22
CD 5 %	NS	NS	NS	NS

Table 9.11 Flowering fruit set and fruit retention in custard apple

Variety	Number of flowers/shoot	Number of flowers/plant	Number of fruit set/plant	% set/ plant	Number of fruit retained/plant	% retention/ plant
Balanagar	42.29	350.68	91.03	26.29	49.33	14.66
Washington 98797	167.61	4133.60	24.00	0.70	6.33	0.19
Seedless atemoya	163.13	3140.00	61.00	1.88	9.33	0.30
Pink Mammoth	168.76	2597.36	493.66	25.64	61.66	2.74
Island gem	145.70	2719.61	33.00	1.21	3.66	0.16
Atemoya × Balanagar	59.21	791.71	404.66	52.26	76.66	9.69
Red Sitaphal	220.38	2095.50	84.00	3.97	13.33	0.78
Local Sitaphal	297.48	2589.64	266.66	10.64	47.66	1.94
CD 5 %	59.61	2400.15	69.12	15.31	17.89	1.89

9.12 Flowering Fruit Set and Fruit Retention

Low marketable yield is a major problem in various annonaceous types. The reasons for low fruit set are dichogamy, lack of pollinizers, and production of flowers without attractive color, poor pollen germination, and environmental and tree condition. Close planting may increase the pollination. Venkataratnam (1979) and Thakor and Singh (1965) reported that stigma remains receptive from one day before anthesis to about 2–3 h after anthesis, but it decreased abruptly so by the time of dehiscence stigma turned almost unreceptive and observed protogynous dichogamy. Similar trend was reported by Limaye (1966) and Rajput (1985). Ahemad (1935) reported that the best time for artificial pollination was from 5 am to 8 am and also reported that drier and hotter period adversely affected the fruit set. Similar result was reported by Karale et al. (1989). Rajput (1985) reported that there are no insect pollinizers for *Annona* because the flowers are unattractive. Pollen germination was found to be poor ranging from 11.5 to 20.00 %. Work done at CHES, Vejalpur, on germplasm evaluation of eight varieties of *Annona* revealed that maximum number of flowers per plant were produced in Washington (4,133.60 no.); however fruit set and retention were the least (0.70 % and 0.19 %, respectively). In cultivar Balanagar though the least of flowers were produced (350.68), fruit set and retention were found to be highest, 26.29 % and 14.66 %, respectively (Table 9.11) (Hiwale 2002).

The time of flowering and fruit development in different varieties varies considerably. Custard apple flowers from March to July and set fruit in June to July and took 4 months for fruit development whereas cherimoya flowers from May to June and July to September. It takes 6 months for fruit development. On the other hand, bullock's heart flowers from August to October and takes 8 months for fruit development.

9.13 Cultural Practices

9.13.1 Planting

Pits of 60 cm × 60 cm × 60 cm size are dug and left open to sun for a week. They are filled with topsoil mixed with 25–30 kg of well-decomposed farmyard manure. Custard apple should be transplanted 5 m × 5 m apart (400 plants/ha). However, setting plants closer may be preferred. Plant spaced at 4 m × 4m (625 plants/ha) not only accommodates over 50 % additional plants/unit area but promotes better fruit set by improving pollination, a problem in annonnas. Due to large canopy, atemoya and bullock's heart require a plant-to-plant spacing of 6 m×6 m and cherimoya and soursop 8 m×8 m spacing. Planting should be carried out preferably in spring so that plants establish roots in summer, start growing as the weather warms up, and put up vigorous growth during rains. However, if adequate irrigation facilities are lacking, monsoon is the optimum time for planting. To keep the graft joint well above the ground is a must. As soon as a young tree is planted, it should be irrigated till it establishes.

9.13.2 Training and Pruning

Annona requires little pruning. It is essential to develop a good crown and better yields over a long period of time. Without pruning, the plants become bushy and their bearing efficiency comes down. Hence, timely removal of misplaced limbs is necessary to build a strong framework. Selective and mild pruning of deadwood and very old branches should be carried out to avoid congestion and encourage well-spaced branching. Severe pruning is detrimental for the plant growth. Yellowing of leaves starts as the harvesting season of fruits ends. The leaves begin drop with onset of winter and fresh growth occurs in spring. Flowering occurs singly or rarely in small clusters mostly on current season's growth and occasionally on old wood. Training to a single stem is the only option when rootstock is employed.

9.13.3 Manures and Fertilizers

Manures and fertilizers application to custard apple is not common but its plants respond very well to fertilization, increasing vigor, yield, and fruit quality. Fertilizer application checks decline and extent longevity of trees. To specify the precise dose common to all soils is difficult, but general recommendations are given. The area below the crown of trees should be cleaned of weeds, loosening of soil in the basin should be carried out, and then fertilizers should be applied in the basin under the tree 30 cm away from the trunk. Subsequently it is desirable to irrigate the trees and incorporate the fertilizers. Fertilizer application should coincide with rapid vegetative growth and fruit development. As fruits are born on new as well as old wood, application of slightly higher dose of N is not harmful (Hayes 1953). Biofertilizer and VAM application can increase the productivity of the plant and soil sustainability. Application of castor cake and bone meal or super phosphate in the ratio of 2:1 was found to be beneficial (Rao 1974) (Table 9.12).

9.13.4 Aftercare

After planting, the young plants must be watered and supported by stacks to keep them erect. To start with a 60 cm×60 cm basin around the plant is adequate. Regular watering during dry periods; occasional hand digging of the basins to check weeds, to keep the soil loose; attending plant-protection measures; removing the sprouts on stock; and building up of a good framework are necessary cultural operations. The basins around the plant should be enlarged, as the plants grow bigger. They should be made little larger than the spread of the plant.

In young orchards, a lot of land remains vacant between the rows for 4–5 years. Hence, short duration vegetables – tomato, onion, chilly, okra, brinjal, radish or cowpea, green gram, horse gram, or any green manure crop – can be intercropped. These crops should not be raised too near the tree, lest they compete with them for nutrients.

Table 9.12 Nutritional requirement in custard apple

	Plant age (years)		
Source (g/plant)	1–2	3–5	Above 5
Nitrogen	75	150	250
Phosphorus	50	100	125
Potash	25	100	125
Organic manure (kg)	25	25–30	50

9.13.5 Irrigation

Most of the *Annona* produce a moderate crop even in the absence of irrigation. Irrigation to plants during flowering and fruit development is essential. Fruit set, yield/plant, and quality are superior in irrigated plants with more edible pulp/segment. Plants receiving regular water grow luxuriantly with each bearing. Pruning, fertilization, and irrigation are quite essential to get maximum yield. In regions having limited water, pitcher, trickle, or drip irrigation systems help in judicious use of water. Fruits are raised in rain-fed areas in low rainfall areas. Land shaping to divert rainwater near the plantation may be taken up. Contour terraces, contour bunds, and micro-catchments also help in efficient water use. Plowing and mulching of the plantation during rainy season helps better conservation of moisture. Kulkarni (1993) reported that eight irrigations at 15 days interval starting from April 15 was best for custard apple in semi arid areas of Rahuri for increasing fruit set and yield of custard apple.

9.13.6 Crop Regulation

Sometimes there is heavy fruit set in custard apple leading to smaller sized fruits or mummified fruits, and for obtaining out of season crop, fruiting in custard apple can be regulated by the use of manual defoliation or chemical defoliation. Both treatments produced higher number of fruits as well as resulted in increased fruit weight (Table 9.13).

9.14 Cropping System

To utilize the interspaces in the young orchards and to maximize productivity per unit area, various intercrops, viz., okra, maize, pigeon pea, green gram, *Sesamum*, and moth bean were grown during kharif season under rain-fed conditions. The vegetative growth parameters were not influenced by the intercropping (Table 9.14). Higher productivity per unit area was obtained in custard apple intercropped with okra with a B. C. ratio of 1:1.54 (Table 9.15). Apart from this the

Table 9.13 Effect of early induction of new growth by defoliation on yield of *Annona*

	Number of fruits			Weight of fruit (kg)		
Variety	Chemical defoliation	Manual defoliation	Control	Chemical defoliation	Manual defoliation	Control
Balanagar	98	99	56	16.17	13.47	7.76
British Guiana	78	125	59	13.55	21.02	9.85
Local Sitaphal	99	85	60	14.32	9.0	8.93
Red Sitaphal	94	81	54	13.15	12.35	7.97
Washington 107005	96	69	38	14.10	7.95	4.70
Mammoth	109	109	61	14.75	14.55	8.80

Table 9.14 Mean morphological parameters of custard apple in agri-horticulture system

			Plant spread (m)	
Name of intercrops	Plant height (m)	Stock diameter (mm)	NS	EW
Okra	2.89	66.35	2.39	2.33
Maize	2.77	67.81	2.26	2.35
Pigeon pea	2.71	66.36	2.20	2.23
Green gram	2.69	64.30	2.15	2.19
Sesamum	2.75	65.04	2.35	2.40
Moth bean	2.73	66.88	2.12	2.29

Table 9.15 Mean yield, net return and B.C. ratio in agri-horticulture production system

Name of intercrops	Main crop (q/ha)	Inter crop (q/ha)	Pure crop (q/ha)	Net return (Rs.)	B: C ratio
Custard apple	26.07	–	–		
Custard apple + okra	23.98	22.26	31.67	13,964	1.54
Custard apple + maize	22.85	9.05	15.22	9,101	1.23
Custard apple + green gram	23.69	1.84	3.06	7,276	0.97
Custard apple + *Sesamum*	23.55	2.68	3.79	11,134	1.53
Custard apple + pigeon pea	24.51	1.37	2.01	7,186	1.02
Custard apple + moth bean	23.19	2.14	2.99	8,245	1.12

Table 9.16 Rainfall, runoff, and soil loss in different cropping system

	Runoff (mm)				% of runoff				Soil loss tones/ha			
Treatment	2002	2003	2004	Mean	2002	2003	2004	Mean	2002	2003	2004	Mean
Cultivated uncropped	115	147	104.7	122.2	32.63	29.28	33.55	31.82	5.3	7.6	6.9	6.60
Cultivated cropped	83.94	98.38	66.24	82.85	23.87	19.59	21.28	21.28	1.6	2.15	4.08	2.61
Staggered trench planting	64.94	76.10	61.51	61.51	18.55	15.16	19.74	17.81	0.6	0.98	1.23	0.93
No tillage	46.44	27.46	40.27	38.05	13.26	5.47	12.90	10.54	0.8	1.24	2.12	1.38

cropping system also resulted in reducing the runoff loss of water as well as reduced soil erosion, there by increasing the sustainability of the soil.

9.15 Rainfall, Runoff, and Soil Loss in Different Cropping System

In custard apple-based cropping system, maximum runoff (31.82 %) was recorded in cultivated uncropped soil. It was minimum in no tillage (10.54 %). Staggered contour trench planting was found to be the best for reducing soil loss effectively (0.93 tones/ha). Runoff and soil loss was higher in custard apple due to total shading of leaves, and new growth took 20–30 bays to cover the soil and due to higher slope of the land and its soil type. Adopting staggered contour trench planting method resulted in reduced runoff of water, and soil erosion losses were also reduced by almost seven times compared to cultivated uncropped, thus resulting in saving the valuable soil and water resources, thereby maintaining the sustainability of the soil (Table 9.16).

9.16 Root Distribution Pattern of Custard Apple

Root distribution varies according to the type of soil and method of propagation. Studies were therefore initiated to find out root distribution patter of seedling type 6-year-old custard apple cv. Balanagar. Root distribution pattern under semiarid rain-fed conditions revealed that the root system is shallow in nature as below 60 cm soil depth not much root activity was recorded and the spread was within the plant canopy. Maximum root activity on fresh weight basis was observed in 0–30 cm radial distance from tree trunk (70 %), which reduced as the radial distance increased. Depth-wise the root distribution was maximum at 0–30 cm depth (56 %) and subsequently reduced as the depth increased (Table 9.17). The root spread was higher depth-wise (up to 120 cm), whereas it was just 85 cm radially.

Table 9.17 Root distribution pattern of custard apple (fresh wt. basis)

Depth/radial	0–30	%	30–60	%	60–90	%	% total
0–30	1.00	40	0.30	12	0.10	4	56
30–60	0.40	16	0.20	8	–	–	24
60–90	0.25	10	0.15	6	–	–	16
90–120	0.10	4	–	–	–	–	4
Total	1.75	70	0.65	26	0.10	4	

Main root wt. (kg)	Lateral root wt. (kg)	Total root wt. (kg)	Aboveground biomass (kg)	Ratio of shoot/root	Rooting depth (cm)	Horizontal spread (cm)
0.90	1.60	2.5	12.00	4.8	120	85.25

Table 9.18 Growth phase

	PAR	EVAP	PN	GS	Carbon intake
	micromoles/m^2/s	millimoles/m^2/s	micromole/m^2/s	millimoles m^2/s	(ppm)
C. apple	301.6	2.46	10.9	268.0	244.0
C. apple + okra	593.0	6.66	23.7	1283.4	522.5
C. apple + *Cajanus cajan*	499.6	6.17	17.4	987	557.6
C. apple + maize	528.8	5.29	26.5	620.2	512.8
C. apple + cowpea	528.4	6.35	16.0	993.6	594.0
C. apple + *Sesamum*	544.6	6.89	19.5	904.8	520.6

9.17 Photosynthetic Efficiency, Conversion Coefficient, Stomatal Conductance, and Transpiration Rate in Different Cropping Systems

Agroforestry systems are more productive than the sole cropping systems because of greater light interception, better moisture utilization, and maximum incidental light, and its interception was recorded in custard apple intercropped with *Sesamum* at both the crop growth stages. Maximum photosynthetic activity was recorded in custard apple with maize cropping system. However, transpiration rate, stomatal conductance, as well as carbon intake was the least in the same combination. Dry matter production was highest in custard apple with maize combination. Rate of photosynthesis is higher than the rate of evapotranspiration; therefore there is net accumulation in the leaves, resulting in higher biomass production (Tables 9.18 and 9.19).

9.18 Natural Resource Management

Energy conservation efficiency is the ratio of output of calories captured by vegetation to the input (solar radiation), in a unit area over a certain period of time. All the intercropping systems resulted in higher production in terms of biomass, thus making better utilization of natural resources like solar energy, available moisture, higher carbon intake, and more stomatal conductance, leading to higher photosynthesis and there by biomass accumulation.

9.19 Harvesting and Post-harvest Management

Custard apple starts bearing fruits at the stage of 4 years. Production declines on the 15th year depending upon the maintenance. Custard apple produces single crop in a year during August–October in south India and September–November

Table 9.19 Fruit set phase

	PAR	EVAP	PN	GS	Carbon Intake
	micromoles/m²/s	millimoles/m²/s	micromole/m²/s	millimoles/m²/s	(ppm)
C. apple	1,011.6	6.3	27.05	506.0	300.0
C. apple + okra	1,644.6	14.22	70.35	2009.0	655.4
C. apple + *Cajanus cajan*	1,953.4	15.22	79.05	2176.0	802.8
C. apple + maize	1,837.0	14.69	82.65	1484.8	547.4
C. apple + cowpea	2,620.8	18.25	75.65	1987.2	569.8
C. apple + *Sesamum*	2,625.8	15.5	62.45	1873.4	555.6

in north India. On maturing, fruits turn light green. The inter-areole space widens, the fruit turn creamy-white. Custard apples are harvested manually when they are fully mature but still firm. About 4–5 picking are required. Occurrence of deformed fruit is due to unfertilized areoles failing to grow. Fruit yield varies widely from tree to tree. Normally a 7-year-old tree produces 100–150 fruits, the total yield being 7 tones/ha.

Bullock's heart and cherimoya are ready for harvesting during December–January, yielding 50–75 fruits/plant. Since both have thick stalks, it is necessary to harvest them with the stalks using secateurs. Atemoyas are somewhat early (September–October) and are shy bearer. Higher yield (50–60 fruits) may be obtained with hand pollination. Soursop, the largest *Annona*, produces about 25 fruits/tree during June–August in south India.

Annona are climacteric fruits. Custard apple takes about 3 days to ripen, while others 4–7 days. Prior to ripening the pulp of matured custard apple is not separated into segments on flakes. It is only during conversion of starch to sugar differentiation occurs. In ripe fruits of bullock's heart and cherimoya, the pulp is more or less homogeneous mass of closely cohering carpels, which cannot be separated easily.

Ripe fruits of custard apple and soursop are very fragile, and with the slightest pressure, the fruits easily get disintegrated into segments. Hence, extra care is necessary while handling. Ripe custard apple can be stored about 2 days, but other annonas can be stored for 3–4 days. This may be partly due to the characteristic feature of the rind. In bullock's heart and cherimoya, the carpels are not associated with the external areole division on the rind and the surface appears to be contiguous or almost fused unlike that in custard apple. Thus, the fruits do not split easily along the deep furrows between the inter-areole spaces, the weakest portions of the rind. Custard apple, atemoya, bullock's heart, and cherimoya are normally used as fresh fruits. Ripe fruits are popular among the poor. The unripe fruits of custard apple are eaten in Andhra Pradesh after backing the roasting. The row fruits of soursop are commonly used to prepare soup or vegetable.

The pulp of custard apple mixed with milk is made into a delightful drink or ice cream. Development of a repulsive off flower on heating beyond 65 °C and presence of gritty cells are major constraint in processing custard apple. But its juice is a potential ingredient to prepare squash, syrup, nectar, and fermented alcoholic beverages. Jam, jelly, conserves, and tarts can also be prepared from the pulp of custard apple. It is also possible to can the pulp. The sweetish sour flash of soursop is fibrous and juicy with pleasant aroma and is amenable for preparation of ice cream, sherbet, syrup.

9.20 Physiological Disorders

9.20.1 Stone Fruits or Mummified Fruits

In certain neglected plants or those under severe moisture or nutritional stress, the dormancy sets well advance and some fruits turn brown, become hard, and without further growth remain on tree for months. Such fruits are termed as "stone fruits" or "mummified fruits." About 20–25 % fruits are found to be mummified in local Sitaphal.

Control

- Clean cultivation
- Application of balanced nutrition
- Thinning of fruits to reduce the load, particularly fruits which are set late

9.20.2 Fruit Cracking

Sudden fluctuation in water supply to the plants may cause cracking of fruits. Irrigating plants after prolonged dry spell may be one of the reasons. An equitable irrigation schedule will help in solving the problem.

9.21 Insect Pest and Diseases

9.21.1 Mealy Bug (*Planococcus pacificus*)

Mealy bug (*Planococcus pacificus*) is seen affecting the fruit and young branches there by influencing the quality and the market price.

- *Control* – Spraying of 0.05 % phosphamidon or dichlorvos (Shukla and Tondon 1984).

9.21.2 Leaf Spot (*Alternaria* spp.)

The affected leaves fall down, causing considerable loss in the production of British Guiana, Island gem, and Red Sitaphal that were found to be moderately susceptible (Anon, 1987).

9.21.3 Anthracnose (*Colletotrichum singulata*)

Anthracnose (*Colletotrichum singulata*) has been attacking Sitaphal in Udaipur area (Anon 1981).

- *Control* – Fortnightly sprays of Benlate 0.05 % or bicor 0.1 % can control both the diseases.

9.21.4 Tree Decline – Caused by Water Stagnation

Trees shrivel and drying of old branches takes place and the plants die suddenly.

Control

- Stagnation of water should be avoided.
- Saline soils with high clay content should be avoided.
- Drainage channel should be laid out to carry out excess water.
- Drenching with copper fungicide 3 g/L.

References

Ahemad MS (1935) Pollination and fertilization in *Annona squamosa* under dry climatic condition in Egypt. Hortic Abstr 4:76

Anonymous (1981) Annual report All India coordinated project on arid Zone fruits, pp 21

Anonymous (1987) All India Coordinated project on Arid Zone fruits. Technical document no. 20

de Smet S, van Damme P, Sheldman X, Romero (1999) Acta Hortic 497:269–288

Gholap SV, Dod VN, Bharad SG, Wankar AM (2000) Crop Res (Hissar) 20:158–159

Hayes WB (1953) Fruit growing in India, 3rd edn. Kitabishtan, Allahabad

Hiwale SS (2002) Annual report. Central Institute for Arid Hort, pp 12–13

Hiwale SS, More TA, Bagle BG (2009) Vegetative propagation in Annona (*Annona squamosa* L.). In: National conference on production of quality seeds and planting material, Health Management in Horticultural Crops, New Delhi from 11–13 March 2009

Jalikop SH, Sampatkumar (1998) Annual report, Indian Institute of horticultural Research. Hessarghtta, Bangalore

Karale AR, Kesar BG, Dhawale BC, Kale PN (1989) The influence of tree condition at the time of hand pollination as a factor on fruit set in Annona species. Maharashtra J Hortic 2:114–116

Khan KF, Rao IKS (1952) A note on vegetative propagation and clonal performance in custard apple (Annona squamosa L.). Indian J Hortic 10:140–144

Kulkarni SS (1993) Effect of cultural and chemical treatments on fruit set, yield and quality of custard apple. M.Sc. (Agri) thesis MPKV, Rahuri, MS
Limaye SP (1966) Hand pollination studies in Atemoya. Agric Coll Mag Parbahni 2(1):41–42
Rajput CBS (1985) Custard apple in fruits of India: tropical and subtropical. Naya Prokash Publication, Calcutta, pp 479–488
Rao SN (1974) Annona the legendary fruit. Indian Hortic 19(3):19–34
Ratan PB, Reddy SE, Reddy YN (1993) South Indian Hortic 41:171–173
Shukla RP, Tondon PL (1984) J Entomon 9:181–183
Thakor DR, Singh RN (1965) Studies on floral biology of Annonas. Indian J Hortic 22(3–4):238–253
Venkataratnam L (1979) Floral morphology and blossom biology on some Annona. Indian J Agric Sci 29(4):68–76

Phalsa (*Grwia subinaequalis* L.) 10

Abstract

India is considered to be the home of phalsa (*Grewia asiatica*). In early literature, phalsa is mentioned for its medicinal qualities. It belongs to the family Tiliaceae. Phalsa is a hardy fruit which is capable of growing even under drought conditions. Once the plant is established, it requires little care. Irrigation is considered necessary in many parts of the country. Phalsa is a small bush which grows all over the country except at higher elevations. In Punjab, Haryana, Uttar Pradesh, and Andhra Pradesh, it is grown commercially. Harvesting of phalsa is labor intensive. Phalsa sherbet is a refreshing drink in hot summer months and has a cooling effect.

10.1 Introduction

The origin of phalsa is India. It has a chromosomal status $2n=18$. Phalsa belongs to the family Tiliaceae. The genus *Grewia* has 140 species out of which 40 occur in India. The wild species *Grewia elastica* grows on the lower hills all over India; other important species are *Grewia glabra, Grewia microcos, Grewia tiliifolia*, and *Grewia villosa.*

It is often a cross-pollinated crop, and insects are the chief pollinating agents.

The plant genetic resource status of phalsa indicated that there are no distinct cultivars available in the country.

Phalsa is a small bush, which bears many small berrylike fruits. Its popularity is restricted by the fact that the fruit is very perishable and the tiny fruits have to be picked from a bush several times during fruiting season.

10.2 Reproductive Biology

In the current year's growth, the axillary flowers are borne with 2–9 peduncles of 2.5 cm length each. Each peduncle has 3–6 pedicellate flowers. There were about 18–47 peduncles on each branch. Flowering started in the third week of November and continued till the last week of February. Randhawa and Dass (1962) reported flowering in phalsa during March–May. Anthesis started at 6 am and continued up to 3 pm with a peak at 10 am. The dehiscence commenced from 7 am and continued till 1 pm with a peak at 10

S. Hiwale, *Sustainable Horticulture in Semiarid Dry Lands*,
DOI 10.1007/978-81-322-2244-6_10, © Springer India 2015

am. Stigma receptivity was maximum on the day of anthesis (Narayan Swamy et al. 1986).

10.3 Varietal Development

Types such as local and sharbati are popular. Two distinct types, tall and dwarf, have been recognized, of which dwarf types were found more productive. Two types on the basis of fruit size (bigger and smaller) have been identified and evaluated at Godhra.

10.4 Soil

It can grow under a wide variety of soil and is considered to be one of the hardiest fruits that is drought resistant. Well-drained loamy soils are preferred. In stagnant water the plant becomes chlorotic. They are moderately sensitive to salinity (up to 6 mmhos/cm), and in calcareous soils Fe chlorosis is a common problem. Rich loam is considered to be an ideal soil for its profitable cultivation.

10.5 Crop Production

10.5.1 Propagation

Phalsa is commonly raised by seed. The seeds are collected from fresh ripe fruits at harvest. They are sown in well-prepared nursery beds at 5–7 cm apart in rows normally kept at 25–30 cm apart. The best time for sowing is July–August. The seedlings thus raised are ready for transplanting in the months of February–March or July–August.

Only a few attempts on asexual method of propagation have been made with success. Shield and ring budding in the month of June have met with 90 and 82 % success, respectively. Significant success was achieved in propagating phalsa by hardwood cuttings during the rainy season. Treating the cuttings with 100 ppm indole butyric acid has been found beneficial in rooting. The cuttings are dipped in indole butyric acid (1,000 ppm) solution prepared in 50 % alcohol for about 5 s and then planted in nursery beds (Singh 1974).

The phalsa is planted at a distance of 3–4 m apart by square system of planting. It should be planted in well-mannered pits measuring $0.6 \times 0.6 \times 0.6$ m. February is considered to be the best month for planting when the seedlings can be lifted without the ball of earth. Seedlings can be transplanted in July–August with a little more care. They should be lifted with the ball of earth.

10.5.2 Nutrition and Nutritional Status

Being a hardy crop, fertilization is hardly practiced. But since the bearing is on new growth, it would definitely respond to fertilization. Nijjar (1969) considered 1 kg N/plant to be sufficient for a good crop. However, Chundawat and Gupta (1974) believed that application of 15 kg FYM after pruning followed by 125 g N/plant after sprouting is optimum for high production. Pundir and Pathak (1981) recorded high yield in phalsa by the application of N, P, and K at 100, 40, and 25 kg/ha, respectively. This plant is very sensitive to Fe^- deficiency; therefore, spray of 0.4 % $FeSO_4$ would benefit the plants. Beneficial effects of $ZnSO_4$ (0.5 %) and $FeSO_4$ (0.4 %) have also been recorded by Singh et al. (1981). These sprays can be applied at prebloom and post-bloom stages.

10.5.3 Manuring

Manuring in phalsa is not commonly practiced as it is grown as a wasteland crop. However, if regular manuring schedule is adopted, phalsa can give better returns. About 10–15 kg farmyard manure is considered as an adequate dose for each plant. Manuring should be done in late January, immediately after the bushes are pruned. Manure should be properly mixed in soil. One irrigation immediately after manuring is beneficial.

10.5.4 Irrigation

Phalsa is regarded as a drought-tolerant crop. However, it has been observed that irrigation is necessary for quality and big harvest. Irrigations at 2–3-week interval during summer are desirable.

10.5.5 Soil Management

During the rainy season green manure crop could be taken which can be plowed in at flowering stage; for the whole winter the field can be left untilled, and with pruning it can be deep plowed to clear off weeds and mixed with FYM during summer. The field has to be kept free of weeds to allow good fruiting.

10.5.6 High-Density Plantation

High-density planting in phalsa was done with Neddlers fan design under semiarid conditions. The results indicated that maximum plant population of 8,333 plants per hectare can be accommodated in 3 m×04 m spacing. Though the yield per plant was the least, maximum B.C. ratio of 2.78 was recorded closely followed by 3×0.8 m spacing. The least B.C. ratio (2.28) was recorded in 3 m×1.2 m spacing which may be due to higher cost of cultivation (Hiwale 2007) (Table 10.1).

- *High-density in phalsa; Neddlers fan design*

Treatments

1. 1.3 m×0.4 m=8333 plants/ha
2. 3 m×0.8 m=4166 plants/ha
3. 3 m×1.2 m=2,722 plants/ha
4. 3 m×1.7 m=1,960 plants/ha
5. 3 m×2.1 m=1,587 plants/ha
6. 3 m×2.4 m=1,388 plants/ha

10.5.7 Pruning and Training

Pruning phalsa to different intensities significantly influenced the production of the number of sprouts/shoots as well as yield per plant. Physicochemical characters were not affected significantly. The maximum number of sprouts

Table 10.1 Yield, net return, and B.C. ratio in high-density plantation of phalsa

Spacing	Plants/ha	Yield kg/plant	Yield q/ha	Cost of cultivation (Rs.)	Gross return (Rs.)	Net return (Rs.)	B.C. ratio
3 m×0.4 m	8,333	0.87	72.49	26,000	72,490	46,490	2.78
3 m×0.8 m	4,166	1.37	57.07	21,000	57,070	36,070	2.71
3 m×1.2 m	2,722	1.40	38.88	17,000	38,880	21,880	2.28
3 m×1.7 m	1,960	1.42	27.83	12,000	27,830	15,830	2.31
3 m×2.1 m	1,587	1.48	23.48	9,000	23,480	14,480	2.60
3 m×2.4 m	1,388	1.53	21.24	8,000	21,240	13,240	2.65

Phlasa planting fan design

Phalsa in bearing

was produced in plants pruned to 50 % of the previous season's growth. Fruit set per shoot as well as yield per plant was also highest in plants pruned to 50 % intensity (1.416 kg/plant). In respect to physicochemical characters, fruit weight and fruit length were significantly influenced by different pruning treatments. The maximum fruit weight and length were recorded in ground-level pruning which may be due to less no fruit set and more leaf area. TSS and acidity were not significantly influenced (Tables 10.2 and 10.3).

Studies carried out on the effect of time of pruning under semiarid conditions of central Gujarat with four treatments, viz., pruning plants in the first week of December and subsequent treatments at 15 days' interval, indicated that plants pruned in the first week of December produced the maximum number of sprouts/plants resulting in higher fruit set (63.01 %) and yield per plant (1.53 kg/plant). Analysis of physicochemical characters revealed that fruit weight was significantly influenced by various treatments. Data on other characters was not significantly influenced by various pruning treatments (Hiwale and Raturi 1990).

10.6 Use of Plant Growth Regulators

Spray of GA 60 ppm plus urea 1 % during flowering reduced the duration of harvest, while GA 60 ppm plus 2,4,5-t ppm was effective increasing the fruit size and yield (Chundawat and Singh 1980). Ethephon 500 ppm at color break stage considerably reduced the number of pickings (Singh et al. 1981) (Table 10.4).

Work carried out at CHES, Vejalpur, under semiarid condition on the effect of growth regulators on growth yield and quality of phalsa indicated that the plants that were sprayed with GA at 75 ppm concentration produced maximum shoot length (53.5 cm) and maximum diameter in GA 100 ppm (5.45 mm). Maximum percent fruit set and yield per plant were recorded in NAA 50 ppm spray (58.4 % and 1.41 kg/pl, respectively). 2,4-D at both 2.5 and 5 ppm restricted the shoot growth resulting in the least fruit set and yield (44.7 % and 0.87 kg/pl), respectively. Thus, application of NAA at the rate of 50 ppm spray immediately after sprouting increased yield per plant as well as fruit weight.

Table 10.2 Effect of pruning intensity on growth, yield, and quality of phalsa

Treatment	No. of sprout/ plant	% set	Yield kg/plant	Fruit weight (g)	Fruit length (mm)	Fruit dia. (mm)	TSS ° Brix	Acidity (%)
25 %	62.00	62.72	1.069	0.47	8.29	9.33	25.10	2.68
50 %	66.75	65.89	1.416	0. 56	9.99	9.31	26.40	2.84
75 %	49.25	60.71	1.25	0.51	9.57	9.06	27.55	2.56
Ground level	29.50	60.15	0.64	0.69	10.11	10.35	28.3	2.67
Control	21.50	59.43	0.40	0.50	8.10	9.24	24.85	2.76
CD at 5%	27.33	1.77	0.29	0.049	0.128	NS	NS	NS

Table 10.3 Effect of pruning time on growth, yield, and quality of phalsa

Treatment	No. of sprout/ plant	% set	Yield kg/plant	Fruit weight (g)	Fruit length (mm)	Fruit dia. (mm)	TSS ° Brix	Acidity (%)
First week of December	103.2	63.01	1.53	0.61	9.06	9.60	25.16	2.24
Middle of December	90.8	61.18	1.44	0.56	8.40	9.04	25.00	2.41
First week of January	53.0	60.77	1.03	0.51	8.34	8.4	22.,64	2.49
Middle of January	77.4	59.9	1.15	0.47	8.03	8.76	23.68	2.54
CD at 5 %	18.84	NS	0.163	0.085	NS	NS	NS	NS

Table 10.4 Effect of plant growth regulators on growth, yield, and quality of phalsa

Treatment	Shoot length (cm)	Shoot dia. (mm)	% set	Yield kg/plant	Fruit weight (g)	Fruit length (mm)	Fruit dia. (mm)	TSS ° Brix	Acidity (%)
GA 50 ppm	49.2	5.02	55.1	1.17	0.48	8.3	9.6	27.0	4.23
GA 75 ppm	55.5	5.17	50.3	1.23	0.42	7.8	8.4	26.1	4.41
GA 100 ppm	51.72	5.45	47.6	1.05	0.49	8.4	9.2	27.1	3.86
NAA 50 ppm	47.12	4.80	58.4	1.41	0.64	8.8	9.9	27.1	3.57
NAA 75 ppm	45.47	4.75	47.2	1.23	0.60	9.0	9.7	26.8	3.64
NAA 100 ppm	48.9	5.0	47.9	1.12	0.65	8.9	9.9	26.9	3.98
2, 4-D 2.5 ppm	37.6	3.95	50.8	1.03	0.62	9.0	9.9	26.7	3.50
2,4-D 5 ppm	38.5	3.97	48.6	0.98	0.58	8.7	8.8	25.2	4.14
Control	40.5	4.20	44.7	0.87	0.41	8.0	7.9	26.5	3.37
CD at 5 %	4.08	0.615	5.45	0.67	0.105	0.778	0.692	NS	NS

References

Chundawat BS, Gupta OP (1974) Effect of growth retardants on vegetative growth and yield of phalsa (*Grewia asiatica*). Haryana J Hortic Sci 3(3 & 4):113–115

Chundawat BS, Singh R (1980) Effect of plant growth regulators on fruit size and yield in phalsa. Indian J Hort 37(2):124–131

Hiwale SS (2007) High density planting in arid and semi-arid fruit crops. In book entitled "Advances in Arid zone fruit culture", Patil RS (ed), pp 61–67

Hiwale SS, Raturi GB (1990) Effect of time of pruning on growth and yield of Phalsa (*Grewia subinaequalis* L.). In: National symposium on fruit crops, IARI, New Delhi during November 1990

Narayan Swamy P, Mokashi AN, Mohan Kumar GN, Nalwadi UG, Dever KV (1986) Floral biology of phalsa (*Grewia asiatica* L.). South Indian Hortic 34(2):115–116

Nijjar GS (1969) Phalsa in Punjab. J Res, P.A.U. Ludhiana 6:848–858

Pundir JPS, Pathak SP (1981) Abstract national symposium on tropical and subtropical fruits, Bangalore, 21–24 January 1981. Horticulture Society of India, New Delhi

Randhawa GS, Dass HC (1962) Studies on floral biology of phalsa (*Grewia asiatica* Linn.). Indian J Hortic 19:10–14

Singh JP (1974) The Phalsa. Indian Hortic 19(3):15–16

Singh R, Bagde TR, Sharma VP (1981) Studies on the effect of some plant growth regulators on the yield and duration of harvesting in phalsa (*Grewia subinaequalis* L.) National symposium on tropical and sub tropical fruit crops, Bangalore. Horticultural Society of India, pp 116

Fig (*Ficus carica*) 11

Abstract
The common fig (*Ficus carica*) is a species of flowering plant of the genus *Ficus*, from the family Moraceae. It is a native of Middle East and Western Asia. It is grown since 4000 B.C. It is a gynodioecious, deciduous tree or large shrub, growing to a height of 7–10 m. The fig growing on the seashore of southern India (*Ficus roxburghii*) is of inferior quality. Fig is richer in iron and copper content than most of the other fruits. Though the fresh ripe fig fruit is very delicious and highly perishable, the great bulk of the world production is dehydrated. Products like fig syrup, jam jelly, spiced or pickled fig spread, and fig meat are also prepared from it.

11.1 Introduction

Fig is grown since the ancient times in the world. It is a native of Middle East and Western Asia. It is grown since 4000 B.C. Turkey, Spain, Italy Greece, Portugal, Algeria, and Turkistan are the countries which are growing fig on a commercial basis. Apart from this, it is grown in countries like Australia, California, New Zealand, and India. Turkey is the leading country producer with 26 % of world production. The area under fig cultivation in the world is around 6.5 lakh ha. In India the area under fig cultivation is 3,000 ha. Out of which, in Maharashtra, the area is around 1,800 ha, and in Karnataka, it is 500 ha.

Fig (*Ficus carica*) belongs to the family Moraceae. The importance of this fruit as food can hardly be overemphasized. The fruits of *Ficus glomerata* are dried and can be made into powder which can be used in various preparations. The fig growing on the seashore of southern India (*Ficus roxburghii*) is of inferior quality. The figs are of four types.

11.2 Major Growing Areas

Fig is cultivated in the dry and arid region of the state. In these areas, the minimum and maximum temperatures vary from 8 to 45 °C, during winter and summer months, respectively. The rainfall in these areas is about 500–550 mm per annum.

Different Fig Types

11.2.1 Capri Fig

The male and female flowers are present on the same plant. It is wild type. The plants bear fruits in all the three seasons. Pollination by

S. Hiwale, *Sustainable Horticulture in Semiarid Dry Lands*,
DOI 10.1007/978-81-322-2244-6_11, © Springer India 2015

Blastophaga wasp is needed for fruit development.

11.2.2 Smyrna

The fruit is a cluster of flowers. At the end of the flower cluster, there is a hole. The flower cluster has both male and female flowers. The fruits drop down if not pollinated; therefore, 3–5 % plants of Capri fig are interplanted or pollinated. Artificial pollination results in seed set.

11.2.3 Adriatic (Common)

There is no need for artificial pollination in this type. The fruit develops parthenocarpically. The seeds are small and are soft. The majority of the varieties under cultivation are from this group.

11.2.4 White San Pedro

The fruits of this type are of Adriatic and Smyrna type, and hence, artificial pollination is needed. They are called as Brewa also. The plant bears fruit in March–April season.

The total mineral content is two to four times higher than most of the other fruits. Fig is richer in iron and copper content than most of the other fruits. Both fresh and dry figs contain appreciable quantities of vitamins A and C and smaller quantity of B and D (Wealth of India 1984). Figs have 162 μg of carotene, 60 μg thiamine, 50 μg of riboflavin, and 600 μg of niacin per 100 g of fruit. The nutritive index of fig is reported to be as high as 11 as compared to 9 of apple, 8 of raisin, and 6 of dates and pears (Condit 1947).

The chemical composition of fresh Poona fig reported by Pawar et al. (1992) is as follows: the fruit contains 78.00 % moisture, 14.77 % reducing sugar, 2.42 % nonreducing sugar, 17.195 % total sugar 0.24 %, crude fat, and 0.88 % ash. Fig has high sugar, calcium, iron, and copper content (Mortenson and Bullard 1968). Fig is also used as a poultice (Parmar and Raushal 1982). The chemical composition of fresh fruits of exotic varieties of fig, viz., Conardia, Dianna, and Excel, has been reported by Gawade and Wasker (2002).

Though the fresh ripe fig fruit is very delicious and highly perishable, the great bulk of the world production is dehydrated. Products like fig syrup, jam jelly, spiced or pickled fig spread, and fig meat are also prepared from it (Woodroof 1975).

Fig is a minor fruit crop. Its commercial cultivation in the state is of recent origin; a few progressive farmers in the northern and northeastern districts introduced this crop on their fields during the early 1990s, and since then, its area and production in the state are steadily increasing; at present, the area under fig in Maharashtra is 1,800 ha. Karnataka has 531 ha with a total annual production of about 5,000 tons.

11.3 Chemical Composition of Fresh Figs

Cheema et al. (1954) reported that figs have a very high food value owing to their carbohydrate and protein content. The average composition of the edible portion of fresh fig indicated 79 % moisture, 1.5 % proteins, and 15.51 % total sugar. The analysis of Poona fig showed that they contain 75 % moisture, 15.20 % reducing sugar, and 2.1 % nonreducing sugar.

Cheema et al. (1954) also reported the analysis of figs with 89.8 % moisture, 1.3 % protein, 0.2 % fat, 0.6 % mineral matter, 17.1 % carbohydrate, 0.06 % calcium, 0.03 % phosphorus, 12 mg iron, 75 calorific value, and 270 IU carotene per 100 g. Norman Franklin (1969) reported that a pound of fig contains 357 cal, 6.4 g protein, 1.89 g fat, 89 g carbohydrates, 245 mg calcium, 145 mg phosphorous, 2.7 mg iron, 360 IU vitamin A, 0.25 mg thiamine, 0.23 mg riboflavin, 2.5 mg niacin, and 7 mg ascorbic acid. He also reported that figs have a definite laxative effect due to the fiber and bulk of the seed together with some specific solvent in juice. They also have an excess alkalinity of ash. Woodroof (1975) reported the composition of raw green fig as 98 % edible portion, 84.6 % water, 2.5 % unavailable carbohydrate, 9.5 % sugar as monosaccharide, and 0.21

% nitrogen. He also reported that fig contains 1.3 % protein and 41 cal per 100 g.

Gopalan et al. (1980) have reported the chemical analysis of cv. Poona fig as edible portion 99 %, protein 1.3 %, fat 0.2 %, minerals 0.6 %, fiber 2.2 %, carbohydrates 7.6 %, calcium 80 mg, phosphorus 30 mg, iron 1 mg, riboflavin 0.05 mg, niacin 0.6 mg, vitamin C 5 mg, and energy 37 kcal per 100 g of edible portion.

Parmar and Kaushal (1982) also reported the chemical composition of the fruit as 42.5 % extractable juice, 80.5 % moisture, 12.1 % TSS, 0.71 % acidity, 5.98 % total sugar, 0.2 % pectin, 3.35 % mg vitamin C, 1.72 % protein 0.924 % ash, 0.034 % phosphorus, 0.295 % potassium, 0.071 % calcium, 0.076 % magnesium, and 0.004 % iron.

Rajput and Singhani (1987) reported the chemical composition of fig fruit per 100 g edible portion as protein 1.3, fat 0.2, carbohydrate 7.6 %, and fiber 2.2 %. Pawar et al. (1992) reported that the pH of Daulatabad fig was less than Poona fig resulting in Daulatabad fig than that of Poona fig which was 0.239 % and 0.144 %, respectively. TSS in Daulatabad fig was 10° Brix, and in Poona fig, it was 21° Brix. Mali (1997) reported the chemical composition of fresh Poona fig fruit as 78 % moisture, 17.19 % total sugar, 14.77 % reducing sugar, 2.06 % nonreducing sugar, 1.46 % protein, 0.88 % ash, 14 % TSS, and 2.2 mg/100 g ascorbic acid.

Postharvest factors affect the chemical composition of fig; Baskaya and Crane (1950) compared the PCPA- and IBA-treated Calimyrna fig with parthenocarpic Kadota and caprified Kadota figs which contained high sugar percent than the caprified one; Creane and Grosi (1960) tried gibberellin concentration of 20, 40, and 80 ppm applied as spray to Mission fig and found that sweetness tended to decrease with these concentrations. In Black Ischia and Brown turkey varieties of fig application of gibberelic acid at 50 ppm concentration showed 6.72 % non reducing sugar which decreased with the increase in concentration of gibberelic acid in both varieties; Ben Yehohua et al. (1970) observed that after application of ethylene or ethrel, the ripe fruit had double TSS and dry weight than the untreated fruits of similar age and obtained normal size, color, texture, and flavor one month before harvest. Crane et al. (1970) found that ethrel 500 ppm applied to the Mission fig stimulated color but produced fruit of poor quality. Singh and Chaurasia (1991) reported that spraying 750 ppm ethephon on mature unripe fruits brought about ripening over five times more fruits in 12 days than those of control. The treatments also caused an increase in TSS and helped in overcoming the problem of pre-ripe fruit drop. Puech et al. (1971) reported that ethephon application to Mission fig at or after the critical stage of development caused ripening within a week and this was associated with a rapid reduction in the epidermal levels of chlorophyll a and b, b-carotene, leutin, neoxanthin, and violaxanthin with an increase in the rate of anthocyanin and accumulation.

Gaikwad (1975) observed that ancymidol at 250 ppm produced fruits having almost equivalent TSS, reducing sugar, and nonreducing sugar to that of control. Phad (1978) found that the CCC treatment at 500, 750, and 1,000 ppm increased the quality of fruit in terms of reducing sugar and nonreducing sugar (Table 11.1).

Opinions differ as to the effect of caprification on flavor and composition of fresh figs. Analysis of three varieties by Lecleark du sablon (1908) showed that caprified figs contain on an average 14.5 % sugar and uncaprified figs 18.7 % sugar. The acid content of fresh figs was lower than that of any other common fruit.

The percentage of acid in fresh Dottato figs was found to range from 0.10 to 0.44. Diseased Adriatic figs contained more than ten times as much free acid as normal fig but less in citric acid besides a small quantity of malic acid. Pectic acid gradually decreased according to variety as follows: Mission, 8.0 to 1.73 %; Adriatic, 7.8 to 2.4 %; and Dottato, 5.9 to 1.73 %. Fruits of the variety Adriatic, Mission, Lob Injir, and Dottato at three stages, immature fruit of first crop fruit, firm ripe stage as well as soft ripe stage of the first and second crop. Reducing sugar increased slowly during the early stages of fruit development and rapidly during the later stages. Sucrose was present in relatively small quantity. Lob Injir figs con-

Table 11.1 Composition of fresh fig

Analyst	Year	Water (%)	Protein (%)	Ash fiber (%)	Crude (%)	Reducing sugar (%)	Acid (%)
G.E. Colby	1894	70–84.6	0.72–2.58	0.36–1.16	–	8.0–20.9	0.06–0.24
A. Girard	1898	59.5	–	0.69	0.41	13.0	1.54
Atwater and Bryant	1906	79.1	1.5	0.80	–	18.8	–
Alice Thomson	1915	–	1.4	0.49	1.16	10.8	0.14
H.W. Wilkey	1917	79.1	1.5	0.58	–	15.5	–
G. Riviere and G. Picherd	1926	78.5	–	0.75	–	15.8	0.34
Twining	1927	71.7	–	0.64	1.55	21.3	0.31
Traub and Fraps	1928	74.8	1.3	0.60	0.90	19.4	–
C. Chatfield and L. McLaughlin	1928	78.4	1.3	0.60	0.90	19.4	–
Ouida B. Abbott	1931	79.0	1.5	0.60	–	15.5	–
J.S. McAlester	1931	79.1	1.5	0.60	–	18.8	–
A.L. Stahi	1935	84.8	0.5	0.51	–	11.2	0.25

tained low percentage of crude fiber 7.42 % when green and 3.95 % when ripe. In comparison, Adriatic fig showed 15.30 % when green and 3.80 % when ripe. All the varieties decreased during the period of fruit development.

As the fig matures and dries, the fallowing changes take place: moisture content decreases to about 80–16 %, sugar content increases from 16 to 60 %, and the pulp becomes more or less coherent, syrup mass enveloping the seeds.

Fig is one of the perishable fruits. The post-harvest life is limited. The importance of this fruit as a food can hardly be overemphasized. Both fresh and dry figs contain appreciable quantities of vitamins and minerals.

11.4 Chemistry and Food Value of Fig

11.4.1 Proximate Composition

A mature fresh fig consists of 84 % pulp and 16 % skin.

11.4.2 Analysis of Fresh Fig

Analyses of fresh figs as reported by various authorities are given below: the water content in fig ranges from 59 to 84 % and sugar content from 12 to 19 %. The total sugar content varies from 6.0 to 13.50 % in the Caucasus varieties of fig and from 8.97 to 15.50 % in Len Koran and Kirovabad varieties. The sugar content of fresh Dottato (Kadota) figs was found to vary from 19 to 24 %.

11.4.3 Enzymes

Various studies on the sap of fig tree have shown that it contains an enzyme capable of dissolving proteins. The enzyme is known as ficin or cradein and is pectic in action. A comparative study of the latex of the fig starch-splitting properties of fig latex is as strong as the latex of mulberry, but its power to coagulate is 100 times as that of mulberry. Subcutaneous injections of fig latex into the pigeon result in fever, local congestions, lesions of necrotic characteristic, convulsions, and, finally, death in a state of coma. The peptolytic enzyme from fig is identical with that of papaya.

On the island of Majorca, the peasants prepare curdled milk by heating the milk and then stirring it with split fig branches; this causes rapid coagulation. The coagulating power of the various parts of fig tree, such as buds, branches, fruit, and foliage at five periods of vegetation from 2 March to

29 July on the last date; when mature leaves were used, the most rapid coagulation was at a temperature of 94 °C.

Mission figs gave a very strong positive test for both catalase and peroxidase; the presence of catalase was indicated by evolution of O_2 upon the addition of H_2O_2. The temperature required for inactivation of peroxidase activity varied with pH. The amylase and other enzymes of the fig have been treated in detail. In 1936, Kunio Okahara extracted from the leaves of fig a chemical substance which is named "ficusin." This substance was precipitated as colorless needlelike crystals having the empirical formula $C_{11}H_6O_3$. It has a bitter taste and a faint odor.

11.4.4 Laxative Properties

Both fresh and dried figs have long been recorded as laxative foods. Figs are said to be mucilaginous, sweet, and slightly alkaline and cause copious bowel movement. The dried figs have higher alkalinity than other alkaline foods. Their total mineral content is two to four times as great as that of most of the fresh foods. Only cheese and nuts have higher calcium content. Dried figs are rich in iron and in copper surpassing in this aspect all the fresh fruits and vegetables and even most of the dried fruits.

11.4.5 Vitamins

Vitamins of fig have received the attention of various investigators. In 1931, Hahn reported that dried figs have less than 16 units of vitamin C as compared with 200 units in orange juice and 100 units in rose hips. Brown Turkey fig grown in Hawaii is a fair source of vitamins A and B and poor source of vitamin C. Figs as good source of thiamine (vitamin B1) and fair source of riboflavin.

Fresh fig contains vitamin C, but it is lost when the fig is dried, though it may be sulfured. Figs normally undergo partial drying on the trees before harvesting, and vitamin C destruction happens during that period. Sulfuring favors the preservation of vitamin A in all but Adriatic variety and destruction of B1 in all the varieties. Vitamin retention was greater when the fruit was dehydrated than when sundried.

11.4.6 Figs for Stock Food

Dried figs especially unsuitable for human consumption have long been used for stock food in ancient times. Figs have been used for fattening hogs. The value of fresh and dried fruits for stock food and the figs will increase weight more rapidly than by grains. The high food value of dried figs is due to the high sugar and protein; even after the figs have been fermented for alcohol, the residue makes a good stock food. Fig leaves have also been used as fodder for cattle. The leaves are harvested just before their natural drop. The fallen leaves serve as sheep pasture.

11.5 Varieties

There are many varieties under cultivation; about 1,000 varieties are listed but many have common characteristics. The varieties suitable to warm climate are Celeste and Brown Turkey followed by Burns week and Marseilles.

11.5.1 Celeste

The fruits are small to medium in size with purplish brown skin. The pulp is white in color or pink with good quality and rich flavor; the fruit is pear shaped.

11.5.2 Brown Turkey

Fruits are medium to large pyriform in shape, without neck, copper colored, with few seeds. The cultivar is well adapted to warm climate. It is mainly grown in Hawaii islands.

The fruits are light red inside with very good quality.

11.5.3 Marseilles

The variety can be grown up to the height of 1,524 m from mean sea level. Plant height is 1.2 m and volume is 2.90 m^3; fruits are round or obovate without neck of medium size. Fruits are yellowish green with streaks of green, pulp white in color, sweet, with large seeds.

11.5.4 Black Ischia

The fruits of this variety are dark purple in color, and hence, it is named as Black Ischia. The variety is cultivated in southern India and in Uttar Pradesh particularly in Lucknow and Saharanpur area. The variety is drought tolerant. It is an Italian variety; fruits are elongated and pear shaped with noticeable ribs. The fruit weighs around 30.5 g. The pulp is red and of good quality; the fruit is small. It is mostly grown as an ornamental plant.

11.6 Varieties Under Cultivation

The varieties of fig predominantly cultivated in the state of Maharashtra are Poona fig, Dinkar, Dianna, and Conardia and in Karnataka, Bellar y fig, Poona fig, and Ganjam fig. Late new varieties are also being introduced, and the noteworthy ones are Dianna and Dinkar.

11.6.1 Dianna

It is best for preparing juice. The variety is having bigger-size fruits compared to other varieties. TSS (%) is 22.8–25.0, acidity (%) 0.11–0.16, skin percent 12.0, pulp percent 82.0, fruit weight (g) 60–75, and calories 75, color golden yellow.

11.6.2 Dinkar

The variety is released by Marathwada Krishi Vidyapeeth, Parbhani. This variety is mostly cultivated in Aurangabad area of Maharashtra. Fruit weight is 40–50 g, TSS 18.0–24.0 %, acidity 0.21–0.26 %, skin (%) 15.00, pulp (%) 75.0, fruit weight (g), calories 40–48, and color brown. Five-year-old plant yields up to 30 kg of fruits per plant.

11.6.3 Poona Fig

The variety belongs to the Adriatic group. The plant height is 1.8 m and volume 3.40 m^3. Fruit weight is 30–40 g, TSS 16.0–22.0 %, acidity 0.18–0.28 %, skin (%) 16.00, pulp (%) 78.0, fruit weight (g) 35–50, calories 70, and fruit color brown. Five-year-old plant yields up to 25–30 kg of fruits per plant. The skin is very thin, and hence, special care needs to be taken while packing. Fruits are bell shaped, medium sized, thin skinned, light purple with red flesh, and sweet with good flavor.

11.6.4 Conardia

The variety is imported from the USA. It is a high yielder and hence popular among the farmers. Average fruit weight is 38.5 g. The fruits are green in color and the pulp is pink in color. The fruits dry early. TSS is 25.5–28.8 %, acidity 0.06–0.12 %, skin (%) 10.00, pulp (%) 85.0, fruit weight (g) 50–60, and calories 78, color light green.

11.6.5 Excel

The variety is imported from America. The average fruit weight is 34 g. The skin color is yellow and the pulp is pink in color. The TSS is around 21° Brix. There is no cracking in the fruits after drying.

11.7 American Fig Varieties

11.7.1 Diredo

The fruits are light green in color, and the pulp is brown in color. Fruit weight varies from 50 to 60 g. This variety is suitable for drying. The fruits crack in adverse atmosphere.

11.7.2 Faldders

The fruits of this variety are very juicy and then become black upon drying. The skin is light yellow and the pedicel is purple in color, and pulp is brown in color; fruit weight varies from 45 to 54 g.

11.7.3 Black Mission

This variety is cultivated in California on a large scale. This variety is old but is highly productive. As per name, the fruits are black in color. The fruits are sweet and tasty and are suitable for drying and preparation of jam. The fruits of the first season are bigger in size compared to the second season.

11.7.4 Yovonne

The fruits are egg shaped. Average fruit weight is 45–55 g. The skin is light yellow and the pulp is light brown in color. It is an early variety.

11.7.5 Salebi

It is also an early variety and heavy yielder. The fruits are round in shape. The skin is light yellow and the pulp is pink in color. Average fruit weight is 48–60 g. The fruits are of export quality.

11.7.6 Tena

The fruits are round in shape. The skin is green and the pulp is light brown in color. Average fruit weight is 45–60 g.

11.7.7 Renia

The skin is green, which is a special characteristic of this variety, and is very attractive. The fruits are round in shape. Average fruit weight is 45–60 g.

11.7.8 Nardinne

The skin is light yellow and the pulp is light brown in color. Average fruit weight is 42–54 g.

11.7.9 Vienna

The fruits are round in shape. The skin is light yellow, and the pulp is light brown in color. Average fruit weight is 45–60 g.

11.7.10 Gulban

The skin is light yellow. Average fruit weight is 65–78 g. It is the highest producer.

11.7.11 Everem

It bears attractive fruits, round in shape. The skin is light yellow. Average fruit weight is 35–55 g.

11.8 Varietal Trial on Fig

Growth observations recorded indicated that maximum plant height and shoot diameter were recorded in cv. Dianna (1.98 m and 37.6 mm, respectively). In cv. Dianna, maximum stems per plant were recorded (24). Data on fruit set indicated that there were significant differences in respect to the number of fruits set per plant; the maximum number of fruits per shoot was recorded in cv. Dianna (253/plant). Maximum fruit weight was recorded in cv. Dianna (65 g/fruit). Fruit diameter was also maximum in Dianna (48 mm); however, fruit length was maximum in Dinkar. TSS was maximum in Dianna (22° Brix). Acidity was found to be maximum in Black Ischia (0.53 %) followed by Dinkar (0.49 %).

Data on the number of leaves on the annual extension growth and leaf area per leaf revealed that the number of leaves in Dianna ranged from 23 to 45 per shoot compared to Dinkar (27–30 no./shoot) followed by Poona fig (range 20–22 no./shoot). However, leaf area per leaf was highest in Poona fig (552 cm^2) and was least in Dianna (246.5 cm^2).

Growth observations recorded indicated that the maximum annual shoot length and diameter were highest in cv. Dianna (2.24 m and 41.3 mm, respectively). Stem diameter was least in Black Ischia. TSS was highest (21.8° Brix) in Dianna and least in Black Ischia (15.5° Brix). The number of fruit set/shoot was highest in Dinkar (16.00 no.) followed by Dianna. Fruit yield per plant was highest in Dianna (11.4 kg/plant) and least in Black Ischia (5.2 kg/plant). The variety Dianna was found to perform better in terms of growth yield and quality. The fruit size is bigger; however, the color of the fruit remains green, but its keeping quality is the best among all the varieties. The plants were pruned in the middle of October. Light pruning was practiced, removing about 1.75 kg of wood on fresh weight basis.

Varietal Trial on Fig (Tables 11.2, 11.3, and 11.4)

11.9 Climate

The fig can be grown to a height of 1,525 m above sea level. In tropical climate, it can be grown at a height of 800–11,200 m. In southern India, the variety Marseilles is found growing on the hills and mountains. The tree can tolerate frost and a temperature range of 10–20 °C. However, it favors dry climate with light early spring rains. Rains during fruit development and ripening are detrimental to the crop, causing fruits to split. The semiarid tropical and subtropical regions of the world are ideal for fig cultivation; however, a source of irrigation is a must. Fig requires hot climate; however, if the temperature goes beyond 39 °C, the quality of the fruit deteriorates. Also the fruit ripens early, and if the humidity is more than, the fruit cracks. The temperature between 15 and 36 °C is ideal for the cultivation of fig. Fig requires hot climate; however, if the temperature goes beyond 39 °C, the quality of the fruit deteriorates. Also the fruit ripens early, and if the humidity is more than, the fruits crack. The temperature between 15 and 36 °C is ideal for the cultivation of fig.

Table 11.2 Mean growth parameters of fig cultivars

	Plant height (m)		Shoot dia. (mm)		Plant spread (m)			
					NS		EW	
Cultivar	2012	2013	2012	2013	2012	2013	2012	2013
Poona fig	1.84	2.06	31.32	34.75	1.38	1.67	1.38	1.52
Dinkar	1.88	2.24	30.47	36.81	1.26	1.62	1.26	1.58
Dianna	1.98	2.17	37.6	41.30	1.72	1.71	1.72	1.68
Excel	1.75	1.96	27.52	32.56	1.55	1.55	1.55	1.48
Conardia	1.67	2.04	28.44	31.62	1.43	1.51	1.43	1.53
Black Ischia	1.43	1.40	15.72	16.50	1.41	1.31	1.41	1.37
CD at 5 %	NS	0.68	3.87	4.32	NS	0.45	NS	0.49

Table 11.3 Mean physicochemical characteristics of fig cultivars

	Fruit wt. (g)		Fruit length (mm)		Fruit dia. (mm)		TSS (° Brix)		Acidity (%)	
Cultivar	2012	2013	2012	2013	2012	2013	2012	2013	2012	2013
Poona fig	32.0	31.7	34.0	33.7	39.2	37.9	18.5	17.4	0.33	0.46
Dinkar	27.0	29.5	39.3	37.6	33.5	34.5	16.4	15.2	0.49	0.56
Dianna	65.0	61.0	36.4	37.2	48.2	47.6	22.2	21.8	0.27	0.32
Excel	29.0	27.7	34.5	35.1	39.4	38.3	17.4	16.5	0.40	0.45
Conardia	31.0	28.5	35.3	34.5	41.3	40.7	18.2	15.8	0.35	0.51
Black Ischia	25.0	15.5	34.1	33.7	33.6	30.2	14.5	15.0	0.53	0.44

Table 11.4 Number of shoot per plant, fruit set, and yield per plant

	Fruit set/plant		No. of shoots/plant		Yield (kg)/plant	
Cultivar	2012		2012	2013	2012	2013
Poona fig	231	264	12.7	23.3	7.39	9.29
Dinkar	156	171	16.0	15.7	4.21	6.48
Dianna	253	287	13.2	24.5	10.12	12.7
Excel	158	183	10.5	12.7	4.47	5.25
Conardia	149	165	11.5	17.4	4.32	5.57
Black Ischia	99	76	7.3	18.2	4.89	3.03

11.10 Soil

The soil for fig cultivation should be well drained. The fig can be grown on a wide range of soils, but the productivity is maximum in clay and heavy soils. The pH of the soil should be in the range of 7–8. Highly acidic soils are not suitable. The fig is fairly tolerant to moderate salinity. The soil in which the fig crop is grown is predominantly red loams, and occasionally, they are grown in black soil also.

Poona fig Plant

Bearing in Cv. Dianna

Fruits of Dianna

Fruits of Cv. Poona fig.

11.11 Plant Propagation

The air layers are the most common planting material used to start the new plantation. The grafts on the wild fig and on *Ficus glomerata*, its wild relative have not fared well in any of the state. Hardwood cuttings from 2- to 3-year-old plant with a thickness of 1.25–2 cm and 8–12 cm long are used. The cuttings can be treated with IBA 2 at 500 ppm to enhance rooting. The cuttings should be planted within 24 h in partial shade. After 70–75 days, they should be transplanted to polythene bags filled with a mixture of soil, sand, and FYM in the ratio of 1:1:1 and after 4–6 months in the field. There is need to identify suitable rootstock, and then the superior varieties can be grafted/budded on them to prevent from the attack of stem borer. *Ficus palmata* can be used as rootstock. Apart from this, the plants of superior type can be multiplied by tissue culture technique also.

11.11.1 Planting

The plants are planted in the pits of size 1 × 1 × 1 m which were filled with 15 kg FYM and 20–40 kg of sand in clay soils before onset of monsoon. The recommended spacing is 5 × 5 m; however, with a view to accommodate more number of plants, 4 × 4 m spacing is being adopted by the farmers. Nowadays, farmers are using grafting/budding technique and adopting a spacing of 2.5 m × 2.5 m.

Plant spacing varies from country to country, for example, in Palestine, fig growth is about 8–10 ft; however, in Africa, it is 40 ft, and in Smyrna-type varieties, it is 60 ft. So the climatic condition and soil type play an important role in the kind of spacing to be adapted.

11.11.2 Training and Pruning

After 1 year of growth, the single-stem plants should be headed back to 40–60 cm from the ground level, and multiple sprouts should be allowed to have more number of shoots so that the plants remain in manageable condition. If the plants are not trained in this way, they grow very tall and become unmanageable over the years.

Fig bears fruits on the current season's growth and hence needs to be pruned every year. The tree after 3 years of age needs to be pruned every year. In fig, two crops can be raised in a year. However, it is advisable to raise only one crop to maintain the sustainability of the trees and reduce the intensity of pest and disease incidence on the crop. Fruits maturing in rainy season, i.e., July–August, are called "Khatta bahar" owing to inferior fruit quality. The second bahar the fruits mature in the months of March to May and are also called as "sweet bahar." Quality is best as the fruits mature in mild hot climate. To take the crop out of this bahar, it is essential that the plants are not watered 1 month before the initiation of bahar. After leaf shedding, the basin and field should be cleaned. The pruning time in semiarid condition is from 15 September to 15 October every year. The dried and diseased branches should also be removed. The main framework with 4–5 main shoots should be cut above 1.5 m in height, and the remaining branches should be pruned by allowing two eyes on the shoot. Paste should be immediately applied to the cut portion to prevent shoot as well as fungal infection. The paste should be prepared by dissolving 4 kg of geru in 10 L of water; allow it to stand overnight; the next day, add linden 20–25 ml or chlorpyrifos 30–50 ml and 50 g copper oxychloride (Kulkarni and Masalkar 2006).

11.11.3 Fertilizer Requirement

Depending upon the variety, soil type, and climate of the area, the fertilizer should be applied. For good-quality fruits, 20 % of nitrogen may be applied through ammonium sulfate and 30 % of K through sulfate of potash.

Micronutrient deficiency can be corrected by application of zinc sulfate at 30 kg/Ha, magnesium sulfate 50 kg/Ha, and boron 12 kg/Ha. It is advisable to test the soils before fertilizer application. The soil should be irrigated after application of fertilizers.

Age of plant (years)	FYM (kg/plant)	Neem cake (kg/plant)	N (g/plant)	P (g/plant)	Kg/plant
1–2	15.0	0.5	75	50	50
3–5	25.0	1.0	150	100	100
6 and above	40.0	2.0	300	200	200

Walunj (2006)

11.11.4 Irrigation

Most of the fig plantations are irrigated through drip in the state. This method is not only convenient but also helps to keep the humidity at a lower level in the microclimate of the orchard and reduces weed growth. Application of fertilizers through drip for the figs is also picking up. If water is applied through drip, the water use efficiency can be increased to 90–100 %. This results in saving of water up to 40–50 % as well as increase in productivity by 20–30 %. Tentative water requirement of the trees has been worked out depending upon evapotranspiration in a particular locality. The requirement will vary according to the climatic conditions prevalent in the area of cultivation, and the water requirement may be worked out accordingly.

11.11.5 Fruit Harvesting

The fruits are ready for harvest after the 15th of April. Big-size fruits of ¾ maturity should be harvested. After grading, the fruits should be packed in corrugated boxes after putting some packing material at the bottom of the box. The fruits should be handled very carefully as they are highly perishable. The fruits can be stored in cold chamber at 0 °C temperature and 85–90 relative humidity for 2 weeks.

11.12 Production Problems

1. Occurrence of leaf rust disease, which spoils the prospects of good crop.
2. Extended rains or cyclonic rains spoil the crop, in certain areas, occasionally.
3. Extensive bird damage to the fruit at the time of ripening.
4. High level of chemical (pesticide) residue in the fruit, putting the consumer at high health risk.
5. Organic packages of pesticides have not been standardized, thus limiting its organic cultivation.

Month	Evaporation (mm)	Crop ratio	Water requirement (mm)	Rate of application (mm)	Time required (hours)
June	4	0.2	0.8	1.28	1.0
July	4	0.2	0.8	1.28	1.0
August	4	0.2	0.8	1.28	1.0
September	5	0.3	1.5	1.28	1.17
October	5	0.3	1.5	1.28	1.17
November	5	0.3	1.5	1.28	1.17
December	5	0.3	1.5	1.28	1.17
January	6	0.3	1.8	1.28	1.40
February	6	0.3	1.8	1.28	1.40
March	6	0.3	1.8	1.28	1,40
April	7	0.3	2.1	1.28	1.64
May	8	0.3	2.4	1.28	1.87

11.12.1 Remedies Suggested

1. Introduction of new varieties which are resistant to leaf rust and are known to produce higher-quality fruits suitable for fresh market, dehydration, and processing, especially for the organic production of figs, is needed to produce figs that are safe for consumption.
2. Mass multiplication of quality planting material for supply to the growers at reasonable price.
3. Restricting the cropping season to dry periods of the year, so that the growers can produce quality fruits with minimum risks.
4. Research work should be undertaken for standardization of dehydration process and also to produce value-added product.

11.12.2 Changes in Chemical Composition of Fruit During Storage

Thonte and Patil (1988) reported that the total sugar content of dried fig decreased from 34 to 27.17 % during storage of one year. A significant variation in the reduction of sugar content of fig fruit has been reported during storage. The reducing sugar content in fig increased with increase in storage period. Thonte and Patil (1988) reported that the nonreducing sugar content in dried fig decreased from 3.14 to 1.50 % during storage.

11.12.3 Analysis of Dried Fig

Analysis of dried fruit harvested at full maturity should show less variation in constituents. Differences in varieties and climatic conditions account for the variability in dried figs as shown in the table below.

Composition of dry fig

Analyst	Year	Water (%)	Protein (%)	Ash (%)	Crude (%)	Reducing sugar (%)	Acid (%)
G.E. Colby	1894	23.0	4.28	2.02	10.14	60.05	0.42
Atwater and Woods	1896	22.5	5.10	2.40	–	70.02	–
A. Plethico	1907	20.7	–	–	–	48.20	
G. Gungielmt	1908	20.4	5.18	2.25	6.83	56.51	–
H.W. Wilkey	1917	28.7	3.58	2.75	6.19	51.43	0.71
Twining Lob Injir	1922	15.7	3.39	2.10	5.80	62.84	0.42
Adriatic		16.8	3.28	2.22	6.95	64.34	0.43
Mission		16.3	3.30	1.94	4.50	64.29	0.43
J.S. McAlester	1931	18.8	4.3	2.4	–	74.20	–
N. Mauri	1942	17.5	3.50	2.50	–	60.0	–

Dried Dottato figs contain reducing sugar 35.2 % when caprified and 28.4 when uncaprified. Analysis of Dottato figs made by Twining Laboratory, Fresno, showed a smaller percentage of sugar in uncaprified fig. confirmed this by stating that caprified fig contains higher percentage of reducing sugar, ash, and protein but a significantly lower percentage of crude fiber.

11.12.4 Effect of Drying on the Chemical Composition of Fruits

11.12.4.1 Moisture

Moisture plays an important role in the growth of microorganism on the product. Pawar et al. reported that dried Poona fig contained 15.3–15.97 % moisture. The moisture content in dried figs was reported to vary from 16.8 to 24 % (Norman and Derosier 1982; Gebhardt et al. 1982). Cheema et al. (1954) reported that a sample of dried Poona fig contained 19.25 % moisture. Salem and Nour (1979) reported that dried fig contained 16.75 % moisture. Mali (1997) reported that the moisture content in dried Poona fig varied from 15 to 24 %.

11.12.5 Total Sugar

Increase in total sugar has been reported in dried fig fruits. Pawar et al. (1992) reported that dried Poona fig contained 49.1–52.02 % total sugar. Gebhardt et al. (1982) reported 52.90 % total sugar in dried fig fruit. Thonte and Patil (1988) reported that sugar-treated dried fig fruits contained 41.50–59.00 % sugars. Salem and Nour (1979) found that dried fig contained 73.32 % total sugars. Mali (1997) found a gradual increase of the total sugar content in dried Poona fig, i.e., 47.00–63.21 %, after 90 days of storage. El-Razik et al. (1997) observed the continuous decrease in total sugar content of dried cv. Kadota during storage.

11.12.6 Reducing Sugar

Pawar et al. (1992) reported that dried Poona fig contained 40.52–42.58 % reducing sugar. Thonte and Patil (1988) reported that ethereal-treated and dried figs contained 32.50 % reducing sugar, whereas control fruits had 33.71 %. The sugar-treated and dried fig fruits contained 40.32–57.00 % reducing sugar. Gebhardt et al. (1982) reported

51.3 % reducing sugar content in the dried fig. (Cheema et al. 1954) reported that a sample of dried Poona fig contained 45.95 % reducing sugar. Salem and Nour (1979) found that dried fig contained 65.50 % reducing sugar. Mali (1997) found a gradual increase in reducing sugar content in dried Poona fig, i.e., 46.01–61.79 %, after 90 days of storage.

11.12.7 Nonreducing Sugar

Pawar et al. (1992) reported that the nonreducing sugar content in dried Poona fig was 6.64–9.82 %. Gebhardt et al. (1982) reported 1.60 % nonreducing sugar content in dried fig. Mali (1997) found a gradual decrease in nonreducing sugar content in dried Poona fig which was 9.95–0.91 after 90 days of storage.

11.12.8 Protein

Protein content in dried figs also increased during drying and dehydration. Pawar et al. (1992) reported that dried Poona fig contained 4.13–4.72 % protein. Bolin and King (1980) reported 2.83–3.0 % protein in dried figs. Mali (1997) reported that dried Poona fig contained 360–4.73 % protein.

11.12.9 Ascorbic Acid

It was reported that significant amount of ascorbic acid is lost during processing. Gebhardt et al. (1982) reported that 1 mg/100 g ascorbic acid was present in dried figs. Gebhardt et al. (1982) reported 2.6–3.6 mg ascorbic acid/100 g fruit. Mali (1997) found that 0.21–2.15 mg/100 g ascorbic acid was present in dried Poona fig. He also reported a decrease in ascorbic acid content after 90 days of storage.

11.12.10 Ash

An increase in ash content has been found in dried figs. Norman and Derosier (1992) reported that ash content in dried fig was 2.4 %. Pawar et al. (1992) reported 2.9–3.13 % ash content in dried fig. Mali (1997) found 2.0 to 3.63 % ash in dried Poona fig. He also reported a gradual increase in ash content.

11.12.11 Total Soluble Solid

Thonte and Patil (1988) reported that TSS contents in oven-dried fig fruits treated with dry sugar and sugar solution were 46 and 44 %, respectively. Mali (1997) reported an increase in TSS content of dried figs during storage.

11.12.12 Organoleptic Properties

Thonte and Patil (1988) reported that organoleptic evaluation of fig with different treatments indicated that the fruits in dry sugar subjected to oven-drying followed by pricked and non-pricked fruits dipped in sugar solution and dried in oven gave good quality of dried fig fruits which were comparable to excellent grade dry fruit from the market. Gouda et al. (1975) revealed that var. Brown Turkey, Black Mission, and Sultani figs were of substandard grade; Britogta blance was of standard; and Britogta Niro and Kadota were of medium standard. And Candearis and Adriatic figs were of fancy grade.

11.12.13 Pest and Diseases

About 50 species of insect pest have been recorded feeding on fig tree (Butani 1979); the major pests include stem-boring beetle, chafer beetles, leaf-eating caterpillars, scales, thrips, bark-eating caterpillars, and fruit flies; minor pests are mites, jassids, psyllids, midges, leaf rollers, mealybugs, and fruit borers.

Fig stem borer (*Batocera rufomaculata* DeGeer): It is the most destructive pest of fig. It is highly polyphagous and fig is the preferred host. The damage is done by the grubs which feed on portion of the bark and xylem.

Management: Spray chlorpyrifos 20EC at 0.1 % in the months of June–July. While regular inspection of plants for pest infestation and destruction of infected portion by cutting and application of kerosin oil and dichlorvos in the tunnels through injection should be practiced.

Jassids (*Velu caricae* Ghauri): It is a sucking pest; it is very active during growth of new leaves. The leaf margin turns yellow, curls, and becomes reddish brown causing typical burn symptoms. The shoots wither and fruits drop down.

Management: Spray dimethoate, monocrotophos, oxydemeton-methyl, or quinalphos each at 0.02 % concentration.

Cockchafer beetle (*Adoratus* spp.): The insect is polyphagous in nature. Eggs are laid in the soil and the grubs feed on the root. The beetles are nocturnal in habit and devour the foliage in the night hours.

Management: Spraying with carbaryl 0.2 % or dimethoate, monocrotophos, oxydemeton-methyl, and/or quinalphos at 0.03 % can control the pest.

Use of parasite fungi like *Eugenia brongniartii* and *Metarhizium anisopliae* has been recommended.

Scales and mealybugs: Several species of scales have been reported infesting fig (Butani 1975); the nymphs and adult females suck the sap from the leaves, shoots, and fruits. In cases of severe infestation, the affected parts are dried and plant growth is affected severely.

Management: Removal and destruction of heavily infested portions.

Spray chlorpyrifos at 0.05%.

Use of fungal pathogen *Verticillium lecanii* at the rate of 2 g/L of water can be tried.

Thrips (*Frankliniella* spp.): Thrips cause damage to the fruits through scarring resulting in quality deterioration.

Management: Spray Rogor at 0.2 % or cypermethrin at 0.005 %.

11.12.14 Diseases

Leaf rust – It is a very common disease in fig. The causal organism is *Cerotelium fici* bringing premature leaf fall, thereby reducing yield. The disease is most prevalent in rainy season.

Fig mosaic is caused by a virus and is incurable. The affected plants should be destroyed.

11.12.15 Postharvest Marketing Problems

1. High degree of perishability of fruits leading to high rate of spoilage after harvest.
2. Low sugar content in the fruit limits the scope of dehydration.
3. In the market, lack of proper facilities (like cold room) forces the growers to hurriedly sell the produce, often leading to improper sell realization.
4. Poor product profile of this fruit limited the scope for processing into different products.
5. Low quality of fruits and the presence of high levels of pesticide residue have pulled down the export prospectus of fig grown in the state.

Fig Jam

2 pints ripe Black Mission figs
or dried Calimyrna figs if fresh figs are not available
1 orange, peeled, seeded, and finely chopped
1/2 cup packed light brown sugar
1 tbsp. finely grated lemon zest
1/4 tsp. kosher salt
1 tsp. vanilla extract

Shortbread Dough

1 cup (2 sticks) unsalted butter, at room temperature
6 tbsp. granulated sugar
2 tbsp. confectioner's sugar
1 egg yolk
1 tsp. vanilla extract
1 cup unbleached, all-purpose flour
1 cup cake flour
1/2 tsp. baking powder
1/2 tsp. kosher salt

To make the filling: Remove any stems on figs, then cut them into quarters, and place in a medium nonreactive saucepan. Add the orange, brown sugar, lemon zest, and salt. Place over medium heat and bring to a simmer. Reduce the heat to medium-low and simmer uncovered, stirring occasionally for about 40 min, or until the figs have softened and lost their shape and the filling is jam-like. If you are using dried figs, add a little water as they simmer to keep the mixture from drying out. Remove from the heat, stir in the vanilla, and let it cool for 1–2 h or to room temperature (if the jam is too watery, drain a little excess liquid before using as filling). If you used dried figs, place the finished cooked figs in a food processor and pulse 5 or 6 times until the figs achieve a jam-like consistency. This can be made in advance up to three days ahead.

To make the shortbread dough: Using a stand mixer fitted with a paddle attachment, cream together butter, granulated sugar, and confectioners' sugar on medium speed for about 5 min or until the mixture is light and fluffy. Stop the mixer a few times and use a rubber spatula to scrape down the sides and bottom of the bowl and the paddle to release any clinging butter or sugar. Beat in the egg yolk and vanilla on medium speed for 2–3 min or until thoroughly combined. Scrape the bowl and paddle to make sure the egg is thoroughly incorporated.

In a small bowl, sift together the all-purpose flour, cake flour, baking powder, and salt. On low speed, slowly add the flour mixture to the butter-sugar mixture and then mix for about 15 s, or until the flour mixture is totally incorporated and the dough is evenly mixed. Stop the mixer and scrape the bowl again to make sure all of the flour is thoroughly incorporated.

Scrape the dough onto a sheet of plastic wrap and wrap entirely, pressing down to form a disk about 6 in. in diameter and 1 in. thick. Refrigerate the dough for about 30 min or until it has firmed up but is still somewhat pliable. Position a rack in the center of the oven and heat to 350 F. Position the rectangle with the long side facing you. Spoon the cooled filling lengthwise along the center of the rectangle, in a strip of about 2 1/2 in. wide. Lifting the edge of the parchment farthest from you, drape the top of the dough rectangle over the jam, covering the top half of it. Gently peel the parchment away from the dough. Repeat with the bottom edge of the parchment, draping the bottom of the dough over the jam. The edges of the dough rectangle should meet in the middle. Gently pinch the edges of the dough together, and then turn the rectangle over, so it is facing seam-side down. Using a pastry brush or your hands, brush any excess flour off the parchment. Cook for 65–70 min or until the shortbread is entirely golden brown. A little fig juice may leak out on the sides. Cool on the baking sheet for at least 2 h or until completely cool. Using a knife, cut on the diagonal into strips about 5 in. long and 1 in. wide.

References

Basakya M, Crane JC (1950) Comparative histology of naturally parthenocarpic, hormone induced parthenocarpy and caprified synconia. Bot Gaz 111:395

Butani DK (1975) Insects pests of fruit crops and their control. Pesticides 9:32–36

Butani DK (1979) Insects and fruits. Periodical Expert Book Agency, Delhi, pp 173–186

Cheema GS, Bhat SS, Naik KC (1954) Commercial fruits of India with special reference to Western India. Macmillan Publishing Co., Bombay, p 380

Childers NF (1969) Modern fruit science. Hort Pub Rutgers University, New Jersey, p 489

Condit IJ (1947) The fig. Chronica Botanica Co., Waltham, p 32

Crane JC, Marei N, Nelson MM (1970) Ethrel speeds growth and maturity of fig. Calif Agric 24(3):8

El-Razik FAAA, Farghai SA, El-Kayati SM (1997) Utilization of solar energy in dying kadota fig

Gaikwad TS (1975) Effect of plant growth regulators on hastening maturity and quality

Gawade MH, Waskar DP (2002) Dehydration of exotic varieties of fig and storage of dried fig under various environments. Accepted for publication in J. FD. Sci. and Tech

Gebhardt SE, Cutrufali R, Mathews RH (1982) Composition of food: fruits and fruit juices, raw processed and prepared. Hand Book no. 89. US Department of Agriculture, Washington, DC

Gopalan CR, Sastri BV, Balsubramanium C (1980) Nutritive value of Indian foods. National Institute of Nutrition, ICMR, Hyderabad

Gouda MS, Zoueil MB, El-Zalaki EM, Mohamed M (1975) Technological studies of fig varieties planted in Arab republic of Egypt. Alex J Agric Rec 25(3):459

Kulkarni S, Masalkar S (2006) Fig season and crop management. In: National seminar on fig production and fig processing industry in India present status and future strategies, Pune, Souvenir, pp 21–22

Mali BB (1997) Effect of preparation and storage on quality of dried figs. M.Sc. (Agri) thesis. Mahatma Phule Krishi Vidyapeeth, Rahuri

Mortenson E, Bullard ET (1968) Handbook of tropical and subtropical horticulture. USDA, Agency for International Development, Washington

Norman W, Derosier JN (1982) The technology of food preservation, 4th edn. AVI Publishing Company, West Port

Parmar C, Raushal MK (1982) Wild fruits of the sub Himalayan region. Kalyani Publisher, New Delhi/Ludhiana

Pawar SG, Kulkarni DN, Shere DM, Kulkarni KD, Patil VK (1992) Effect of pretreatments on chemical composition and drying rates of solar dried figs. Indian Food Pack 1:39

Phad VS (1978) Effect of various growth regulators on size and quality fig variety Daulatabad. M.Sc. (Agri) thesis, Andhra Pradesh Agricultural University, Hyderabad

Puech AA, Rebeiz CA, Crane JC (1971) The effect of Z-chloroethylphosphonic acid on pigment changes in the mission fig. Plant Physiol 47

Rajput HS, Singhani AR (1987) The heart-cook. Book Orient Paperback Publication, New Delhi

Salem SA, Nour A (1979) Sugars and amino acids in dried Iraqi fig. J Food Agric 30:620

Singh SP, Chaurasia SN (1991) Induction of early ripening in fig (*Ficus carica*) by ethephon. J Hill Res 4(2):110

Thonte GT, Patil VK (1988) Studies on drying of fig fruits. Indian Food Pack 42(4):94

Walunj C (2006) Fig growing and water management. In: National seminar on fig production and fig processing industry in India present status and future strategies, Pune, Souvenir, pp 23–25

Woodroof JG (1975) Storage life of canned frozen dehydrated and preserved fruits. Commercial Fruit Processing. AVI Publications, 595

Yehoshua B, Iwahori S, Lyons JM (1970) Role of ethylene and ethrel in the development of fig fruits. Israel J Agric Res 20:143

Bael (*Aegle marmelos* Correa.)

12

Abstract

Bael (*Aegle marmelos* Correa) is an important indigenous fruit of India. Like many other species of Rutaceae family, bael also has fragrant flowers. They are also highly valued in *Ayurvedic* medicines. All the parts of the tree, whether it is stem, bark, root, leaf, flower, seed oil, or fruits of any stage of maturity and ripening, are used in various *Ayurvedic* medicines. There is lot of variability as the stray plants are of seed origin. There are no regular orchards of bael. It can tolerate salinity to some extent and can be grown successfully once the orchard is established under semiarid rainfed conditions. In situ softwood budding/grafting can be used to establish an orchard. The gestation period is reduced by adopting vegetative propagation. Varieties NB-5 and NB-9 are performing better in semiarid areas, and a yield of 40–50 kg can be obtained per plant.

It is the most suitable crop for value addition. The pulp is diluted with water and add the requisite amount of sugar to tamarind juice to prepare a delicious cooling drink. The bael pulp is used as a base of various fruit products like squash, jam, slab, toffee, powder, nectar, RTS, etc. Green bael fruits are used for preparing *Murabba* (preserve).

12.1 Introduction

India is the second largest producer of fruit (63.50 million tons) obtained from 5.78 million ha area accounting for 10.00 % of the world production of fruits. Bael (*Aegle marmelos* Correa) is an important indigenous fruit of India. It is grown in various parts of South Asian countries including India, Sri Lanka, Pakistan, and Bangladesh and most of the Southeast Asian countries including Myanmar and Thailand. In India, bael is being grown throughout the country and is also known by other vernacular names like bael, bili, bilva, belo, sriphal, and Bengal quince (Johns and Stevenson 1979). The bael tree has found mention in mythological treatises. It is grown near the temple of the Lord Shiva in India. According to Hindu customs, the leaves of the tree are considered sacred and offered to the Lord Shiva. In history, the mention of bael tree has been traced to Vedic times (C 2,000 B.C.–C 800 B.C.) in the Yajurveda. Om Prakesh (1961) recorded mentions of bael in early Buddhist and Jain literature (C 800 B.C.–C 325 B.C.) describ-

S. Hiwale, *Sustainable Horticulture in Semiarid Dry Lands*,
DOI 10.1007/978-81-322-2244-6_12, © Springer India 2015

ing various methods of ripening of the bael fruit along with some other fruits. In the "Ramayana" period, the bael fruit was known and its trees were reported to be growing in "Chitrakuta" hills and "Panchavati." In the "Upavana Vinod," a Sanskrit treatise on silviculture (Majumdar 1935), and in the "Brihat Samhita," mention had been made of bael fruit (Aiyer 1956); as the legend goes, in the forest, Lord Rama performed religious rites by offering various fruits including bael (Aiyer 1956). The bael fruit has been portrayed in the painting of Ajanta Caves along with other fruits (Om Prakesh 1961). Like many other species of the Rutaceae family, bael also has fragrant flowers. It is believed that this tree acts as an indicator plant for tracing of underground water (Singh and Roy 1984).

12.2 Importance

The fruits are official in the Indian Pharmacopoeia. They are also highly valued in *Ayurvedic* medicines. The peripheral part just within the rind is fleshy and thick, has a pleasant resinous odor. The walls separating the chambers have a light yellow tint which becomes yellowish brown on exposure and have slightly acrid bitter taste. The chambers are full of amber- or honey-colored viscous very sticky or glutinous, translucent pulp, which is slightly sweet and feebly aromatic. The gummy substance surrounding the seeds serves as a good adhesive and is added to water paints to improve strength and brilliancy. It is more abundant in young fruits. The gum has been used for the stabilization of drilling fluids. The stem also contains a gum similar to gum arabic.

The importance of bael fruit lies in its curative properties, which make the tree one of the most useful medicinal plants of India (Kirtikar and Basu 1935). Its medicinal properties have been dealt with in the "Charaka Samhita," an early medical treatise in Sanskrit (Aiyer 1956). All the parts of the tree, whether it is stem, bark, root, leaf, flower, seed oil, or fruits of any stage of maturity and ripening, are used in various *Ayurvedic* medicines.

The demand of bael fruit is not much higher as other major fruits like mango, banana, guava, apple, etc. The ripe fruit is eaten as fresh and has a more demand among those who wants it for therapeutic use. Sometimes the pulps are diluted with water, and the requisite amount of sugar and tamarind was added to prepare a delicious cooling drink. The bael pulp is used as a base of various fruit products like squash, jam, slab, toffee, powder, nectar, RTS, etc. Green bael fruits are used for preparing *Murabba* (preserve), which is generally taken for stomach ailments. The green bael fruit slices often are dried and stored for future use. The unripe or half ripe fruit is regarded as astringent, digestive stomachic, and good for heart and brain (Kirtikar and Basu 1935). The fruit is used in chronic diarrhea and dysentery and to act as tonic for the heart and brain. It is a useful adjuvant, as it helps to remove constipation which hinders the healing ulcerated surfaces of the intestine. Besides the fruits, the root is an ingredient of *dasamula* (ten roots) used as Ayurvedic medicine. The roots as well as bark are used in the form of a decoction as a remedy in melancholia, intermittent fever, and palpitation of the heart. The root has anti-amoebic and hypoglycemic properties. The young leaves and shoots are used as fodder for cattle, sheep, and goats. The leaves are bitter and used as febrifuge; poultice made of leaves is used for ophthalmia and ulcers. Fresh leaves are also used as remedy for dropsy and beriberi associated with weakness of the heart. The astringent rind of ripe fruits and bark are employed in dyeing and tanning. The timber is commonly used for making pestles of oil and sugar mills and for posts, shafts, axles, and naves of cart. The shells of smaller fruits are used as snuff boxes. A yellow dye is extracted from the rind of unripe fruit. The diluted leaf juice is used for catarrh. The alkaloid aegelin present in the leaves is efficacious in asthma. Utilization of bael fruit has been reported by Roy and Singh (1979b).

12.3 Chemical Composition

Various chemical constituents, viz., alkaloids, coumarins, steroid, etc., have been isolated and identified from the different parts of the plant. The bael fruit is one of the most nutritious fruits. Analysis of the bael fruit gave the following values: 61.5 g moisture, 1.8 g protein, 0.39 g fat, 31.8 g carbohydrates, 1.7 g minerals, 55 mg carotene, 0.13 mg thiamine, 1.19 mg riboflavin, 1.1 mg niacin, and 8.0 mg vitamin C per 100 g of edible portion (Gopalan et al. 1961). No other fruit has such a high content of riboflavin. Tannic acid is the only phenolic substance detected from bael fruits. The fruit contains allo-imperatorin, marmelosin identical with imperatorin, and β-sitosterol. Chakraborty et al. (1978) reported that marmolide, an isomer of imperatorin which exhibits tyrosinase-accelerating and tryptophan pyrolase-inhibiting furocoumarin, was isolated from ripe fruits. The fruit yields 2 % water-soluble dried gum. Hydrolysis of the gum gave galactose, 20.4 %; arabinose, 10.7 %; D-GALACTURONIC acid, 25.2 %; and traces of rhamnose (Haskar and Kendurkar 1961). Compounds like auroptin, marmin, umbelliferone, and lupeol have also been found from the bark of bael (Patra et al. 1979). Occurrence of auropotins, umbelliferone, marmin, lupeol, and skimmianine has also been reported in roots (Chatterjee and Choudhury 1960). Besides these compounds, psoralen, xanthotoxin, scopoletin, and tembamide have also been isolated from the roots of bael (Shoeb et al.1973).The presence of aegelin, an alkaloid, has been reported in the leaves of bael fruit (Chatterjee and Roy 1957). Analysis of the leaves gave the following values (dry basis): crude protein, 15.13; ether extract, 1.54; crude fiber, 16.45; N-free extract, 52.83; ash, 14.05; calcium, 5.93; and phosphorus, 0.69 %. Marmelosin is probably the most therapeutically active compound present in bael fruit (0.03–0.37 %) and varies according to variety and locality (Dixit and Dutt 1932). The mature bark contains marmin, auropotins, umbelliferone, lupeol, and skimmianine (Chatterjee and Bhattacharya 1959). The bael fruit mucilage on hydrolysis shows the presence of three reducing sugars, galactose, arabinose, and rhamnose (Parikh et al.1958). The wood contains a furoquinoline alkaloid, dictamnine, marmesin, and neutral compound. The seed yields oil 34.4 % on dry basis and the fatty acid composition of oil as follows: palmitic, 16.6; stearic, 8.8; oleic 30.5; linoleic, 30.0; and linolenic, 8.1 %. Bael seed contains 62 % protein and 3 % each carbohydrate and ash (Banerjee and Maiti 1980).

12.4 Area and Distribution

So far, there is no organized orcharding of bael in our country; hence, exact data on acreage and production is not available. However, in recent years, concerted efforts have been made for the collection of elite genotypes of bael from all over the country and the evaluation and establishment of germplasm block at ICAR Institutes/Regional Stations and State Agricultural Universities. The bael is grown in India and in neighboring countries, namely, Nepal, Sri Lanka, Pakistan, and Bangladesh and most of the Southeast Asian countries, Myanmar and Thailand. In India, it is distributed throughout the country, but concentrated area under bael is in the eastern parts of the Gangetic Plains and nearby areas particularly in Uttar Pradesh, Bihar, West Bengal, and Orissa. Its trees are also available in wild state in sub-Himalayan tract from Rajasthan to West Bengal and central and southern India. In Gujarat, bael trees are found growing naturally in the forest with great diversity (Singh et al. 2008b). Most of the genotypes available in Gujarat are having small-size fruits (Singh et al. 2008a). Apart from systematic orchards, bael trees are also planted in nutritional gardens, parks, temple gardens, and roadsides for various purposes. It was introduced into Europe from India in 1759 (Johns and Stevenson 1979).

12.5 Soil and Climate

The bael tree is very hardy and deciduous and can thrive well in swampy, alkaline, and stony soils having pH range from 5.0 to 10.0 (Jauhari and Singh 1971). Its trees are cold hardy and found to be grown up to an altitude of 1,200 m above mean sea level. It has a wide range of adaptability to adverse soil and climate. Under hot semiarid ecosystem, the extent of hardiness of bael plants has also been observed, and the plants are giving good yield in rainfed conditions. The extent of hardiness of bael plants under *Thar* desert has also been observed that the plant even after being buried under sand for 2–3 months is capable of rejuvenating itself (Anon 2001). Marked reductions in the contents of leaf NPK and Ca were observed in response to increase in salinity and sodicity level in the soil in which plants were grown. Salinity caused significant increase in leaf Mg, while sodicity decreased it. Leaf Na was at toxic levels in both saline and sodic soils (Shukla and Singh 1996b).

Bael is bestowed with a natural characteristic for being tolerant to the extremes of temperature and soil moisture stress by shedding its leaves during summer. However, young plants need to be protected from low temperature (4 °C) and desiccating hot winds. It can also be successfully grown in saline, sodic, and sandy wasteland provided the soil is treated with gypsum and pyrite before plantation.

12.6 Botany and Morphology

Aegle marmelos Corr. belongs to the family Rutaceae. Other members of the family Rutaceae are *Citrus, Casimiroa, Clausena, Eremocitrus, Limonia, Feroniella, Fortunella, Poncirus, Triphasia*, etc. The genetic name *Aegle* is of Greek origin and the species *marmelos* is of Portuguese origin. Its chromosome number is $x=9$ and $2n=36$. The tree is medium to tall, deciduous, and slow growing up to the height of 5–10 m. Its leaves are green aromatic trifoliate; often branches have spines and trunk is strong and stout. The leaves are divided into three leaflets, i.e., a pair and a terminal one; the terminal one is usually the largest. Some leaf abnormalities of *Aegle marmelos* have been noticed. The branches are unusual with long, straight spines. The bark is shallowly furrowed and corky. The bisexual flowers are nearly 2 cm wide and borne in clusters, sweet scented, and greenish white. The calyx is shallow with five short, broad teeth, pubescent outside. There are five petals which are oblong oval, blunt, thick, pale greenish white, and dotted with glands. Stamens are numerous, sometimes coherent in bundles. The ovary is oblong ovoid and slightly tapering, the axis being wide; cells are numerous 8–20, small, and arranged in a circle, with numerous ovules in each cell. The ripe fruits are woody, large, spherical, up to 23 cm in diameter, oblong, or pear shaped, with a more or less smooth or slightly tuberculate surface. The fruit is usually globose with a pericarp nearly smooth, grayish yellow, 1.6–2.7 mm thick, hard, and filled with soft, yellow and orange, very fragrant, and pleasantly flavored pulp. Botanically, the fruit is berry with hard pericarp. The number of cells in the fruit, arranged in a circle, is equal to the number of cells in ovary. Seeds are numerous, compressed, and arranged in closely packed tiers in the cell surrounded by very tenacious, slimy, transparent mucilage, which becomes hard when dry. The testa is white with wooly hairs and the embryo has large cotyledons and a short superior radical. Pollination is usually through honey bees. The nectar secreting disk found beneath the ovary is the main source of attraction for the insects (Reuther et al. 1967; Srivastava and Singh 2000; Singh 1989; Pal and Mishra 2005; Singh and Mishra 2004; Singh et al. 2007).

Flower buds

Flower bud emergence, flowering duration, time of anthesis, dehiscence of anther, stigma receptivity, and pollen viability vary according to variety and locality (Srivastava and Singh 2000).

12.7 Flowering and Fruiting

Generally, flower bud emerges in the month of April and full bloom in the month of May under hot semiarid ecosystem of western India. The seedling tree requires 7–8 years to bear fruits, while budded plants start bearing fruits at the age of 3–4 years after planting, and fruits are ready for harvest after 10–11 months by March–April. However, this varies according to genotypes and agroclimatic conditions.

12.8 Improvement

In popularization of bael, improvement in terms of higher productivity per unit area and quality of fruits in terms of percentage of edible portion, color, shape, taste, and flavor are important. Generally, bael is raised through seeds leading to great variability in morphological characteristics; bearing behavior, size, and shape of fruit; physical composition of fruit, viz., pulp, mucilage, fiber, and peel; and chemical composition in terms of TSS, total sugar, phenolics, and vitamin C, etc. There is a wide genetic diversity in the existing population. This provides ample scope for selection of genotypes for improvement.

In view of the limitations of the conventional breeding techniques, biotechnological approaches for bael improvement hold great potential. Strategies emanating out of advances made in plant biotechnology include embryo rescue, soma clonal variation, haploidy, protoplast fusion, in vitro conservation of germplasm, and recombinant DNA technology (genetic engineering). Genetic maps of species may be prepared with the use of molecular markers.

12.9 Varietal Wealth

Flowers are in clusters, sweet scented, and greenish white. The calyx is shallow with five short, broad teeth, pubescent outside. There are five petals which are oblong oval, blunt, thick, pale greenish white, and dotted with glands. Stamens are numerous, sometimes coherent in bundles. The ovary is oblong ovoid, slightly tapering, the axis being wide; cells are numerous 8–20, small, and arranged in a circle, with numerous ovules in each cell. The ripe fruits are woody, large, spherical, up to 23 cm in diameter, oblong, or pear shaped, with a more or less smooth or slightly tuberculate surface. The fruit is usually globose with a pericarp nearly smooth, grayish yellow, 1.6–2.7 mm thick, hard, and filled with soft, yellow and orange, very fragrant, and pleasantly flavored pulp. Botanically, the fruit is berry with hard pericarp. The number of cells in the fruit, arranged in a circle, is equal to the number of cells in ovary. Seeds are numerous, compressed, and arranged in closely packed tiers in the cell surrounded by very tenacious, slimy, transparent mucilage, which becomes hard when dry. The testa is white with wooly hairs, and the embryo has large cotyledons and a short superior radical. Pollination is through entomophily usually by honeybee. The nectar secreting disk found beneath the ovary is the main source of attraction for the insects (Reuther et al. 1967; Srivastava and Singh 2000; Singh 1989; Pal and Mishra 2005; Singh and Misra 2004; Singh et al. 2007).

Flower bud emergence, flowering duration, time of anthesis, dehiscence of anther, stigma receptivity, and pollen viability vary according to variety and locality (Srivastava and Singh 2000). Size and shape of floral organs in terms of bud size, flower size, petal size, etc. of the varieties evaluated at CHES, Godhra, under rainfed condition of semiarid ecosystem by Singh et al. (2008c).

12.9.1 Narendra Bael-5

The plants are small with a semi-spreading growth habit and precocious and prolific in bearing. The budded plants start fruiting in the 4th year. The average fruit yield of a 6-year-old plant is 28.78 kg. The fruits are medium in size (12.50 cm × 11.50 cm), round with a smooth surface and very thin rind (0.16–0.17 cm), straw yellow at maturity, low in mucilage, and moderately fibrous, having an attractive yellow pulp, with low seed content. Excellent in taste and flavor, the fruits have a total soluble solid of 33° Brix in pulp and 48° Brix in mucilage and ascorbic acid 18.63 mg/100 g of edible portion. The fruit weight ranged from 0.8 to 1.0 kg under rainfed conditions of semiarid ecosystem during the sixth year of orchard life. Its taste is good and can be used as fresh as well as processed into various value-added products.

12.9.2 Narendra Bael-7

Plants are tall and semi-spreading. They are sparse in bearing with large size fruit. The budded plants start fruiting in the 4th year. The average fruit yield of a 6-year-old plant is 32.10 kg (6th year). The fruits are medium in size (18.25 cm × 22.50 cm), round with smooth surface and very thick rind, yellow at maturity, low in mucilage, and fibrous, having an attractive yellow pulp, with low seed content. Fruits are good in taste and flavor, having a total soluble solid of 30° Brix in pulp and 42° Brix in mucilage, slight acidity, and 19.78 mg/100 g ascorbic acid. It is highly suitable for processing.

12.9.3 Narendra Bael-9

The plants are semi-vigorous and semi-spreading, having compact canopy. The variety is precocious and prolific bearer. The average fruit yield of a 6-year-old plant is 56 kg. Fruits are medium to large in size (16.00 cm × 13.50 cm), roundish oblong with smooth surface and thick rind (0.31 cm), light yellow at maturity, average in mucilage, and moderately fibrous, slightly golden yellow pulp with low seed content. The fruits are good in taste, containing a total soluble solid of 38° Brix in pulp and 41° Brix in mucilage, slightly acidic and ascorbic acid 19.20 mg/100 g of edible portion. It can be used as fresh as well as processed into various value-added products. Its keeping quality for storage is very good.

12.9.4 Pant Aparna

Its trees are dwarf with drooping foliage, almost thornless, precocious, and heavy bearers. The leaves are large, dark green, and pear shaped. The fruit has a globose shape with an average size of 13.00 cm × 12.00 cm and weight of 0.8–1.25 kg. The fruit pulp is yellow and rind is thin. TSS is 34° Brix in pulp and 47° Brix in mucilage, titratable acidity 0.67 %, and ascorbic acid 17.15 mg/100 g of pulp. Mucilage, seed, and fiber are low. Mucilage and seeds are enclosed in separate segments. Flavor and taste are very good. Yield during the 6th year is 40.25 kg/plant.

12.9.5 Pant Shivani

It is an early midseason variety. Trees are tall, vigorous, dense, upright growing, precocious, and heavy bearer. The fruit shape is ovoid and oblong and the size is 18.50 cm × 15.00 cm. The fruit weight ranges from 2 to 2.4 kg. The color of the fruit is lemon yellow and its storage quality is good. Rind is medium thin, pulp is lemon yellow with pleasant flavor, and mucilage, seeds, and fiber are low to medium. The taste is very good. It has 69 % pulp, TSS 36° Brix in pulp, mucilage TSS 48° Brix, total titratable acidity 0.47 %, and ascorbic acid 19.55 mg/100 g of flesh.

12.9.6 Pant Sujata

It is an early midseason variety but has problem of fruit splitting. So far, it has not been reported under rainfed conditions of hot semiarid ecosystem of western India. Trees are medium dwarf with drooping and spreading foliage, dense, precocious, and heavy bearers. Thorns are stout and bigger. The fruit is globose shaped and depressed at both ends with average size of 14.50 cm × 13.50 cm, and weight varied from 1.12 to 1.40 kg under rainfed condition of hot semiarid ecosystem of western India. The fruit and pulp are light yellow. The storage life is better, rind is thin, and seeds, mucilage, and fiber are low. Its flavor is pleasant and taste is very good. The flesh is 72 %, TSS 32° Brix in pulp and 42° Brix in mucilage, acidity 0.44 %, and ascorbic acid 17.10 mg/100 g of flesh.

12.9.7 Pant Urvashi

It is a midseason variety. Trees are tall, vigorous, dense, upright growing, precocious, and heavy bearers. The fruit is ovoid oblong with average size of 14.50 cm × 17.20 cm, and the fruit weight ranges from 1.5 to 2.50 kg. The fruit is yellow, rind is medium to thin, and pulp is light yellow. The fruit has 62.35 % pulp with pleasant flavor. Seeds and mucilage are medium, fiber content low, TSS 33° Brix in pulp and 41° Brix in mucilage, titratable acidity 0.49 %, and ascorbic acid 17.15 mg/100 g pulp.

12.9.8 CISHB-1

It is an early maturing variety. The plants are semi-tall and have a spreading growth habit. The budded plants start fruiting in the 4th year. The average fruit yield of a 6-year-old plant is 42.64 kg. The fruits are medium in size (16.50 cm × 12.00 cm), oval oblong with smooth surface, yellow at maturity, low in mucilage, and fibrous, having an attractive yellow pulp, with high seed content. Excellent in taste and flavor, the fruits have a total soluble solid of 32° Brix in pulp and 43° Brix in mucilage. The fruit weight varies from 0.8 to 1.40 kg.

12.9.9 CISHB-2

The plants are dwarf and spreading. The average fruit yield of a 6-year-old plant is 38.45 kg. The fruits are medium in size (16.00 cm × 14.00 cm), round with smooth surface and thick rind, yellow at maturity, low in mucilage, and fibrous, having an attractive yellow pulp, with low seed content. It is good in taste and flavor and the fruits have a total soluble solid of 31° Brix in pulp and 38° Brix in titratable acidity (0.41 %). The fruit weight ranges from 1.7 to 2.6 kg/fruit.

Bael Pant Sujata

Bael Pant Shivani

Bael NB-5

Bael Pant Urvashi

Bael NB-7

CISH Bael-1

12.10 Plant Propagation

With the standardization of vegetative propagation method for bael, the commercial nursery has got a boost. Traditionally, bael was propagated by seeds; however, seed propagation is limited for the raising of rootstocks only. True-to-type planting materials can be produced through vegetative means only. Methods of vegetative propagation have been reported by Singh (1954). The bael fruit can be grafted on a number of related species, such as *Aegle fraegle-gabonensis, A. chevalieri, A. paniculata, and Swinglea glutinosa* (Hays 1957).

12.10.1 Seed

Seed propagation is the most common method of propagating bael. Bael seeds have no dormancy; hence, fresh seeds can be sown 2–3 cm deep in the nursery within 10–15 days. The fresh bael seeds germinate in 8–15 days after sowing. The seedlings become ready for transplanting in spring or the next monsoon. The orchard raised by seedlings is not true to type and exhibits variability. Therefore, vegetative propagation techniques are recommended for commercial orcharding of bael.

12.10.1.1 Raising Rootstock Seedlings

For raising of seedlings in the nursery, beds are thoroughly cleaned and plowed. For better germination, higher survival, and establishment, well-rotten FYM should be mixed with the soil before sowing of seedlings. Fresh seeds can directly be sown because of no dormancy. Young seedlings should be protected from frost during winter under arid ecosystem. Polyethylene tubes can also be used to raise the seedlings. FYM, sand, and soil (1:1:1) should be mixed before filling in the polythene tubes. Seeds are sown in the polythene tubes or bags during rainy season for better germination and survival. Delayed and poor seed germination and reduced plant growth were observed in response to increased sodicity (Shukla and Singh 1996a). Performance of bael with respect to seed germination and plant growth was observed satisfactorily in sodic soils up to 29.0 ESP without application of any chemical amendments. The media requirement for seed germination and seedling establishment has also been suggested by Chattopadhyay and Mahanta (1989).

The foliar sprays of plant growth regulators, i.e., gibberellic acid (GA_3) and IBA (both at 250, 500, 750, and 1,000 ppm) and potassium nitrate (250, 500, 750, and 1,000 ppm), improve seedling vigor by means of improved growth of stem and roots.

12.10.1.2 Raising Rootstocks in Polyethylene Bags

Deshi rootstock seeds are collected in April–May are extracted from fruit was dried and sown in polythene bags. The seeds germinate after 10–15 days. To reduce the time for raising rootstock and to avoid damage during handling and transportation, polyethylene tubes and polyethylene bags may be used on commercial scale. Generally, polyethylene bags (25 cm × 10 cm) are used for raising the rootstocks. Small holes are made in the bottom and sides of polythene bags for drainage and aeration and filled with porous rooting medium or pot mixture for raising rootstocks. Generally, 1–2 seeds are sown in each polythene bags and then placed in trench bed, so that it can be irrigated easily. Sometimes coiling of root becomes a problem; hence, root pruner is also used for trimming of roots. Growth of the rootstocks can be improved with the application of 1 g/L urea solution. About 8–12-month-old seedling of uniform size having stem of pencil thickness are used as rootstock for budding and grafting. Plants raised in the polythene bags can easily be transported to distant places with higher planting success.

12.10.1.3 Rootstock

Rootstock selection for vegetative propagation of bael is important, as it controls the vigor and equilibrium between yield and quality. Dwarfing rootstocks induce dwarfness and facilitate easy management of the orchard. Generally, seeds of *deshi* plants are used for raising rootstocks in nursery, but it can also be grafted on the *Aegle fraegle-gabonensis, Aeglopsis chevalieri*, and *Aegle paniculatum.* However, it is not commonly used for raising of bael plant.

12.10.2 Vegetative Propagation

12.10.2.1 Patch Budding

Rectangle incision is made on the rootstock by placing the bud on the rootstocks to mark the exact size of the bud on them and after removing the bark of the rootstock and tying with white polythene strip (200 gauge thickness and 2 cm wide). In case the cuts on rootstock are wider, at least one side bark of scion and stock must be matched properly. The rootstock is cut about 10 cm above the bud to facilitate bud sprouting. The time of budding influences the survival of plant in different varieties. Singh et al. (1976) reported 100 % bud take during the month of June or July. Effect of scion genotypes on patch budding in bael has been reported by Mishra and Jaiswal (2001).

12.10.3 In Situ Patch Budding

In arid and semiarid regions, in situ budding is the most successful method for establishing a bael orchard. This is done by sowing 2–3 seeds directly in the field or by planting seedlings. After 1 year, budding is done in the field. In bael, the tap root system is very vigorous. The root system is, therefore, disturbed during the process of planting of grafts, which ultimately affects growth and establishment adversely in the field conditions. The plants propagated by in situ patch budding in the months of June–July recorded maximum success (94.14 % and 90.82 %, respectively). Budding in bael in June–July from a 1-month-old scion gave 80 % success, and patch budding is an ideal method of bael multiplication (Singh 1954; Moti and Chaturvedi 1976; Singh et al. 1976; Kumar et al. 1994; Chadha 2001).

12.10.3.1 Softwood Grafting

Shoots of 3–4 months old are defoliated 10–12 days prior to grafting operation. For this, seedling rootstock is cut at 10–15 cm height. With the help of a knife, a 5 cm long vertical downward incision is made in the center of the rootstock. A sharp cut of 5 cm is made on both sides on the base of the scion shoot to make wedge shape, and the graft is tightly securedusing a 200 gauge thick and 2 cm wide polythene strip. Maiti et al. (1999) reported that bael can be propagated successfully, and among different grafting methods tried, whip grafting had given good response (70 % success).

Soft wood grafted plant

12.10.3.2 In Situ Softwood Grafting

The deshi rootstock is raised at desired spacing directly in the field; the seeds should be sown directly in the field during rainy season under rainfed condition.

In situ softwood grafting through wedge method is done in the months of June–July on a 1-year-old seedling. The growth below the graft union is removed regularly to encourage sprouting and subsequent growth of the scion shoots. The bud sprouts within 15–20 days of grafting. The polythene strips are carefully removed after completion of the union. The plants are given support with the help of stakes to protect them from stormy winds. High temperature and relative humidity during June–July have helped in early sprouting and better graft success, because of fast establishment of vascular connection with rootstock.

12.10.3.3 Root Cuttings

Bael can be propagated successfully by root cutting. Separation and planting of root suckers can be done during monsoon. To ensure establishment, suckers are planted in nursery beds for about 2 years after uprooting and are then shifted to the main field. Some root suckers, which arise from the roots of the bael trees having profuse roots, can be separated during monsoon and directly planted into the field. But, the success percentage is very low. It is not recommended for arid and semiarid conditions.

12.10.3.4 Stem Cutting

Bael can also be propagated through cutting. Ray and Chatterjee (1996) reported that growth regulator and etiolation treatments were significantly effective in inducing roots in ringed stem cuttings of *A. marmelos*. An invigoration treatment (the production of water shoots following removal of large branches) accompanied by growth regulator and etiolation treatment significantly increased the root quality of cuttings. The highest rooting rates of 75–80 % were achieved by using 5,000 ppm IBA etiolation and invigoration treatments. Rooting rates of 45 and 40 % can be obtained with the application of 100 ppm IBA and 100 ppm IAA, respectively. It was observed that tip cuttings of bael rooted well under intermittent mist and treatment with IBA at 5,000 ppm produced 100 % rooting.

12.10.3.5 Layering

Air layering is very successful in bael provided that mother trees are given invigoration treatment by heading back to few of the thick branches during April. Air layers are prepared in the second week of August by bark ringing and application of IBA at 10,000 ppm in lanolin paste.

12.10.3.6 Micropropagation

Recently, micropropagation techniques have also been found successful in bael. True-to-type and disease-free plants can be generated from a very small piece of plant in aseptic condition in artificial growing medium rapidly throughout the year. Regeration can be done from explant nucellus (Hossain et al. 1993) and cotyledons leaf (Islam et al. 1993). Multiplication of shoots can be done by using micro shoots. Arumugam and Rao (1996) reported that cotyledonary node explants excised from 15-day-old seedlings of bael were placed in MS medium supplemented with BAP [benzyl adenine], IBA, IAA, or NAA. BAP induced the best production of multiple shoots and subsequent plant regeneration. Rhizogenesis of shoots was achieved in the presence of IAA (Varghese et al. 1993). Elongated shoots were rooted on half strength MS medium supplemented with 0.1 mg IBA liter^{-1} (Islam et al. 1993). The maximum survival (90 %) was recorded when medium was supplemented with 0.5 mg/l BAP and 0.5–1.0 mg/L kinetin from the plantlet regeneration from axillary bud. Similarly, maximum number of shoots (4.70) was developed in culture flask containing MS medium supplemented with 2.0 mg/L BAP and 1–0 mg/L kinetin (Bhargava et al. 2008).

12.11 Agro-techniques

12.11.1 Orchard Establishment

The land may be prepared by usual plowing, harrowing, and leveling. There should be gentle slope to facilitate proper irrigation and proper drainage to avoid the harmful effects of water stagnation during rainy season particularly in black cotton soil. Well-decomposed organic matter is mixed with soil and pits are filled. Planting is done during rainy season when the soil in pits has already settled. While planting, one should be careful that the earth-ball does not break and graft union remains well above the ground level. Soil all around the stem should be pressed properly to avoid the formation of air pockets. The plants should be irrigated immediately after planting. In the initial 2–3 years, it is advised to protect plants against low and high temperature injury and from hot desiccating winds by covering plants with some short of cover, leaving the one side open. Shelter belt and windbreaks around the orchard protect the tree from hot desiccating wind during summer. For this, 2–3 rows of fast growing drought hardy tree species should be planted in staggered manner. In India, bael is planted at the distance either 8 m×8 m or 10 m×10 m. No systematic work has yet been taken up on nutritional requirement of bael tree.

12.11.2 Planting

The planting of bael seedling is done at 10 m×10 m or 8 m×8 m depending upon the agroclimatic conditions. Under rainfed condition of hot semiarid ecosystem, planting of vegetatively propagated saplings can be done at 5 m×5 m spacing to maximize the productivity. The pits of 1 m×1 m×1 m are dug and exposed for solarization to kill harmful soil organisms. The pits are filled with top soil mixed with 20–25 kg FYM after drenching with chlorpyriphos at 3 ml/L to avoid the attack of termite during the early phase of plant growth. The ideal time of planting under rainfed condition is July. In situ planting is found more suitable for orchard establishment under rainfed condition of semiarid ecosystem.

12.11.3 Canopy Management

Canopy management of the crop deals with the development and maintenance of their structure in relation to the size and shape for the maximum productivity and quality. Tree vigor, light, temperature, and humidity play a vital role in production of quality fruits. The crux of canopy management lies in the fact as how best we manipulate the tree vigor and maximum use of available sunlight and temperature to increase the productivity and minimize the adverse effect of weather. Pruning is done to improve and regulate tree size and shape to achieve the desired architecture of the canopy and also to reduce the foliage density by removing unproductive branches of the tree.

12.11.4 Planting System

No proper research has been done on plant geometry of bael orchard. Generally, bael plantation is being done in square system. Seedling of bael is planted on boundary of orchard as windbreak. Planting of the bael at 6 m × 6 m in square system and at 5 m × 7 m in rectangle system has been recommended. The main objective to follow particularly in planting system is to accommodate the maximum number of trees per unit area without affecting the yield efficiency and fruit quality adversely. Some of the popular systems of planting in vogue are the square, rectangular, quincunx, hexagonal, contour, hedgerow, double hedgerow, paired, and cluster planting. In India, most of the farmers are poor and have less resource like landholding, irrigation facility, etc., and high-density planting is suitable to increase their productivity by accommodating more number of plants per unit area. The layout of square, hedgerow, double hedgerow, cluster, and paired systems is as follows:

12.11.5 Irrigation

It promotes better growth during establishment and the early stages of growth, especially during the summer. In the early age, plants require 8–10 irrigations in a year, while bearing trees require 4–5 irrigations during the time of fruit development and ripening. In dry areas, the use of water harvesting techniques during the rainy season will be useful for ensuring proper irrigation to improve subsequent growth and yield. In dry areas, the use of water harvesting techniques during rainy season and mulches should be adopted. Nevertheless, bael can successfully be grown under rainfed conditions of hot semiarid and arid region.

12.11.6 Mulching

Continuous use of organic mulches is helpful in improving the soil physicochemical properties, microbial flora, and soil aeration which ultimately resulted into better growth and yield of plant. Under rainfed condition, application of organic mulch in tree basin is very beneficial for successful cultivation of bael. It reduces the loss of moisture from the soil, enhances the rate of penetration of rainwater or irrigation in the soil, and controls the growth of weed. Mulching can be done with black polythene or any suitable organic material. Mulching with paddy straw, maize straw, grasses, and rice husk reduces the weed population and conserves the moisture in the soil. Mulches should be applied in the tree basin (20 cm thick) after rainy season, and undecomposed organic mulches should be incorporated and mixed with soil of tree basin in the forthcoming monsoon.

12.11.7 Integrated Nutrient Management

No systematic work has yet been taken up on manuring and fertilization of bael, a minor fruit crop. Generally, bael trees are not manured. However, an annual dose of about 20 kg of FYM during the pre-bearing period and 50–80 kg per tree at bearing stage is considered beneficial. It is suggested to apply 10 kg farmyard manure and 50, 25, and 50 g NPK, in a 1-year-old plant, respectively. This dose should be increased every year in the same proportion up to the age of 10 years. Sometimes in rich soils, the trees have a tendency to put on more vegetative growth with the result that the fruiting is delayed. Singh and

Mishra (2000) have reported seasonal variation in root distribution pattern of bael.

12.11.8 Training

Training helps in avoiding difficulty in intercultural operation. The plants are trained to straight central stem in which branches are not allowed up to 60 cm. Young plants should be allowed with 4–6 well-spaced branches in all direction. Young plants are trained with the help of stake, if needed, so that they can grow erect. Suckers appearing from rootstocks should be removed.

12.11.9 Pruning

Generally pruning is not recommended in bael, but pruning is a tool to regulate tree size and shape to achieve a desired architecture of the canopy and also to reduce the foliage density by removing the unproductive branches to make the tree open. Generally, bael plant is not pruned once the tree starts fruiting, because the branches of such plant are self-oriented, even though, in case of rosette growth, few branches should be removed from its place of origin to have well-spaced scaffolds. Regular pruning in bael plant is not required because fruiting takes place on new shoots as well as old shoots. However, dried, criscross, weak, and diseased branches should be removed as and when required. It will facilitate easy harvesting of the fruits. Pruning of few growing branches becomes necessary particularly when orchard has been established under high-density planting system.

12.11.10 Weed Management

Most weeds although complete their life cycle in a shorter period but compete with plants for light, water, and nutrients and thereby reduce the yield. In the orchard, hoeing, hand weeding, and plowing of the land 2–3 times a year are done to suppress the weed growth. Intercropping and mulching also help in controlling weeds in tree basin.

12.11.11 Intercropping

Intercropping is intended to maximize land and space use efficiency to generate supplemental income particularly during the initial unproductive phase of the orchard to protect the interspaces from losses through weeds, erosion, impact of radiation, temperature, wind, and water, enriching it by nitrogen-fixing legume crops. During the early phase of orchard establishment, interspaces left between tree rows can efficiently be utilized by raising suitable crops which not only enrich the soil but also generate additional income.

12.11.12 Fruit Growth and Development

The growth and development of the bael follow a single sigmoid curve. The growth rate of bael has three distinct phases: the initial slow increase for 1 month followed by rapid increase for 3 months and then more or less a stationary phase until the fruits are harvested. The moisture content of bael fruit decreases during development and ripening. With the decrease in peel moisture, the hardness of starch appears and increases steadily till harvest, but disappears with ripening. The rate of respiration in bael fruit at early stage of development is rapid and it declines with growth. However, an upsurge in respiration is noticed after picking the fruit from the plant. Based on the respiratory studies, the bael fruit can be classified as a climacteric fruit. With the fall in the rate of increase in mucilage, the starch appeared and continued to increase with fruit development but disappeared while ripening. The total and nonreducing sugars of the fruit show a rising trend during development. The total phenolic content of the bael fruit decreased during development and ripening. The inherent low acidity of the fruit shows decreasing trend during development and ripening. However, the fruit ripens normally only in April–May. Studies on biochemical changes in fruit reported an increasing trend in both reducing and nonreducing sugars. Fruit pulp contained very low acid levels which did not vary much with fruit development. The ascorbic acid content increased with fruit maturity. A sharp increase in pectin, tannin, and marmelosin contents in the pulp was recorded until January, and thereafter, a gradual decline was noticed (Roy and Singh 1980; Pandey et al. 1986).

There is a progressive fall in crude protein content during fruit development. However, a slight increase is noticed during ripening. The pectin expressed as calcium pectate increases during development and ripening of bael fruit. It has been observed that the bael fruit matures in December–January under rainfed conditions of western India. However, the fruit starts ripening from March onward because of prevailing high temperature under rainfed conditions of hot arid ecosystem.

12.11.13 Pest and Diseases Management

12.11.13.1 Pests

Generally, bael is free from serious pests and diseases, but the insect pest damages the crop considerably, especially when environmental conditions are very conducive. Termite attack on new saplings is a major problem. Application of chlorpyriphos at 2–3 ml/L/plant has been found effective to control the termite. Chafer beetles or leaf-eating caterpillars cause damage to the plant, and it can be controlled by 2–3 sprays of dimethoate at 1.5–2.0 ml/L at 15 days interval.

12.11.13.2 Diseases

A new leaf spot disease of bael caused by *Fusarium roseum* Link has been reported by Kore and Dhande (1973). Bacterial shot hole and fruit canker of bael are caused by *Xanthomonas bilvae.* The symptoms on the leaves are characterized by round, water-soaked spots (0.5 mm) surrounded by a clear halo. Gradually the spots increase in size (3–5 mm) and form brown lesions with saucer-like depression in the center surrounded by oily raised margin. The primary localized lesions all over the leaf are always followed by falling-out of the necrotic dead tissues, leaving circular or slightly irregular perforation of shot holes. The pathogens also infect the fruit, twigs, and thorns. The disease can effectively be controlled by 2–3 sprays of 500 ppm streptomycin at 15 days interval.

12.11.13.3 Fruit Rot (*Aspergillus nidulans)*

Internal rotting of fruit is a serious problem, which is mainly caused by damage to fruits during harvesting, storage, transportation, or harvesting. To avoid such, maintain proper ventilation during storage and avoid storage especially in polythene enclosures. Such disorders can be avoided by harvesting of fully mature fruits and also to avoid damage to the fruit can be wrapped/stored with newspaper or phenol papers.

12.11.13.4 Physiological Disorders

12.11.13.4.1 Fruit Cracking

Fruit cracking has been observed as a major physiological disorder, and its degree of damage depends according to genotypes/varieties and locality. Fruit cracking takes place twice in a year, i.e., winter season (December–January) while developing fruits are immature and during summer season (March–April) when the fruits are mature and in ripening phase. The cracking at later stage is more severe than the former one. The cracking can be minimized by maintaining optimum soil moisture regime and by provision of windbreaks against hot desiccating wind side of orchard. Organic mulches like paddy straw, maize straw, and *subabul* lopping can effectively be utilized in maintaining soil moisture of tree basin particularly during summer under rainfed conditions of hot semiarid ecosystem (Saini et al. 2004).

12.12 Harvesting

Bael fruits are likely to get damaged if proper care is not taken during harvesting. The tree is in leafless condition during harvesting particularly in late maturing varieties, while early maturing varieties do not shed their leaves at the time of harvesting under rainfed conditions of semiarid ecosystem of western India. Mature bael fruits are harvested individually from the tree along with the portion of fruit stalk (2–3 cm) to avoid infection, and it also helps to judge the ripening. The stalk is easily separated while pressing the fully ripened fruit which is indication of ripening. Proper care is required for harvesting of bael fruits. Harvesting by shaking the trees should be discouraged, as the fruits are likely to develop cracks on impact because the peel of fruit is highly brittle which invites infection and can cause heavy loss during storage.

12.13 Yield and Yield Attributing Characteristics

The number of fruits per tree may vary from 30 to 45 at the age of 6–7 years depending on genotype, soil, and climatic condition. However, a seedling tree at 30–40 years age can yield 500–800 fruits. Physical characteristics of fruit in terms of fruit size, fruit weight, and shell thickness varied in different varieties/genotypes.

12.14 Ripening

Fruit setting in bael takes place in early May, and ripe fruits are available in March–June. However, this may vary in different agroclimatic zones of the country. It takes 18–24 days for the fruits to be artificially ripened. The composition of bael fruit, whether ripened artificially or naturally, does not vary much; the sugar accumulation in natural ones is slightly more than artificially ripened. The specific gravity of the bael fruit is initially high, which falls gradually up to 3 months, and thereafter, it increases and remains more or less constant. No climacteric rise in respiration is noticed as long as the fruit is attached to the plant. However, rapid upsurge in fruit respiration, total sugar, and reducing sugar and decline in moisture, phenolics, and acidity have been observed after harvesting the fruit till complete ripening which coincides with the optimum ripening condition of fruit (Roy and Singh 1981).

12.15 Quality Attributes

Results of the study on the varieties evaluated for their quality attributes revealed that the physicochemical attributes differed significantly among the evaluated varieties. The physical composition in terms of peel, pulp, mucilage, fiber, and seed percent in fruits and chemical composition, i.e., TSS, total sugar, reducing and nonreducing sugar, acidity, phenolics, and vitamin C content, varied in different varieties. Differences in physicochemical characteristics in the bael genotypes have been reported by Teotia et al. (1963), Jauhari et al. (1969), Majumdar (1975), Pandey et al. (1986), Ram and Singh (2003), Singh et al. (2000), and Singh et al. (2008a).

12.16 Grading and Packing

Bael fruits have different shapes and sizes; hence they should be graded accordingly. At the time of harvest, generally but not always, the tree is in leafless condition at the time of harvesting, and the fruits are completely exposed particularly in late maturing cultivars/genotypes. There is no recommended practice for packing bael fruits. At present, the fruits are packed in gunny bags, baskets, or wooden crates, and sometimes, they are transported without any packing. It is highly essential that some cushioning material, namely, straw paper, sawdust, newspaper liner, etc. should be used while packing bael fruits. The fruit should not develop any crack or damage during packing, transportation, marketing, and storage; otherwise, it may cause spoilage by fungal infection.

12.17 Storage

Fruits harvested at full maturity for preserve making can be stored up to 15 days, and fruit harvested at ripe stage can be stored up to 7–9 days at room temperature. Fruit can be stored up to 3 months at about 9 °C and 85–90 % humidity under cool storage. It is sensitive to low temperature injury like other subtropical fruits. At low temperature, spoilage is caused mainly due to chilling injury, i.e., appearance of brown spots on the fruit surface during storage below 8–9 °C, while at high temperature, spoilage is mainly due to fungal attack. During storage, an increase in total sugars and greater accumulation of reducing sugars are observed.

During storage of bael fruit products, there is reduction in nonreducing sugars and increase in reducing and total sugars. Addition of SO_2 not only improves the initial quality of the bael fruit slab, toffee, and powder but also prevents nonenzymatic browning reaction during storage of all the bael fruit products. The optimum relative humidity for the storage of bael fruit slab, toffee, and powder is found to be 63, 58, and 5 %, respectively. Practically no change in organoleptic quality is noticed in frozen pulp after 6 months, and in case of other products stored at 37 °C, the organoleptic quality remained up to acceptable point (Roy and Singh 1979a).

12.18 Processing

The bael fruit is not popular as dessert fruit due to its hard shell, mucilaginous texture, and numerous seed and fiber contents. Bael fruit has been used widely from time immemorial for processing in the mature green form to prepare preserves, but recently methods have also been standardized to process the ripe fruit. Roy and Verma (1950) mentioned the process of manufacturing bael squash and bael jam. In the same year, Agnihotri (1950) published the method for preparing and preserving syrup from ripe fruits. However, an early report of Singh and Dutt (1941) stated that although the fruit is rich in pectin, it cannot form jelly due to the excess of gummy substances. Verma and Ahmed (1958) reported that bael fruit powder could also be manufactured successfully. Fruits can be processed into a number of acceptable products like slab, powder, toffee, squash, jam RTS, etc. which are briefly mentioned below in flow chart.

12.19 Flow Sheet Chart of Various Value-Added Products of Bael

12.19.1 Flow Chart of Preserve

Pulp graded + Washed mature green bael fruits → Break fruits → Scoop out pulp along with seeds and fibers → Discard seeds → Cut pulp in 2.5 cm thick slices → Wash in water → Prick with a fork → Soak overnight in cold water → Blanch → Prepare 40 % sugar solution → Impregnate with sugar by gradually raising the syrup to 78°Brix → Pour into jars and seal

12.19.2 Flow Sheet for Extraction of Bael Fruit Pulp

Ripe bael fruit → Washing → Breaking → Scooping of pulp with seed and fiber (Discarding) → Addition of water equal to the weight of pulp → Addition of citric acid (titratable acidity 0.5 %) → Kneading → Heating at 80 °C for 1 min → Passing through a pulping machine or stainless steel sieve of 20 mesh (discarding seeds and fibers) → Bael fruit pulp →

- → Canning
- → Freezing
- → Preserving by SO_2

12.19.3 Flow Chart for Squash

Pulp → Dissolve citric acid (25 g/kg) in water and add → Dissolve potassium meta bisulfite (2.5 g/kg) in water and add → Filter and add → Prepare syrup by mixing 1.41 water and 1.6 kg sugar per kg of pulp → Mix → Pour into bottles and seal → Heat pasteurize in bottles at 80–95 °C → Cool in cold water.

12.19.4 Flow Chart of Jam

Pulp → Mix 1 kg sugar and 10 g pectin per kg pulp and add → Heat mixture while stirring until the weight of pulp is reduced to half → Dissolve citric acid (5 g/kg) in water and add to pulp → Continue cooking until total sugar content is 68.50 % → Pour into jars and seal → Cool at room temperature

12.19.5 Flow Sheet RTS

Bael fruits (ripe) → Breaking of shell → Removal of pulp with seed and fiber → Addition of water (1:1) → Mixing of pulp with water → Passing through the pulper → Pulp → Mixing with syrup solution (sugar + water + acid) according to recipe → Homogenization → Addition of preservative → Bottling → Crown corking → Pasteurization → Cooling → Storage

12.20 Future Research Needs

Some of the research gaps have been identified which are as follows:

1. A wide range of genetic diversity is available throughout the country especially in the states of UP, Bihar, Uttarakhand, Jharkhand, Chhattisgarh, MP, Gujarat, etc. Existing elite genotypes need to be exploited for improvement.
2. Development of varieties which have less seed and fiber and have more TSS, vitamins, etc.
3. Emphasis should be given on postharvest technology to develop value-added and export-oriented processed products. Small-scale processing units should be established and promoted for commercialization of this fruit crop.
4. Screening of genotype for abiotic especially drought resistant and moisture stress and biotic stress is essential and characterized them for various agroclimatic conditions.
5. Development of integrated crop management strategies for sustainable fruit production of bael.
6. Development of suitable varieties for high-density orcharding.
7. Bael-based cropping systems and cropping models should be developed to provide stability in income to the farmers.

References

Agnihotri BN (1950) The Bael (*Aegle marmelos* Correa.). Indian Fd Packer 4(9):29–31

Aiyer AKYN (1956) The antiquity of some field and forest flora of India. Bangalore Printing and Publishing Co. Ltd, Bangalore

Anon (2001) Annual report. AICRP on AZF, CIAH, Bikaner

Arumugam S, Rao MV (1996) *In-vitro* production of plantlets from cotyledonary node cultures of *Aegle marmelos* (L.) Corr. Adv Plant Sci 9(2):181–186

Banerjee N, Maiti M (1980) Some physico-chemical studies on a protein rich fruit seed (*Aegle marmelos* Correa.). J Indian Chem Soc 57(9):945–946

Bhargava R, Kumar R, Chauhan S, Singh RS, Sharma BD, More TA (2008) Establishment of contamination free culture in bael (*Aegle marmelos* Correa). Indian J Arid Hortic 3(1):11–14

Chadha KL (2001) Bael. In: Chadha KL (ed) A handbook of horticulture. ICAR, New Delhi

Chakraborty DP, Bhattacharyya P, Chattopadhyaya KK (1978) Marmelida, a tyrosinase accelerating and tryptophan pyrrolase inhibitory furanocoumarin from *Aegle marmelos* Correa. Chem Ind 2:848

Chatterjee A, Bhattacharya A (1959) Isolation and constitution of marmin, a new coumarin from *Aegle marmelos*. J Chem Soc, June 1922

Chatterjee A, Choudhury B (1960) Occurrence of auraptene, umbelliferone, marmine, lupel and simmianine in the root of *Aegle marmelos* Correa. J Indian Chem Soc 37:334–336

Chatterjee A, Roy SK (1957) Agelenine, a new alkaloid of leaves of *Aegle marmelos*. Sci Cult 23:106–107

Chattopadhyay RK, Mahanta SK (1989) Media requirement for seed germination and seedling establishment. Indian Hortic 36(3):27

Dixit BBL, Dutt S (1932) The constitution of marmelosin. J Indian Chem Soc 9:271–79

Gopalan C, Rama Sastri BN, Balsubramanian SC (1961) Nutritive value of Indian food. I.C.M.R, Hyderabad

Haskar CN, Kendurkar SG (1961) Uses of bael gum. Printindia 11(1):135

Hays WB (1957) Fruit growing in India. Kitabistan, Allahabad

Hossain M, Karim MR, Islam R, Joarder OI (1993) Plant regeneration from nucellar tissues of *Aegle marmelos* through organogenesis. Plant Cell Tissue Org Cult 34(2):199–203

Islam R, Hossain M, Joarder OI, Karim MR (1993) Adventitious shoot formation on excised leaf explants of in-vitro grown seedlings of *Aegle marmelos* Corr. J Hortic Sci 68(4):495–498

Jauhari OS, Singh RD (1971) Bael-valuable fruit. Indian Hortic 16(1):9–10

Jauhari OS, Singh RD, Awashi RK (1969) Survey of some important varieties of bael (*Aegle marmelos* Correa). Punjab Hortic J 9:48–53

Johns L, Stevenson V (1979) The complete book of fruit. Angus and Roberson Publications, Sydney

Kirtikar KR, Basu BD (1935) Indian medicinal plants, vol I. L. M. Basu, Allahabad

Kore SS, Dhande GW (1973) A new spot of bael (*Aegle marmelos* Correa) caused by *Fusarium roseum* link from India. Punjab Rao Krishi Vidyapeeth Res J 1(2):223–24

Kumar D, Pathak RK, Ali W (1994) Studies on the effect of duration and methods of budding in bael. Indian J Hortic 15(2):127–129

Maiti CS, Nath A, Sen SK (1999) Studies on the propagation of bael (*Aegle marmelos* correa) by different grafting methods in West Bengal. J Appl Hortic 1(2):131–32

Majumdar CP (1935) Upvana-Vinoda (A Sanskrit literature on arborihorticulture). The Indian Research Institute, Calcutta

Mazumdar BC (1975) Physico-chemical analysis of some types of bael (*Aegle marmelos* Correa). I. Fruits growing in west Bengal. Indian Agric 19(3):295–98

Mishra KK, Jaiswal HR (2001) Effect of scion genotypes on patch budding in bael (*Aegle marmelos* Corr.). Adv Hortic Forestry 8:61–66

Moti DL, Chaturvedi OP (1976) Propagation of some tropical and subtropical fruits by budding. Punjab Hortic J 16(1–2):33–38

Om Prakesh (1961) Food and drinks in ancient India. Munshi Ram Manohar Lal Orient Book Sellers and Publishers, Delhi

Pal M, Mishra KK (2005) Effect of pollen on fruit set, fruit drop and final fruit retention in self-and cross pollination in bael (*Aegle marmelos*). Indian J Agric Sci 75(12):800–804

Pandey NC, Singh AR, Maurya VN, Katiyar RK (1986) Studies on bio-chemical changes in bael (*Aegle marmelos* Correa) fruit. Prog Hortic 18(1–2):29–32

Parikh VM, Ingle TR, Bhide BV (1958) Studies in carbohydrates. Bael fruits (*Aegle marmelos*) mucilage. J Indian Chem Soc 35:125–29

Patra A, Mukhodhyay AK, Ghosh A, Mitra AK (1979) Composition of *Aegle marmelos*. Carbon-13 NMR spectra of aurapten and marmin. Indian J Chem B 17(4):385–387

Ram D, Singh IS (2003) Physico-chemical studies on bael (*Aegle marmelos* Correa) fruits. Prog Hortic 35(2):199–201

Ray DP, Chatterjee BK (1996) Effect of different concentrations of growth regulators, etiolation and vigoration treatment on the rooting of stem cuttings of bael (*Aegle marmelos* Correa). Orissa J Hortic 24(1&2):36–41

Reuther W, Webber HJ, Bachelor LD (1967) The citrus industry. Univ Calif Div Agric Sci I:407–409

Roy SK, Singh RN (1979a) Studies on utilization of Bael fruit (*Aegle marmelos* Correa.) for processing II. Extraction of bael fruit pulp. Indian Food Packer 33(1):5–9

Roy SK, Singh RN (1979b) Preliminary studies on storage of bael fruit (*Aegle marmelos* Correa). Prog Hortic 11(3):21–28

Roy SK, Singh RN (1980) Studies on changes during development and ripening of bael fruit. Punjab Hortic J 20(324):190–97

Roy SK, Singh RN (1981) Studies on induced ripening of bael fruit (*Aegle marmelos* Correa). Punjab Hortic J 21(1220):74–82

Roy RS, Verma UP (1950) Bael products. Indian Food Packer 14(8):13–14

Saini RS, Singh S, Deswal RPS (2004) Effect of micronutrients, plant growth regulators and soil amendments on fruit drop, cracking and quality of bael (*Aegle marmelos* Correa) under rainfed conditions. Indian J Hort 61(2):175–76

Shoeb A, Kapil RS, Popli SP (1973) Coumadin's and alkaloids of *Aegle marmelos*. Phytochemistry 12(8):2071–72

Shukla SK, Singh GN (1996a) Seed germination and growth of bael (*Aegle marmelos* Correa.) seedlings as influenced by soil sodicity. J Trop For 12(4):293–294

Shukla SK, Singh GN (1996b) Nutrient imbalances in bael (*Aegle marmelos* Correa) leave as induced by salinity and sodicity. Indian J Plant Physiol 1(4):293–94

Singh LB (1954) Propagating bael (*Aegle marmelos* Correa.) vegetatively. Sci Cult 19:485–486

Singh RD (1989) Studies on bael (*Aegle marmelos* Correa). Morphology of flower, anthesis, dehiscence and receptivity of stigma. Prog Hortic 32:15–24

Singh BN, Dutt S (1941) Studies on formation of jellies from some Indian fruits. Indian J Agric Sci 11:1006–21

Singh AK, Mishra KK (2000) Seasonal variation in root distribution pattern in bael (Aegle marmelos). Indian J Agric Sci 70(10):717–718

Singh VP, Misra KK (2004) Estimation of variability parameters for floral traits and yield in bael (*Aegle marmelos* Correa). Prog Hortic 36(2):253–258

Singh RN, Roy SK (1984) The bael. DIPA/ICAR, New Delhi

Singh VR, Pandey IC, Upadhyay NP, Prasad RS (1976) Propagation of bael (*Aegle marmelos*) by budding. Punjab Hortic J 16:57–59

Singh R, Misra KK, Jaiswal HR (2000) Studies on physico-chemical characters of fruits of bael genotypes. Indian J Hortic 57(4):314–317

Singh AK, Singh S, Bagle BG, Dhandar DG (2007) Studies on flower biology of bael (Aegle marmelos Correa) under semi-arid ecosystem. In: Proceedings of national symposium on production, utilization and export of under utilized fruits with commercial potentialities held at B CKV, Kalyani, Nadia, West Bengal during November 22–24, pp 65–69

Singh AK, Singh S, Bagle BG, More TA (2008a) Varietal evaluation of bael under semi arid ecosystem of western India. National seminar on emerging trends in plant sciences and herbal medicines held at N.D.U.A & T, Kumarganj, Faizabad, U.P. on 17th and 18th March 2008, pp 13

Singh AK, Singh S, Bagle BG, More TA (2008b) Evaluation of bael genotypes under semi-arid environment of western India. Paper presented in third Indian horticulture congress on new R & D initiatives in horticulture for accelerated growth and prosperity, organized by The Horticultural Society of India, New Delhi, during 06–09 November, Bhubaneswar, pp 79

Singh AK, Singh S, Joshi HK, Bagle BG, More TA (2008c) Evaluation of bael genotypes for growth, behaviour and floral traits under semi-arid ecosystem of western India. The Hortic J 21(30): 140–142

Srivastava KK, Singh HK (2000) Floral biology of bael (*Aegle marmelos*) cultivars. Indian J Agric Sci 70(11):797–798

Teotia SS, Maurya VN, Agnihotri BN (1963) Some promising varieties of bael (*Aegle marmelos* Correa) of eastern districts of Uttar Pradesh. Indian J Hortic 20:210–14

Varghese SK, Inamdar JA, Kalia K, Subramanian RB, Nataraj MA, Kalia K (1993) Micro propagation of *Aegle marmelos* (L.) Correa. Phytomorphology 43(1–2):87–92

Verma UP, Ahmed S (1958) Drying of bael (*Aegle marmelos* Correa.) pulp. Indian Food Packer 12(7):7–9

13 Tamarind (*Tamarindus indica* L.)

Abstract

Tamarindus indica L. is a member of the dicotyledonous family Fabaceae of subfamily Caesalpinioideae having five to nine tribes based on morphology. Cultivation of tamarind started in Egypt as early as 400 BC. The tree is also mentioned in the Indian Brahma Samhita scriptures between 1200 and 200 BC and in Buddhist sources from around the year 650 AD.

In recent times identification of superior types of varieties has led to regular plantation of the crop all over India.

India is the world's largest producer of tamarind products. In India, it is naturally regenerated on wastelands and forest lands. Since ancient times, India has been exporting processed tamarind pulp to Western countries, mainly the European and Arab countries and more recently the USA.

13.1 Introduction

Cultivation of tamarind occurred in Egypt as early as 400 BC. The tree is also mentioned in the Indian Brahma Samhita scriptures between 1200 and 200 BC and in Buddhist sources from around the year 650 AD (Salim et al. 1998). The name tamarind is derived from the Arabic Tamar-u'l-Hind because the dark brown pulp of the fruit was thought to resemble dried dates. It was therefore called the Tamere-Hindi or the "date of India." According to Coates-Palgrave, this inspired Linnaeus in the eighteenth century to name the tree as *Tamarindus indica*. According to a Native American legend, the tree existed in Ecuador in the pre-Colombian era (FAO 1988).

13.2 Area

At present tamarind is cultivated in 54 countries of the world; 18 countries in its native range and 36 other countries where it has become naturalized. The major areas of production are in the Asian and American continents. In most countries, tamarind is a subsistence tree crop mostly meeting local demands, although some of it is also exported. The major product is the fruit which is used for culinary purposes, making juices, chutneys, sauces, etc., while the seed is the main component used in industrial applications. The timber is hard having little commercial value but is used at the local level for making furniture and tools and used as fuelwood.

S. Hiwale, *Sustainable Horticulture in Semiarid Dry Lands*,
DOI 10.1007/978-81-322-2244-6_13, © Springer India 2015

India is the world's largest producer of tamarind products. In India, it is naturally regenerated on wastelands and forestlands. Since ancient times, India has been exporting processed tamarind pulp to Western countries, mainly the European and Arab countries and more recently the USA.

More recently, Thailand has become a major producer of tamarind, with sweet and sour varieties in production. Thailand is particularly prominent due to the availability of the sweet-tamarind types. Many other countries have minor production areas of tamarind and depend on wild trees established through natural regeneration.

13.3 Taxonomy

Tamarindus indica L. is a member of the dicotyledonous family Fabaceae of subfamily Caesalpinioideae having five to nine tribes based on morphology. It is a diploid species with chromosome number of $X=12$ and $2n=24$ (Purseglove 1987). According to Leonard (1957), the *Amherstieae* has a total of 25 genera, 21 in tropical Africa, 2 in tropical America, and 2 in Asia. Both *Tamarindus* and *Amherstia* are derived, with zygomorphic and showy flowers and stamen filaments connate in a sheath, but have many differences in the floral structure, leaves, fruits, and seeds. *Tamarindus* is said to have some resemblance to *Heterostemon Desf.* from the upper Amazon region of South America. No other tree bearing any resemblance to tamarind has been reported in other countries.

13.4 Vegetative Characters

Tamarind is a long-lived, large, evergreen, or semi-evergreen tree, up to 30 m tall. The trunk can be up to 8 m in circumference, with a spreading, rounded crown, up to 12 m in diameter. The tree has a poor form, the bole splitting at about 1 m from the ground and often multi-stemmed, though it has fairly good coppicing ability. The trunk is short fissured with bark light gray to brown, very rough and scaly; the inner bark is 1–2 cm thick and green white in color. The branches are often crooked, thick, wide spreading, and drooping at the ends to form a dense crown (Storrs 1995). The blood-red-colored gum exudes from the bole and branches when they are damaged. It has a maximum height of 30 m (Allen and Allen 1981) and maximum tree diameter of 4.1 m (Gamble 1922). Long-lived tree reported 150 years in India.

13.4.1 Wood

Bright yellow sapwood and brown violet heartwood; it is dense, very hard, and close grained and takes a fine polish. Resistant and durable but susceptible to termite attack.

13.4.2 Root

In most locations tamarind produced a deep taproot, except in poorly drained or compacted soils, and has an extensive lateral root system. The main root is wiry and flexuose, while the lateral roots are numerous, moderately long, and distributed down the main root. In its seedling, the hypocotyl is distinct from and thicker than the root, 3.8–8.9 cm long, slightly compressed and finely tomentose.

13.4.3 Leaves

These are alternate, even, and paripinnately compound, 7–15 cm long with a pulvinus at the base and 6–20 pairs of opposite leaflets. The leaflets are $12–32 \times 3–11$ mm in size, narrowly oblong, unequal and entire, rounded or slightly emarginate at the apex, and rounded at the base; they are glabrous or slightly puberlous, glaucous beneath and darker green above with reticulate venation, and are sessile to sub-sessile. The apex is rounded to almost square, slightly notched and asymmetric with a tuft of yellow hairs. A prominent scar is seen after leaf fall. Tropism is observed in tamarind; the leaflets fold after dark.

13.4.4 Inflorescences and Flowers

The inflorescence racemes are small, 5–10 cm long, terminal and lateral drooping, and often panicled. The flowers are bisexual and 2–2.5 cm in diameter. Pedicels are about 5 mm long, nodose and jointed at the apex.

13.4.5 Pollen and Seed Dispersal

The presence of nectar in the tamarind flower suggests that pollination is carried out by insects (Prasad 1963). The structure of the flower however does not exclude the possibility of self-pollination, which can also result in seed set. The large, hard seeds are dispersed by both animals and man.

13.4.6 Pollination and Fruit Set

Thimmaraju et al. (1977) reported that tamarind is protogynous, entomophilous, and highly cross-pollinated. They also reported that flower bud development takes about 20 days from first visible initiation.

13.4.7 Fruit

The fruit is pendulous; pods are 5–10 cm × 2 cm, oblong or sausage-shaped, curved or straight with rounded ends. The shell is light grayish or scurfy brown and filled with soft pulp surrounding the seed cavities (Coronel 1991).

13.4.8 Seed

Pods contain 1–10 seeds, each 3–10 × 1.3 cm, irregularly shaped, flattened and rhomboid. The seeds are very hard, shiny, reddish or purplish brown, and non-arillate exalbuminous. There are about 720 seeds per kg of fruit (Hong et al. 1996).

13.5 Climate

Tamarind is adapted to a wide range of ecological conditions, reflecting its wide geographical distribution in the sub- and semiarid tropics. It grows wild in many countries and is also widely cultivated. According to Dalziel (1937), the tree has an abundant leaf fall and usually has no undergrowth. Although the tree is often planted for its pleasant shade in villages, people in India object to sleeping under the tree, as they fear the "harmful acid exhalations." The leaves bear acid exudation droplets, which refresh the air at hot times of day (Jansen 1981). Joshi (1985) has suggested that tamarind is a weather-indicating plant of tribals in Rajasthan, India.

13.6 Temperature

Essentially a tree of tropical climates, tamarind thrives under a maximum annual temperature ranging from 33 to 37 °C and a minimum of 9.5–20 °C. Trees are very sensitive to fire and frost and require protection when small. Older trees are resistant to extremes of temperature than young trees and can withstand temperatures as high as 47 °C and as low as −3 °C without serious injury. Tamarind is more cold tolerant than mango, avocado, and lychee (Verheij and Coronel 1991).

13.7 Light

It is a light-demanding tree. The ground is usually bare around the tree due in part to the dense shading by the canopy.

13.8 Soil Requirements

It is growing in a wide range of soils and has been suggested that no other specific requirements needed. With little or no cultivation, it can flourish in poor soils and on rocky terrain. In India, it tolerates sodic and saline soils where it grows in ravines and on degraded land. The optimum pH for tamarind is 5.5–6.8, which is slightly acidic, though it also grows well in alkaline soils.

13.9 Uses

13.9.1 Domestic Uses

Production of tamarind for home consumption in home gardens and small waste areas would assist in providing nutrition security, especially in drought-prone regions where fresh fruit is a small component of the daily diet.

13.9.1.1 Pulp

It is the most valuable and commonly used part of the tamarind. The pulp constitutes 30–50 % of the ripe fruit (Purseglove 1987), the shell and fiber account for 11–30 %, and the seed about 25–40 % (Chapman 1984). Fruit composition is variable, depending on locality. Tamarind has low water content and a high level of protein, carbohydrates (60–72 %), and minerals. The soluble content varies from 54 to 69.9° Brix. The pulp contains oil, which is greenish in color, and liquid at room temperature. The major constituents of it include furan derivatives and carboxylic acids. It contains tartaric acid ranging from 12.2 to 23.8 %, which is uncommon in other plant tissues. It contains 25–45 % reducing sugars, of which 70 % is glucose and 30 % is fructose. Ascorbic acid content of tamarind is very small and varies from 2 to 20 mg/100 g. Other organic acids are also present like oxalic acid, succinic acid, citric acid, and quinic acid. According to Lakshminarayan Rao et al. (1954) about 55 % of the total nitrogen in the tamarind pulp was non-protein N or soluble in 10 % trichloroacetic acid, and 70 % of this is contributed by free amino acids. It is also rich in minerals.

13.9.1.2 Seeds

It comprised the seed coat or testa (20–30 %) and the kernel or endosperm (70–75 %) (Coronel 1991; Shankaracharya 1998). Tamarind seed is a raw material used in the manufacture of tamarind seed kernel powder, polysaccharide, adhesive, and tannin. They are alternative source of protein, rich in some essential amino acids. Unlike the pulp, the seed is a good source of protein and oil. Whole tamarind seed and kernels are rich in protein (13–20 %), and the seed coat is rich in fiber (20 %) and tannins (20 %). Panigrahi et al. (1989) reported that whole tamarind seed contains 131.3 g/kg crude protein, 67.1 g/kg crude fiber, 48.2 g/kg crude fat, and 56.2 g/kg tannins. About 70 % of the proteins were extractable. The protein isolated was relatively high in lysine, phenylalanine, tyrosine, and leucine. Albumins and globulins constitute the bulk of the seed proteins. The seed is rich in cystine and methionine but threonine and tryptophan are limiting. It has a very good balance of essential amino acids. The seed kernel is rich in phosphorus, potassium, and magnesium.

13.9.1.3 Seed Kernel Oil

The seed oil is golden yellow, semidrying oil, which in some respects resembles groundnut oil. Andriamanantena et al. (1983) extracted the oil with hexane and a mixture of chloroform and methanol; the yield was 6.0–6.4 % and 7.4–9.0 %, respectively.

13.9.1.4 Leaves and Flowers

The leaves are used as vegetable by endemic peoples in producing countries. They contain 4.0–5.8 % proteins, while the flowers contain only 2–3 %. The leaves are also a fair source of vitamin C and beta-carotene and the mineral content is high, particularly in potassium, phosphorous, calcium, and magnesium. The leaves are also used as fodder for domestic and wild animals, including elephants.

13.9.1.5 Wood

Tamarind wood is used for many purposes including making furniture, wheels, mallets, rice pounders, mortars, pestles, plows, well construction, tent pegs, canoes, side planks for boats, cart shafts and axles, and naves of wheels, toys, hubs, oil presses, sugar presses, tools and tool handles, turnery, etc.

The wood also makes good fuel with a calorific value of 4,850 kcal/kg, producing great heat, which is required in brick making. The wood makes excellent charcoal, is valued for making gun powder (NAS 1979; Chaturvedi 1985), and used as a major fuel to produce gas units that powered Indian cars and trucks during World War 2.

13.9.2 Medicinal Uses

Tamarind products, leaves, fruits, and seeds have been extensively used in traditional Indian and African medicine.

13.9.2.1 Pulp

It is used in Africa, Asia, and America. In former times, the fruit pulp was used as a gentle laxative under the name "pulpa tamarindorum." Tamarind pulp alone or in combination with lime juice, honey, milk, dates, spices, or camphor is used as a digestive even for elephants.

Tamarind pulp is also said to aid in the cure of malarial fever (Timyan 1996). The pulp is also effective in ridding domestic animals of vermin in Colombia through the application of pulp with butter and other ingredients (Morton 1987).

13.9.2.2 Seed

The seed is usually powdered and often made into a paste for the treatment of most external ailments. In Cambodia and India, it has been reported that powdered seeds have been used to treat boils and dysentery.

Seed powder has also been externally applied on eye diseases and ulcers; boiled, pounded seeds are reported to treat ulcers and bladder stones and powdered seed husks are used to treat diabetes (Rama Rao 1975). The seed can also be used orally, with or without cumin seed and palm sugar, for treatment of chronic diarrhea and jaundice.

13.9.2.3 Leaves

Tamarind leaves are usually ground into powder and used in lotions or infusions. The leaves, mixed with salt and water, are used to treat throat infections, cough, fever, intestinal worms, urinary troubles, and liver ailments. Leaf extracts also exhibit antioxidant activity in the liver. Also taken internally, the leaves are used in cardiac and blood sugar reducing medicines.

The leaves are also used to treat ulcers, and the juice of the leaves, boiled with oil, is applied externally to treat rheumatism and external swellings in the Philippines and West Africa (Jayaweera 1981; Rama Rao 1975).

13.9.2.4 Bark, Flower, and Root

The medicinal properties of the bark, flower, and root are similar in many respects to the pulp. Treatments for digestive tract ailments and indigestion have been reported from Cambodia, India, and the Philippines. The bark is the usually the most effective method of administration. The bark has also been used to recover loss of sensation due to paralysis. Gargling the ash with water has been used in the treatment of sore throat. The bark is astringent and used as a tonic and in lotions or poultices to relieve sores, ulcers, boils, and rashes in the Philippines and eastern Sudan (Dalziel 1937). The poultice of flower is used in the treatment of eye diseases and conjunctivitis in the Philippines (Brown 1954).

13.9.3 Industrial Uses

Tamarind pulp is used as a raw material for the manufacture of several industrial products, such as tamarind juice concentrate (TJC), tamarind pulp powder (TPP), tartaric acid, pectin, tartarates, and alcohol (Anon 1982a).

13.9.3.1 Tamarind Kernel Powder

The major industrial use of the seeds is in the manufacture of tamarind kernel powder (TKP). It is prepared by decorticating the seed and pulverizing the creamy white kernels. The decorticated seed is ground to the required mesh size by machines to obtain a yield of 55–60 %.

Mixing with sodium bisulfate before packing will prevent enzymatic deterioration. It should have flavoring characteristics when dissolved in water and be free of any burnt or other undesirable flavors.

13.9.3.2 Pectins

Polysaccharides obtained from tamarind seed kernels form mucilaginous dispersions with water and possess the characteristic property of forming gels with sugar concentrates, like fruit pectin. However, unlike fruit pectin, tamarind polysaccharide can form gels over a wide pH range, including natural basic conditions. Jellose is prepared on a large scale by adding TKP to 30–40 times its weight of boiling water, containing citric or tartaric acid at a concentration of 0.2 %.

13.9.4 Other Uses

Tamarind pulp mixed with sea salt has been reported to polish brass, copper, and silver in Sri Lanka (Jayaweera 1981), India (Benthall 1933), West Africa (Morton 1987), South Africa, and Somalia. The seed testa contains 23 % tannin, which when suitably blended is used for tanning leather and imparting color-fast shades to wools. The fruits are reported to have antifungal and antibacterial properties (Guerin and Reveillere 1984). The seed husk has also been found to be an effective fish poison (Roy et al. 1987). Tamarind extract has also been reported to have an inhibitory effect on plant virus diseases. In India, extracts obtained from tamarind plant parts have completely inhibited the activity of both cowpea mosaic and the mung bean mosaic viruses (Singh et al. 1989). Antioxidant activity of tamarind seed was investigated by Osawa et al. (1994). The seed is also used as filler for adhesives in the plywood industry.

13.10 Genetic Resources

13.10.1 Genetic Resources of Tamarind

Plant genetic resources are basic raw materials for improvement of any crop plant. The wider genetic diversity is fundamental for the development of new varieties with good quality and higher yields (Frankel and Hawkes 1975). The genetic variation of tamarind has been based on the phenotypic variation observed. This is primarily based on fruit characteristics, such as length of pod, pod weight, seed number, pod color, and sweetness of pulp. Recent evidence also indicates the existence of variation of characteristics of fruit pulp in different countries, some with less acid or sweet fruits. The variations have also been reported for tolerance to drought, wind, poor soils, waterlogging, high and low pH, and grazing. Phenological diversity also exists in tamarind, and tree to tree variations are common in flowering and in maturing fruits (Mahadevan 1991), which may reflect either genetic variation or genotype x environmental interactions and/or both. Wide phenotypic variation in tamarind germplasm has been attributed to geographic isolation and gene mutation (Feungchan et al. 1996a). The origin of a sweet tamarind has been attributed to a point mutation.

13.10.2 Evaluation of Germplasm (Table 13.1)

Table 13.1 General characteristics useful to distinguish tamarind cultivars

Plant part	Character
Flower	Color of petals (dark pink to whitish cream)
	Flowering pattern (early-late)
Pod	Pod form, length, breadth, curvature, shape, pod size, pod weight, color of shell
Pulp	Color (red/whitish), pulp/shell ratio
	Real pulp value, pulp yield/pulp recovery %
	Fiber content, ease of pulp extraction
Seed	Color, size, weight, number of seeds per pod, seed/pulp ratio
Yield	Total yield of pods, alternate/regular bearing habit
Biochemical	Sweetness of pulp (ratio of tartaric acid/sugar/protein and mineral constituents, amino acid composition)
Others	Resistance to salinity, drought, degraded soils, waterlogging, high pH, low pH, grazing diseases and pests, tree form

13.10.3 Germplasm Evaluation in India

In India, most of the area under tamarind cultivation is planted with unselected, inferior cultivars. The Bharata agricultural and industrial foundation (BAIF), Pune, India, has attempted to supply improved planting material to smallholders by selecting superior trees from among the existing natural populations. The parameters for selection of superior trees have been based entirely on pod characteristics such as pod length, pod color, and pulp yield per pod.

The most preferred cultivar is Periyakulam 1 (PKM-1), a sour type, a clonal selection from a local variety of a village named Endapalli near Periyakulam. It has a pulp recovery percentage of 39 % compared to the local cultivars which is 28 %. The yield is also higher. Studies were also undertaken to select high yielding varieties based on their flowering pattern. In seedling populations, early, mid, and late flowering tamarind types have been identified. Recent developments in biochemical and molecular markers could be effectively used to identify the genetic variations within and between tamarind populations.

13.11 Cultivars

- *India*: Pratishthan, Periyakulam (PKM 1), Urigam. Under semiarid conditions of Gujarat, India, promising genotypes (24) of tamarind were collected and evaluated for growth, flowering, fruiting, and fruit quality attributes. The maximum number of fruits per panicle was recorded in Pratishthan (4.00), closely followed by Goma Prateek (3.50), Sweet Type (3.20), and T-263 (3.00). Peak period of ripening time in majority of genotypes was March. Maximum fruit yield per plant (85.00 kg) was recorded in Goma Prateek during the 12th year of orchard life under rainfed conditions of hot semiarid ecosystem, closely followed by T-10 (43.00 kg/plant), while minimum was recorded in PKM-1 (12.00 kg/plant).

13.11.1 Brief Characteristics of Tamarind Cultivars

13.11.1.1 Tamarind-13

Plant height is 4.7 m, stem girth 63.66 cm, plant spread N-S 7.52 m, and E-W 7.43 m. Tree form spreading type, leaves opposite, mostly firm and glossy, elliptic, pinnately veined. Flowering initiation of bloom starts from the last week of May, full bloom in the first week of June, and end bloom in the fourth week of June. Mean length of panicle is 14.81 cm. Ripening time starts from the first week April. Mean fruit yield/plant is 49.83 kg. Pod weight is 26.70 g. Seed weight per fruit is 7.58 g. TSS is 71.00 ° Brix and acidity 14.06 %. Total sugar is 55.81 % and reducing sugar 27.27 %. Protein content is *3.29* g/100 g pulp, calcium 177.70 mg/100 g pulp, magnesium 44.17 mg/100 g pulp, and phosphorus 70.83 mg/100 g pulp.

13.11.1.2 Tamarind T-263

Plant height is 4.66 m, stem girth 6.5 cm, plant spread N-S 7.3 m, and E-W 7.29 m. Tree form spreading type, leaves opposite, mostly firm and glossy, elliptic, pinnately veined. Flowering initiation of bloom starts from the last week of April, full Bloom in the last week of May, and end bloom in the fourth week of June. Mean length of panicle is 10.11 cm. Ripening time starts from the first week April. Mean fruit yield/plant 11.13 kg. Pod weight is 14.94 g. Seed weight per fruit is 7.58 g. TSS is 68.00 ° Brixand acidity 11.15 %. Total sugar is 52.49 % and reducing sugar 25.28 %. Protein content 2.55 g/100 g pulp, calcium 164.31 mg/100 g pulp, magnesium 35.24 mg/100 g pulp, and phosphorus 42.83 mg/100 g pulp.

13.11.1.3 CHEST-10

It has semi-spreading growth habit, thick trunk, and drooping branches. Peak period of ripening time starts the first week of April. It recorded 43 kg fruit yield per plant, pulp percent 52.20 %, and TSS 71.30 °Brix.

13.11.1.4 CHEST-11

It has semi-spreading growth habit, thick trunk, and drooping branches. Peak period of ripening time starts the last week of March. It recorded 22 kg fruit yield per plant, pulp percent 52.10 %, and TSS 71.10° Brix.

13.12 Ex Situ Conservation

Conservation of tamarind germplasm is an important prerequisite to prevent the loss of genetic diversity of this species. Ex situ management is considered to be one of the best and economically and socially acceptable methods for conservation of tamarind germplasm (Singha 1995). In this method, phenotypically superior germplasm is collected from different countries and multiplied through vegetative propagation to be established in clonal orchards. Seed germplasm banks can be maintained since seeds may be stored for longer periods. A low cost method of conserving tamarind germplasm by the establishment of protected areas could also be effective as practiced for the conservation of forest genetic resources in some parts of the world (Collins et al. 1991).

13.13 Genetic Improvement

Recently, genetic improvement using superior clones has been described by Kulkarni et al. (1993). Tamarind has a relatively long generation time and is believed to be primarily outcrossing. In the short term, those trees having superior characteristics could be selected for vegetative propagation by air layering or grafting methods to produce fast-growing trees for local (home gardens) and commercial purpose.

13.13.1 Selection

Therefore, the selection of plus trees is an important step in tamarind tree improvement programs. Plus trees can be selected using the following characteristics: acidity of the pulp, content of tartaric acid and sugar, real value of pulp, pod bearing ability (flowering and fruit maturing), pod size, pulp, fiber and seed weights, and number of seeds. Some of the trees that are selected on this basis are reported to have pods 25 cm long and 5 cm wide (Jambulingam and Fernandes 1986).

7 Year old tamarind tree

Profuse flowering

Bumper fruit harvest

Promising genotype

13.13.2 Ideotypes

It is essential to identify different ideotypes for different purposes, localities, environments, and cultural practices. There is no universal ideotype suitable for all sites and end users.

1. Earliness.
2. Deep root system.
3. Fruit size long.
4. The fruiting season could be longer.
5. Heavy yielder, 600–800 kg/tree/year.
6. Regular bearer.
7. Fleshy, sweet/acidic pulp, high pulp recovery.
8. High in nutrients and fodder value of leaf.
9. Wider adaptability and resistance to pests and diseases.
10. Short statured plants suitable for high density.
11. Very good pulp quality for value addition and export.

13.14 Cultivation

Nowadays with identification of high yielding and regular bearing varieties, the area under tamarind is increasing fast. Regular orchards are being planted on marginal and wasteland. Apart from this, demand from market is also increasing, and at the same time, there is a huge export market available to this fruit.

13.15 Propagation

Tamarind is traditionally grown from seed, though vegetative and tissue culture propagation methods have been developed.

13.15.1 Seed Germination and Propagation

The seed of tamarind is orthodox (Riley 1981 quoted by Hong et al. 1996). Fresh seeds retain viability for at least 6 months when kept at ambient temperature in dry conditions. Seed germination begins within a week following sowing and may take a month to complete. On average, tamarind seeds begin to germinate about 13 days after sowing (Padolina 1931; Galang 1955). The germination capacity of fresh or well-stored tamarind seed is reported to vary from 65 to 75 %. Coronel (1991) stated that, depending on the conditions, germination might vary from 30 to 70 %. Some studies (FAO 1988) indicate that germination can be improved by storing the seed for 6 months before planting.

Experiments have been conducted to improve the germination of tamarind seeds. Masano (1994) compared three seed pretreatments with an untreated control. The result indicated that slicing the seed was the best, with 92 % germination compared to the control (82 %). Slicing the seed also hastened germination. Germination begins after 7 days and completed at 9 days. The other treatments had no positive effect.

Tamarind is normally grown from seed. Seeds should be collected from high yielding trees with well-formed, rounded, fully ripe pods, although they may not come true to type due to outcrossing. Under natural conditions, the seed pods fall from the trees and the seeds germinate naturally on the onset of monsoon. When propagating by seed, mature ripe pods should be collected from the trees in March–April. The best medium for seed germination is sand or soil mixed with cow dung. Seeds may be germinated in nursery beds, seed boxes, pots, or plastic bags. When grown in nursery beds, the recommended spacing is 20–25 cm in both directions. After planting in the field, seedlings should be protected from browsing animals.

13.15.2 Vegetative Propagation

Vegetative propagation is preferable to seed propagation, as seed propagation does not produce true-to-type progenies. It has advantage of producing true-to-type progeny which can be taken from selected, superior, mother trees. The methods include stem cuttings, shield and patch budding, or grafting onto the seedling rootstocks and air layering or marcotting.

13.15.2.1 Stem Cuttings

A technique using softwood terminal cuttings has been developed, and the protocol standardized, by the forest research station at Maddimadugu, Andhra Pradesh, India (Srivasuki et al. 1990).

They are immediately dipped for 10 s in 1,000 ppm of indole butyric acid (IBA) and in 50 % isopropyl alcohol, before being planted in polypropylene tubes containing vermiculate/perlite (1:1) and placed in a mist propagator with 70–80 % humidity. Soft or semi-soft stem cuttings, 15–20 cm long, taken from 1- to 2-year-old branches can also be rooted (Swaminath et al. 1990).

13.15.2.2 Budding and Grafting Methods

Vegetative propagation methods like, shield and patch budding, cleft grafting, whip grafting, approach grafting, and air layering are reliable methods (Purushotham and Narasimharao 1990). For patch budding, seedling rootstocks should be grown in raised beds and transplanted to polythene bags of 25 × 10 cm size, immediately after germination and seedlings may be budded when they are 6–9 months old. This is a suitable method for large-scale multiplication of tamarind. Pathak et al. (1992) reported 96 % and 94 % success, respectively, of patch budding and modified ring budding of 9-month-old seedling rootstocks. Approach grafting is a very reliable method and up to 95 % success can be obtained (Swaminath and Ravindran 1989). For veneer grafting, rootstocks about 6 months old and of uniform size should be selected. It is reported to give about 50 % success (Amin 1978). For softwood grafting, the rootstock seedlings are defoliated and their tops cut off at 15 cm high immediately before grafting. It was shown to be the best grafting method in terms of successful unions and survival rates (Navaneetha et al. 1990).

13.15.2.3 Micropropagation Techniques

Tamarind may be propagated by tissue culture techniques. Shoot tips, cotyledons, and cotyledonary nodes have been used successfully as explants for tamarind tissue culture (Splittstoesser and Mohamed 1991). Tissue culture raised plants have been reported to show better growth in height, branching habit, spread of branches, early flowering and fruiting cycle, with the start of flowering at an average height of 3.7 m. Although these micropropagation techniques have shown promise, none of them has reached the stage of commercialization.

13.15.2.4 In Situ Seed Sowing

Direct sowing of the seed of tamarind can be can be done in the field (Chaturvedi 1985). Seeds may be seen directly to establish plantations, hedgerows, or home gardens. The seed should be planted up to 1.5 cm deep. The planting holes should however have been previously prepared and filled with well-decomposed manure or compost.

13.15.2.5 Spacing

There are no regular orchards of tamarind; the stray plants of seed origin are found in the villages and in the fields. The recommended spacing between plants is 10 m apart. However, looking at the long life span, the spacing can be varied. In parts of India, tamarind is established at 8 m×8 m, 8 m×12 m, or 12 m×12 m (Jambulingam and Fernandes 1986).

13.15.2.6 Time of Planting

The best time for field planting is at the beginning of the rainy season, particularly in seasonally dry regions. This will reduce the need for frequent watering until the plants are firmly established in the soil. Seedling growth in the field is initially fast (about 1.2 m in the first two years) but slows later.

13.16 Management

13.16.1 Pruning and Training

Tamarind requires minimal care except in the very early stages of growth. Tamarind is a compact tree and produces symmetrical branches. Young trees should be trained to allow 3–5 well-spaced branches to develop into the main scaffold structure of the tree. Usually no pruning is done in tamarind; however, old and dried branches need to be removed every year. Apart from this, the very old trees which have become nonproductive should be headed back keeping 15–20 scaffold branches.

13.16.2 Intercropping

In the initial years, vegetatively propagated plants up to 5 years of age and seed origin orchard up to 10 years of age can be intercropped with traditional crops to increase production potential of the unit area. Intercropping may be practiced in order to obtain some income during the early stages of tree growth and until the trees starts bearing. It is good practice to plant short season crops which grow in the rainy season.

13.17 Nutrition

13.17.1 Fertilizers and Manures

No recommendation on fertilizer dosages is available. Trees are known to fruit well even without fertilizer application, due to their deep and extensive root system. In India, inorganic fertilizers are not applied to tamarind trees, but 5 kg of farmyard manure is applied to the planting hole at the time of planting. In Thailand, the commercial growers use inorganic fertilizers, mostly urea.

13.17.2 Nitrogen Fixation

Tamarind being leguminous tree fixes atmospheric nitrogen. Leguminous species are associated with rhizobium bacteria in root nodules which fix atmospheric nitrogen. Very little information is available on nitrogen-fixing bacteria in tamarind compared to other cultivated legumes. Some evidence suggests that it does form a symbiotic association with rhizobium bacteria enabling the tree to fix atmospheric nitrogen under appropriate conditions (Postgate 1979).

13.17.3 Mycorrhizal Associations

Tamarind seedlings inoculated with 13 vesicular arbuscular mahycorhize (VAM) fungi from various sources around the world have been demonstrated to exhibit increased leaf number, plant height, stem girth, biomass, and phosphate and zinc content. Container-grown seedlings of several leguminous species including tamarind were inoculated with three arbuscular mycorrhizal fungi: *Glomus fasciculatum*, *G. mosseae*, and *Gigaspora margarita*.

13.17.4 Irrigation

Irrigation is not normally practiced in tamarind cultivation, but it promotes better growth during establishment and the early stages of growth, especially during the dry seasons (Yaacob and Subhadrabhandu 1995).

13.18 Pest and Disease Management

13.18.1 Pests

Tamarind trees are liable to be attacked by a large number of insect pests. In India alone, 40 insect pests have been reported as attacking tamarind, causing severe economic losses (Joseph and Oommen 1960). Some of the most serious pests in India are the hard scale insects (Timyan 1996). Mealybugs also attack tamarind trees. Nymphs and the females suck the sap on the ventral surface of the leaflets, the base of leaf petioles, tender shoots, and even the mature shoots. Some thrips are reported to attack tamarind flowers. *Scirtothrips dorsalis* Hood is a polyphagous thrip and the adults live for 10 or 15 days and complete 25 overlapping generations per year (Raizada 1965). The other minor pests in India include the bruchid beetle, *Pachymerus gonagra*, is the most serious pest of tamarind in India and Pakistan (Beeson 1941). The lac insect, *Kerria lacca* kerr. while not considered a major pest, is a widely distributed polyphagous insect in India which attacks tamarind stems, twigs, and leaves and many other cultivated and wild plants. Larvae of *Achaea janata* Linn. are reported to cause heavy losses when epidemics of the moth infest flowers in tamarind plantations in Tamil Nadu (Ahmed 1990). Aphids *Toxoptera aurantii* are serious pests that attack tamarind and many other plants. It is a major pest which sucks the sap of tender shoots and leaflets causing them to become distorted and covered with molds growing on the secreted honeydew. Nematodes are also reported to attack tamarind. The major nematodes are *Radopholus similis*, the burrowing nematode, and *Meloidogyne incognita*, a common root knot nematode.

13.18.2 Diseases

Several diseases have been reported to infect tamarind in India, including various tree roots and bacterial leaf spots. The major diseases reported are leaf spot. In Karnataka state, India, stony fruit disease caused by the fungal pathogen *Pestalotia macrotricha* Syd. makes the fruits hard and stony with fibrous structures. A mildew caused by *Oidium* sp. Is a common occurrence in nursery seedling. The disease causes defoliation and early growth is severely retarded.

13.19 Harvesting

Trees grown from seed may take more than 7 years to start bearing and up to 10 or 12 years before an appreciable crop is produced. Well-tended trees grown in open areas will come into bearing early, in about 7 years and less. Grafted trees will however come in to bearing in 3–4 years. Tamarind seed pods fill at maturity, the pulp becomes brown to reddish brown, and the skin becomes brittle and cracks easily. The pods are gathered when ripe and the hard pod shell is removed. The pulp is preserved by placing it in casks and covering it with boiling syrup or packing it carefully in stone jars with alternate layers of sugar.

13.19.1 Harvesting Stage

In the Philippines, fruits are harvested at two stages, green for flavoring and ripe for processing. The fruits of sweet types are also harvested at two stages, half-ripe and ripe. At the ripe stage, the pulp shrinks, due to loss of moisture, and changes to reddish brown and becomes sticky. In most countries, the sour tamarind fruits are harvested by shaking the branches and the pods are collected on a mat. In sweet tamarind, the pods fetch a high price in the local market and are carefully harvested by handpicking. Sometimes bamboo ladders are used to pick the fruits. Generally, the fruits are left to ripen on the tree before harvesting, so that the moisture content is reduced to about 20 %.

13.20 Yield

The yield of tamarind varies considerably in different countries, depending on genetic and environmental factors. Feungchan et al. (1996a) reported that the fruit yields are influenced by environmental and genetic factors, but the age of the tree is not correlated with fruit yield. Recently, Usha and Singh (1996) reported that cross-pollination results in higher fruit set and retention in tamarind than when open or self-pollinated. Fruit set was only 36 % with open pollination and increased to 56 % with cross-pollination. The effect of growth regulators on fruit setting of sweet tamarind indicated that 4 CPA at 15 ppm concentration helped in maximum fruit set of 216 fruits compared to control (81 fruits) (Feungchan et al. 1996b).

13.21 Processing and Storage

The fresh fruits are often dried using small-scale dehydrators; however, in most countries, rural households dry pods in the sun. The shells, fibers, and seeds are then removed and the pulp stored in plastic bags or earthenware pots. The dry ripe pods can easily be cracked and the pulp and fibers separated from the broken shells. The pulp is then processed by peeling and removing the fiber strands from the pulp. After separating the pulp from the fibers, the seeds, and shells, it is then compressed and packed in palm leaf mats, baskets, corn husks, jute bags, or plastic bags for storage and marketing. In most of the tamarind-growing countries, such as India and Africa, the pulp is pressed and preserved in large masses and sold in small shops and bazaars by weight. The quality and condition of the pulp and the selling price in the market are often related to the care taken during storage. In dry conditions the pulp remains good for about 1 year. Feungchan et al. (1996c) attempted several methods to prevent change of pulp color including powdered salt, steam, sun drying, hot air incubation, and cold storage.

13.22 Economics of Production

The cost involved in production is mostly with labor for field establishment, digging pits, weeding, and the purchase of seedlings or grafts. There is no income in first 3 years but intercropping with annual crops could compensate the labor and other costs. The orchards established through clones are of uniform size, the fruits are of high quality, and the trees yield well. Each tamarind will give an average yield of 100 kg of fresh tamarind pulp per year at maturity. The national and international commerce of tamarind is more limited for the following reasons:

- Tamarind has not received sufficient research attention over the years, and in most countries, unimproved trees are cultivated.
- Fresh fruits of the sour type have village level markets.
- Most of the fruits and seeds go to waste due to lack of technologies for processing and storage.
- Processed pulp is of low quality
- Alternate low cost processing technologies are not available.
- Products are restricted to domestic markets of producing countries.
- Most of the exports are from only a few of the major producing countries.
- Lack of both local and international market information restricts expansion of crop and product diversification.

13.23 Marketing

Tamarind is a delicacy in the producing countries and is used for various culinary purposes. It is consumed in fresh, dried, and other processed forms. The ripe pods are collected from trees grown in the wild or in home gardens; they are processed in households and sold in the village markets.

13.24 Export

International trade in tamarind has been in existence for a long time, though it is limited to the major producing countries. Only a small proportion of the total production is exported and this is mostly in the dried form. Tamarind is also exported as pulp, fresh fruit, paste, and industrial products including tamarind seed kernel powder. Demand for these products varies according to quantity and the country.

13.25 Post Harvest

13.25.1 Tamarind Juice Concentrate

It is a convenient product, as it is easy to dissolve and reconstitute in hot water (Anon 1982b). It can also be stored for long periods. It is prepared by extracting cleaned pulp with boiling water using the countercurrent principle, where dilute extracts are used for extracting fresh batches of the pulp.

13.25.2 Tamarind Pulp Powder

It is prepared by concentrating, drying, and milling the pulp into a powder form. Depending on the manufacturing process, wide variations in the physicochemical characteristic are reported. On a small scale, the fruit pulp is made into a refreshing drink after dissolving in water and squeezing by hand. The extraction and processing techniques of the pulp for the preparation of canned tamarind syrup, clarified tamarind juice, and other soft drinks have been reported by Bueso (1980).

13.25.3 Tamarind Pickle

Pulp is used commercially to prepare tamarind pickle. The pickles are commonly used in Asia. Pickles are hot, spicy, and of salty-sour taste and can be preserved for several months.

13.25.4 Jam

In making tamarind jam, ripe fruits are shelled and the pulp is boiled for 10 min. The pulp is then drained and separated from the seeds. For every cup of pulp, two cups of brown sugar are added. The mixture is then cooked and constantly stirred while boiling until it becomes thick in consistency. The resulting jam is cooled, packed in dry, sterilized jars, and sealed.

13.25.5 Syrup

Tamarind syrup is made by boiling immature fruit pulp until it is soft and then strained through cheesecloth. To every cup of juice, a half-teaspoon of baking soda is added. The mixture is boiled down to one half the original quantity, removing the rising scum in the process. The juice is again strained and for every cup obtained; a quarter cup of sugar is added. The mixture is boiled again for 20 min. The cooled syrup is poured in to sterilized bottles and sealed.

13.25.6 Candy

Sweetened tamarind fruit is made by peeling whole ripe fruits and pouring boiling sugar syrup over the fruits and placing in a deep enamel basin. Boiling syrup is prepared by mixing three parts of sugar with one part water. After soaking for 3 days, the fruits are drained of the old syrup and again covered with freshly prepared syrup. The process is repeated until the fruits are sweet enough.

References

Ahmed M (1990) Deviation in feeding behavior of *Achaea janata* Linn. (Lepidoptera: Noctuidae). Ann Entomol 8(2):15–17

Allen ON, Allen EK (1981) The leguminosae: a source book of characteristics, uses and nodulation. University of Wisconsin Press, Madison, 812 p

Amin RS (1978) 'Scion wood grafting' a new technique for hardwood plants. Curr Sci 47:468–469

Andriamanantena RW, Artuad J, Gaydofn EM, Iatrides MC, Chavalier JL (1983) Fatty acid and sterol composition of Malagasy tamarind kernels. J Am Oil Chem Soc 60(7):1318–1321

Anon (1982a) Some recent developments. Central Food Technological Research Institute, Mysore, India

Anon (1982b) Tamarind juice concentrate plant starts in Mysore. Indian Food Ind 1:43–44

Beeson CFC (1941) The ecology and control of forest insects of India and the neighbouring countries. Government of India Publication, New Delhi,767 p

Benthall AP (1933) The trees of Calcutta and its neighbourhood. Thacker Spink & Co., Calcutta, 513 p

Brown WH (1954) Useful plants of the Philippines. Republic of the Philippines, Department of Agriculture and Natural Resources, Technical Bulletin. Bur Printer, Manila

Bueso CE (1980) Soursop, tamarind and chironka. In: Nagy S, Shaw PE (eds) Tropical and subtropical fruits. AVI Publishing, Westport, p 375

Chapman KR (1984) Tamarind. In: Tropical tree fruits for Australia. Complied by Page PE. Queensland Department of Primary Industries, Information series Q 183018, Brisbane, pp 83–86

Chaturvedi AN (1985) Firewood farming on the degraded lands of gangetic plain. Utter Pradesh forest bulletin no.50, Lucknow, India. Government of India press (1), 286 pp

Collins NM, Sayer JA, Witmore TC (1991) The conservation atlas of tropical forests of Asia and the Pacific ICUW. Macmillan Press, London

Coronel RE (1991) *Tamarindus indica* L. in plant resources of South East Asia. Wageningen; Pudoc. no. 2. In: Verheij EWM, Coronel RE (eds) Edible fruits and nuts. PROSEA Foundation, Bogor, Indonesia, pp 298–301

Dalziel JM (1937) The useful plants of West Tropical Africa. Crown Agents for Overseas, London. Governments and Administrations, 612 p

FAO (1988) Fruit bearing trees. Technical notes. FAO-SIDA For Paper 34:165–167

Feungchan S, Yimsawat T, Chindaprasert S, Kitpowsong P (1996a) Tamarind (*Tamarindus indica* L.) plant genetic resources in Thailand. Thai J Agric Sci, Special Issue (1):1–11

Feungchan S, Yimsawat T, Chindaprasert S, Kitpowsong P (1996b) Effect of plant regulators in fruit setting. Thai J Agric Sci Special Issue (1): 48–51

Feungchan S, Yimsawat T, Chindaprasert S, Kitpowsong P (1996c) Studies on the colour change of tamarind pulp in storage. Thai J Agric Sci Special Issue (1):52–56

Frankel OH, Hawkes JG (1975) Crop genetic resources. Today and tomorrow. Cambridge University Press, Cambridge

Galang FG (1955) Fruits and nuts grown in the Philippines. Rizal AIA Printing Press, Malabon

Gamble JS (1922) A manual of Indian timbers. Sampson Low, Marston & Co, London, 866 p

Guerin JC, Reveillere HP (1984) Active antifongique d'extraits vegetaux a usage therapeutique. 1. Etude de 41 extracts sur 9 souches fonguques (Antifungal activity of plant extracts used in therapy). 1. Study of 41 plant extracts against 9 fungal species). Allalles Pharmaceutiques Francaises 42(6): 553–559 (English summary)

Hong TD, Linnington S, Ellis RH (1996) Seed storage behaviour: a compendium. Handbook for genebanks no.4. International Plant Genetic Resources Institute, Rome

Jambulingam R, Fernandes ECM (1986) Multipurpose trees and shrubs in Tamil Nadu State (India). Agrofor Syst 4(1):17–32

Jansen PCM (1981) Spices, condiments and medicinal plants in Ethiopia, their taxonomy and agricultural significance. Centre for Agricultural Publishing and Documentation (PUDOC) Wageningen Agricultural Research Reports No. 906

Jayaweera DMA (1981) Medicinal plants (Indigenous and exotic) used in Ceylon. Part III. Flacourtiaceae-Laytharecae. A publication of the National Science Council of Sri Lanka, pp 244–246

Joseph KV, Oommen P (1960) Notes on some insect pests infesting dry tamarind fruits in Kerala State. Indian J Entomol 22(3):172–180

Joshi p (1985) Weather indicating plants of tribals in Southern Rajasthan. Bot Bull India 27(1–4):100–104

Kulkarni RS, Gangaprasad S, Swamy GSK (1993) Tamarind: economically an important minor forest produce. Minor For Prod News 3(3):6

Lakshminarayan Rao MV, Subramanian N, Srinivasan N (1954) Free amino acids in tamarind pulp. J Sci Ind Res 13B: 377

Leonard J (1957) Genera des Cynometreae et des Amherstieae africaines (Leguminosae-Caesalpinioideae). Memoire Acad Royale Belgique 30(2):1–314

Mahadevan N p (1991) Phenological observations of some forest trees as an aid to seed collection. J Trop For 7(3):243–247

Masano M (1994) Effect of pre-treatment on the germination process of Asam (*Tamarindus indica* L.) seeds. Bull Penelitian-Hitam 563:33–42

Morton J (1987) Fruits of warm climates, J.F. Morton, Miami, pp 115–121

NAS (1979) Tropical legumes: resources for the future, NAS, Washington, DC, pp 117–121

Navaneetha N, Palniswamy KP, Abdul-Khader M, Kumar N (1990) Studies on rooting of cuttings cuttings of tamarind. South Indian Hortic 39(4):220–224

Osawa T, Tsuda T, Watanabe M, Ohshima K, Yamamoto A (1994) Antioxidative components isolated from the seeds of tamarind (*Tamarindus indica* L.). J Agric Food Chem 42(12):2671–2674

Padolina F (1931) Vegetative propagation experiments and seed germination. Philipp J Agric 2(4):347–355

Panigrahi S, Bland B, Carlaw PM (1989) Nutritive value of tamarind for broiler chicks. J Anim Feed Sci Technol 22(4):285–293

Pathak RK, Ojha CM, Dwivedi R (1992) Adopt patch budding for quicker multiplication in tamarind. Indian Hortic 36(2):17
Postgate J (1979) Nitrogen fixation. The Camelot Press Ltd, Southampton
Prasad A (1963) Studies on pollen germination in *Tamarindus indica* L. Madras Agric J 50L:202–203
Purseglove JW (1987) Tropical crops. Dicotyledons. Longman Science and Technology, London, pp 204–206
Purushotham K, Narasimharao SBS (1990) Propagation of tamarind by veneer and soft wood grafting. South Indian Hortic 38(4):225
Raizada U (1965) Life history of *Scirtothrips dorsalis* Hood. with detailed external morphology of the immature stage. Bull Entomol 6:30
Rama Rao M (1975) Flowering plants of Travancore. Bishen Singh Mahendra Pal Singh, Dehra Dun, 484 p
Roy RN, Maiti SS, Mondal CR (1987) Tamarind seed husk as an effective fish poison. Environ Ecol 5(3):467–471
Salim A, Simons A, Waruhin A, Orwa C (1998) Agroforestry tree database: a tree species reference and selection guide and tree seed suppliers directory. International Council for Research in Agroforestry, PO Box 30677, Nairobi, Kenya
Shankaracharya NB (1998) Tamarind-chemistry, technology and uses-a critical appraisal. J Food Technol 35(3):193–208
Singh M, Singh AK, Singh AK (1989) Screening of extracts of higher plants for virus inhibitory action. Geobios (India) 16(2–3):74–77
Singha RK (1995) Biodiversity conservation through faith and tradition in India. Some case studies. Int J Sustain Dev World Ecol (UK) 2(4):278–284
Splittstoesser WE, Mohamed Yaseen Y (1991) In vitro shoot regeneration of tamarind (*Tamarindus indica* L.) and carob (*Ceratonia siliqua*) with thidiazuron. In: Proceedings of the Inter-American society for tropical horticulture, 35–37th annual meeting, Vina-del-mar, Chile, 7–12 October 6–8
Srivasuki KP, Reddy RD, Reddy KK (1990) Rooting of terminal cuttings of *Tamarindus indica* Linn. Indian Forester 116:984–985
Storrs AEG (1995) Know your trees. Some common trees found in Zambia. Regional Soil Conservation Unit (RSCU), Zambia
Swaminath MH, Ravindran DS (1989) Vegetative propagation of fruit yielding tree species. My For 25(4):57–360
Swaminath MH, Ravindra DS, Mumtaz J (1990) Propagation of tamarind through stem cuttings. My For 26(2):207–208
Thimmaraju KR, Narayana Reddy MA, Swamy N, Sulladmath UV (1977) Studies on the floral biology of tamarind (*Tamarindus indica* L.). Mysore J Agric Sci 11:293–298
Timyan J (1996) BWA. Yo: Important trees in Haiti. Southeast Consortium for International Development, 1634, 1 Street N.W. Suite 702, Washington, DC, 20006
Usha K, Singh B (1996) Influence of open and cross pollination on fruit set and retention in tamarind (*Tamarinds Indica* L.). Rec Hortic 3(1):60–61
Verheij EMW, Coronel RE (eds) (1991) Edible fruits and nuts. Plant Resources of Southeast Asia, Handbook 2, Wageningen
Yaacob O, Subhadrabhandu S (1995) The production of economic fruits in South East Asia. Oxford University Press, Oxford, pp 133–142.1

14 Guava (*Psidium guajava*)

Abstract

Guava (*Psidium guajava*) is becoming a popular fruit crop in India because of its wider adaptability to soil and climatic conditions apart from its ability to produce fruits throughout the year. Guava, the "poor man's fruit" or "apple of the tropics," is a popular fruit tree of the tropical and subtropical climates and was native to the tropical America. It is very hardy in nature and is least attacked by insects and pests. It grows on different soils and in both humid and dry climates. The largest producer countries are Mexico, Brazil, India, and Thailand. The fruits are eaten fresh or are processed into juices and jams.

14.1 Introduction

Guava (*Psidium guajava*) is becoming a popular fruit crop in India because of its wider adaptability to soil and climatic conditions apart from its ability to produce fruits throughout the year. It is very hardy in nature and is least attacked by insects and pests. It has been cultivated in India since the early seventeenth century. Apart from fresh consumption it is a potential crop for value addition, and many of its products (pulp concentrate, jelly, etc.) have export potential. Guava is a rich source of vitamin C, which is 2–5 times than fresh oranges, and fair source of vitamin A, calcium, phosphorous, pantothenic acid, riboflavin, thiamin, and niacin. It is also a good source of pectin. Guava seeds contain iron as well as oil. The leaves yield dye and tannin. The leaves have also medicinal value for curing diarrhea. It is an ideal tree for home gardening.

Guava (*Psidium guajava* L.; $2n=22$) belongs to the genus *Psidium* of the family Myrtaceae and is native to the tropical regions where it grows on different soils and in both humid and dry climates. The largest producer countries are Mexico, Brazil, India, and Thailand. The fruits are eaten fresh or are processed into juices and jams. In some areas the fruits are also used to make pickles and soups (Landrum et al. 1995). Moreover, the positive effects of guava extracts on human ailments have been described (Lozoya 1999). The pharmacological actions and the medicinal uses of aqueous extracts of guava leaves in folk medicine include the treatment of various types of gastrointestinal disturbances such as vomiting, diarrhea, and gastric pain (Lutterodt 1992; Lozoya et al. 1994).

S. Hiwale, *Sustainable Horticulture in Semiarid Dry Lands*,
DOI 10.1007/978-81-322-2244-6_14, © Springer India 2015

14.2 Area and Production

It ranks fourth in area and production. UP has about half of the total area under guava, and district Allahabad has the reputation of growing the best guava in the country as well as in the world (Mitra and Bose 1985).

Since 1991 the area and production under the crop have more than doubled; however, the productivity has stagnated (Table 14.1). The contribution of guava in the overall fruit production varied from 3 to 4 % with a production increase of 1.5 times that of 1991.

Data on state statistics in respect indicated that in 2008 and 2009 area was highest in UP, whereas it was reduced to almost half during 2010. In MS, however, it showed a steady increase. There was a lot of variation among the states in respect to productivity per ha. It was maximum in UP followed by Punjab (Table 14.2)

Table 14.1 All-India area, production, and productivity of guava

Year	Area (in 000' ha)	% of total FRT. area	Production (in 000' mt)	% of total FRT. production	Productivity (in mt/ha)
1991–1992	94.0	3.3	1,095.1	3.8	11.7
2001–2002	154.6	3.9	1,715.5	4.0	11.1
2002–2003	154.6	4.1	1,793.0	4.0	11.6
2003–2004	166.4	3.6	1,830.7	4.0	11.0
2004–2005	161.0	3.2	1,682.8	3.4	10.5
2005–2006	166.5	3.1	1,736.6	3.1	10.4
2006–2007	176.0	3.2	1,831.0	3.1	10.4
2007–2008	179.0	3.1	1,981.0	3.0	11.1
2008–2009	204.0	3.3	2,270.0	3.3	11.1
2009–2010	219.7	3.5	2,571.5	3.6	11.7
2010–2011	205.0	3.2	2,462.0	3.3	12.0

Bijay Kumar et al. (2011)

Table 14.2 State-wise area, production, and productivity of guava

	Area in 000' ha			Production 000' mt			Productivity mt/ha		
	2008–2009			2009–2010			2010–2011		
State	Area	Production	Pdy.	Area	Production	Pdy.	Area	Production	Pdy.
Maharashtra	32.5	257.3	7.9	33.5	258.0	7.7	36.0	311.0	8.6
M.P.	–	–	–	8.2	238.5	29.0	9.7	280.8	29.0
UP	34.0	412.6	12.1	39.9	486.7	12.2	14.6	241.4	16.5
Bihar	29.1	229.2	7.9	29.2	231.5	7.9	29.4	235.2	8.0
West Bengal	12.5	170.5	13.7	13.4	175.7	13.1	13.6	178.8	13.2
Punjab	8.0	160.5	20.0	8.0	169.3	21.2	7.8	171.0	21.8
Gujarat	9.3	144.2	15.4	9.8	156.6	15.9	10.2	150.7	14.7
Karnataka	7.0	137.5	19.6	7.2	138.8	19.3	7.3	141.6	19.4
A.P.	10.8	162.0	15.0	10.1	150.8	15.0	8.5	128.2	15.0
Chhattisgarh	–	–	–	–	–		13.4	102.8	7.7
Orissa	–	–	–	14.1	100.0	7.1	14.2	102.1	7.2
Tamil Nadu	9.2	101.7		7.0	92.5	13.2	7.6	100.8	13.2
Others	52.2	494.8	11.0	39.242	372.0	9.5	32.4	317.9	9.8
Total	203.7	2,270.1	9.7	219.7	2,571.5	11.7	204.8	2,462.3	12.0

Bijay Kumar et al. (2011)

14.3 Climate and Soil

Guava can be a very hardy fruit tree and can be grown in tropical as well as subtropical climate up to 1,500 m above mean sea level. It can withstand high temperatures of the North and has ability to withstand drought but is susceptible to severe frost. Superior quality guava can be obtained if night temperature is below 10 °C that prevails during winter season. A well-distributed rainfall of 100 cm is sufficient for raising a good crop of guava. However, rains during harvesting can spoil the quality of fruits. It requires dry atmosphere at the time of flowering and fruiting; high temperature at the time of fruit development causes fruit drop (Mitra and Bose 1985).

Guava can be cultivated on various types of soils – heavy clay, very light sandy soils. Nevertheless, very good quality guavas are produced in river basins. It tolerates a soil pH of 4.5–8.5 (Elbaradi 1975). Maximum concentration of its feeding roots is available unto 25 cm soil depth. Thus, the topsoil should be quite rich to provide enough nutrients for accelerating new growth, which bears fruits. It does not stand waterlogging. Some plantations of guava are coming up well in river beds in Tamil Nadu and Uttar Pradesh.

14.4 Varieties

The varietal characteristics in guava are not as distinct as found in the majority of other fruits. Its propagation through seeds reduces the distinctive characteristics of a variety in commercial cultivation. Important guava varieties are shown below.

Variety	Shape	Size	Skin color	Taste	Keeping quality
Lucknow 49	Roundish ovate	Medium	Primrose yellow	Sweet	Excellent
Allahabad Safeda	Roundish	Medium to large	Yellowish white	Sweet	Excellent
Banarasi	Round	Medium	Yellow	Sweet	Good
Chittidar	Subglobose	Medium to small	Straw yellow with a few red dots on skin	Sweet	Poor
Red fleshed	Roundish ovate	Medium	Saffron yellow with a few red dots on skin	Sweet	Good
Apple color	Spherical	Medium	Dawn pink with deep minute dots on skin	Sweet	Good
Behat coconut	Round	Medium	Aureolin	Sweet	Good
Pear shaped	Pyriform	Medium	Straw yellow with large dots	Sweet	Good
Seedless	Oblong to globose	Medium to large	Straw yellow; flesh is creamy white	Sweet	Excellent
Baruipur	Round	Medium	Dresden yellow; flesh is white	Sweet	Good
Hafshi	Spherical	Medium	Yellow; flesh is red	Sweet	Good
Safeda Jam (Allahabad Safeda x Kohir)	Roundish	Medium	Yellowish white	Sweet	Good
Arka Mridula	Roundish	Medium	Yellowish white with few soft seeds	Sweet	Excellent

Table 14.3 Growth characteristics of guava varieties

Variety	Plant height (m)	Stock dia. (mm)	Scion dia. (mm)	Spread NS (m)	EW (m)
Apple color	2.005	68.03	67.54	2.38	2.46
Sardar	2.57	88.90	79.65	3.42	3.42
Allahabad Safeda	2.12	73.12	73.17	2.81	2.66
Selection-8	1.99	67.06	66.44	2.44	2.30
Chittidar	2.02	64.19	60.03	2.35	2.30
Red fleshed	2.29	73.74	67.59	2.39	2.42
SEM±	0.007	3.08	3.24	0.15	0.134
CD at 5 %	0.020	9.29	9.75	0.456	0.404

Table 14.4 Fruit retention, yield, and physicochemical characteristics of guava

Treatment	% retention	Yield/plant (kg)	Fruit wt. (g)	Fruit length (mm)	Fruit dia. (mm)	TSS °Brix	Acidity (%)	100-seed wt (g)	Seed no./ fruit
Apple color	76.24	8.95	70.12	51.52	50.0	12.55	0.945	1.47	328
Allahabad Safeda	67.69	13.72	64.22	48.42	48.9	16.05	0.50	0.89	566
Selection-8	60.32	12.73	87.38	55.52	51.53	13.25	0.427	0.772	485
Red fleshed	70.34	10.26	67.83	44.31	51.3	17.25	0.542	1.502	530
Sardar	40.26	11.72	172.75	65.66	68.7	18.00	0.553	1.775	504
Chittidar	70.53	10.14	83.22	51.86	51.9	18.85	0.528	1.03	517
SEM±	NS	0.483	6.37	0.755	1.254	0.82	0.108	0.022	–
CD at 5 %	–	1.455	19.23	2.27	3.78	2.46	0.326	0.066	–

Trial on six varieties of guava was carried out under semiarid areas of Gujarat. The performance of varieties indicated that plant height was maximum in cv. Sardar and minimum in Selection-8 (Arka Mridula). Stock and scion diameter were maximum in cv. Sardar and minimum in Chittidar. North–south and east–west plant spread were highest in cv. Sardar and least in Chittidar (Table 14.3).

The highest percent fruit retention was observed in apple color and minimum in cv. Sardar, but fruit weight was maximum in cv. Sardar and minimum in Allahabad Safeda followed by Selection-8 and was least in apple color. Maximum TSS was recorded in Chittidar and minimum in apple color. Acidity was least in Selection-8 and highest in apple color. 100-seed weight per fruit was least in Selection-8 followed by Allahabad Safeda indicating a small seed size in the fruits and was maximum in cv. Sardar which have bold seeds. Looking in to higher yield potential and soft seededness, cv. Allahabad Safeda was found to be the best followed by Selection-8 which therefore can be recommended for commercial cultivation under semiarid areas (Table 14.4).

14.5 Propagation

Guava is propagated both by seeds and vegetative means. But vegetative propagation is commercially followed.

14.5.1 Seed Propagation

The propagation of guava through seeds should not be encouraged because the seedlings have a long juvenile phase, give lower yields, and bear poor-quality fruits. But the seedlings serve as rootstock material for grafting or budding. The seeds should be sown as soon as possible after extraction from the ripe fruits. Soaking of seeds in water for 12 h. or in hydrochloric acid for 3 min gives about 90 % germination. About 1-year-old seedlings become ready for grafting or budding. For plantation seedling, seeds should be collected from the plants producing high-quality fruits and high yield. Guava seeds lose viability within a short period after extraction; however, viability can be prolonged with potassium nitrate at 1 % and parahydroxybenzoic acid at 10^{-3} M (Ghosh 1978).

14.5.2 Vegetative Propagation

In northern India, guava is propagated by inarching, giving a very high percentage of success during the rainy season. But inarching is cumbersome and gives a limited number of plants from the mother plant.

14.5.2.1 Budding

Budding has been adopted only on a limited scale in some parts of the country where atmospheric humidity is high. The main problem encountered in this method is disbudding of rootstock making it labor intensive. Among the various methods of budding – shield, Forkert, patch, and chip – the patch budding is ideal, giving the highest percentage of success. However, the best time of budding differs from locality to locality. The Forkert method of budding can be followed in July which is superior to patch and shield budding (Srivastava 1962).

14.5.2.2 Layering

Layering is being commercially followed in southern and western India with very good results. After bending the plant, its branches are covered with soil leaving the terminal portion open. In a few months the rooting of branches takes place which are then separated from the mother plants and planted in the nursery for further sale. Layering is a labor-intensive method. A limited number of plants can only be multiplied from a mother plant. When mother plants are very tall, air layering of shoots is done during the rainy season using polythene and moist sphagnum moss. The use of root-promoting plant growth regulator, IBA (3,000 ppm), promotes the rooting of air layers up to 100 %. The main limitation of air layering is the poor establishment of air layers in the nursery after detachment from the mother plant. Further, the method is very cumbersome and labor intensive.

14.5.2.3 Cutting

Propagation of guava by cuttings under ordinary conditions may not be done. It is only successful under intermittent mist conditions with the aid of rooting hormones like IBA and NAA. A very high percentage (93.3 %) of rooting has been obtained in semi-hardwood cuttings under mist p-hydroxybenzoic acid (200 ppm) and IBA (5000 ppm) treatment as quick dip (Dhua, et al. 1982). However, under a plastic house without mist, a high relative humidity was maintained by trapping the water vapor produced after irrigation, and IBA 2500 ppm-treated hardwood cuttings of guava cv. 'Allahabad Safeda' planted in pots containing sand resulted in 87.5% rooting and 62.86% survival of rooted cuttings (Reddy and Singh 1988).

14.5.2.4 Stooling

Stooling is the easiest and cheapest method of guava propagation. The self-rooted plants (cutting or layers) are planted 0.5 m apart in the stooling bed. These are allowed to grow for about 3 years. Then these are cut down at the ground level in March. New shoots emerge on the beheaded stumps. A 30 cm wide ring of the bark is removed from the base of each shoot rubbing the cambium of the exposed portion in May. All the shooters are mounded with the soil to a height of 30 cm. The soil is covered with mulch to conserve the moisture. After a period of 2 months upon the onset of monsoon, the shoots are detached from the mother plant at ringed portion and planted in the nursery. The shoots are headed back to maintain the root and shoot balance before planting in the nursery. By following the technique of ringing and mounding of the shoots, second time stooling is done on the same mother stools in the first week of September. The rooted stool layers are detached in the first week of November. Thus, stooling is done twice on the same mother stools in a year. The stooling of a mother stool can be done for many years. With the advancement in its age, the number of stool layers also increases every year. The growth and development of stool layers are better than seedlings. The application of rooting hormone is not required. However, it is reported (Rathore 1984) that in Punjab, propagation of guava by stooling has become a commercial practice.

A semidwarfing rootstock for guava Aneuploid No. 82 has been developed.

14.6 Cultivation

14.6.1 Planting

The field for planting is prepared during the summer season by plowing, leveling, and removing weeds. The pits of 1 m × 1 m × 1 m size are dug and filled with a mixture of farmyard manure and soil. If the soil is good and irrigation facilities are available, the preparation of land digging of pits is not required. The planting is done during the rainy season by adopting square planting system. Guava is commercially planted at a distance of 5–8 m. The exact planting distance is, however, decided according to variety, soil fertility, and availability of irrigation facilities. Guava Lucknow 49 needs more spacing than apple color and Allahabad Safeda. Under irrigation and high soil fertility, the plants become very vigorous requiring more spacing. In normal conditions, a planting distance of 7 m is optimum; high-density planting reduces total soluble solids, sugars, and ascorbic acid but increases titratable acidity. The lower plant population results in the spread of the crown, while higher-density planting causes erect growth of branches making the plant tall and compact. High-density planting gives higher yield/unit area in early years of fruiting.

14.6.2 Manure and Fertilizers

Although guava is grown without the application of any manure and fertilizer, it responds very well to their application by giving higher yield and better-quality fruits. For guava-growing regions of the country, different fertilizer schedules – 600 g N and 400 g K in the northern region; 260 g P and 260 g K in the eastern region; 900 g N, 600 g P, and 600 g K in the southern region; and 600 g N, 300 g P, and 300 g K/plant/year in the western region – have been recommended.

The fertilizer application should be based on the leaf nutrient status of an orchard, wherever feasible. The time of fertilizer application depends on the crop taken and the region. In north India, fertilizer is given in the first week of May for the rainy season crop and in the first week of July for the winter season crop. In West Bengal, fertilizers are applied in two equal split doses, one in January and the other in August. At Bangalore, full K and 70 % N are applied in June and full P and 30 % N in September. Since 48 % of feeder roots of guava are found in the surface soil up to 25 cm depth, the fertilizer should be placed in 25 cm trenches 1 m away from the trunk for better uptake. During June FYM and fertilizers are applied followed by irrigations till rains are received.

Sometimes guava suffers a deficiency, which is characterized by reduction in leaf size, interveinal chlorosis, suppression of growth, and dieback of leaders. It can be corrected by spraying of ZnSO4 (0.45 kg) and hydrated lime (0.32 kg) in water (33 L). Bronzing is another common problem in guava. It is caused by the deficiency of B, Zn, N, P, and K due to low soil pH. The soluble P level of leaves is a better index for bronzing. Guava Lucknow 49 is more susceptible than Allahabad Safeda. Improving the soil pH and treating the soil with N, P, K, and Zn at 200, 80, 150, and 80 g/year, respectively, or fortnightly foliar spraying of these nutrients each at 2 % for 4 months can reduce it.

14.6.3 Irrigation

Guava is mostly grown under rainfed condition, and irrigation is rarely practiced wherever this facility available. However, irrigation enhances the yield of guava by making the plant more vigorous and increasing the fruit set. Irrigation is especially desirable after planting for survival of the plants and thereafter for 2–3 years to obtain early good growth. Irrigation of fruiting plants depends upon the adoption of a particular cropping pattern. For the whole year, irrigation, a cropping pattern which is commercially adopted all over the country except the northern region, is given during the summer season. Normally, the winter season cropping pattern is adopted in north India which requires fortnightly irrigation during October–November. Irrigation is given to make the soil of the root zone moist; thus, heavy irrigation is unnecessary. The fruit quality of guava is adversely affected by high moisture content during harvesting.

14.6.4 Training/Pruning

Traditionally, no pruning is done in guava because the plant bears heavily even without it. But no pruning results in the formation of narrow crotches, limb breakage due to heavy fruit load, and overcrowding. Therefore, training of plants in young stage to build strong framework and to avoid weak crotches is necessary, whereas fruiting trees should be pruned to check overcrowding in the orchard. The plants should be trained as low-headed trees to facilitate multiple handpickings. The open center or delayed open center system may be adopted.

The scaffold branches in young plants are to be tipped back to encourage secondary branching. The root suckers, water sprouts, and crisscross branches are to be removed altogether. In Maharashtra, bending of horizontal branches is practiced to some extent by tying the branches of 2 adjoining plants to increase fruiting in young plants, but it is labor intensive and creates hindrances in cultural operations.

In every growing season, a large number of new shoots emerge in guava, the majority of which are lateral and a few are terminal. These shoots produce fruits. After, 1 year most of the lateral shoots dry out, while terminal shoots put forth the extension growth. Hence, to check the overcrowding and to control the plant height, the terminal shoots on the periphery may be headed back at about 40 cm level in alternate years. Pruning also takes place during harvesting as the fruit is plucked along with the shoot on which it is borne. Pruning is usually recommended after harvesting or in spring. Summer pruning may damage the plant by sunburning.

14.7 High-Density Planting

Guava Aneuploid No. 82, when planted at 3 × 3 m with 1111 plants per ha., yielded 8–20 tons of guava per ha (Sharma et al. 1992). Gorakh Singh (2005) reported that cv. Allahabad Safeda when planted at 2 × 1 m spacing 5000 plants per ha., with suitable pruning management, can yield 48 tons of guava per ha (5 years old).

14.8 Aftercare

Guava plants do not require much are after planting. The weeds are removed by shallow cultivation. Green manuring should be done during the rainy season and clean cultivation during the rest of the year. Leguminous crops can be grown as intercrops during the first 3 years of planting to obtain more income and to increase the N content of the soil. Both rainy and winter season crops are very heavy compared with spring crop. Fruit quality of the winter crop is best. Therefore, winter crop is preferred over the rain season crop. In northern India, normally hot and dry summers along with low soil moisture do not allow summer flowers to set the fruits. But in mild summer and normal soil moisture, the summer flowers set fruits for rainy season crop which is known for its poor-quality fruits and severe incidence of fruit fly and fruit borer. The practice of taking winter crop instead of rainy season crop is known as crop regulation. The rainy season crop can be removed by spraying of urea (10 %) on Allahabad Safeda and 20 % on Lucknow 49 at the time of peak flowering in the summer season. Other methods of removing rainy season crop such as hand removal of flowers and fruits, spraying of bioregulators, root exposure, withholding irrigation, and pruning of ¾ of the flower-bearing shoots are either costlier or impractical or ineffective.

14.9 Time of Flowering and Fruiting in Guava

Under natural conditions, a guava tree produces flowers and fruits twice in a year in northern India, but it is thrice, i.e., almost throughout the year, in western and southern India which results in a no rest period, and ultimately a guava tree bears small crops at different times of the year; this pattern of flowering and fruiting is not desirable for commercial cultivation. The well-defined periods are:

1. *Ambe bahar*: When a guava tree flowers during February–March or spring season, this flowering period is known as *ambe bahar*. The fruits ripe from July to September in the rainy season. The fruits obtained during this season are insipid, watery, and poor in quality (i.e., taste and keeping quality).
2. *Mrig bahar*: When a guava tree produces flowers in June–July or the monsoon season, this flowering period is known as *mrig bahar*. The fruits ripe from November to January in the winter. The fruits obtained during winter are excellent in quality, and, therefore, the guava trees are made to produce the *mrig bahar* flowering only.
3. *Hasth bahar*: Sometimes, a guava tree produces flowers in October, this is known as *hasth bahar*. The fruits ripe from February to April. The quality is good, but yield is very low. However, it fetches good prices. *Hasth bahar* in guava is not very common. It is easy to induce. It is mostly a chance crop. *Hasth bahar* is observed in western and southern India.

14.10 Effect of Different Mulches on Yield, Quality, and Water Content of Guava Allahabad Safeda

At CHES, Vejalpur, under rainfed conditions, guava fruit development takes place during the period when moisture depletion is very fast. There is shortage of water, thereby affecting the yield and quality of guava. An experiment was initiated at the Central Horticultural Experiment Station, Godhra, Gujarat, in a randomized maize block design. The results of the study revealed that relative water content of cv. Allahabad Safeda was highest under maize straw (84.45), closely followed by black polythene. However, after two months of rainfall, it was black polythene mulch under which maximum RWC was recorded. It was on par with maize straw (80.07 %).

The highest yield per plant was obtained in mulching with maize straw (17.60 kg) closely followed by black polythene mulch. It was least in control (10.57 kg). All the treatments had non-significant effect on the physicochemical characteristics of fruits. The results thus clearly indicated that mulching with maize straw and black polythene helps substantially in conserving soil moisture resulting in increased fruit yield (Hiwale and Raturi 1997) (Tables 14.5 and 14.6).

14.11 Harvesting and Postharvest Management

Guava is harvested throughout the year (except during May and June) in one or other regions of the country. However, peak harvesting periods in north India are August for rainy season crop, November–December for winter season crop, and March–April for spring season crop. In the mild climatic conditions of other parts of the country, the peak harvesting periods are not so distinct. Guava fruits develop best flavor and aroma only when they ripen on tree. However, seeds from fruits only 8 weeks old become fully mature and viable indicating that seeds from hand-pollinated fruits can be sown only 8 weeks

Table 14.5 Effect of mulching on yield and quality of guava cv. Allahabad Safeda

Treatment	Yield per plant (kg)	% retention	% increase over control	Fruit weight (kg)	Fruit length (kg)	Fruit dia. (mm)	TSS (Brix)	Acidity (%)
Maize	17.61	69.86	66.6	50.00	51.35	47.27	19.0	0.54
Paddy	16.98	50.89	60.64	47.75	54.37	46.55	18.2	0.55
Subabool	15.72	63.46	48.72	40.50	46.92	46.65	19.4	0.59
Polythene	17.32	74.21	63.86	48.75	50.77	46.65	19.6	0.57
Jal shakti	11.80	44.29	11.63	41.25	47.50	46.75	18.0	0.55
Control	10.57	45.32	–	48.50	49.72	46.47	17.4	0.53
SEM	1.32	5.94	–	–	1.35	–	–	–
CD at 5 %	2.81	17.87	–	–	4.07	–	–	–

Hiwale and Raturi (1997)

Table 14.6 Effect of mulching on moisture content of guava cv. Allahabad Safeda

	RWC	RWC	RWC	
Treatment	Immediately after rainfall	One month after rainfall	Two months after rainfall	% moisture over control
Maize	90.28	84.45	80.07	6.19
Paddy	89.16	81.86	78.23	3.75
Subabool	87.39	83.16	77.22	2.41
Polythene	88.68	82.34	81.96	8.70
Jal shakti	88.61	83.88	75.96	4.74
Control	87.82	79.43	75.40	–
SEM	1.15	1.02	1.18	–
CD at 5 %	3.47	3.08	3.57	–

Hiwale and Raturi (1997)

after pollination (Welgemoed and Preez 1986). Fruits attaining maturity show signs of changing color from dark green to yellowish green (Singh et al. 1967).

In most of the commercial varieties, the stage of fruit ripeness is indicated by color development, which is usually yellow. For local market, fully yellow but firm fruits are harvested, whereas half yellow fruits should be picked for distant markets. The fruits are harvested selectively by hand along with the stalk and leaves. The plants begin bearing at an early age of 2–3 years, but they attain full bearing capacity at the age of 8–10 years. The yield of a plant depends on its age, cropping pattern, and the cultural practices. A 10-year-old plant yields about 100 kg of fruits every year. If both rainy and winter season crops are taken, more yields may be obtained in the rainy season.

Ripening of guava starts on the tree and continues even after harvest. It is accelerated in the rainy season due to high temperature and slows down in winter season due to low temperature. The fruits are packed in baskets made from locally available plant material. For distant markets, wooden or corrugated fiberboard boxes are used along with good cushioning materials – paddy straw, dry grass, guava leaves, or rough paper. Good ventilation is necessary to check for buildup of heat.

Guava is a delicate fruit requiring careful handling during harvesting and transporting. The fruits should reach consumers in a firm condition. Because of their perishable nature, guavas are disposed off immediately after harvesting in the local market, and a very small quantity is sent to distant markets. The shelf life of guava fruits has also been found to increase for about 10–12 days when they are dipped for 5 min in a solution of

NAA at 150 ppm and packed in 150-gauge polythene bags with vents (Dhoot et al. 1984). Since fruits are sold at a cheaper price and are available for a very long period of the year, they are not kept in cold storage. However, the shelf life of guava can be extended up to 20 days by keeping them at low temperature of 50 °C and 75–85 % relative humidity. It can also be stored for about 10 days at room temperature (180–230 °C) in polythene bags providing a ventilation of 0.25 % (Tandon et al. 1984).

Six varieties of guava, viz., Selection-8, Chittidar, Allahabad Safeda, apple color, red fleshed, and Sardar, have very short shelf life. Therefore, in order to extend their shelf life, the fruits of these varieties were treated with various growth regulators and packed in polythene bags both perforated and unperforated along with refrigerated storage. The data indicated that percent weight loss was highest in fruits treated with GA50 ppm and least in refrigerated storage followed by polythene bags (200 gauge). There was improvement in total soluble solids in refrigerated storage compared to other treatments. Therefore, storing fruits of six guava varieties under refrigerated condition was the best for extending the shelf life of guava fruits up to 8 days (Tables 14.7, 14.8, and 14.9).

Table 14.7 Postharvest evaluation (fruit fresh wt, g)

Variety	Selection-8	Chittidar	Allahabad Safeda	Apple color	Red fleshed	Sardar
GA50 ppm	74	90	74.6	57.8	76.8	143.0
BA50 ppm	67	73	70.8	54.6	73.2	115.2
Polythene bags 2% perforation	67.2	83	68.3	51.4	69.4	88.0
Polythene bags	66.3	81.3	63.6	60.3	69.3	90.6
Refrigerator	52.3	84.3	73.	54.6	63.6	92.0

Table 14.8 Percent weight loss after 4 days

Variety	Selection-8	Chittidar	Allahabad Safeda	Apple color	Red fleshed	Sardar
GA50 ppm	15.62	20.54	11.86	12.82	14.2	10.06
BA50 ppm	13.43	10.88	10.72	12.45	11.19	8.68
Polythene bags 2% perforation	9.82	10.84	4.66	2.33	2.59	7.27
Polythene bags	1.05	1.30	1.23	2.01	1.01	1.10
Refrigerator	0.57	0.36	0.94	1.65	0.94	1.08

Table 14.9 Total soluble solids after 4 days' storage

Variety	Fresh	GA (50 ppm)	BA (50 ppm)	Polythene bags 2% perfo.	Polythene bags	Refrigerator
Selection-8	16.8	13.0	13.8	13.6	17.0	20.4
Chittidar	19.0	15.2	16.0	16.4	15.0	20.8
Allahabad Safeda	18.0	15.8	16.2	15.4	13.0	17.2
Apple color	19.0	11.8	15.0	14.2	16.0	18.0
Sardar	17.8	14.6	15.2	14.6	19.0	18.6
Red fleshed	20.0	12.6	14.0	14.0	19.8	27.6

Table 14.10 Export of guava from India

Country	2008–2009		2009–2010		2010–2011	
	Qty	Value	Qty	Value	Qty	Value
UAE	252.4	56.9	44.6	15.5	82.3	22.1
UK	23.9	14.9	38.0	19.1	19.0	12.6
Maldives	15.9	5.5	10.0	3.2	14.8	4.9
Oman	112.6	23.4	78.0	17.9	23.9	4.2
Bahrain	36.7	8.4	66.5	7.1	42.9	3.0
Nepal	268.7	19.8	59.7	4.4	37.2	2.8
Saudi Arabia	363.5	75.5	15.8	3.2	4.7	2.7
Mexico	0.0	0.0	0.0	0.0	17.4	2.5
Gambia	0.0	0.0	0.0	0.0	12.5	2.1
Malaysia	61.7	8.0	1.7	0.4	9.4	1.9
Others	556.1	92.2	202.0	42.7	25.2	4.4
Total	1,691.4	304.4	516.3	113.4	289.3	63.3

Bijay Kumar et al. (2011)

14.12 Pest and Diseases

14.12.1 Guava Fruit fly (*Dacus dorsalis* Hendel)

This becomes a dangerous pest during the monsoon. The fly lays eggs on the surface of the fruit. On hatching which takes a few days, the young white maggots enter into the fruit and feed on the soft pulp. In most cases fruits drop off.

14.12.2 Guava Wilt

Guava plants are attacked by wilt, which causes heavy losses. It is very difficult to find a guava orchard more than 30 years in age because most of its plants die at about 20 years of age due to wilt. The incidence is more severe during the rainy season and in alkaline soils (Mehta 1951). Various fungi causing wilt are *Fusarium roseum*, *F. oxysporum f. sp. pisi*, *F. solani*, *Macrophomina phaseolina*, and *Gliocladium roseum*. Resistant rootstock is the only solution. The planting material should not be obtained from a wilt-infected region or nursery (Table 14.10).

The major importers of guava are Saudi Arabia followed by the UAE in 2008–2009. However, there seems to be a declining trend in guava export in 2009–2010 and 2010–2011. It has dropped by three times that of 2008–2009 in 2009 and seven times that of 2008 in the year 2010. The situation needs to be carefully investigated as during these years the production has declined.

References

Bijay Kumar, Mistri NC, Singh B, Gandhi CP (2011) Indian horticulture data base, pp 79–80

Dhoot LR, Desai UT, Rane DA (1984) J Maharashtra Agric Univ 9(2):185–188

Dhua RS, Mitra SK, Sen SK, Bose TK (1982) Effect of ethephon and IBA on rooting of guava. Sci Cult 48:444–445

Elbaradi TA (1975) Guava. Abstr Trop Agric 1(3):19–26

Ghosh SN (1978). M.Sc. AG. thesis submitted to Bidhan Chandra Krishi Viswavidyalaya, Kalyani

Gorakh Singh (2005) Meadow orchard system in guava orchard. Indian Hortic 17

Hiwale SS, Raturi GB (1997) Effect of different mulches on yield, quality and water content of Guava Allahabad safeda. In: Paper presented in national conference on impact of biotechnology on modern horticulture on rural development at Jadavpur U. Calcutta on 20–22 November 1997

Landrum LR, Clark WD, Sharp WP, Brendecke J (1995) Hybridization between *Psidium guajava* and *P. guineense* (*Myrtaceae*). Econ Bot 49:153–161

Lozoya X (1999) Xiuhpatli, Herba officinalis. SSA/ UNAM, Mexico, p 127

Lozoya X, Meckes M, Aboud-Zaid M, Tortoriello J, Nozolillo C, Arnason JT (1994) Quercetin glycosides

in *Psidium guajava* L. leaves and determination of a spasmolytic principle. Arch Med Res 25:11–15
Lutterodt GD (1992) Inhibition of Microlax-induced experimental diarrhea with narcotic-like extracts of *Psidium guajava* leaf in rats. J Ethnopharmacol 37:151–157
Mehta PR (1951) Plant Prot Bull New Delhi 3:7–12
Mitra SK, Bose TK (1985) Guava. In: Bose TK (ed) Fruits of India-tropical and subtropical. Naya Prokash, Calcutta, pp 277–297
Rathore DS (1984) Punjab Hortic J 24(1–4):75–78
Reddy KM, Singh RN (1988) Indian J Agric Sci 58(1):81–82
Sharma YK, Goswami AM, Sharma RR (1992) Effect of dwarfing Aneuploid guava rootstock in high density orcharding. Indian J Hortic 49(1):31–36
Singh S, Krishnamurthy S, Katyal SL (1967) Fruit culture in India. ICAR, New Delhi
Srivastava RP (1962) Further studies in budding of guava. Sci Cult 28:433–434
Tandon DK, Adsule PG, Kalra SK (1984) Indian J Hortic 41(2):88–92
Welgemoed C, Preez RDu (1986) Inf Bull Citrus Subtrop Fruit Res Inst 162–164.

15 Wood Apple (*Feronia limonia* Linn.)

Abstract

The wood apple, *Feronia limonia* (Linn.) *Swingle*, synonyms *Limonia acidissima* L., *Feronia elephantum correa*, and *Schinus limonia* L., is the species of its genus, in the family Rutaceae. Besides wood apple, it may be called elephant apple, monkey fruit, curd fruit, kath bel, and other dialectal names in India. The plant is a native and common in the wild in dry plains of India and Ceylon. The plants are hardy with deep penetrating root system, which makes them survive harsh soil and climatic condition. In India, the fruit was traditionally a "poor man's food" until processing techniques were developed in the mid-1950s. The fruit is much used in India as a liver and cardiac tonic and, when unripe, as a means to halt persisting diarrhea and dysentery and effective treatment for hiccup, sore throat, and diseases of the gums. In a Poultice, the pulp is used to help with bites and stings of venomous insects, as is the powdered rind.

15.1 Introduction

The wood apple, *Feronia limonia* (Linn.) *Swingle*, synonyms *Limonia acidissima* L., *Feronia elephantum correa*, and *Schinus limonia* L., is the species of its genus, in the family Rutaceae. Besides wood apple, it may be called elephant apple, monkey fruit, curd fruit, kath bel, and other dialectal names in India. In Malaya, it is gelinggi or belinggi; in Thailand, Ma-khwit; in Cambodia, Kromsang; and in Laos, Ma-fit. In French, it is pomm-d-elephant, pomm de bois, or citron des mois, with chromosome no. $2n=18$ (Morton 1987).

15.2 Origin and Distribution

The plant is a native and common in the wild in dry plains of India and Ceylon. It is found growing; it is also frequently grown throughout Southeast Asia, in Northern Malaya, and in Penang Island. In India, the fruit was traditionally a "poor man's food" until processing techniques were developed in the mid-1950s. It occurs wild or cultivated, up to an elevation of 1,500 ft; in Western Himalayas, it is more common in the Deccan andThane and Chandrapur districts of Maharashtra. It is also reported to occur in parts of Hazaribagh, Palamau, and Chota Nagpur in

S. Hiwale, *Sustainable Horticulture in Semiarid Dry Lands*,
DOI 10.1007/978-81-322-2244-6_15, © Springer India 2015

Jharkhand. It is often cultivated on borders of fields and as a roadside tree near villages and is sometimes planted in orchards.

15.3 Composition

Analysis of the edible part (55–58 %) of the fruit gave (food value per 100 g edible portion) the following value.

Component	Fruit pulp	Seeds
Moisture (%)	64.20	4.00
Protein (%)	7.10	26.18
Fat (%)	3.70	27.10
Carbohydrates (%)	18.10	35.49
Ash (%)	5.00	5.03
Calcium (%)	13.00	1.58
Phosphorus (%)	110.00	1.43
Iron (%)	0.60	0.03
Tannins (%)	1.03	0.08
Ether extract (g)	0.60	–
Mineral matter (g)	1.90	–
Fiber (g)	5.00	–
Riboflavin (μg/100 g)	0.17	–
Vitamin C (mg/100 g)	3.00	–
Carotene (μg)	6.10	–
Thiamin (μg)	0.04	–
Niacin (μg)	0.80	–
Energy (k/cal)	134.00	–

Gopalan et al. (1984) and Morton (1987)

15.4 Food Uses

The pulp represents 55–56% of the whole fruit. The pectin content of the pulp is 3–5 % (16 % yield on dry weight basis). The seed contains bland, non-bitter oil high in unsaturated fatty acids. The rind must be cracked with a hammer. The scooped-out pulp, though sticky, is eaten raw with or without sugar or is blended with coconut milk and palm-sugar syrup and drunk as a beverage or frozen as an ice cream. It is also used in chutneys and for making jelly and jam. The jelly is purple and much like that made from black currants. A bottled nectar is made by diluting pulp with water, passing through a pulper to remove the seeds and fiber, further diluting, straining, and pasteurizing. A clear juice for blending with other fruit juices has been obtained by clarifying the nectar. Pulp sweetened with syrup of canes or palm sugar has been canned and sterilized. The pulp can be freeze-dried. A stiff jelly can be made from the wood apple, but the flavor is somewhat harsh, so it is seldom used alone, but more frequently mixed with such fruits as the guava. Anon (1956), however, speaks of the jelly as having an exceeding agreeable flavor. He also reported syrup and chutney as other possible products. He refers to two types of wood apple, one with fruit larger and sweeter than the other, and states that the ripe fruit pulp contains 2.3 % acid and 7.25 % sugars.

15.4.1 Medicinal Value

The fruit is much used in India as a liver and cardiac tonic and, when unripe, as a means to halt persisting diarrhea and dysentery and effective treatment for hiccup, sore throat, and diseases of the gums. In a poultice, the pulp is used to help with bites and stings of venomous insects, as is the powdered rind.

The juice of young leaves is mixed with milk and sugar candy and given as a remedy for biliousness and intestinal troubles of children. The powdered gum, mixed with honey, is given to overcome dysentery and diarrhea in children. Oil derived from the crushed leaves is applied on itch and the leaf decoction is given to children as an aid to digestion. Leaves, bark, roots, and fruit pulp are used against snakebite. The leaves are aromatic, carminative, and astringent.

15.4.2 Other Uses

15.4.2.1 Gum

From the trunk and branches of the tree exudes a gum resembling gum arabic in properties. The exudation is profuse after the rainy season. It is considered to be a good substitute for gum arabic and commercial samples of the latter are utilized as substitutes for, or adulterant of, gum arabic and are also used in making artists' watercolors, ink, dyes, and varnishes. It consists of 35.5 % arabinose and xylose, 42.7 % D-galactose, and traces of rhamnose and glucuronic acid.

15.4.2.2 Wood

The wood is yellowish gray to grayish white and hard and heavy (sp.gr. -0.83). It is somewhat refractory to season. It is durable both under cover and in exposed situations. It is used for house building, hanes of wheels, and oil crushers. It can be used as wood pattern wood, for shoe lasts, penholders, and similar articles. It is also used for agricultural implements and ornamental carving (Anon 1956).

15.4.2.3 Pectin

The pectin has potential for multiple uses in India, but it is reddish and requires purification. Fruit pulp contains 2.66 % pectin on fruit weight basis (Roy and Majumdar 1988).

15.4.2.4 Rind

The fruit shell is fashioned into snuffboxes and other small containers.

15.5 Area and Production

There are no regular plantations; however, plants are usually found along the border of fields, roads, railway lines, and banks of the river. Normally 20-year-old plants are found to bear 1,000–1,500 fruits per trees. However, the size of the fruit is varying. The fruit weight varies from 175 to 350 g. There is tremendous scope for increasing the area under this crop considering its production potential, which is as high as 1.75–3.50 q per tree per year, and multifarious uses.

15.6 Soil and Climate

The plants are hardy with deep penetrating root system, which makes them survive harsh soil and climatic condition. Wood apple can be grown on saline marginal lands, which are unsuitable for cultivation of other crops or trees. Once established they can withstand salinity and drought in a better way.

Mature plants can tolerant low temperature (0–15 °C) as well as a temperature as high as 47.7 °C (Troup 1921). Wood apple is found to be growing in high as well as low rainfall areas of tropical and subtropical conditions. It can also be grown in semiarid regions of India. The warm season appears conducive for the initiation of floral buds.

15.7 Taxonomy

The slow-growing tree is erect, with a few upward-reaching branches bending outward near the summit where they are subdivided into slender branchlets drooping at the tips. The tree grows up to an elevation of 450 m in the Western Himalayas. The bark is rigid, fissured, and scaly and there are sharp spines: 2–5 cm long on some of the zigzag twigs; Leaves pinnate, 8–10 cm long, The deciduous, petiole and rachis, flat, often narrowly winged, glabrous, entire, alternate Leaves, dark green, leathery, often minutely toothed, blunt or notched at the apex, are dotted with oil glands and slightly lemon-scented when crushed. The leaflets 3–9, opposite, sessile, or short petiolate, obovate with crenate tip and cuneate base and glabrous. Flowers are dull red or greenish, 1.25 cm wide, and are borne in small, loose, terminal, or lateral panicles. They are usually bisexual. The calyx is very small with 5–6 lobs, deciduous; petals 5–6, elliptic-oblong, spreading or bent downward. The fruit is round to oval, 5–12.5 cm wide, with a hard, woody, grayish white, scurfy rind about 6 mm thick. The pulp is brown, mealy, odorous, resinous, astringent, and acid or sweetish, with numerous small, white seeds scattered through it. It grows wild and is considered an indigenous fruit. It is found in gardens, but is not grown systematically.

Warm season appears conducive for the initiation of floral buds. The tree sheds its leaves and branches are bare for a short period during January, and flowering starts in February–March. Flowering is mainly on new shoots; emergence of panicles commences in the middle of February and continues up to the 3rd week of May. Opening of flowers starts in the 2nd week of

March. The flowers are mainly staminate and hermaphrodite. Ovary, style, stigma are in both hermaphrodite and male flower but rudimentary in the latter. Average fruit weight is 350 g with rind thickness of 0.3–0.6 cm, acidity 1.04–4.50 %, and total sugar 4.08–4.47 %.

15.8 Germplasm Conservation and Improvement

Breeding and improvement is not studied so far systematically. The plants growing so far or of seed origin and found to have lot of variability which can be used for making selection of superior types. There are two forms: one with large, sweetish fruits and the other with small, acidic fruits.

15.9 Ideotypes of Wood Apple

High yielding (more than 500 fruits on 10-year-old plants)
Earliness in bearing
Dwarf nature
Thin shell (thickness less than 0.3 cm)
Less fiber
High TSS (more than 15 °Brix)
Pulp color dark brown
Strong aroma

15.10 Evaluation

Twenty germplasm lines collected from different places were evaluated at Central Horticultural Experiment Station, Vejalpur (Gujarat). Seed germination ranged from 15.56 to 62.40 %. Survival percentage ranged from 80.00 to 94.00 %. It is observed that germination percentage was higher in ACC No.-2 followed by ACC No.-4 and percent survival was higher in ACC No.-5 followed ACC No.-6 (Table 15.1).

There was significant variation in various parameters like fruit length, fruit diameter, pulp weight, skin weight, TSS, number of seeds, and seed weight. The following were recorded: fruit weight ranged from 187 to 350.53 g, fruit length 71.50–95.06 mm, fruit diameter 69.84–86.00, pulp weight 83.23–155.69 g, skin weight 82.93–164.48 g, TSS ranged from 10.67 °Brix to 14.33 °Brix, number of seed (239.67–505.00), and seed weight 3.00–15.00 g, indicating that there is lot of variability in the species, which needs to be collected and evaluated for identifying high yielding cultivars for commercial cultivation.

Table 15.1 Germplasm evaluation of wood apple

ACC. no.	Germination (%)	Survival (%)	Fruit wt. (g)	Fruit length (mm)	Fruit dia. (mm)	Pulp wt. (g)	Skin wt. (g)	TSS °Brix	Seed no.	Seed wt. (g)
1	22.88	87.00	289.83	81.90	80.33	106.72	125.86	12.67	409.67	9.00
2	62.40	90.00	257.16	74.73	82.33	83.23	116.84	12.33	262.00	7.00
3	15.56	80.00	217.83	73.06	79.53	92.47	87.72	14.33	505.00	15.00
4	33.47	83.00	215.16	75.13	77.33	92.03	82.93	11.33	224.00	13.07
5	25.16	94.00	350.53	86.20	84.16	155.69	164.48	14.33	451.33	12.67
6	21.14	91.00	328.67	95.06	86.00	148.15	158.42	10.67	239.67	3.99
7	–	–	187.00	71.50	69.84	86.63	86.30	12.50	247.00	7.33
CD at 5 %	–	–	34.13	3.87	7.90	18.45	16.05	1.79	28.45	1.53

Annon (2003)

15.11 Vegetative Growth Characteristics of Wood Apple Germplasm

There were significant differences in respect to all the growth parameters of wood apple during the year except stock diameter. Maximum plant height was recorded in line CHES-10 (7.37 m) and least in CHES-11 (5.5 m). Stock and scion diameter was highest in line CHES-1 and least in CHES-11. Plant spread in both North south and east-west direction was maximum in line CHES-8 (5.56 and 5.30 m) and least in line CHES-10 (3.71 and 3.8 m) (Tables 15.2 and 15.3).

There were significant differences among the various characteristics, viz., yield per plant, fruit retention, fruit weight, skull weight, pulp weight, pulp skull ratio, and pulp seed ratio. The clone CHES-2 retained maximum fruit per plant and highest yield per plant (90.40 kg/plant). However, fruit weight was highest in line CHES-4 (627.33 g) fallowed by CHES-2. The least fruit retention (46.67) and yield per plant (12.97 kg) were recorded in CHES-10 and least fruit weight in CHES-9 (174 g). The least skull weight and pulp weight in were recorded in line CHES-9 (48.13 g and 55 g, respectively). It was highest in CHES-4 (179.33 and 319 g, respectively). Similarly pulp skull ratio and pulp seed ratio were highest in the selections. Based on fruit retention, fruit weight, yield per plant in CHES-2 and CHES-4, they were found to outperform (Table 15.4).

Table 15.2 Vegetative growth characteristic of wood apple germplasm

Line no.	Plant height (m)	Stock dia. (cm)	Scion dia. (cm)	Plant spread (m) NS	Plant spread (m) EW
CHES-1	6.12	29.5	26.46	4.38	4.53
CHES-2	6.44	25.36	21.86	4.12	3.91
CHES-3	6.05	23.36	23.43	4.54	4.10
CHES-4	5.95	23.53	23.66	4.72	4.41
CHES-5	6.12	26.36	22.43	4.18	4.10
CHES-6	6.26	26.43	22.43	5.49	5.18
CHES-7	6.37	26.00	25.23	4.65	4.31
CHES-8	7.16	25.46	23.00	5.56	5.30
CHES-9	6.44	26.23	23.73	5.46	5.13
CHES-10	7.37	24.40	23.43	3.71	3.8
CHES-11	5.5	21.53	21.33	4.32	5.2
CD at 5 %	0.297	NS	1.46	0.185	0.194

Hiwale (2013)

There were significant differences among the various characteristics, viz., seed number per fruit, seed weight per fruit, reducing sugar, total sugar, TSS, pulp skull ratio, and pulp seed ratio. The highest seed number and seed weight per fruit were recorded in line CHES-8 (638.33 and 19.33 g, respectively). The superior type CHES-4 and CHES 2 had medium composition. The least seed number and seed weight per fruit were recorded in line CHES-5 (388.33 and 11.67 g, respectively). Reducing sugar and total sugar were highest in superior clones CHES-4 (1.42 and 3.07 %) fallowed by CHES-2. The results clearly indicated the superiority of clones CHES-4 and CHES 2 in respect to the above

Table 15.3 Wood apple yield per plant/fruit retention and physical characters of fruit

Line	Fruit no/ plant	Fruit weight g	Yield kg/ plant	Fruit length mm	Fruit dia. mm	Skull weight g	Pulp wt. g	Pulp skull ratio	Pulp seed ratio
CHES-1	106.67	423.33	43.45	85.00	92.66	116.00	178.33	1.54	10.69
CHES-2	206.67	440.00	90.40	88.33	89.33	107.33	185.67	1.73	12.11
CHES-3	87.00	421.00	35.57	87.66	90.00	150.33	211.33	1.41	11.95
CHES-4	134.00	627.33	85.25	96.66	103.66	179.3	319.00	1.78	27.33
CHES-5	181.00	266.33	48.60	79.33	80.66	88.67	98.33	1.11	8.44
CHES-6	141.00	370.33	52.11	83.00	83.67	124.67	156.00	1.25	12.31
CHES-7	114.67	426.00	60.52	82.33	91.90	149.33	217.00	1.46	15.14
CHES-8	72.67	227.00	20.88	76.33	75.00	95.33	151.67	1.59	7.84
CHES-9	402.00	174.00	67.75	69.00	65.00	48.33	55.00	1.14	3.88
CHES-10	46.67	283.33	12.97	78.17	75.67	74.00	126.67	1.71	7.31
CHES-11	73.33	299.33	21.98	75.33	82.67	98.00	136.33	1.88	11.36
CD5%	23.35	53.33	2.37	3.43	4.35	8.12	7.24	–	–

Hiwale (2013)

Wood apple tree in bearing

Bunch bearing in the selection

Fruits of CHES selection

wood apple germplasm

Table 15.4 Fruit analysis

Line	Seed no./fruit	Seed wt. (g)	TSS ° Brix	Acidity (%)	pH	Reducing sugar (%)	Total sugar (%)
CHES-1	516.33	16.67	7.60	2.31	3.47	1.01	2.59
CHES-2	493.67	17.67	11.60	3.25	3.31	1.39	2.17
CHES-3	488.87	15.33	11.00	2.54	3.62	1.11	1.95
CHES-4	570.67	16.67	9.60	2.08	3.48	1.42	3.07
CHES-5	388.33	11.67	9.60	2.31	3.28	1.12	2.90
CHES-6	450.33	12.67	11.00	2.33	3.70	0.99	2.79
CHES-7	574.00	16.00	9.00	2.62	3.72	0.65	2.06
CHES-8	638.33	19.33	6.00	2.54	3.57	1.01	2.83
CHES-9	490.33	14.33	7.30	4.81	3.28	1.03	1.91
CHES-10	449.33	17.33	11.0	3.08	3.51	0.69	2.21
CHES-11	451.33	12.00	5.30	3.65	3.51	0.65	1.77
CD at 5 %	37.41	1.74	1.27	0.196	0.208	0.22	0.20

Table 15.5 Nutritional composition of wood apple germplasm

Line	N (%)	P (%)	K (%)	Ca (%)	Mg (%)	Protein
CHES-1	5.31	0.074	1.38	0.34	0.53	33.58
CHES-2	3.58	0.048	1.87	0.13	0.58	23.45
CHES-3	3.27	0.071	1.61	0.196	0.64	20.56
CHES-4	4.30	0.071	1.76	0.39	0.71	27.36
CHES-5	3.10	0.017	1.55	0.27	0.49	20.41
CHES-6	3.54	0.030	1.36	0.113	0.35	21.63
CHES-7	3.16	0.059	1.62	0.22	0.38	25.42
CHES-8	3.79	0.049	1.76	0.12	0.52	20.84
CHES-9	4.21	0.034	1.41	0.12	0.45	26.41
CHES-10	3.76	0.053	1.29	0.29	0.45	22.61
CHES-11	3.04	0.044	1.26	0.12	0.57	19.69
CD at 5 %	0.204	0.0073	0.14	0.196	0.063	1.701

characteristics. Maximum TSS was observed in line CHES-2 (11.6° Brix) and least in CHES-8 (6° Brix). Percent acidity was highest in line CHES-9 (4.81 %) and least in CHES-4 (2.08 %). The pH of the juice was maximum in line CHES-3 (3.62) and least in CHES-5 and CHES-9 (3.28) (Table 15.5).

15.12 Nutritional Composition of Wood Apple Germplasm

Nutritional composition of wood apple indicated that there were significant differences among the various characteristics, viz., N, P, K, Ca, Mg, and protein content. The maximum P, Ca, and Mg content was recorded in clone CHES-4 (0.071 %, 0.39 %, and 0.71 %, respectively). However, the maximum N, K, and protein content was recorded in clone CHES-2 (5.31, 1.87, and 33.58 %). The least N, K, Ca, and protein content was recorded in line CHES-11 (3.04 %, 1.26 %, 0.12, and 19.69 %, respectively), P in line CHES-5 (0.017 %), and Mg in line CHES-6 (0.35 %). The superior clones CHES-4 and CHES 2 had better nutritional composition.

15.13 Fruit Drop in Wood Apple

Data on fruit drop at weekly interval showed that heavy fruit drop was recorded in line CHES-5 (no. 7530); least fruit drop was recorded in line CHES-2 which resulted in higher fruit retention and yield per plant fallowed by CHES-4. The initiation of fruit drop started immediately after 10 days of fruit set from last week of March. It was at peak during the middle of May and continued till June. The fruit drop continued due to continuous growth during the hot summer months. Most of the drop occurred at pinhead stage of the fruit.

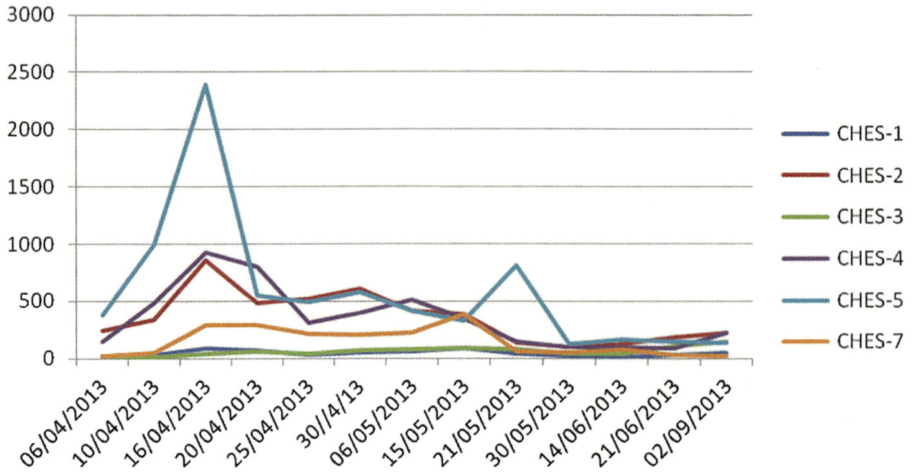

Fruit drop in wood apple

15.14 Plant Propagation

Wood apple is mostly propagated from seed, although root cuttings or layers are said to be successful. Multiplication may also be by budding onto self-seedlings to induce dwarfing and precariousness. Recent work done at CHES, Vejalpur, showed that there is scope for softwood grafting for large-scale multiplication of the plants (Table 15.6).

To reduce the gestation period of 12–15 years required for fruiting in wood apple, a successful method of softwood grafting has been standardized. Plants raised from seeds of same species (1 year old) were defoliated 10 days prior to grafting. These scions were grafted (on the 1-year-old) local rootstock seedling. Percent success in grafting recorded on monthly basis indicated that it was highest when grafting was performed in the months of January and February (48.88 and 53.33 %, respectively) and lowest in December and April (30.24 % and 34.56 %, respectively).

Table 15.6 Soft wood grafting/budding in wood apple

	Percent success	
Time of soft wood grafting	In grafting	In budding
December	30.24	20.3
January	48.88	27.17
February	53.33	29.83
March	44.28	29.73
April	34.56	24.13
CD at 5 %	13.56	5.21

Hiwale (2008)

Local root stock

Grafted plant

15.14.1 In Situ Softwood Grafting

Wood apple plantation can be established by in situ grafting method. The seeds of local type are collected in October–November and during June are sown in the marked pits in the field (2–3 seeds/pit). The seeds germinate on the onset of monsoon. The growth is slow which takes 8–9 months for the rootstock seedlings to become ready for grafting. The plants are grafted in situ during March when there is leaf fall on the mother tree. The graft starts sprouting after 15 days after grafting. The percent success was as high as 80 %. Thus, the orchard can be established under rainfed condition in semiarid areas (Hiwale 2006).

In situ softwood grafting

15.15 Agro-techniques

15.15.1 Orchard Establishment

Planting is done during rainy season. Plantation can be done at 8–10 m spacing, which can vary according to soil and climate and growth habit of plant. Alley plantation is possible in wood apple where in it can be cropped with various arable crops to obtain additional income.

15.15.2 Irrigation Management

The tree can be grown without any irrigation under rainfed condition. However, irrigating the crop will increase its bearing, fruit size, and ultimately yield. The plants need to be watered in the first year of plantation. After harvesting the fruit in October–November, water should be withheld and plant should be irrigated only after flowering in March. Its water requirement is least.

15.15.3 Integrated Nutrient Management

No work so far has been done on this fruit tree species. Leaf analysis done at Central Horticultural Experiment Station, Godhra, revealed that the plant removes substantial quantity of these major nutrients (nitrogen (2.016 %), phosphorus (2.963 ppm), and potassium (2.325 %) content) from soil which needs to be replaced.

15.15.4 Canopy Management

The flowering is mainly on new shoots. Old non-bearing trees can be made productive by heading back the branches. As such the shape of the tree is very compact. Infected and dried branches should be regularly removed.

15.15.5 Weed Management

Land where the wood apple is to be planted should be free from *Cynodon dactylon* and *Cyperus rotundus*. For controlling these weeds, four spray of glyphosate at 3-month interval should be given. Manual digging and removal of the weed biomass from the field is the best control measure but is expensive. Earthing up of soil at frequent interval will help in controlling other seasonal weeds and also in improving aeration of the soil.

15.16 Harvesting and Yield

Fruit is tested for maturity by dropping it onto a hard surface from a height of 1 ft. Immature fruits bounce, while mature fruits do not. After harvest, the fruit is kept in the sun for 2 weeks to fully ripen. Ripened fruits are known for their typical aroma.

Fruit takes about 210–240 days to mature from flowering. The fruits ripen in 10–15 days after harvesting. Single fruit should be picked. If the fruit is dropped on soil, there is every possibility of fruit cracking and subsequent infection by fungi leading to rotting of fruit during storage. The fruits have very hard skin making it suitable for long storage of 15–20 days without any treatment. It is also easy to handle.

15.17 Marketing and Economics

Mature fruits are harvested and sold in the market as such. The fruits fetch remunerative prices in the market. A 20-year-old seedling plant gives a yield of 500–700 fruits per plant at 10×10 m spacing, there will be 100 plants/ha, which will yield 50,000 fruits and a yield of 12.5 tons per hectare. An income of Rs.25,000–30,000/- can be obtained from the plantation.

There is tremendous export potential for this crop as it can be made into different value-added products and very profitable processing industries could be established.

15.18 Value Addition

A good quality stiff jelly of attractive color can be prepared from the pulp. Similarly, wood apple toffee can be prepared by boiling the pulp with sugar. It is also possible to prepare syrup and chutney from wood apple.

15.18.1 Wood Apple Syrup

Wood apple syrup can be prepared from pulp. Pulp is extracted from fruit by cracking it and water is mixed with it and is boiled two times. Sugar syrup is added to the extraction to raise the TSS to 13 Brix. The prepared mixture after boiling for 30 min is filled in cans while it is still hot. The cans are sealed and stored at room temperature for about 1.5 years.

15.18.2 Jelly

Select fresh, disease-free half ripen fruits and remove the hard cover of the fruit; pulp should be used for making jelly.

Extract juice by adding 1.5 times water to the weight of fruit pulp; use stainless steel and aluminum vessel for storing juice. To preserve juice and its color heat the juice unifomly. Overheating converts pectin into pectic acid; also the color and aroma is lost. Fix the time for which the juice has to be heated. To make jelly transparent, the juice should be extracted by tying in cloth and allowing it to drop on its own. Do not press it by hand. Filter the juice through muslin cloth to maintain the acidity and pectin content. Addition of sugar, acid, and pectin will depend upon the natural sugar, acid, and pectin present in the fruit. For wood apple, 1.5 kg juice, 1.5–1.75 kg sugar is required when it is heated at 104–105 °C. The resultant produce is 1.5–1.75 kg jelly.

Mix the sugar only in boiling juice before adding sugar; prepare its syrup and mix with juice. The juice should be heated till the sugar fully dissolves and acid sugar and pectin blend is achieved.

15.18.3 Chutney

Select matured fruits; keep the fruits for 8–10 days so that color of pulp and its aroma develops to its best. Crack them and remove the pulp with seed. To the 1 kg fruit pulp should be mixed with 350 g sugar, 30 g chili powder, 5 g jeera, 5 g pepper, 55 g, 20 g ginger, add pulps with seeds, to the mixture grinder and pack in a well-sterilized bottle or any suitable container. Taste of the chutney develops when storage period increase; its aroma and taste also increases.

15.18.4 Plant Protection

In wood apple, there were no serious pests and diseases observed, due to its hardy nature. An infection of the leaves, fruits, twigs, and thorns by *Xanthomonas bilvae* was found, which is also capable in infecting the wood apple (Anon 1956). Other diseases like gummosis caused by aspergillus are also observed but it is not common.

References

Anonymous (1956) The wealth of India. A dictionary of Indian raw materials and industrial products, vol IV. CSIR Publication, New Delhi. pp 18–20

Anon (2003) Annual report. Central Institute for Arid Horticulture, Bikaner, Rajasthan, India

Gopalan C, Ramashastri BV, Balsubramanium SC (1984). Nutritive value of Indian foods. National Institute of Nutrition, ICMR, Hyderabad, India (reprinted)

Hiwale SS (2006) Wood apple (*Feronia limonia* Correa) book chapter on Advances in Arid Horticulture, vol II, pp 435–442

Hiwale SS, Dhandar DG, Bagle BG (2008) Vegetative propagation of wood apple (*Feronia limonia* Correa). Indian J Agrofor 10(2):58–61

Hiwale SS (2013) Annual report. CIAH, Bikaner

Morton JF (1987) Fruits of warm climate, Wood apple, pp 190–191

Roy P, Majumdar BC (1988) Sci Cult 55:110–111

Troup RS (1921) In the silviculture of Indian trees, vol III. Govt. of India Publication, New Delhi, pp 101–103

Jamun (*Syzygium cuminii*) 16

Abstract

Jamun, *Syzygium cuminii*, is one of the most hardy fruit crops and can easily be grown in neglected and marshy areas where other fruit plants cannot be grown successfully. The fruit is a good source of iron, sugars, minerals, protein, carbohydrate, etc. Jamun originated in India. Small Jamun fruits, which are not suitable for table use, can be used in the beverage industry as they contained a high amount of acidity, tannins, and anthocyanin. Leaf extract of Jamun reduces radiation-induced DNA damage in the cultured human peripheral blood lymphocytes. Therefore, Jamun fruits have high value in terms of therapeutics and nutrition. Jamun fruits are used as an effective medicine against diabetes. The area under the crop is increasing fast owing to tremendous potential for commercialization. However, the shelf life of the fruit and value addition needs to be worked out to sustain the increase in area, because of identification of superior varieties like Vengurla seedless and Lucknow seedless.

16.1 Introduction

Jamun, *Syzygium cuminii* Skeels, is a nutritious fruit with a variety of uses. It is one of the most hardy fruit crops and can easily be grown in neglected and marshy areas where other fruit plants cannot be grown successfully. The fruit is a good source of iron, sugars, minerals, protein, carbohydrate, etc. Fully ripened fruits are eaten as fresh fruit and can be processed into beverages like jelly, jam, squash, wine, vinegar, and pickles. A little quantity of Jamun fruit's syrup is much useful for curing diarrhea. Small Jamun fruits, which are not suitable for table use, can be used in the beverage industry as they contained a high amount of acidity, tannins, and anthocyanin (Anonymous 1986). Jamun seeds contain alkaloids like jambosin and glycoside, which reduce the diastatic conversion of starch into sugars. Jamun fruits are used as an effective medicine against diabetes and heart and liver trouble (Singh 2001). The powdered seeds have reputation for being useful in the treatment of diabetes. Oral administration of an aqueous Jamun seed extract for 6 weeks caused a significant decrease in lipids and thiobarbituric acid reactive substances and an increase in catalase and superoxide dismutase in the brain of diabetic rats (Prince et al. 2003). Leaf extract of Jamun reduces radiation-induced DNA damage in the cultured human peripheral blood

S. Hiwale, *Sustainable Horticulture in Semiarid Dry Lands*,
DOI 10.1007/978-81-322-2244-6_16, © Springer India 2015

lymphocytes. Therefore, the Jamun fruits are having high value in terms of therapeutics and nutrition.

16.2 Origin and Distribution

Jamun originated in India. It is also grown in other countries like Thailand, the Philippines, and Madagascar. Jamun has been successfully introduced into many other tropical countries like West Indies, California, Algeria, and Israel. In India, a maximum number of Jamun trees are scattered throughout the tropical and subtropical regions. It also occurs in the lower range of Himalayas up to an elevation of 1,300 m and in the Kumaon hills up to 1,600 m. It is widely grown in the larger part of India from the Indo-Gangetic Plain in the north to Tamil Nadu in the south.

16.3 Soil and Climate

Jamun adapts well to a wide range of soils. For high yield potential and good plant growth, deep loam and well-drained soils are needed. Such soils have sufficient soil moisture retention capacity, which is very beneficial for optimum growth and good fruiting. It can flourish in poor soils with little or no cultivation. It tolerates sodic and saline soils particularly in ravines and degraded land. It can also be grown in waterlogged conditions. Jamun survived and grew better under lower salinity and shallower water table condition (Hebbara et al. 2002). Jamun plants survived in alkali soils up to 10.5 pH (Singh et al. 1997).

It adapts well to a wide range of ecological conditions, reflecting its wide geographical distribution in the sub- and semiarid tropics. It can be grown in semiarid subtropical regions with an annual rainfall ranging from 350 to 500 mm (Vaishtha 1991). Jamun requires dry weather at the time of flowering and fruit setting. In subtropical areas, early rain is considered to be beneficial for fruit ripening and proper development of size, color, and taste.

16.4 Taxonomy of the Crop

It belongs to the important genus *Syzygium* and family Myrtaceae having chromosome number $2n=40$. There are about 400–500 species, of which a few are considered as edible fruit bearers (Chundawat 1990). Some of the species are described below.

Syzygium jambos (Rose Apple or Safeda Jamun)
The trees are medium in size, evergreen, and ornamental. The leaves have very small petiole. The fruits are light yellow to white in color. They are rose scented and have a persistent calyx. The seeds are polyembryony, and it is being grown in Assam, Bihar, Andhra Pradesh, Tamil Nadu, West Bengal, coastal areas of Maharashtra, and Gujarat.

Syzygium fruticosum
The trees are suitable for windbreak. Fruits are edible and small in size.

Syzygium javanica (Water Apple)
This species is found in south India and West Bengal.

Syzygium densiflorum
It is suitable for use as a rootstock for *Syzygium cuminii* as it shows resistance against termites.

Syzygium uniflora (Surinam Cherry or Pitanga Cherry)
It is also a small tree and bears small-size fruits having bright red color and aromatic flavor. The tree is grown in south India.

Syzygium zeylanica
The tree is small and bears edible fruits and grown in Western Ghats of India.

Jamun is an evergreen tree, attaining medium to large in size. The trunk is thick and grayish white in color. Leaves are opposite, simple, mostly firm and glossy, elliptic, and pinnately veined with lateral veins close together; flowers are in a few flowered panicles, hermaphrodite, light yellow in color, and borne in the axils of leaves on the branchlet and have 4 calyx lobes, calyx tube not extending beyond the summit of ovary, and 4 petals and are white, spreading and have many stamens, with ovary inferior, 2 celled (Bailey and Baily 1978).

16.5 Reproductive Biology

New vegetative shoots in the Jamun emerge in two distinct flushes from February to May and from August to October; the number of inflorescences per shoot is found maximum in February flush and least in those produced in October (Mishra and Bajpai 1971). The maximum anthesis (18.71–43.08 %) and dehiscence have been recorded between 10 am and 12 noon. Pollen fertility is higher in the beginning of the season. Honeybees, houseflies, and wind do the pollination. The highest level of pollen germination in Jamun may be obtained in 20 % sucrose solution (Singh 1978). The maximum fruit set (32.6–36.0 %) can be obtained when pollination is done 1 day after anthesis, and thereafter, a sharp decline could be observed in fruit set. Fruit is berry, purplish red, ovoid, edible, and polyembryonic.

16.6 Reproductive Biology of Jamun Under Semiarid Ecosystem of Western India

The month of February is the peak period for panicle emergence. Peak period of flowering and fruit set is observed in the month of March. The highest panicle length (16.24 cm) and fruit set per panicle (25.00) have been recorded in GJ-19. The time taken for complete development of flower bud ranges from 20 to 26. The peak period of anthesis is observed between 8 am and 1 pm. Anther dehiscence commences after opening of flowers, i.e., at 7 am, and continued till 4 pm. Peak period of dehiscence may be observed between 9 am and 2 pm. Pollen viability ranges from 90.20 to 98.50 %. Pollen germination and pollen length range from 24.00 to 37.50 % and 29.00 to 48.90 μm, respectively.

16.7 Status of Genetic Diversity

Little work has been done to exploit the genetic resources of Jamun. Now, the crop is facing severe genetic erosion of several species as a result of activities related to urbanization and intensive agriculture.

A number of seedling strains having a lot of variation in respect to fruit shape and size, pulp color, TSS, acidity, and earliness particularly in Uttar Pradesh, Gujarat, and Maharashtra provide a good scope for selection of better varieties. The survey in Pune and Ahmednagar districts of Maharashtra revealed wide variation in fruit weight (3.5–16.5 g), pulp content (54.29–85.71 %), TSS (4.5–17 %), and acidity (0.16–0.55 %) where promising types No. 15, 4, 14, and 13 were identified (Keskar et al. 1989). Ashraf (1987) reported that fruit shape in Jamun varied from round to oblong and apex of fruits from flat to pointed. He also observed great variability in the physicochemical characteristic of fruits offering possibility of selecting a variety suitable for fresh market and processing. Small seed size and high pulp content with better chemical properties are considered ideal characteristics. Singh et al. (1999) evaluated eight genotypes of Jamun under Faizabad conditions and reported that oblong types had more fruit weight and relatively less seed weight. Among the locally available genotypes of Jamun in West Bengal, Selection No. 1 (oval-shaped large fruit) and Selection No. 2 (cylindrical-shaped medium-size fruit) proved to be better on the basis of yield and fruit quality attributes (Kundu et al. 2001). A survey was undertaken in the Karnataka state to investigate the nature and extent of variability present in Jamun seedling progenies for the morphological characteristics of the tree. It was observed that there was a high degree of variability regarding plant girth, leaf area, petiole length, and leaf length to petiole length ratio (Prabhuraj et al. 2002). A survey was carried out in North Goa to identify the elite genotypes of Jamun, and individual fruit weight ranged from 3.42 to 13.67 g, length from 3.31 to 5.26 cm, girth from 5.21 to 9.82 cm, pulp percentage from 58.57 to 84.55, TSS from 12.00 to 26.80 %, titratable acidity from 0.59 to 1.63 %, total sugars from 6.87 to 25.31 %, and sugar-to-acid ratio from 15.39 to 27.92 (Devi et al. 2002). Patel et al. (2005) made a survey in Uttar Pradesh and Jharkhand and identified the elite genotypes among its population. The genotypes RNC-26 and RNC-11 were found to be promising and had higher pulp and fruit weight.

Singh and Singh (2005) also made a survey in Gujarat to identify the elite genotypes among its population. Flowering, fruiting, and the physicochemical attributes of ripened fruits from 17 selected genotypes were studied. The study revealed that there was a wide variation among the genotypes. The earliest flowering (mid-February) took place in GJ-24 and latest (late March) in GJ-23, GJ-30, GJ-31, GJ-32, and GJ-33. Maximum panicle length (15.24 cm) and number of fruits per panicle (26.00) were found in GJ-19. Collection numbers GJ-19, GJ-21, GJ-22, and GJ-25 have been found to have the earliest (first week of May) ripening period, while GJ-17, GJ-28, and GJ-32 ripened at the last (late June). Maximum yield per plant was recorded in GJ-19 (175.00 kg) followed by GJ-23 (170.00 kg), GJ-22 (160.00 kg), and GJ-18 (150.00 kg). Individual fruit weight ranged from 10.10 to 22.50 g, length from 1.99 to 3.24 cm, and pulp percentage from 73.66 to 85.68. There was a wide variation in chemical characteristics also. Total soluble solids varied from 10.30 to 12.34 %, total sugar from 8.58 to 9.13 %, and vitamin C from 32.12 to 46.37 mg/100 g. On the basis of overall performance, GJ-18, GJ-19, GJ-23, GJ-24, and GJ-25 were found to be promising among the genotypes studied based on its yield and physicochemical characteristics.

16.8 Varietal Wealth

Common cultivar grown under north Indian condition is "Ra-Jamun". It produces big-size (length 2.5–3.5 cm and diameter 1.5–2.0 cm), oblong fruit, deep purple color at fully ripe stage, juicy and sweet in taste. Fruit ripens in June–July and possesses small stone. Another late maturing variety bears a small-size (length 1.5–2.0 and diameter 1–1.5 cm), slightly round fruit, deep purple or blackish in color at full ripe stage. The stone present in these cultivars is comparatively large in size. Fruits ripen in the month of August. Jamun is a fruit which offers a good scope for selection of better varieties out of the innumerable seedling strains that are in existence in this country.

Flowering in Jamun

Jamun fruits

Clonal selection

16.9 Plant Propagation

16.9.1 Seeds

The most common method of propagating this fruit tree is through its seeds. Jamun seeds have no dormancy period; hence, fresh seeds can be sown (within 10–15 days) 4–5 cm deep at a distance of 25 cm × 15 cm. The seeds germinate 10–15 days after sowing. The seedlings become ready for transplanting in spring or in the next monsoon. If the seeds are sown too deep, emergence of seedling is delayed, and there may be poor rooting due to lack of aeration. Seeds may also be sown in polythene bags as it facilitates easy handling of rootstocks and grafted plants. There is occurrence of polyembryony in Jamun up to 20–50 %; hence, nucellar seedlings may be used to produce true-to-type plants (Singh and Thakur 1977). The seeds of Jamun take 24–61 days for total germination under Bihar conditions (Singh and Thakur 1977). Sasthri et al. 2001 recorded that large-size seeds had higher germination rate (99–98 %) than small-size seeds (89–79 %).

16.9.2 Vegetative Propagation

16.9.2.1 Budding

In India, shield, patch, and Forkert methods of budding are generally employed. Patch budding has been found successful when performed in the month of June under Gujarat conditions (Chovatia and Singh 2000). Singh and Singh (2006) reported that patch budding was successful in the month of March under semiarid environment of western India.

16.9.2.2 Softwood Grafting

Softwood grafting method was proved to be better for in situ grafting under rainfed conditions. Madalageri et al. (1991) reported that softwood grafting was a successful method for the multiplication of Jamun under Karnataka conditions. Chovatia and Singh (2000) recorded a success rate of 41.67 % through softwood grafting during the month of June under Gujarat conditions. Softwood grafting was found successful in the months of July–August under semiarid environment of western India (Singh and Singh 2006).

16.9.2.3 Use of Polycontainers

Raising of rootstock in nursery beds and lifting budded plants with earth-ball in highly sandy soils are practically not feasible. Transportation of plants from a long distance may also cause high mortality, particularly under semiarid and arid environment. To reduce the time for raising rootstock and to avoid damage during handling and transportation, polyethylene tubes and polyethylene bags may be used on a commercial scale. Generally polyethylene bags (25 cm × 15 cm) are used for raising the rootstocks. Small holes are made in the bottom and sides of polyethylene bags for drainage and aeration and filled with porous rooting medium or pot mixture for raising rootstocks. Generally 1–2 seeds are sown in each polythene bag and then placed in trench bed, so that it can be irrigated easily. Sometimes coiling of root can be a problem; hence, root pruner is also used for trimming the roots.

16.10 Cultivation

Once the climate and soil requirements of Jamun have been taken into consideration, the next step is to layout the orchard. The land may be prepared by usual plowing, harrowing, and leveling.

There should be a gentle slope to facilitate proper irrigation and prompt drainage to avoid the harmful effects of water stagnation during rainy season. Jamun may be grown under various cropping systems, i.e., as an orchard crop in a pure land, as an agroforestry species in mixed cropping systems, or as a hedgerow tree. After marking the places for the plants, pits of 90×90×90 cm are usually dug out during summer months. Digging of pits is very essential for heavy type of soil or soils with shallow hardpan. While digging, it is necessary to keep the topsoil and subsoil separately in two heaps near each pit for about 2–4 weeks. This helps in exposing harmful soil organisms to weathering agencies, providing better aeration in the future rooting zone and making provision for the nutritional requirements for the healthy development of the plants. Well-decomposed organic matter is mixed with soil and filled in pits. Planting is done during the rainy season after settling the soil in the pits. While planting, one should be careful that the earth ball does not break and graft union remains well above the ground level. The planting should preferably be done during cloudy weather and in the evening. The plants should be irrigated immediately after planting. In the initial 2–3 years, it is advisable to protect plants against low temperature injury by covering plants with some short of cover, leaving the eastern side open for entrance of light. Grafted saplings of Jamun may be planted at a distance of 8×8 m to achieve optimum yield. It is frequently planted along land borders, canal banks, on field boundaries, in village groves, or in home gardens as individual tree.

16.11 Irrigation

Irrigation is not normally practiced in Jamun cultivation, but it promotes better growth during the establishment and the early stages of growth, especially during the dry seasons. In the early age, plants require 8–10 irrigations in a year, while bearing trees require 4–5 irrigations during the time of fruit development and ripening. In dry areas, the use of water-harvesting techniques during the rainy season should be considered as it encourages subsequent growth and fruiting.

As soil moisture is one of the limiting factors for successful cultivation of Jamun, use of mulch is very beneficial. It reduces the loss of moisture from the soil, enhances the rate of penetration of rainwater or irrigation in the soil, and controls the growth of weed, thus eliminating the competition between weeds and the fruit trees. Mulching can be done with black polythene or any organic materials. In general, mulching with paddy straw reduces the weed population and conserves the moisture in the soil. However, it needs to be standardized under various agroclimatic conditions of the country.

16.12 Fertilizers

The exact doses and type of manures and fertilizers for Jamun is yet to be worked out. Generally Jamun trees are not manured; however, an annual dose of about 20 kg of FYM during the pre-bearing period and 50–80 kg per tree to bearing trees is considered beneficial (Bose et al. 2000). Apart from manure, 125 g N, 50 g P_2O_5, and 50 g K_2O per plant per year may be applied. This dose should be increased every year in the same proportion up to the 8th year, after which the fixed dose should be applied each year in two splits during June and September. It may vary according to the soil conditions.

16.13 Canopy Management

Canopy management of the crop deals with the development and maintenance of their structure in relation to the size and shape for maximum productivity and quality. The basic concept in canopy management of a perennial tree is to make the best use of the land and the climatic factors for increased productivity in a three-dimensional approach. Tree vigor, light, temperature, and humidity play a vital role in the production and quality of the fruits. The crux of the canopy management lies in the fact as to how best we manipulate the tree vigor and use the available sunlight and temperature to increase the productivity and quality and minimize the adverse effects of weather parameters. Some of the basic principles in canopy management are (1) maximum utilization of light, (2) avoidance of the buildup of microclimate congenial for the disease and pest (3), convenience in carrying out cultural operations, and (4) maximizing the productivity and quality. Basically the training is a potential tool to manage the canopy architecture of the plant. Young plants should be allowed 3–5 well-spaced branches to develop into the main scaffold structure of the tree. The framework of the branches is allowed to develop above 60–100 cm from the ground level. Pruning is a tool to regulate tree size and shape to achieve a desired architecture of the canopy and also to reduce the foliage density by removing the unproductive branches of the tree. Regular pruning in Jamun plant is not required; however, dry, week, and diseased branches should be removed.

16.14 Weed Management

Productivity of Indian orchards can be increased only when all the aspects of production technology including weed management are given due consideration. Weeds injure crops very slowly in a subtle way. Most weeds complete their life cycle in a shorter time compared to fruit trees and compete for light, water, and mineral nutrients and reduce yields. In the orchard, hoeing, hand weeding, and plowing the land 2–3 times a year are done to suppress weed growth. Intercropping and mulching may also be followed to control weeds.

16.15 Maturity Indices

Optimum stage of maturity has been worked out based on the physical and chemical composition of the fruit. In Jamun, Shukla (1979) recorded three distinct phases of fruit growth, viz., during the first phase, the rate of growth was slow (15–52 days after fruit set); in the second phase (52–58 days after fruit set), the rate of development was quite rapid; and the third and last phase (58–60 days fruit set) comprised slower growth with little increase in fruit weight, moisture content, and total and reducing sugars. Shukla and Prasad (1980) studied the respiration rate in developing fruits of Jamun at frequent intervals from 15 to 61 days after fruit set. Initially respiration rate (15–30 days) declined, rose again at 56 days (climacteric peak), and again declined until harvest maturity at 60 days after fruit set. Geetha et al. (1992) reported that Jamun fruit followed a sigmoidal curve with the first phase (7–14 days after fruit set), the second phase (14–35 days after fruit set), and the last phase (35–42 days after fruit set). They also confirmed that fruit length, fruit weight, fruit volume, and moisture content increased during maturation of Jamun fruit. The total soluble solids and total and reducing sugars showed a continuous increase as the fruits developed with a marked increase during ripening. A similar trend was also exhibited by the moisture content and acidity. There was a gradual decrease in tannins, whereas pectin rose and then fell during the growth period. TSS/acid ratio and anthocyanin content followed an increasing trend with advancing maturity and markedly increased during ripening (Garande et al. 1998). Singh et al. (2006) observed that fruit growth was faster at the initial stage and slowed down while reaching toward maturity stage and followed a sigmoid growth pattern in all the genotypes of Jamun. The specific gravity showed an increasing trend (more than one) in all the genotypes during development. Total soluble solids, total sugar, and vitamin C increased as the fruits reached toward maturity. Titratable acidity showed declining trend while reaching toward ripening stage. Deep purple color on the fruit surface was observed in all the genotypes during ripening.

16.16 Harvesting and Yield

Seedling of Jamun plants starts bearing after 8–10 years of planting while grafted ones after 4–5 years. The fruit ripens in the months of June–July. The main characteristic of ripe fruit at full size is the development of deep purple or black color. The Jamun fruit should be picked immediately when it ripens, because it cannot be retained on the tree at the ripe stage. The ripe fruits are picked by hand, and in all cases, care should be taken to avoid all possible damage to the fruits. Bamboo pegs are used for harvesting of the fruits. The fruits of Jamun are generally harvested daily and sent to the market on the same day. The average yield of fruit from a full-grown seedling Jamun tree is about 80–100 kg and from a grafted one 60–70 kg per year.

16.17 Postharvest Handling

16.17.1 Sorting

On arrival in the packing area, the Jamun should be sorted to cull out the undersized, damaged, and bruised fruits to avoid further spoilage loss.

16.17.2 Grading and Packaging

Considerable variation exists in the quality of harvested fruit due to genetic, environmental, and agronomic factors and therefore requires grading to get suitable returns from the market. Systematic grading coupled with scientific packaging and storage reduces postharvest losses and marketing costs substantially, which enables the producer to fetch a competitive price. Jamun should be graded on the weight basis to fetch better price in the market.

Packaging is a vital component of postharvest management to assemble the produce in convenient units and to protect it from deterioration during handling and marketing. Adequate packaging protects the fruits from physiological, pathological, and physical deterioration in the marketing channels and retains their attractiveness. Jamun fruits are highly perishable and are normally packed in bamboo baskets for transport to the local market. Physiological loss in weight, shriveling, discoloration, and rotting were minimum when packed in bamboo baskets using leaves as liner. Fruits prepacked in leaf cups covered with perforated polythene show less loss in weight and shriveling and better maintenance of appearance as compared with conventional prepackaging only in green leaf cups (Singh and Pathak 1988). Depending upon the distance to be covered, Jamun fruits are transported as head loads, in trucks and train. Among the different types of containers, shallow plastic crates are better during transportation of the fruits from the field to the market to prevent bruising losses. Gunny bag should not be used as they do not protect the fruits from mechanical damage. Care should be taken not to overfill or underfill the boxes to avoid damage during transport. It is advisable to transport it in the cooler part of the day to avoid heat damage. Small-size (2–3 kg capacity) corrugated fiberboard boxes may be used for transportation to distant markets.

16.17.3 Storage

The Jamun fruit is highly perishable and can be kept in good condition for about 2–3 days under ordinary condition. Shukla (1979) observed that precooled fruits packed in perforated polythene bags could be stored well up to 3 weeks at low temperature 8–10 °C and 85–90 % relative humidity. Ethanol-treated fruits stored in polyethylene bags at 10 °C were completely deastringent after 9 days of storage (Mohammed and Wickham 1999).

16.18 Insects

16.18.1 Leaf-Eating Caterpillar (*Corea subtilis*)

Caterpillars attack the leaves and the tree becomes defoliated. To control this pest, spray Rogor 30 EC 0.1 % or malathion 0.1 %.

16.18.2 White Flies (*Dialeurodes eugenia*)

It damages the tree all over India. Affected fruits get wormy appearance on the surface. It can be controlled by maintaining sanitary situation in the orchard, which consist of picking up the affected fruits and burying them deep in the soil, and the area under the tree should be dug, so that the maggots in the affected fruits and the pupae hibernating in the soil may be destroyed.

16.18.3 Bark-Eating Caterpillars (*Indarbela tetraonis* and *Indarbela quadrinotata*)

The larva feeds on live bark tissues and shelters under the covering of silken webbing during night. The latter makes a tunnel into the branch and stem and remains in the hole during daytime. As a result, the tree loses vitality and yield declines. It can be controlled by maintaining sanitary situation in the orchard, and petrol should be injected in the hole and be plugged. Foliar sprays with dimethoate (0.05 %) or monocrotophos followed by endosulfan (0.07 %) at triweekly interval control the pest effectively.

16.18.4 Jamun Leaf Miner (*Acrocercops syngramma* and *Acrocercops phaeospora*)

This pest causes damage during the reproductive phase, i.e., from April to September. The newly hatched caterpillar mines a narrow threadlike silvery gallery on the leaf along the midrib upward or latterly. The larval mine later on is transformed into a tubular blister-like swelling on the dorsal surface of the leaf. The pest can be controlled by clipping and burning the affected leaves followed by spraying of dimethoate 30 EC (1.2 ml/L water).

16.18.5 Jamun Leaf Roller (*Polychorosis cellifera*)

The larvae web the leaves by folding the tip downward on both margins parallel to the midrib and feed inside. In case of severe attack, 1/4th of the lamina is eaten up. The pest undergoes 3–4 generations between March–April and September–October in north India and the second generation is most harmful. Regular clipping and burning of affected leaves can keep the population under control. In case of severe attack, spraying with chlorpyrifos 20 EC (2 ml/L) or endosulfan 35 EC (2 ml/L) is recommended.

16.18.6 Leaf Webbers (*Argyroploce aprobola* and *Argyroploce mormopa*)

The newly hatched larvae in large numbers web together the tender leaves at shoot tips and feed within. Regular clipping and burning of affected leaves can keep the population under control. In case of severe attack, spraying with chlorpyrifos 20 EC (2 ml/L) or endosulfan 35 EC (2 ml/L) is recommended. Birds also damage the Jamun fruits. For keeping them away, beating of drum or flinging stones is found useful.

16.19 Diseases

16.19.1 Anthracnose

The fungus incites leaf spot and fruit rot. Affected leaves show small scattered spots, light brown or reddish brown in color. Affected fruits show small, water-soaked, circular, and depressed lesions. Fruits rot and shrivel. This disease can be controlled by spraying of Dithane Z-78 at 0.2 %.

References

Anonymous (1986) Research highlight, 1985. ICAR, New Delhi, p 27

Ashraf SM (1987) Studies on post harvest technology of jamun fruits. Ph.D. thesis, Department of Horticulture, NDUAT, Kumarganj, Faizabad

Bailey LH, Baily EJ (1978) Hortus third- a concise dictionary of plants cultivated in the United States and Canada. MacMillan Publishing Co., New York, pp 457–458

Bose TK, Mitra SK, Sanyal D (2000) Fruits: tropical and subtropical, vol II. Pub. Naya Udyog, 206, Bidhan Sarani, Calcutta, India, pp 643–656

Chovatia RS, Singh SP (2000) Effect of time on budding and grafting success in jamun (*Syzygium cuminii* Skeels). Indian J Hortic 57:255–258

Chundawat BS (1990) Arid fruit culture. Pub. Oxford and IBH Publishing Co. Pvt. Ltd., New Delhi, pp 165–171

Devi SP, Thangam M, Desai AR, Adsule PG (2002) Studies on variability in physico- chemical characters of different *jamun* (*Syzygium cuminii* Skeels) accessions from Goa. Indian J Hortic 59:153–156

Garande VK, Joshi GD, Wasker DP (1998) Studies of changes in chemical composition during growth and development of jamun fruit. Orissa J Hortic 26:76–78

Geetha CK, Babylata AK, Methew KL, George ST (1992) Fruit development in jamun (*Syzygium cuminii* Skeels). South Indian Hortic 40:350–351

Hebbara M, Manjunatha MV, Patil SG, Patil DR (2002) Performance of fruit species in saline- water logged soils. Karnataka J Agric Sci 15:94–98

Keskar BG, Karale AR, Dhawale BC, Choudhary KG (1989) Improvement of Jamun (*Syzygium cuminii* Skeels) by selection. Maharastra J Hortic 4:117–120

Kundu S, Ghosh DK, Maiti SC (2001) Evaluation of some local types of jamun (*Syzygium cuminii* Skeels) of West Bengal. Environ Ecol 19:872–874

Madalageri MB, Patil VS, Nalawadi UG (1991) Propagation of jamun (*Syzygium Cuminii* Skeels) by soft wood wedge grafting. My For 27:176–178

Mishra RS, Bajpai PN (1971) Vegetative growth studies in the jamun. Indian J Hortic 28:273–277

Mohammed M, Wickham L (1999) Effect of modified atmosphere packaging and ethanol on the deastringency process in Jamun (*Syzygium cuminii* Skeels) fruit. J Appl Hortic 1:105–107

Patel VB, Pandey SN, Singh SK, Bikash D (2005) Variability in jamun(*Syzygium cuminii* Skeels) accessions from Uttar Pradesh and Jharkhand. Indian J Hortic 62:244–247

Prabhuraj HS, Swamy GSK, Athani SI, Patil BR, Hulamani NC, Patil PB (2002) Variability in morphological characteristics of Jamun (*Syzygium Cuminii* Skeels) trees. My For 38:187–189

Prince PSM, Kamalakkanan N, Menon VP (2003) *Syzygium cuminii* Skeels seed extracts reduce tissue damage in diabetic rat brain. J Anthropol 84:205–209

Sasthri G, Srimathi P, Malarkadi K (2001) Effect of seed size on seed quality in jamun (*Syzygium cuminii* Skeels). Madras Agric J 88:524–526

Shukla JP (1979) Ph.D. thesis, Department of Horticulture, CSA University of Agriculture and Technology, Kanpur

Shukla JP, Prasad A (1980) Changing pattern of jamun fruit during growth and development III. Changes in respiratory activity. Prog Hort 12:71–73

Singh G, Dagar JC, Singh NT (1997) Growing fruit trees in highly alkaline soils- a case study. Land Degrad Dev 8:257–268

Singh HL (1978) Studies o pollen germination in jamun (*Syzygium cuminii* Skeels). Plant Sci 10:151–152

Singh IS (2001) Minor fruits and their uses. Indian J Hortic 58:178–182

Singh IS, Pathak RK (1988) Packing of Jamun, Aonla and Bael fruits a Souvenir on packing of fruits and vegetables in India. Agri- Horticultural Society Publication, Hyderabad, pp 108–111

Singh IS, Srivastava AK, Singh V (1999) Improvement of some underutilized fruits through selection. J Appl Hortic 1:34–37

Singh RK, Thakur S (1977) Seed germination and seedling growth of jamun (*Syzygium cuminii* Skeels) types. Proc Bihar Acad Agric Sci 25:139–142

Singh S, Singh AK (2005) Genetic variability in Jamun (Sizygium cuminii Skeells)types. Proc Bihar Acad Agric Sci 25:139–142

Singh S, Singh AK (2006) Standardization of method and time of propagation in Jamun (Sizigium cuminii) under semiarid environment of western India. Indian J Agric Sci 76:242–245

Singh S, Singh AK, Joshi HK, Bagle BG, Dhandar DG (2006) Developmental pattern and standardization of maturity indices in jamun (Syzygium cuminii Skeels) under semi arid ecosystem of western India. National symposium on production, utilization and export of under utilized fruits with commercial potentialities, 22–24th November, 2006 at BCKV, Kalyani, West Bengal, p 114

Vaishtha BB (1991) Lecture delivered in summer institute on fruit production and utilization in wastelands; NDUAT, Faizabad, 24th June–12 July 1991

Chironji (*Buchanania lanzan* Spreng.)

17

Abstract

Chironji is a common tree species found in deciduous forests throughout India. Chironji or charoli belongs to the family Anacardiaceae and was first described by Francis Hamilton in 1798. It is rich in protein and mineral content. It bears fruits which contain a single seed which is popular as an edible nut, known as chironji. It is used as fuel, fodder especially for buffalos, and alternative host for Kusmi lac insect and its oil for cosmetic and soaps. It attains the height up to 20 m and stem girth to 1.5 m. Roots are used in treatment of diarrhea, leaves are used in skin diseases, and fruits are used in treating cough and asthma. The bark furnishes a natural varnish and is used in tanning in Kerala. The leaves are reported to be valued for their tonic and cardiotonic properties, and their powder is a common medicine for wounds. In wasteland development and dry land horticulture, it assumes great significance due to its multifarious uses and capacity to withstand adverse climatic conditions.

17.1 Introduction

Chironji or charoli belongs to the family Anacardiaceae and was first described by Francis Hamilton in 1798. It a common tree species found in deciduous forests throughout India. It attains the height up to 20 m and stem girth to 1.5 m. It is used as fuel, fodder specially for buffalos, and alternative host Kusmi lac insect and its oil for cosmetic and soaps. The kernels are highly nutritive and are used in confectionary. Chironji oil is used as a substitute for almond and olive oil in both confectionary and medicine for glandular swelling of the neck, and chironji fruits are laxative. Kernels of fruit are used as ointment in skin diseases (Das and Agrawal 1991). It is an excellent tree of agroforestry and social forestry for wasteland development and dry land horticulture; it assumes great significance due to its multifarious uses and capacity to withstand adverse climatic condition.

At present, it is growing under forest condition as an underexploited fruit and gives monitory reward to the tribal community of the country and seems to be boon for them. It is a valuable species found in dry deciduous forest throughout the country excluding eastern Himalayan forest (Singh 1982; Singh et al. 2006). The flesh of the

S. Hiwale, *Sustainable Horticulture in Semiarid Dry Lands*,
DOI 10.1007/978-81-322-2244-6_17, © Springer India 2015

ripe fruit is palatable and is largely eaten raw or roasted, and the oily kernels are the most important part. Mesocarp of the ripe fruit is edible and cherished by children (Munde et al. 2003). Very good juice may be prepared from the pulp of chironji fruits. The kernel is highly nutritious and rich in protein (25.0–30.0 %) and yields sweet oil. The kernel contains 33.50 % oil, 1.90 % of which is unsaponifiable. The saponifiable part contained 20.00 % of linolenic acid. Chironji oil is non-repellant and nontoxic and is suitable for human consumption (Banerjee and Jain 1988). The kernels are reported to be used by tribals of Gujarat as brain tonic. An ointment made out of the kernels is used to cure itch of the skin and to remove blemishes from the face. It is also used in diarrhea and intercostal pains. The bark contains 13 % tannin and therefore used in tannin industry. It is also used for coating tablets for delayed action. Roots are used in treatment of diarrhea, leaves are used in skin diseases, and fruits are used in treating cough and asthma. The bark furnishes a natural varnish and is used in tanning in Kerala. The leaves are reported to be valued for their tonic and cardiotonic properties, and their powder is a common medicine for wounds. Chironji provides quality timber wood for various uses. The alcoholic extract of *Buchanania lanzan* is commonly used for wound healing Chitra et al. (2009).

17.2 Present Status in Distribution

Information regarding area and production of this fruit in India is not available because it is not grown on plantation scale. They can be seen growing in forest areas. The production in India is mainly concentrated in the drier states, and the produce is collected by the villagers and sold in the local market. Kernels are very rich in protein, and it is being used as a dry fruit for preparation of various food products. Looking into the importance of the plant, it has tremendous potential for dry land horticulture. Its cultivation may spread to arid-semiarid areas, resource-poor areas, and wastelands.

17.3 Origin and Distribution

This tree is found in the north, west, and central India, particularly in the dry regions of the country. Plants may be seen in forest area of Uttar Pradesh, Madhya Pradesh, Chhattisgarh, Maharashtra, Bihar, Jharkhand, Orissa, Andhra Pradesh, and Gujarat.

17.4 Soil and Climate

Chironji is a very hardy plant and thrives well on rocky and gravelly red soils. Though it is a very hardy tree, plants do not survive under waterlogged conditions. It can grow even in pockets of soil between crevices of barren rock. Trees are grown on degraded rocky area including salt-affected soils. However, for its better growth and productivity, well-drained deep loam soil is ideal. It prefers tropical and subtropical climate and can withstand drought admirably.

17.5 Botany

Chironji or charoli (*Buchanania lanzan* Spreng.) belongs to the family Anacardiaceae. It is highly heterozygous, cross-pollinated crop, and as such, seedlings exhibit a wide range of variations. It is a dicot woody plant of deciduous nature. Flowers are typical of family Anacardiaceae. Flowering starts in the months of January–February on the well-developed panicles, which have hermaphrodite flowers, and they are located on the annular prominent disk, and the fruit is drupe. Fruits become ready for harvest during the months of April and May. Harvesting of fruits done in the last week of May when the fruits are fully mature with a fruit size of 9 to 12 mm black in color with 35 g weight of 100 fruits, 2857 no. of fruits/kg, 100 kernel weights of 6.4 and 182 g kernel weight/kg fruit. The seed moisture content of the fruits harvested in the third week of May is the least, i.e., 7.13 %, and therefore retains its viability up to 52 % with the highest seed germination of 38.5 (Sharma 2012). Early harvesting results in low seed setting and poor germination potential. If the fruits are harvested at the right time, the oil is high as 62 % with a protein content of 48 %.

17.6 *Buchanania lanzan*

The tree is having 13–18 m height with straight trunk, young branches clothed with silky hairs. The leaves are thickly coriaceous, broadly oblong, obtuse, sometimes emarginated, glabrescent above, more or less villous beneath, reticulate veined, and the nerves and veins impressed on the upper surface, base rounded, the main nerves having 10–20 pairs, and petioles long (about 1.2 cm). Flowers are small, sessile, and greenish white, in terminal and auxiliary pyramidal ferruginous-pilose panicles which are shorter than the leaves, and bracts are small and caduceus. Calyx lobes are short, broadly ovate, and ciliate. Petals are long, ovate oblong, subacute, and disk fleshy. Stamens are 10 and a little shorter than the petals, filaments are flattened, and anthers are about as long as the filaments. One ovary is perfect, conical, and villous, and the other 4 are reduced to cylindrical filaments. Drupes are obliquely lentiform in the long diameter, black, stone hard, and 2-valved.

17.7 *Buchanania angustifolia*

It is a glabrous tree. Leaves are thinly coriaceous, linear oblong, elliptic or elliptic lanceolate, obtuse or rounded, very often emarginated, quite glabrous, reticulate veined, and the nerves and veins slightly prominent on the upper surface, base rounded or acute, the main nerves having 12–15 pairs and petioles are oblong and slender. Flowers are in glabrous branched panicles about equaling the leaves. Calyx lobes are semi-orbicular. Petals are long and oblong. Fertile ovary is pilose. Drupes are obliquely globose and slightly compressed.

Diversity-rich area of chironji in Madhya Pradesh, Chhattisgarh, Orissa, Maharashtra, Andhra Pradesh, Karnataka, and Gujarat was surveyed to select elite genotypes among its population. Promising genotypes were earmarked and multiplied through vegetative propagation and planted at our experimental field.

17.8 Reproductive Biology

The reproductive biology of the crop was studied at the Central Horticultural Experiment Station, Vejalpur, Gujarat. It was observed that the peak period of panicle emergence and flowering were recorded in the months of January and February, respectively. The peak period of fruit set was recorded in the month of February in majority of genotypes. The highest panicle length (35.11 cm) and fruit set per panicle (36.20) were noted in CPT 1. The time taken for complete development of flower ranged from 16 to 20 days. The peak period of anthesis was recorded 6–11 am in all the genotypes. None of the genotypes showed anthesis before 4 am and after 2 pm. Anther dehiscence commenced after opening of flowers, i.e., at 7 am, and continued till 3 pm. The peak period of dehiscence was recorded between 8 am and 12 noon in all genotypes. The flower diameter varied from 5.12 to 6.30 mm. The stamen and carpel length varied from 1.97 to 2.12 mm and 1.22 to 1.38 mm, respectively. Pollen viability ranged from 54.55 to 70.38 %. Pollen germination ranged from 20.00 to 35.00 %. Maximum stigma receptivity was recorded on the day of anthesis. Fruit set/panicle was found to be positively and significantly associated with panicle length, and it may be observed while selecting elite genotypes.

17.9 Genetic Diversity

There are no standard cultivars of chironji available since little work was initiated to exploit the genetic resources of chironji in India. Due to the denudation of forests, already a lot of genetic base has been eroded. The seedling origin trees were identified for their variability and were marked for their special characteristics in Gujarat, Maharashtra, Utter Pradesh, Bihar, Orissa, Chhattisgarh, and Andhra Pradesh. The existing variability was collected at the station from the seedling progenies on the basis of yield, fruit size, fruit weight, TSS, etc. Flowering starts in the first week of February and continues till the end of March. The panicle length varied from 15.34 to 32.10 cm, pollen germination from 24.10 to 35.0 %, and pollen viability from 55.0 to 70.38 %. Fruit weight ranged from 0.97 to 1.35 g, pulp percent from 44.04 to 61.60, TSS from 19.00 to 22.20 %, stone weight from 0.38 to 0.69 g, kernel weight from 0.08 to 0.14 g, and protein content from 24.00 to 30 70 %.

17.10 Flowering and Fruiting

Chironji flowers in the months of January–February in the semiarid areas of Gujarat. It takes the flower 15–20 days to set the fruits. The plants start flowering in the 4–5 years of plantation in the grafted plants. In the seed-originated plants, it takes anywhere between 12 and 15 years. The fruit takes 4–5 months for ripening.

Flowering

Fruit set

Fruits

kernels

17.11 Fruit Maturity and Harvesting

The fruits take about 3–4 months for their full development; harvesting starts from the end of April and continues till the third week of May. A 5-year-old plant gives 25–30 kg fruits. After processing getting 1.0–1.2 kg charoli kernels obtained. The process of extracting kernels is labor intensive.

17.12 Plant Propagation

There was no standard method for propagation of chironji. The seedling trees are coming up over the ages from seed dropped from the dropping of the birds. Seed germination was found to be 83 % after scarification. Seed treatment with sulfuric acid 5–7 % was also suitable for enhancing the seed germination. The seeds are extracted from the ripened fruits during April–May. The seeds are sown in raised beds or polythene bags of 5 × 10 cm size. The seeds take 20–25 days to initiate germination. The seedlings become ready for grafting after 1 year.

17.12.1 Vegetative Propagation

Softwood grafting method was standardized for propagation of chironji; a 1-year-old rootstock was grafted through wedge grafting in the month of June when the rootstock seedling put forth new growth. Defoliation of scion was undertaken 10–15 days prior to grafting by removing leaves and allowing the petiole attached to the stem. The union is tied with polythene strip of 200 gauge thickness; 60–70 % success can be obtained during the months of July–August.

In situ softwood grafting method can be used for raising the orchards directly in the field. The rootstock seeds are sown on the onset of monsoon in the pits of 1 × 1 × 1 m size prepared as per the layout. The rootstock seedlings thus raised were softwood grafted after 1 year. The scion material was defoliated 10–15 days prior to grafting by allowing the petiole to remain attached to the scion shoot. The scion is ready as the apical bud swells and the petioles drop down.

Month	% success
February	3.33
March	34.33
April	29.00
May	16.33
June	39.00
July	66.66
August	40.33
September	24.00
October	17.00

The percent success was highest in July and was below 40 % in the rest of the months except November, December, and January where it is negligible (Singh et al. 2007).

17.12.2 Root Cutting

This is not the commonly practiced method for propagation of chironji; however, due to poor seed germination and slow growth of the seedling, the method is practiced by some nurserymen. Rooting can be obtained by treating the root cutting of 3.6–5.5 cm thickness with auxin like IBA 1500 which recorded as high as 67.66 % success (Singh et al. 2002).

17.12.3 Tissue Culture

Chironji is a vulnerable species. A tissue culture technique for rapid clonal multiplication has been standardized by Shende and Rai (2005). The decorticated seed was used for culturing on MS medium with various concentrations of auxin and cytokinin alone or in combination. The combination of benzyl amino purine (BAP) and naphthalene acetic acid (NAA) was found to be the best at 22.2 μm and 5.37 μm, respectively. For the formation of the maximum no. of shoots, further MS media containing 23.3 μm kinetin induced profuse rooting.

17.12.4 Rootstock

Selection of rootstock is important in the sense that it is a slow-growing plant if some fast-growing rootstock can be identified which will impart vigor to the plant and is also hardy in nature can withstand salinity having tolerance to drought. Work on this aspect needs to be undertaken. Since it belongs to the family Anacardiaceous, rootstock belonging to the same family may be tried.

17.13 Planting

It is observed that the chironji trees as old as 100 years are seen growing in the forest in India even as on date. They grow as huge trees with a height of 15–20 m. And hence at least 10 × 10 m spacing may be kept between the rows and plants. The plants are planted during the start of rainy season as they can be established well. The pits of 1 × 1 × 1 m size are opened during the summer and filled before the onset of monsoon. Solarization will kill most of the disease-causing organism, and also while filling the pits FYM and sand in appropriate proportion depending on the soil type is added so that there is sufficient aeration in the soil, which helps in plant establishment.

17.14 Training and Pruning

The tree is found growing as a single stem. There is no standard training method developed so far. Training plant with single stem up to 60–90 cm above ground level and then allowing 3–4 scaffold branches will be ideal for exposure of the plant to sunlight on all the four sides. This will also facilitate harvesting and other farm operations. As the plant bears on the current season's growth, no pruning is advocated. But to keep the plant dwarfing, there is a need to control the size of the plant where pruning in pre-bearing years will be useful.

17.15 Manuring and Fertilization

There are no standardized fertilizer dosages worked out for chironji as it is found growing mostly in the jungle. However, with the introduction of systematic cultivation, there is a need to standardize the fertilizer doses to obtain maximum production/unit area.

17.16 Intercropping

As the plant is very slow growing, there is very good scope for intercropping the orchard in the initial years. This will help in increasing the potential of the land in the initial years. To meet the basic requirement of food, fodder, and fuel, traditional crops of the area should be grown as intercrops during the rainy season. This will also help in reducing runoff losses and thereby soil erosion.

17.17 Irrigation

The trees are found growing in the jungle without any irrigation. It is found to be drought hardy and hence should be grown in rainfed areas. However, to establish the orchard in the initial years, irrigation should be given for the first 2–3 years at 10–15 days interval during the summer season.

17.18 Processing and Value Addition

Seeds are separated by rubbing in the water and dried. After that, kernels are taken out by breaking the hard seed coat mechanically and packed either in glass jars or polyethylene bags.

There is tremendous scope for preparing beverages from ripened fruit of chironji. Kernels are being used or the preparation of different kinds of sweets. The products like squash, RTS, and nectar may be prepared from the pulp of the fruit.

17.19 Marketing

After harvesting ripened fruits, the pulp is removed by eating and the kernels are then dried in the sun, and the seeds are taken out by grinding on the deshi grinding mill of stone by the ladies. The farmers are forced to sell the produce to the local business man at a very low rate due to ignorance of the farmer and lack of organized market in the local area. Once the regular plantation comes up, then there is scope to develop organized market also.

17.20 Pest and Disease Management

Hoppers: It is sucking pest causing damage during the flowering season. If uncontrolled, the entire tree becomes black. The fruit set is adversely affected due to the development of sooty mold. The pest can be controlled by spraying dimethoate (0.03 %) or phosphamidon (0.5 %) two to three times during vegetative growth and at flowering.

Bark-eating caterpillar (*Inderbella* spp.). The larvae feed on the bark of the stem particularly in the crotches of the stem. The infection if uncontrolled can remove the entire skin of the branch resulting in drying. Pest is active from March to June. The damage can be seen during regular visit of the farm in the form of webs and holes filled with excreta and wood powder. The damage can be controlled by removing the webs and filling the holes with dichlorvos (0.1 %) and closing them with wet soil. Regular spraying with dimethoate at 15 days interval can reduce the further attack of the pest.

17.21 Diseases

Gummosis: The disease is caused by the fungi *Fusarium* spp. In case of severe infection, the stem is covered with gum oozing out from the crotches. The stem becomes dark and branches start drying down. Application of Bordeaux paste and spraying it on the plant is the cheapest way to control this disease. Spraying of copper fungicides 2 g/L of water or bavistin at 1 g/L of water can help in controlling the disease.

17.21.1 Powdery Mildew

The disease appears in the form of white powder on the flowering bunches affecting the fruit set adversely during flowering season. This is fungal disease caused by *Erysiphe polygoni*. The disease can be controlled by spraying with wettable sulfur 80 % at 2 g/L of water two to three sprays at 15 days interval.

References

Banerjee A, Jain M (1988) Investigation on *Buchanania lanzan* Spreng, seed oil. Fitoterapia 59:406

Chitra V, Dharani Prasad P, Pavan Kumar K, Narayan Rao A (2009) Wound healing activity of alcoholic extract of *Buchanania lanzan* in albino rats. Int J Chemtech Res 1(4):1021–1031

Das D, Agrawal VS (1991) Fruits drug plants of India. Kalyani Publisher, New Delhi, p 250

Munde VM, Shinde GS, Sajindranth AK, Prabu T, Machewad PM (2003) Correlation and path analysis studies in charoli (*Buchanania lanzan* Spreng). South Indian Hortic 50:517–521

Sharma A (2012) Scientific harvesting for quality seed collection of *Buchanania lanzan* Spreng for its conservation and sustainable management. Int J Bio-Sci Bio-Technol 4(1):65–74

Shende S, Rai M (2005) Multiple shoot formation and plant regeneration of a commercially useful tropical plant *Buchanania lanzan* Spreng. Plant Biotechnol 22(1):59–61

Singh RV (1982) Fodder trees of India. Oxford and IBH Publication, New Delhi, p 663

Singh J, Banerjee SK, Francis A (2002) Vegetative propagation of *Buchanania lanzan* Spreng. root cuttings. Indian For 128:700–704

Singh S, Singh AK, Bagle BG (2006) Chironji (Buchnania Lanzan Spreng.). A boon for tribals. Intensive Agriculture, March–April, pp 35–36

Singh S, Singh AK, Bagle BG (2007) Propagation of chironji by in situ soft wood grafting. Indian Hortic 52(2):35–36

Mahua (*Bassia latifolia* Roxb.) 18

Abstract

Mahua, botanically known as *Bassia latifolia* Roxb., is an economically multipurpose tree. Its flowers, fruits, and oil are used in various ways. The flowers are used for the preparation of distilled liquor and as feed for livestock. The seed is a good source of oil. A wide range of variability occurs in nature with regard to fruit size and quality owing to its seed propagation, which needs to be conserved and exploited. In recent years, there is considerable awareness about nutritional security and food safety. *Mahua*, a characteristic tree of the dry region, is found in almost all parts of India growing in the forests. It is a multipurpose tree of which every part is used even today by the tribals of India. Its flowers are used in extracting liquor for ages. The dry flowers are a big business. Its seeds are used for extraction of oil which is edible and is used for lighting lamps. Its wood is used in house construction as well as fuelwood and preparation of agricultural implements.

18.1 Introduction

Madhuca longifolia (Koenig) var. *latifolia* (Roxb.) commonly called mahua is attached with the life of the tribals. It is a member of Sapotaceae. The family is known for its being drought hardy, salt tolerant, and surviving the natural rains once established. It is a large deciduous tree. Its importance as a medicinal plant is known to tribals since time immemorial. It is antipyretic, hematoprotective, anti-inflammatory, analgesic, and antitumor and treats rheumatism, ulcer, bleeding, and tonsillitis. It is highly respected like a mother by the tribals and is a large evergreen tree distributed throughout India, Sri Lanka, and Nepal (Saluja et al. 2011). Phytochemical studies indicated the presence of starch, terpenoids, proteins, mucilage, anthraquinones, glycosides, saponin, and tannins (Gopalkrishnan and Shraddha 2012) (Table 18.1).

The elements present in *Madhuca longifolia* play an important role in Ayurvedic treatment. It indicated the association of 5 and 6 microelements. The composition of the leaf indicated a higher concentration in respect to the macroelements P and Mg. In respect to microelements Z, Cu, Na, and Fe are found in higher concentration than the macroelements which might be helpful in hastening the enzyme activity in the human system.

S. Hiwale, *Sustainable Horticulture in Semiarid Dry Lands*,
DOI 10.1007/978-81-322-2244-6_18, © Springer India 2015

Element	%	Conc. (ppm)
Phosphorus	1.43	42.5
Calcium	0.535	665.8
Potassium	0.345	112.9
Magnesium	0.819	140.4
Chlorine	0.478	93.2
Sulfur	0.623	35.6
Sodium	3.88	490.7
Copper	11.8	71.8
Zinc	13.2	64.9
Ferrous	2.61	101.2

Annalaxmi et al. (2012)

Table 18.1 Ethnomedicinal uses of the bark of *Madhuca latifolia*

Use	Worker
Antidiabetic, itch, swelling, fractures, and snakebite	Kumar et al. 2011
Wound healing	Smitha et al. 2010
Antibacterial, rheumatism, bleeding, and spongy gums	Tambekar and Khante 2010
Antioxidant activity	Roy et al. 2008
Astringent, stimulant, emollient, piles, nutritive, and antiepileptic	Khond et al. 2009
Rheumatism, ulcer, and tonsillitis	Prasanth et al. 2010
Skin diseases, epilepsy, pneumonia, and piles	Khare 2007

Akshatha et al. (2013)

It is a multipurpose tree as its fresh flowers are used for eating as well as for making local wine. Oil is extracted from fruits and seeds which till today is used by the adivasi people for eating as well as lighting lamps. The flowers after wine extraction can be used as animal feed. The seed oil is used for soap, lubricating grease, and candles. Mahua seeds can also be used as defatted flour for making bakery products. The cake obtained from oil extraction has industrial as well as commercial application either as manure or as insecticide. Apart from this it provides good-quality timber. The tree is cherished as a prized possession by the adivasi people even until today. Seventy five percent of the tribal households engaged in mahua flower collection, meaning a population of 7.5 million are into this livelihood activity. An estimate indicates that 28,600-man days/year of employment is generated in mahua flower collection (FLGG India 2008) (Table 18.2).

18.2 Mahua Seed Oil

A mahua seed contains 33–43 % oil, which is a pale yellow semisolid fat. It has multifarious uses as cooking and lamp oil, paints, and varnishes and is even mixed with other vanaspati oils (Table 18.3).

18.3 The Major Constraints of the Business Are

1. Fast depletion of resource base
2. Distress selling and low price realization
3. Absence of quality standards

New leaves sprouting

Flowering in Mahua

Close view of flowering

Table 18.2 Nutritional composition of mahua flower

Component	Percent
Moisture	19.8
Protein	6.37
Fat	0.5
Reduced sugar	50.62
Total sugar	54.06
Ash	4.36
Calcium	8.0
Phosphorus	2.0

Kureel et al. (2009)

Table 18.3 Composition of mahua oil

Component	Quantity
Refractive index	1.452–1.462
Saponification value	187–197
Iodine value	55–79
Palmitic	24.5 %
Stearic acid	22.7 %
Oleic acid	37.0 %
Linolenic acid	14.3 %

Kureel et al. (2009)

4. Failed initiatives by state agencies for procurement
5. No commercial breakthrough in alternative product development

There are three major hurdles, namely, knowledge, storage, and credit. Technology relating to standardization of harvesting methodology and postharvest treatments needs to be standardized. Apart from this grading at harvester and trader level, there is the identification of slandered varieties with good-quality flowers and yield per plant. To stop the loss of resources, plantation programs need to be taken up on large scale. A centralized storage infrastructure with a large number of mini godowns at vantage places should be created in PPP mode. A microcredit should be created for meeting the cash needs of mahua collectors, and mahua procurement is assisted through this fund. A decentralized procurement mechanism should be created to get better price and avoid distress sale.

18.4 Origin and Distribution

It is found growing wild throughout India, but it prefers hot and dry climate. It is found growing throughout Madhya Pradesh, Chhattisgarh, Maharashtra, Utter Pradesh, Orissa, Andhra Pradesh, and Gujarat in the jungles. The tree is surviving today because the adivasi people cut any other tree except the mahua trees.

18.5 Area and Production

There are no regular plantations of mahua trees in India, and they are found growing spread all over the forest area; hence, exact statistics regarding area is not available. Chhattisgarh, Madhya Pradesh, and Orissa account for nearly 80 % of mahua trees in India. There are three to five million trees of mahua in undivided M.P. The estimated collection is around 85,000 t out of the potential 4,90,000 t. The difference is due to dwindling tree population. Ranchi is the biggest trade center in India. With the liberalization and opening up of the international market under the WTO regime, the Government of India is trying to promote wines from India. Agriculture Processed Foods Export Development Authority has been entrusted to develop a strategy and find ways to export the product since mahua has the potential of becoming an excellent wine.

18.6 Soil and Climate

It is found growing on the dry lands of semiarid areas. It can be grown on wastelands without much care. The seedlings of the trees are found growing during the rainy season. The birds are the main source of their spread. The birds eat the fruits, and the seeds come out in the form of excreta drop throughout the forest and germinate on the onset of monsoon.

Mahua likes tropical and subtropical climate. The trees of *Bassia latifolia* and *Bassia longifolia* are found growing in an altitude of 1,000–1,200 m.; the trees of the species *Bassia butyracea* are found growing in an altitude of 4500 m in Himachal Pradesh. *Bassia malabarica*species is found growing in the Western Ghats and down south in Karnataka, Kerala, and Tamil Nadu.

18.7 Distribution and Habitat

The species is distributed in central, northern, and southern India, Sri Lanka, and Burma. Of the two varieties *longifolia* is distributed in Sri Lanka, southern India extending toward Maharashtra, and Gujarat. The variety *latifolia* is found growing in some parts of central and northern India and Burma. It is found growing at an altitude of 1,200 m, is a strong light demander, and is suppressed under shade.

18.8 Botany

Mahua belongs to the family Sapotaceae. The species name is *Madhuca longifolia* (Koenig) J.F. Macb. var. *latifolia* (Roxb.) Cheval. Its synonyms are the following: *Madhuca latifolia* Macb., *Madhuca longifolia* Koenig, *Madhuca indica* J.F. Gmel. It is known locally as honey tree or butter tree in English, as mahua in India, and as mi or illuppai in Sri Lanka.

18.9 Few Species of the Genus Bassia Are Described Here

18.9.1 *Bassia Latifolia*

The trees of the species are found growing 12–15 m in height. Trees as old as 50–60 years are very common. The bark is dark in color, with crevices, and the inner cambium is red and if cut than a milky substance oozes out from numerous branches. The tree is a spreading type. The leaves are elliptic or oblong, acuminate, and pubescent or tomentose. The petioles are long. The flowers are in dense fascicles near the end of the branch. The pedicels are long, dropping, fulvous, and rusty pubescent or tomentose. The calyx is long, divides nearly at the base, and is usually 4, rarely 5, in number. The corolla is cream in color, with long fleshy tube and 7–14 lobes. Stamens have 20–30 anthers that are in three series, acuminate and hairy. The ovary is hirsute, with a long style that is hairy at the base. The fruit is a berry which is long, fleshy, ovoid, and greenish. Flowering starts in January and continues up to April. It varies from place to place.

18.9.2 *Bassia longifolia*

The species is commonly found in the tropical forests of south India. The trees are large with multiple branching; the bark is thick dark brown with scales. The leaves are thin and lanceolate in shape or globose, tapered toward the base. The petiole and stipules are long and slender. Flowering starts with the emergence of new growth in clusters. The calyx is long divided nearly up to the base, usually with 4 pubescent segments, ovate-acuminate. The corolla is long divided nearly up to the base with a fleshy inflated globose tube, but rugose when dried. and has 6–12 lobes that are lanceolate and glabrous. Stamens are 16–20 in number arranged in two rows, one above the other, and have subsessile hairy anthers that are cordate at the base. The ovary is densely hairy. The fruit is an oblong berry. It has 1–2 seeds compressed straight on one side. Flowering takes place in November–January.

18.9.3 *Bassia malabarica*

The tree is medium in height. The bark is dark colored. The leaves are scaly, oblong to lanceolate, subobtuse, dark green, and shiny. Venation is reticulate. The petiole is long. Flowering is at the end of the branch in dense fascicles. The calyx is long, divided at the base, with 4 segments. The corolla is long, with densely hairy tube slightly shorter than the lobes which are six in number, oblong, and hairy outside. Stamens are 16 to 18 in number in two rows, with hairy filament and glabrous anther with small tuff of hairs between the basal lobes. The ovary is glabrous. The fruit is a glabrous oblong berry. Flowering takes place in November–January.

Characteristics of *Madhuca longifolia.* It is a large evergreen tree with short stems 3–6 m in length. Its bark is scaly, yellowish gray to dark brown and red inside. The leaves are clustered near the apex; linear lanceolate, tapering toward the base; and glabrous when mature (Anonymous 2012).

The flowers are pale yellow in color, small, many, and aromatic, in dense clusters near the end of the branches, with rusty, pubescent calyx and fleshy corolla. The fruits are berries ovoid in shape, 5 cm long with 1–2 seeds, and yellow when ripe. The seed is yellow or light brown in color, shiny, and smooth. Mahua is a cash crop for the tribals and also provides food security.

18.10 Reproductive Biology and Diversity

Under the semiarid rainfed conditions of India, the period of flowering is from March–April. The maximum number of flowers is 42, and the number of fruits recorded was 9.5 per fascicle. The peak period of anthesis was from midnight till early morning. The ovary and style length showed marked variation. It varied from 4.50 to 6.25 mm and 25–35 mm, respectively. Pollen viability ranged from 90.0 to 98.50 % and pollen germination from 26.50 to 40.20 %. Fruit set per panicle was found to be a promising trait for the selection of high-yielding varieties. The germplasm can be classified as early, midseason, and late based on maturity of the fruit. The mahua is a highly heterozygous, cross-pollinated fruit, and therefore the existing seedling population exhibits a lot of variability. Therefore, a wide range of variability in growth, flowering fruit set, and physicochemical characteristics can be observed in the individual plants. Physicochemical attributes showed considerable variability. Seed weight varied from 4.20 to 12.10 g. The oil content ranged from 41.0 to 45.10 %. Protein content ranged from 20.00 to 24.00 %. TSS of the fruit varied from 24 to 27.50 % and acidity from 0.09 to 1.13 %. Ascorbic acid content showed variability from 43 to 64 mg per 100 g pulp. Total sugar content ranged from 20.00 to 24.24 and reduced sugar from 9.24 to 13.13%.

The production of mahua flowers at different ages of the tree showed an increasing trend. A 10-year-old tree yields 10 kg of flowers, and a 60-year-old tree yields 140 kg of flowers. Though all the parts of the plant are useful either one or another, the flowers are the real income generators as well as a subsidiary food to the farmers as well as the cattle (Table 18.4).

18.11 Flowering and Fruit Set

Flowers appear from February to April in different agroclimatic zones. The flowers take 20–30 days to develop completely. The number of flowers per fascicle varied from 10 to 60. Grafted plants start bearing from 7 years. Though there is profuse flowering, the fruit set is very poor. It ranges from 1.6 to 4.0 and fruit retention ranges from 8 to 13%. The low fruit

Table 18.4 Evaluation of germplasm done at NDUAT, Faizabad, has resulted in some selections, viz, NM-2, NM-4, NM-7, and NM-9

	NM-2	NM-4	NM-7	NM-9
Flowering time	First week of April	Fourth week of April	Second week of April	Fourth week of April
No. of flowers/fascicle	66	54.6	64.30	54.6
Dry flower yield (kg)	9.9	8.5	10.00	6.8
Total sugar	25.50	24.50	22.70	21.80
Fresh flower (%)				
Reducing sugar (%)	18.6	19.60	19.5	20.00
Ripening	Second week of June	Third week of June	Third week of June	Third week of June
Seed weight (g)	7.20	6.0	9.50	5.20
Kernel weight	5.50	4.70	7.20	4.0
Oil content (%)	47.40	45.20	46.70	48.50
Protein (%)	23.60	25.70	22.40	16.80
Flower yield (dry) kg/plant	9.9	8.5	10.0	6.80

Singh (1998)

set may be due to self-incompatibility. There are three distinct phases of flower drop locally known as shuru, bhawari, and knawa. In the initial stage which lasts for 5–6 days, flowers posses a shrunken appearance. In the first stage of drying, the flowers collected yielded 25 % by weight of the total collected flowers. The next stage lasts around a week. The quality of flowers that drop is maximum with a yield as high as 50 % post drying. They possess a bold and succulent appearance. In the third stage of flower drop, the flowers are of original size in appearance as well as in yield.

References

Akshatha KN, Mahadeva Murthy S, Laxmidevi N (2013) Ethnomedicinal uses of *Madhuca longifolia*-a review. Int J Life Sci Pharma Res 3(1):44–53

Annalaxmi R, Uma R, Subhash Chandran GG, Savairaj Sahayam C, Charls A (2012) Evaluation of elemental content of leaves of *Madhuca longifolia* by X-ray fluorescence spectroscopy. J Nat Prod Plant Resour 2(4):490–493

Anonymous (2012) Report on the study on Madhuca sub section. Chhattisgarh State Information and Publication, Raipur, pp 1–30

Gopalkrishnan B, Shraddha N (2012) Pharmacological studies on stem bark of *Madhuca longifolia* *koen. 9. Indian J Prod Resour (3):232–236

Khare CP (2007) Indian medicinal plants; an illustrated dictionary. Springer Science, New York

Khond M, Bhosle JD, Arif T, Mandal TK, Padhi MM, Dabur R (2009) Screening of some selected medicinal plants extracts for in vitro antimicrobial activity. Middle East J Sci Res 4:271–278

Kumar PK, Vidhyasagar G, Ramakrishna D, Reddy MI, Atyam VSSG, Raidu SCH (2011) Screening of *Madhuca indica* for antidiabetic properties in streptozotocin and Streptozotocin–Nicotinamide induced diabetic Rats. Indian J Pharma Tech Res.

Kureel RS, Kishor R, Dev Dutt, Ashutos P (2009) Mahua–a potential tree born oilseed. National oilseed and development Board. Ministry of Agriculture Govt. of India, Gurgaon, pp. 1–27

Prasanth S, Kumar AA, Madhu B, Yennamaneni PK (2010) Antihyperglycemic and antioxidant activity of ethanolic extract of *Madhuca longifolia* bark. Int J Pharma Sci Res 5:89–94

Roy SP, Shirode D, Patel T, Prabhu K, Shetty SR, Rajendra SV (2008) Antiulcer activity of 70 % ethanolic extract of bark of *Madhuca longifolia*. Indian J Nat prod 24:8

Saluja MS, Sangameshwaran B, Hura IS, Sharma A, Gupta SK, Chaturvedi M (2011) In vitro cytotoxic activity of *Madhuca longifolia against* Ehrlich Ascites Carcinoma cell lines (EAC). Int J Drug Discovery Herbal Res IJDDHR 1:55–57

Singh IS (1998) Mahua an oil bearing tree Technical bulletin pp. 3–11. N. D University of Agriculture and Technology Faizabad, Uttar Pradesh

Smitha S, Mukesh CS, Kohli DV (2010) Wound healing activity an formulation of ether benzene 95 % ethanol extract of herbal drug *Madhuca longifolia* leaves in albino rats. J Optoelectron Biomed Mater 1:13–15

Tambekar DH, Khante BS (2010) Antibacterial properties of traditionally used medicinal plants for enteric infection by adivasi's (bhumka) in melghat forest (Amravati Dist, M.S.). Int J Pharmaceut Sci Res 1:120–128

Non Traditional Crops: Cashew (*Anacardium occidentale*)

19

Abstract

Cashew nut is one of the most important items of export from India. The place of origin is Brazil. Other major cashew nut-producing countries are Mozambique, Tanzania, Brazil, Kenya, and Madagascar. Each and every part of cashew nut is useful. The kernels are rich in carbohydrate; unsaturated fats; minerals, like calcium, phosphorus, and iron; and vitamins. The wood is used for firewood, charcoal making, etc. Wood pulp is used to fabricate corrugated and hardboard boxes. It has many medicinal values. Cashew apple can be brewed into tasty wine (fenny), and the remains can be used as cattle feed. There is scope to introduce this crop into the nontraditional area, and its performance in semiarid areas on marginal land is found to be very encouraging.

19.1 Introduction

A cashew tree is a tropical evergreen that produces the cashew nut and the cashew apple. Officially classed as *Anacardium occidentale*, it can grow as high as 14 m (46 ft), but the dwarf cashew, growing up to 6 m (20 ft), has proved more profitable, with earlier maturity and higher yields. The cashew nut is served as a snack or used in recipes, like other nuts, although it is actually a seed. The cashew apple is a fruit, whose pulp can be processed into a sweet, astringent fruit drink or distilled into liqueur. The shell of the cashew nut yields derivatives that can be used in many applications from lubricants to paints, and other parts of the tree have traditionally been used for snakebites and other folk remedies.

Originally native to northeastern Brazil, the tree is now widely grown in tropical regions, Vietnam and Nigeria being major producers, in addition to India, Côte d'Ivoire, and Indonesia. Cashew nut belongs to the family Anacardiaceae and it originated from Brazil. Cashew nut is one of the most important export items from India. In India, cashew nut was introduced by the Portuguese in Malabar Coast in the sixteenth century and planted to check soil erosion. Other major cashew nut-producing countries are Mozambique, Tanzania, Brazil, Kenya, and Madagascar. Table 19.1 gives an idea about its area and production in India.

Its English name derives from the Portuguese for the fruit of the cashew tree, caju, which itself is derived from the indigenous. The name *Anacardium*, originally from the Greek, actually

S. Hiwale, *Sustainable Horticulture in Semiarid Dry Lands*,
DOI 10.1007/978-81-322-2244-6_19, © Springer India 2015

Table 19.1 Cashew producers

Country	Production (Mt)	Yield (Mt/ha)
Nigeria	650,000	1.97
India	613,000	0.66
Côte d'Ivoire	380,000	0.44
Vietnam	289,842	0.85
Indonesia	145,082	0.25
Philippines	134,681	4.79
Brazil	104,342	0.14
Guinea-Bissau	91,100	0.38
Tanzania	80,000	1.0
Benin	69,700	0.29
Total	**2,757,598**	**0.58**

Food and Agriculture Organization (2011)

refers to the nut, core, or heart of the fruit, which is outwardly located (*ana* means "upward" and *-cardium* means "heart"). The fruit of the cashew tree is an accessory fruit (sometimes called a pseudocarp or false fruit). What appears to be the fruit is an oval- or pear-shaped structure, a hypocarpium, that develops from the pedicel and the receptacle of the cashew flower (Varghese and Pundir 1964). Called the cashew apple, it ripens into a yellow and/or red structure about 5–11 cm long. It is edible and has a strong "sweet" smell and a sweet taste. The pulp of the cashew apple is very juicy, but the skin is fragile, making it unsuitable for transport. In Latin America, a fruit drink is made from the cashew apple pulp which has a very refreshing taste and tropical flavor that can be described as having notes of mango, raw green pepper, and just a little hint of grapefruit-like citrus. While the cashew plant is native to northeast Brazil, the Portuguese took it to Goa, India, between 1560 and 1565. From there, it spread throughout Southeast Asia and eventually Africa.

Nigeria was the world's largest producer of cashew nuts with shell in 2010. Cashew nut production trends have varied over the decades. African countries used to be the major producers before the 1980s; India became the largest producer in the 1990s, followed by Vietnam which became the largest producer in the mid-2000s. Since 2008, Nigeria has become the largest producer. Cashew nuts are produced in tropical countries because the tree is very frost sensitive; they have been adapted to various climatic regions around the world between the latitudes of 25°N and 25°S.

Peru reported the world's highest production yields for cashew nuts in 2010, at 5.27 metric tons per hectare, nearly nine times the world average.

Cashew apples have a sweet but astringent taste. This astringency has been traced to the waxy layer on the skin that contains a chemical, urushiol, which can cause minor skin irritation to areas that have had contact with it. It is almost identical to the astringency caused by the skin of a mango, which also contains urushiol. The astringency from mango skin can be mildly tasted in the flesh of mango fruit, just as the astringency of cashew apple skin can be mildly tasted in the flesh of cashew apples. In cultures that consume cashew apples, this astringency is sometimes removed by steaming the fruit for 5 min before washing it in cold water; alternatively, boiling the fruit in salt water for 5 min or soaking it in gelatin solution also reduces the astringency (Azam-Ali and Judge 2004).

19.2 Uses

1. Each and every part of cashew nut is useful. The kernels are rich in carbohydrate; unsaturated fats; minerals, like calcium, phosphorus, and iron; and vitamins. The cashew apple juice is rich in vitamin C (261.5 mg/100 g). It contains 10.15–12.5 % sugars (mostly reducing) and about 0.35 % acid (as malic acid).
2. The wood is used for firewood, charcoal making, etc.
3. Wood pulp is used to fabricate corrugated and hardboard boxes (Nambiar and Pillai 1986).
4. Medicinal uses: Many parts of the plant are used in the traditional medicine of the Patamona of Guyana. They grind the seeds into a poultice for treating snakebites, apply nut oil to cracked heels or as an antifungal agent, and use the fruits, bark, and leaves for many other purposes including antifungal activity, for sores and rashes, as an antipyretic, and for antidiarrheal applications. The leaf

extracts with petroleum ether and ethanol inhibited the growth of several species of bacteria and fungi (Dahake et al. 2009). Chemicals identified in cashew shell oil have been assayed against *Streptococcus mutans*, a bacterium responsible for many dental cavities, and found to have activity in vitro against this and other Gram-positive bacteria.

Cashew apple can be brewed into tasty wine (fenny) and the remains can be used as cattle feed. In Goa, the cashew apple (the accessory fruit) is mashed; the juice is extracted and kept for fermentation for a few days. Fermented juice then undergoes a double distillation process. The resulting beverage is called *feni* or fenny. *Feni* is about 40–42 % alcohol. The single-distilled version is called *arrac*, which is about 15 % alcohol.

19.3 Climate

Cashew is a tropical plant, which requires a moist, mild tropical climate. Plants cannot tolerate frost and extreme cold for a long time and the dry hot summer. It can be grown up to 1,000 m elevation. It grows in areas with rainfall ranging from 100 cm to over 400 cm/annum. It is a sun-loving tree and does not tolerate excessive shade. When temperature exceeds 39–42 °C during the marble stage of fruit development, it causes fruit drop.

19.4 Soil

Cashew is found growing well on a wide range of soil types. It is generally grown as a wasteland crop. It is grown widely on laterite, red, and coastal soils in Kerala, Konkan region of Maharashtra, Goa, Karnataka, Tamil Nadu, Andhra Pradesh, Orissa, and West Bengal. However, the best soil to grow is deep, well-drained sandy loam.

Studies reported from the Cocoa Research Institute, Ibadan, Nigeria, on variability in soils and cashew tree size indicated that tree growth is affected by the soil type in 3 ecological localities. Profile pits were dug, soil samples collected for analysis, and tree girth and canopy diameter measured. The relationship between tree size and soil properties depended on the locality. Generally, high clay content, high pH, high base saturation, poor drainage or seasonally high water table, and the presence of stones or hardpan within 100 cm of soil depth reduced tree size. Only high sand content and high rainfall appeared favorable (Tables 19.2 and 19.3).

Table 19.2 Area and production of cashew nut in different states

	Area (.000 ha)				
State	1991–1992	2008–2009	2009–2010	2010–2011	Yield (kg/ha)
Kerala	155.5	70.0	72.0	78.0	920
Karnataka	74.4	107.0	118.0	119.0	359
Andhra Pradesh	71.2	182.0	183.0	183.0	566
Tamil Nadu	96.0	131.0	133.0	135.0	132
Goa	45.1	55.0	55.0	56.0	321
Mahabharata	22.9	170.0	175.0	181.0	1,398
Orissa	60.1	137.0	143.0	149.0	530
Pondicherry	0.4	6.0	6.0	6.0	789
Tripura	1.1				44
West Bengal	6.9	11.0	11.0	10.0	530
All India	533.6	893.0	923.0	963.0	572

Kumar et al. (2011)

Table 19.3 Production (.000tons)

	Production (.000 tons)			
State	1991–1992	2008–2009	2009–2010	2010–2011
Kerala	143.2	75.0	66.0	71.0
Karnataka	26.8	60.0	53.0	57.0
Andhra Pradesh	40.4	112.0	99.0	107.0
Tamil Nadu	12.7	68.0	60.0	65.0
Goa	14.5	30.0	26.0	24.0
Mahabharata	32.0	225.0	198.0	208.0
Orissa	31.8	95.0	84.0	91.0
Pondicherry	0.3	–	–	–
Tripura	0.1	–	–	–
West Bengal	3.7	11.0	10.0	11.0
All India	305.5	695.0	613.0	674.0

Kumar et al. (2011)

In India the area under cashew was 533.6 thousand hectares which increased to 963.0 thousand hectares with a production of 305.4 thousand tons to 674 thousand tons and 572 kg yield/ha. Highest acreage and production of cashew was recorded in Kerala during the years 1990–1991; however, over the years, Maharashtra has taken a lead in respect to area, productivity, and yield kg/ha. It was maximum in Maharashtra (1,398 kg/ha).

19.5 Varieties

A large number of germplasm have been collected and assessed at different cashew research centers. From Anakkayam/Mannuity, promising selections are BLA-139-1 and BLA-273-1; from Vengurla, Vengurla 1–8; from Vridhachalam, M 25/1, M 262/2, and m3/4; and from Bapatla, Selection No. 129, Selection 56, Selection 1, and Selection 273.

Vengurla 8 in bearing

Cashew Local collection

Variability in Cashew apple

19.6 Performance of Vengurla Selections Under Semiarid Conditions

To explore the possibility of cultivating cashew in semiarid areas, few local collections from the forest areas of Panchmahal was made and Vengurla selections were introduced from KKV, Dapoli. The selection Vengurla 8 was outperformed in respect to yield parameters. The number of clusters per plant was highest in Vengurla 8; however, the number of fruit set per cluster was highest in the local collection. The number of fruits harvested per tree however lot of variation in was maximum in Vengurla 8 followed by the local collection. Seed weight and apple weight was highest in Vengurla 1 followed by Vengurla 8. Yield per plant was maximum in Vengurla 8 followed by the local collection under a semiarid climate (Table 19.4).

19.7 Hybrids

H-3-19 (T.30X Brazil-18) Anakkayam; 2/11 (T. No.1x T.No. 273) Bapatla; Hybrid No.11 (Midnapore Red x Vector 56) Vengurla.

19.8 Propagation

Propagation of cashew by vegetative methods is suggested for higher yields. However, seed is the most important for raising seedlings for its use as rootstocks.

Table 19.4 Performance of cashew varieties at CHES, Vejalpur

Cultivar	No. of clusters/ plant	No. of fruit set/cluster	No. of fruits harvested	Seed wt. (g)	Apple wt. (g)	Yield of nut/ kg plant
Local collection	68	9.46	483	4.66	27.52	2.25
Vengurla 1	79	4.2	40	8.66	92.37	0.34
Vengurla 3	47	3.8	198	5.25	46.60	1.04
Vengurla 4	58	4.4	73	5.72	53.15	0.42
Vengurla 7	65	6.4	56	6.13	39.60	0.34
Vengurla 8	87	6.7	585	7.83	49.75	4.58

Hiwale (unpublished)

19.8.1 Layering

Air Layering: Air layering is one of the most common methods of cashew propagation. Pencil thickness shoots of the previous season's growth are selected, and a ring of bark, 3 cm long, is removed. The cinctured portion is applied with IBA 500 ppm in lanolin paste, and a lump of well-soaked sawdust is placed and wrapped with polythene tape (25 × 25 cm); the ends of the tape are carefully tied and left for rooting. When roots formed sufficiently (60–80 days), which are seen from polythene wrapper, the layers are severed and planted in pots. They are kept under partial shade till transplanting.

19.8.2 Budding

Patch and Forkert method of budding are successful methods of propagation. The operations in patch budding consist of removing a rectangular patch of bark with a bud from the budwood by giving two transverse and two vertical cuts. A corresponding patch of bark is removed from the stock plant. The bud is placed on the stock and wrapped with polythene tape in such a way that a small patch of bud is left uncovered. Within 25–30 days, sprouting takes place. The bud take is about 90 % during May–June.

In the case of Forkert method of budding, the bark is loosened by two vertical incisions, connecting the two with a transverse cut at the bottom. The bud is inserted into the flap by lifting the flap upward, and the bud, covered with the flap, is wrapped with polythene tape. The wrapper and the flap are removed after 25–30 days of budding. The exposed bud starts developing further.

19.8.3 Grafting

Cashew can be successfully propagated by veneer, side, cleft, and whip grafting. Recently, epicotyl and softwood grafting are emphasized much and becoming popular.

19.9 Planting

On well-cleaned field, the pits 50 cm^3 size are dug at 6 × 6 to 8 × 8 m distance. The pits are filled with a mixture of 15 kg farmyard manure and sufficient soil to fill the pit.

Planting is done at the beginning of monsoon, i.e., May–June.

19.10 Manure and Fertilizers

Based on long-term experiments on the nutritional requirements of cashew, the Central Plantation Crops Research Institute recommends 250 g N, 125 G P_2O_5, and 125 g k2o per tree per year in two split doses pre- and post-monsoon. However, under climatic conditions of the Konkan region of Maharashtra, cashew tree responded to N at 125 kg in the presence of p205 at 50 and k2o at 100 kg per hectare (Sawake et al. 1985). At the Agricultural Research Station, Utkal, the trees receiving N:P:K at 25,000:250:250 kg/ha had the highest yield (Hanamashetti et al. 1985). In the laterite tract of West Bengal, high nut yield (3.7 kg/tree) in cashew has been recorded in trees receiving fertilizers at the age of 3 years at the rate of 200 g N, 75 g P, and 100 g K per tree (Ghosh and Bose 1986).

19.11 Irrigation

Mostly, cashew is grown under rainfed conditions. However, it responds well to irrigation during flowering and fruiting; 2–3 irrigations increase yields considerably.

19.12 Interculture and Intercropping

Hoeing is done to keep down weeds and to conserve soil moisture during the early stages of plantation.

It is advisable to grow some intercrops that will enable higher return during the initial years for up to 4 years or so. Seasonal crops like horse gram, cowpea, ground nut, sweet potato, ginger, and black gram can be intercropped. Sometimes paddy crop is taken as intercrop in cashew plantation. Long-living crops like coconut and, sometimes, *Casuarinas* are also grown. Intercropping is not desirable after the cashew trees commence flowering and fruiting.

19.13 Training and Pruning

Young trees are trained by providing support at the time of planting and by removing any water sprouts coming out on the trunk below the union. The branches which are hanging and touching the ground are removed. The dead and diseased twigs and dried branches should be removed to reduce losses through diseases like dieback and also to minimize chance of serious losses due to forest fires (Singh et al. 1967).

19.14 Flowering, Fruiting, and Fruit Development

The cashew tree starts flowering at the age of 3–5 years and attains full bearing by the 10th year and continues to yield usually for 30–40 years. The inflorescence (panicle) is an indeterminate panicle of polygamomonoecious type. More number of hermaphrodite flowers are in the middle portion of the panicle. The flowers are white to light green at the time of opening, later turning to pink. The flower is typically pentamerous. Number of stamens is 10–11, one or two are large, and the remaining is small. The ovary is superior and is unilocular and contains a single anatropous ovule.

Flowering commences in the month of November and extends up to February, and the fruit ripens from March to May.

Growth patterns of the apple, pericarp, and kernel are of different types. Initially, the nut grows much faster than the apple, but in later stages, the apple increases in size much more rapidly and soon outgrows the nut. Both pericarp and kernel reach their maximum size about 30 days after fruit set. Therefore, the pericarp hardens and its size decreases. The apple becomes sweet when fully mature due to accumulation of nonreducing and reducing sugars, and protein content of the kernel increases steadily up to 40 days after fruit set and remains high until harvest (Chattopadhyay et al. 1983).

Fruit setting is poor because the percentage of hermaphrodite flower is less. About 3–10 % fruit set has been reported depending upon the climatic conditions. Fruit set can be increased and fruit drop controlled with the use of growth regulators like 2, 4-D at 10 ppm; spraying twice during the flowering period increases fruit set (Murthy et al. 1975). Two sprays during flowering with Ethrel, naphthoxyacetic acid, and GA each at 50 ppm are most effective in increasing fruit set and fruit retention (Chattopadhyay 1985; Pappiah et al. 1980).

Fruits are ready after 60 days of pollination. The shell of the nut attains full size in 3–4 weeks after flowering, after which the pedicel swells to produce the cashew apple which is juicy and sweet when ripe. It is a good source of vitamin C, sugar, and minerals. The nut is a kidney-shaped seed. Cotyledons are remilunar in shape.

19.15 Harvesting

The fully ripe fruits are harvested. They are allowed to fall on the ground and collected daily and dried in the sun.

19.16 Yield

Yield varies considerably from tree to tree, climatic conditions, soil type, age of the tree, management practices, etc. The average yield in India is 2.1 kg per tree per year. However, it varies from 1 to 45 kg per tree/year. It is providing considerable scope for improvement.

19.17 Processing

Processing consists of moisture conditioning, roasting, shelling, drying, peeling, grading, and packing.

19.18 Insect Pests

19.18.1 Root and Stem Borers (*Plocaederus ferrugineus*)

The borers make holes in the collar region of the tree, gum oozes out from the holes, and the leaves become yellow and start drying. Twigs start drying, and ultimately, the whole tree dies.

Control Measures

1. Plug the holes with cotton soaked in creosote.
2. Remove all dead twigs from the tree.
3. Spray 0.1 % BHC wettable.
4. Take all possible phytosanitary measures.
5. Inject carbaryl 1 % solution into the holes.

19.18.2 Leaf Miner (*Acrocercops syngramma*)

They mine through the tender leaves and severely damage them.

Control Measures

1. Spray new flush with 0.05 % fenitrothion.
2. Spray endosulfan at 0.05 %.

19.18.3 Leaf and Blossom Webber (*Lamida moncusalis* and *Orthaga exvinacea*)

The caterpillars web the shoots and inflorescences together, remain inside, and feed on them.

Control Measures

1. Spray endosulfan at 0.05 % at the time of emergence of new flushes.
2. Spray carbaryl or malathion at 0.05 % (Ayyanna et al. 1977).

19.18.4 Tea Mosquito (*Helopeltis antonii*)

They cause severe damage to the tender vigorously growing shoots and inflorescence. Both the adults and nymphs of the bug suck sap from the tender shoots, floral branches, tender apples, and immature nuts. The affected panicles and shoots dry up.

Control Measures

1. Spray endosulfan at 0.05 % or carbaryl at 0.15 % phosphamidon at 0.03 % or quinalphos at 0.05 % (Nair and Abraham 1982) in the months of October till January at 14–21 days intervals depending upon the severity of the infestation.

19.19 Diseases

19.19.1 Dieback or Pink Disease (*Corticium salmonicolor, Pellicularia salmonicolor*)

Affected shoots dry up from the tip downward. During monsoon season, a film of silky thread of mycelium develops on the branches, and later, the fungus develops a pink growth which represents the spore form (Nambiar and Pillai 1986).

Control Measures

1. Remove all affected parts and destroy them.
2. Paste the cuts with Bordeaux paste.
3. Spray with Bordeaux mixture 1 % in premonsoon, i.e., May–June, and post-monsoon, i.e., September–October.

19.19.2 Damping Off of Seedlings (*Fusarium* sp., *Pythium* sp., *Phytophthora palmivora*, and *Cylindrocladium scoparium*)

The fungi attack either the root or the collar or both of the seedlings. It commonly occurs under poor drainage in the nursery beds.

Control Measures

1. Have provision of draining excess water from the nursery beds.
2. Drench nursery beds with Ceresan 0.1 % or Bordeaux mixture 1 %.

19.19.3 Anthracnose (*Colletotrichum gloeosporioides*)

The causal pathogen (fungus) enters the fruit through the stigma in the flower stage. Rainy season is favorable for the spread of the disease.

Control Measures

1. Remove affected parts and destroy them.
2. Plant thick windbreak of tall growing trees like *Casuarina*, *Eucalyptus*, subabul, etc. The windbreak helps in preventing the spread of the disease by windblown spores.
3. Spray with Bordeaux mixture (3:3:50).

References

Ayyanna T, Subbaratnam GV, Rao BHK (1977) Cashew Bull 14:8–10

Azam-Ali and Judge (2004) Small-scale cashew nut processing. FAO, United Nations

Chattopadhyay RK (1985) Udyanika 5(1–2):45–47

Chattopadhyay PK, Pal B, Roy RN, Sadhu MK, Bose TK (1983) Indian Agric 27(2):149–154

Dahake AP, Joshi VD, Joshi AB (2009) Antimicrobial screening of different extract of *Anacardium occidentale* Linn. leaves. Int J Chemtech Res 1(4):856–858

Fao.org. (2011) Major food and agricultural commodities and producers – countries by commodity

Ghosh SN, Bose TK (1986) Indian Cashew J 18(1):11–16

Hanamashetti SI, Hedge M, Hiremath IG, Khan NM (1985) South Indian Hortic 33(3):190–192

Kumar B, Mistri NC, Singh B, Gandhi CP (2011) Indian horticulture data base, pp 14

Murthy KN, Kumaran PM, Nayar MM (1975) J Plant Crops 3:81–82

Nair GM, Abraham CC (1982) Agric Res J Kerala 20(2):41–48

Nambiar MC, Pillai PKT (1986) Cashew. In: Bose TK (ed) Fruits of India-tropical and subtropical. Naya Prakash, Calcutta, pp 409–438

Pappiah CM, Vijayakumar M, Hameed AS (1980) South Indian Hortic 28(1):1–4

Sawake DP, Gunjate RT, Limaye VP (1985) Acta Hortic 108:95–99

Singh S, Krishnamurthi S, Katyal SL (1967) Fruit culture in India. ICAR, New Delhi

Varghese T, Pundir Y (1964) Anatomy of the pseudocarp in *Anacardium occidentale* L. Proc Plant Sci 59(5):252–258

Non Traditional Crops: Manila Tamarind (*Tamarindus indica* L.) 20

Abstract

The manila tamarind is a fast-growing, long-lived tree that can grow to a height and spread of 20 m. It is known by the name "Madras thorn," but it is not native to Madras. The name "Manila tamarind" is misleading, since it is neither closely related to tamarind nor native to Manila. Genus *Pithecellobium* derives from the Greek words "pithekos" (ape) and "lobos" (pod) and the species name "*dulce*" from the Latin "dulcis" meaning sweet (Anonymous, Taxon: *Pithecellobium dulce* (Roxb.) Benth. Germplasm resources information network. United States Department of Agriculture, 23 Aug, 1994). Manila tamarind is a tree 5–18 m high, with pendulous branches, with short, sharp stipular spines. In India, the bark of the plant is used as astringent in dysentery, febrifuge, and also for dermatitis and eye inflammations. Leaves are used as abortifacient. The leaves are the source of fodder and the fruits are consumed fresh. It is a rich source of protein and fixes atmospheric nitrogen.

20.1 Introduction

Kingdom	Plantae – plants
Subkingdom	Tracheobionta – vascular plants
Super division	Spermatophyta – seed plants
Division	Magnoliophyta – flowering plants
Class	Magnoliopsida – dicotyledons
Subclass	Rosidae
Order	Fabales
Family	Fabaceae – pea family
Genus	*Pithecellobium* Mart. – blackbead
Species	*Pithecellobium dulce* (Roxb.) Benth. – monkeypod

S. Hiwale, *Sustainable Horticulture in Semiarid Dry Lands*,
DOI 10.1007/978-81-322-2244-6_20, © Springer India 2015

20.2 Other Names

It is known by the name "Madras thorn," but it is not native to Madras. The name "Manila tamarind" is misleading, since it is neither closely related to tamarind nor native to Manila. It is called "seema chintakaya" in Telugu. The name "monkey pod" is more commonly used for the rain tree (*Albizia saman*). Other names include black bead, sweet Inga, *cuauhmochitl* (Nahuatl), *guamúchil/cuamúchil/huamúchil* (Mexico, Spanish), *guamá americano* (Puerto Rico), *Makham thet* (Thai), *'opiuma* (Hawaiian), *kamachile* (Filipino), *kodukkappuli* (Tamil), *vilayati ambli* (Gujarati), *jungle jalebi* or *ganga imli* (Hindi), *tetul* (Bengali), *seeme hunase* (Kannada), *vilayati chinch* (Marathi), and *seema chinta* (Telugu) (Grandtner 2005).

20.3 Etymology

It is referred to as manila tamarind because of the sweet-sour tamarind-like taste. Genus *Pithecellobium* derives from the Greek words "pithekos" (ape) and "lobos" (pod) and the species name "*dulce*" from the Latin "dulcis" meaning sweet.

20.4 Botany

Manila tamarind is a tree 5–18 m high, with pendulous branches, with short, sharp stipular spines. The leaves are evenly 2-pinnate, 4–8 cm long. The flowers are white, in dense heads, 1 cm in diameter. Pods are turgid, twisted, and spiral, 10–18 cm long, 1 cm wide, and dehiscent along the lower suture. Seeds are 6–8, with an edible, whitish, pulpy aril. The arillus is sweet when the fruit is ripe.

Pithecellobium dulce is a species of flowering plant in the pea family, Fabaceae, that is native to Mexico, Central America, and northern South America. It is introduced and extensively naturalized in the Caribbean, Florida, Guam, and Southeast Asia like the Philippines. It is considered an invasive species in Hawaii.

20.5 Description

The manila tamarind is a fast-growing, long-lived tree that can grow to a height and spread of 20 m. The trunk grows with very sharp needles which help to protect itself from animals while it is a seedling. The needles fall of from the main trunk when the tree gets large but remain on the branches. The leaves are small for such a large tree, they grow in pairs, and each is oval shaped, 2–3 cm long and 1–2 cm wide. The flowers are small, about 2 cm diameters, and the fruit is a beanlike bulged pod curled up in a spiral that changes from green to red when ripe. Each pod contains from 6 to 10 black seeds each surrounded by white flesh. The white flesh tastes sweet and musky.

Manila tamarind plant

Fruit bearing branch

Seeds with edible cover

Harvested fruits

20.6 Edibility

Pulp around the seed is edible.

20.7 Uses

The manila tamarind is usually eaten out of hand. In some countries it is used to make a beverage (ILDIS 2005).

- Frequent bowel movements: decoction of bark taken as tea
- The leaves, when applied as plasters, used for pain and venereal sores
- Salted decoction of leaves, for indigestion and also used as abortifacient
- Bark used in dysentery, dermatitis, and eye inflammation
- In Brazil, *P. avaremotem*, used as a cancer elixir
- In Mexico, decoction of leaves for earaches, leprosy, toothaches, and larvicide
- In India, bark of the plant used as astringent in dysentery and febrifuge. Also used for dermatitis and eye inflammations. Leaves used as abortifacient
- In Guiana, root bark used for dysentery and febrifuge

20.8 Climate

Manila tamarind requires full sun. Trees can tolerate exceedingly hot conditions (above 40 °C) and also cold conditions (less than 5 °C) provided it is not prolonged.

20.9 Soil

Manila tamarinds, like the sweet-sour tamarind, will tolerate a great diversity of soil types but does best in deep, well-drained soils which are slightly acidic. Trees will not tolerate cold, wet soils.

20.10 Propagation

Manila tamarind seeds remain viable for months and will germinate in a week after planting.

20.11 Properties

- Considered abortifacient, anodyne, astringent, larvicidal, antibacterial, anti-inflammatory, febrifuge, and antidiabetic
- Bark considered astringent
- Leaves considered astringent, emollient, antidiabetic, and abortifacient
- Roots reported to be estrogenic

20.12 Medicinal Value

- *Anti-inflammatory/antibacterial:* study of the fresh flowers of *Pithecellobium dulce* yielded a glycoside quercetin. The activity of the flavonol glycoside confirmed its anti-inflammatory and antibacterial properties.
- *Phenolics/antioxidant:* free radical scavenging activity of folklore, *Pithecellobium dulce*

Benth (Sugumaran 2008). Leaves: study of the aqueous extract of *Pithecellobium dulce* revealed phenolics including flavonoids and showed potent free radical scavenging activity.

- *Anti-inflammatory triterpene:* anti-inflammatory triterpene saponins of *Pithecellobium dulce*, characterization of an echinocystic acid bisdesmoside. A new bisdesmodic triterpenoid saponin, dulcin, was isolated from the seeds of PD (Sahu and Mahato 1994).
- *Genotoxicity:* mutagenic and antimutagenic activities in the Philippine Medicinal and Food Plants; in a study of 138 medicinal plants for genotoxicity, *Pithecellobium dulce* was one of the 12 that exhibited detectable genotoxicity in any system (Lim-Sylianco and Shier 1985).
- *Anti-tuberculosis/antimicrobial:* hexane, chloroform, and alcoholic leaf extracts were studied for activity against *Mycobacterium tuberculosis* strains. The alcoholic extract showed good inhibitory activity and antimicrobial activity against secondary pathogens.
- *Antidiabetic:* the study of ethanolic and aqueous leaf extract of *P. dulce* in STZ-induced diabetic model in rats showed significant activity, aqueous more than the alcoholic extract, comparable to glibenclamide (Sugumaran 2009).
- *Antiulcer/free radical scavenging:* the study of the hydroalcoholic extract of PD was found to possess good antioxidant activity and suggests possible antiulcer activity with its free radical scavenging and inhibition of H, K-ATPase activities comparable to omeprazole (Megala and Geetha 2010). Phytochemical screening yielded flavonoids – quercetin, rutin, kaempferol, naringin, and daidzein.
- *Hepatoprotective:* the study of an aqueous extract of *P. dulce* in a murine model showed hepatoprotection against CCl4-induced oxidative impairments probably through its antioxidative property. Results were supported by histological findings (Manna et al. 2011).
- *CNS depressant:* the study of evaluating the locomotor activity of aqueous and alcoholic extracts of PD in albino mice showed significant CNS depression, the alcoholic extract exhibiting greater effect when compared to chlorpromazine. The activity was attributed to an increase in the concentration of GABA in the brain (Sugumaran et al. 2008).
- *Analgesic/anti-inflammatory:* the study of methanol extract showed significant anti-inflammatory and analgesic effects comparable to standard drugs (Arul and Muthukumaran 2011).

20.13 *Pithecellobium dulce* Variegated

(Variegated Madras thorn, Manila tamarind): family, Mimosaceae/Fabaceae; species, *Pithecellobium dulce*; flower color, cream; life form, tree.

20.13.1 Distinguishing Features

This hardy ornamental tree, a native of Central America, should not be grown near bush land, and any seedlings should be regularly removed before they have a chance to form a thicket. The leaves of the variegated variety have attractive green and white leaflets with a mix of pink. Flowers are in heads cream in color. Pods are curved, green to red on the outside with a white edible pulp on the inside surrounding the shiny black seeds. These seeds germinate readily.

References

Arul SS, Muthukumaran P (2011) Analgesic and anti-inflammatory activities of leaf extract of Pithecellobium dulce Benth. Int J PharmTech Res 3(1):337–341

Grandtner MM (2005) Elsevier's dictionary of trees: with names in Latin, English, French, Spanish and other languages, vol 1. Elsevier, Amsterdam, pp 670–671. ISBN 978-0-444-51784-5

International Legume Database & Information Service (ILDIS) (2005) *Pithecellobium dulce*. Version 10.01, November 2005

Lim-Sylianco CY, Shier WT (1985) Mutagenic and antimutagenic activities in Philippine medicinal and food plants. Toxin Rev 4(1):71–105
Manna P, Bahttacharyya S, Das J (2011) Phytomedicinal role of Pithecellobium dulce against CCl4-mediated hepatic oxidative impairments and necrotic cell death. Evidence-based complementary and alternative medicine
Megala J, Geetha A (2010) Free radical-scavenging and H+, K+-ATPase inhibition activities of Pithecellobium dulce. Food Chem 121(4), 15:1120–1128
Sahu NP, Mahato SB (1994) Anti-inflammatory triterpene saponins of Pithecellobium dulce: characterization of an echinocystic acid bisdesmoside. Phytochemistry 37(5):1425–1427
Sugumaran M (2008) Free radical scavenging activity of folklore: Pithecellobium dulce benth. Leaves. Ethnobot Leafl 12:446–451
Sugumaran M (2009) Antidiabetic potential of aqueous and alcoholic leaf extracts of Pithecellobium dulce. Asian J Res Chem 2(1), Jan–Mar 2009
Sugumaran M, Vetrichelvan T, Darline Quine S (2008) Locomotor activity of leaf extracts of Pithecellobium dulce Benth. Ethnobot Leafl 12:490–493

Part III

Agro Forestry Species

21 Neem (*Azadirachta indica*)

Abstract

Neem is a versatile tree of Indian origin. Since ancient times, its various plant parts have been used to control domestic insects, stored grain pests, and crop pests and in human and livestock medicine. Chemicals extracted have shown potential as pesticides and as nitrification inhibitors and do not pose environmental pollution and related health hazards in human and livestock. It is a favorite garden tree with delicate foliage and lilac-colored flowers in profuse axillary panicles. The Malabar Neem is also found in India. Neem is cultivated throughout India. Nineteen meliaceous genera are represented in Indian subcontinent by one or more species. Neem is the most important of the fourteen Indian timber trees of this family confined to meliaceous genera. Its wood is a good source of fuel wood and used in preparation of agricultural implements. Neem kernels are used in manufacturing cake which is rich in NPK content; apart from this, it acts as insecticide when applied to soil. Seeds are also used in preparation of insecticidal formulations. The leaves are a source of fodder for camels.

21.1 Introduction

Neem is a versatile tree of Indian origin. Since ancient times, its various plant parts have been used to control domestic insects, stored grain pests, and crop pests and in human and livestock medicine. Recently, these properties have been attributed to hundreds of chemicals present in this golden tree. Neem chemicals have shown potential as pesticides and as nitrification inhibitors and do not pose environmental pollution and related health hazards in human and livestock. The Neem tree/margos/Indian lilac is botanically known as *Azadirachta indica*. It belongs to the family Meliaceae. It is a large family consisting 50 genera and 1,400 species. It is a favorite garden tree with delicate foliage and lilac-colored flowers in profusion axillary panicles. The Malabar Neem is also found in India. Chinaberry, *Melia toosendan* Sieb. etZucc., is found in China. The Thai Neem tree is *Azadirachta indica* var. siamensis Valeton.

S. Hiwale, *Sustainable Horticulture in Semiarid Dry Lands*,
DOI 10.1007/978-81-322-2244-6_21, © Springer India 2015

21.2 Habitat

Neem is a native of Burma and is habituated in Southeast Asia, India, Andaman, Pakistan, Bangladesh, Sri Lanka, Thailand, Tarai regions of Nepal, Malaysia, Indonesia, Japan, tropical USA, South America, and Africa.

Neem is cultivated throughout India. Neem was introduced in Africa in the last century as an ornamental avenue tree. It was first introduced into the northern Bornu Province, Nigeria, in 1928 from Ghana. Neem has also been introduced in the Philippines.

21.3 Botany

Neem is moderate to large in size and usually an evergreen tree with a fairly dense rounded crown. Leaves are glabrous, imparipinnate, alternate, exstipulate, and 22.5–37.5 cm long.

Leaflets are 7–17, alternate or opposite, very shortly stalked, 6.25–7.5 cm long, ovate-lanceolate, attenuate at the apex, and unequal at the base. The bark is moderately thick and reddish brown inside. The heartwood is red, hard, and durable. The white flowers are small and numerous.

The calyx is 3, 4, or 5, widespreading, rounded, blunt, and ciliate, with smooth thin sepals. There are 10 stamens situated at the base of the hypogynous disk. Gynaeceum tri-penta-carpellary, superior, as many locular with 2 ovules in each locule.

Style is about the length of the stamina tube and stigma 5-lobed. The gametophytes develop in the usual way, and the embryo sac is formed after polygonum type. The seed is exalbuminous and germination is epigeal (Watt 1891).

21.4 Floral Anatomy and Morphology

Nair (1959) studied the floral anatomy of *Melia azedarach* and described its placenta as parietal. Nair and Kanta (1961) observed that in the floral morphology, there are –2–3 rows of sporogenous cells in a longitudinal section of the anther. Microsporogenesis proceeds in the usual way. The cells of glandular tapetum are binucleate. Pollen grains are shed at the three-celled stage. The number of germinal furrows varies from 3 to 9, the frequent condition being tetraporate. A few cases of polyspory have been observed. A many-celled hypodermal archesporium is distinguished in bitegmic, crassinucellate, and anatropous ovule. Generally one cell develops further. The development of the embryo sac conforms to the polygonum type. The antipodals degenerate before fertilization. Twin embryo sacs and triplets have been observed. Fertilization is porogamous. Syngamy and triple fusion have been observed in several cases. The division of primary endosperm nucleus is earlier than the zygote. The endosperm is free nuclear and later becomes cellular. In the mature seed, three to five layers of the endosperm cells are present. Several cases of endosperm nodules have been observed. The zygote is divided by a transverse wall. Further divisions are very irregular. The mature embryo is dicotyledonous. The additional embryo is developed from the synergid. A seed develops in each ovary. The outer seed coat consists of a thick-walled epidermis and three layers of loosely arranged cells. The inner seed coat is two to three layered. Nair and Yunus (1980) studied the petal surface of *Melia azedarach* and *Azadirachta indica*. In *Melia azedarach*, there are clear differences between the two species with regard to micromorphology of the petal surface, particularly of the tubercles, which are cone capped and restricted to wall ridges in *Melia azedarach*, while in *Azadirachta indica*, the tubercles are flat topped and are restricted to the cell floor.

Studies on its seed biology have revealed the existence of tendency of polyembryony, embryo abortion, and premature fruit a fall. The samples used for the present studies were screened cytologically ($N=14$) and biochemically for protein type and oil contents. Floral initiation which occurs during leaf shedding takes about 2 weeks to reach the blooming stage (end of April). Blooming lasts for 5 weeks during which petals open in succession. Flowers are typically

Table 21.1 Flowering time of Neem in India

Time of flowering	States
First week of January	Kerala
First week of February	Karnataka
First week of March	Andhra Pradesh, Madhya Pradesh, Gujarat
First week of April	West Bengal, Orissa, Bihar, Uttar Pradesh, Delhi, Haryana Punjab
First week of May	Sub-Himalayan States

Randhawa (1975)

pentamerous, but variation in the number of petals, i.e., 4, 6, and 8 occurs in some cases affecting the number of 5 sepals and 10 stamens. Anthesis is mostly nocturnal, i.e., from 17.00 to 4.00 h but anthers' dehiscence is precocious in bud. There is distinct protandry and stigma becomes receptive during evening from 16.00 to 18.00 h (Table 21.1).

In general, microsporogenesis is normal but some polyed formation seems to be the consequences of cytomixis. Even then, the pollen fertility is from 85 to 96 %. Pollen grains fail to germinate in the medium even when supplemented with sucrose. Pollen grains are 39.0–41.8 μm long which are smooth, thin walled, mostly 4-zonocoporate, and subporate, are indicative of anemophily. Whereas creamish color and scented nature of flowers which are visited by moths during night and honey bees and butterflies during day shows entomophily, there is no self-incompatibility. The species is predisposed toward cross-pollination but geitonogamy occurs as usual. There is considerable amount of infructescence which is evident from the limited fruit setting, i.e., 4.2 %.

The chromosome number in Neem is 14 (Deshmukh 1957). In India, it sheds its old leaves in February–March and produces glossy young leaves and fragrant white flowers in the first week of April. The flowers and fruits have foul odor after a shower.

On account of its large number of leaves, the leaf area is considerably big and the rate of photosynthesis is high. It gives more oxygen during the day as compared with many other trees.

21.5 Uses

The Neem oil is consumed by women during pregnancy (MacMillan 1935), and the gum is medicinally esteemed as a stimulant. The twigs are largely used as toothbrush and for oral hygiene.

Kautilya and Charakya recorded the extraction of oil from seeds of nimbi (Neem) in Arthashastra. Nimbi is synonymous with arishta meaning relieving sickness and pichumarda, another name of the plant being equivalent to destroyer of leprosy (Watt 1891).

Neem bark is regarded as a bitter tonic, astringent, and useful in fevers, thirst, nausea, vomiting, as well as skin diseases. The active part is the inner layer of the bark.

The tree is held sacred by the Hindus. Its various parts are used in many of their rituals and ceremonies. It is believed that a few drops of heavenly nectar fell on the Neem. So on New Year's Day (Ghudi Padwa), Hindus eat its leaves and take a bath in water with boiled Neem leaves to be free from diseases.

In the Indian festival Ghatasthapana, villagers gather together at a public place and keep a pot filled with water, on which they put five branches of Neem and coconut. This pot is covered with flowers and worshiped. Sacrifices are also made before the Neem tree to avert ill-luck and diseases.

Neem oil is used for burning the lamps and as a lubricant in machinery. The leaves and twigs are used as cattle feed and for manure. Silk dyers use the Neem gum for preparing their colors. The bark fiber is used to manufacture the ropes.

Neem fruit pulp provides a useful substrate for methane gas production. Neem timber is excellent for making furniture, boards, panels, toys, and plows. Neem tree leaves are used to reclaim wasteland oil cake as manure.

Neem is successfully used as windbreak and shade tree for roadside plantations. Neem is considered a sacred and health-giving tree chiefly due to its insecticidal and medicinal properties.

21.6 Wood and Timber

Nineteen meliaceous genera are represented in the Indian subcontinent by one or more species. Neem is the most important of the 14 Indian timber trees of this family confined to meliaceous genera.

21.7 General Characteristics of Neem Wood

Sapwood is grayish white and heartwood is red when first exposed, fading to reddish brown and resembling mahogany dull to somewhat lustrous, especially on the radial surface.

It is aromatic with characteristic taste, moderately heavy (specific gravity, approximately 0.68), rather uneven, and narrowly interlocked, even grained and medium to somewhat coarse textured (2) (Table 21.2).

The timber is durable even in open situations. It is not usually attacked by insects, due to the oil content in the wood. It is easy to work by hand or machines almost like teak. It lends itself to broad carvings (Mitra 1963) (Table 21.3).

Table 21.2 Mechanical properties of Neem timber

Character	Value
Compression (parallel to grain)	6,680 lbs/sq.inch
Shear (parallel to grain)	1,326 lbs/sq.inch
Transverse strain	11,480 lbs/sq.inch
Young's modulus	10,08,800

Table 21.3 Fuel value of Neem tree

Character	Value
Calorific value (Cal/g)	4,322.82
Ash content (5)	0.97
Moisture content (5)	11.20
Ignition time (second)	20
Prurniry rate (g/min.)	12.50

Shahen and Harode (1987)

21.8 Neem-Based Agri-silvi Production System on Marginal Lands

Five- to six-year-old Neem (10×5 m) was intercropped with various intercrops to maximize production potential of marginal lands. The effect of different intercrops on the growth parameter of the main crop showed that it varied from 5.66 m in the case of *Stylosanthes* to 6.31 m in the case of fodder jowar. However, differences were nonsignificant. The stem diameter increased from 17.29 cm in the fourth year to 20.98 cm in the fifth year indicating an increase of 21.34 %. The effect of different intercrops on the stem diameter of the main crop revealed that it varied from 17.92 cm (*Stylosanthes*) to 20.71 cm (fodder jowar). The N-S plant spread ranged from 3.80 m in the fourth year to 4.53 m in the fifth year showing an increase of 21.34 %, whereas it was maximum (4.55 m) in moth bean and minimum (3.93 m) in *Stylosanthes*. E-W plant spread increased from 3.67 m (5th year) to 4.67 m (6th year), indicating an increase of 26.43 %. The effect of different intercrops on the effect of plant spread (E-W) of the main crop revealed that it varied from 4.03 m in the case of anjan to 4.52 m in the case of fodder jowar (Tables 21.4 and 21.5).

The mean Neem dry biomass production is found to increase from 54.48 q/ha in fifth year to 56.23 q/ha indicating the biomass addition of 1.75 q/ha. Biomass production in different

Table 21.4 Morphological parameters of Neem in agri-silvi production system

Name of Neem intercrops	Plant height (m)	Stem diameter (cm)	Plant spread (m) (N-S)	Plant spread (m) (E-W)
Sunhemp	6.25	19.32	4.21	4.20
Fodder jowar	6.31	20.71	4.14	4.52
Cenchrus	6.09	19.12	4.15	4.03
Stylosanthes	5.66	17.92	3.93	4.27
Moth bean	6.15	19.12	4.55	4.21
Soya bean	5.88	19.50	4.49	4.38

Hiwale (2004)

Table 21.5 Biomass production of Neem in agri-silvi production system

Neem intercrops	Dry biomass produced by main crops (q/ha)	Fresh biomass produced by intercrops (q/ha)	Dry biomass produced by intercrops (q/ha)
Neem + sunhemp	54.32	79.77	24.92
Neem + fod. jowar	57.47	98.63	29.93
Neem + *Cenchrus*	55.65	89.75	27.52
Neem + *Stylosanthes*	56.52	56.96	13.46
Neem + moth bean	53.10	29.55	16.31
Neem + soya bean	55.05	63.45	14.86

intercrops was maximum 57.47 q/ha in Neem+fodder jowar. Raising of various intercrops in the interspaces revealed that substantial quantity of carbon in the form of biomass could be harvested making maximum utilization of precipitation and sunlight. The mean maximum biomass production was in the Neem+fod. jowar (98.63 q/ha).

Intercropping in Neem

Run off and soil erosion in Neem

Neem root system under semi arid rainfed conditions

Table 21.6 Net return and B:C ratio of Neem-based intercropping system

Intercrops	Net return (Rs.)	B:C ratio
Neem + sunhemp	11,812.00	2.720
Neem + fod. jowar	14,543.00	3.315
Neem + *Cenchrus*	13,415.50	3.010
Neem + *Stylosanthes*	4,436.50	0.990
Neem + moth bean	8,642.00	1.935
Neem + soya bean	9,421.50	2.155

Hiwale (2004)

Economics of Neem-based cropping system indicated that maximum mean net return of Rs. 14,543/- can be obtained in Neem + fodder jowar-based cropping system with a benefit-cost ratio of 1:3.32 averaged over the past 2 years followed by Neem + *Cenchrus* (1:3.01) and sunhemp (1:2.72) (Table 21.6).

21.9 Effect of Leaf Litter Decomposition in Respect to Tree Species Involved in the System

Leaf litter fall and its decomposition were studied in fruit tree species, viz., aonla (*Emblica officinalis*) and custard apple (*Annona squamosa*), and silvicultural tree species, viz., Neem (Azadirachta indica) and subabul (*Leucaena leucocephala*). Organic recycling of nutrient through leaf litter fall and its decomposition plays an important role in the sustainability of agroforestry system in cultivated lands. Litter fall is the source of nutrient recycled to the soil from the plants. The decomposition of litter adds substantial quantity of nutrient to the soil through nutrient recycling. There is variation in the rate of decomposition and nutrient release pattern, which is governed by chemical composition of leaf and prevailing weather condition. Addition of leaf litter to the soil over the years helps in the improvement of soil fertility and aggregate stability.

All the four tree species were found to be deciduous in nature, and large amount of leaf litter is added by all the tree species to the soil. The studies on leaf litter fall and nutrient recycled carried out indicated that in all the tree species, leaf litter production increased over the years (on dry wt. basis). However, in case of Neem though mean leaf litter fall/tree was just 4.809 kg/tree and on per hectare basis, it was highest 961.867 kg/ha, which was due to closer spacing adopted in the tree species (Table 21.7).

Recycling of N, P, and K varied with different tree species. However, on hectare basis, it was in Neem leaf litter (21.495 kg N/ha, 1.154 kg P/ha, and 3.087 kg K/ha, respectively) due to less number of plants/ha.

21.10 Leaf Litter Decomposition and Its Nutrient Composition

Maximum leaf litter decomposition was noted in subabul (44.15 %) and minimum in Neem (25.27 %). The decomposition rates vary with substrate quality, climate, and quality and quantity of decomposer organism. Neem leaf litter (decomposed) had N content (1.802 %). Higher N content in subabul can be attributed to atmospheric N fixation capacity of tree species (Table 21.8).

Phosphorus content of decomposed leaf litter was minimum in Neem leaf litter (0.290 %) (Table 21.8). The phosphorus content of fresh leaf litter collected was low. The absolute amount of P in the residue showed biphasic pattern with initial low content followed by slow buildup phase.

Table 21.7 Leaf litter fall (kg/tr. and kg/ha) and nutrient recycled (estimated g/tr. and kg/ha)

Name of tree species	Year	Dry leaf litter fall		Nitrogen		Phosphorus		Potassium	
		kg/tree	kg/ha	g/tree	kg/ha	g/tree	kg/ha	g/tree	kg/ha
Neem	2001	3.700	740.000	52.68	10.536	4.440	0.880	11.914	2.383
	2002	4.888	977.600	69.716	37.318	5.900	1.18	15.695	3.139
	2003	5.840	1,168.000	83.162	16.632	5.900	1.402	18.688	3.738
	Mean	4.809	961.867	68.519	21.495	5.783	1.154	15.432	3.087

Hiwale (2004)

Table 21.8 Leaf litter decomposition rate and its nutrient content

	Year	% leaf litter decomposed	Nitrogen (%)		Phosphorus (%)		Potassium (%)	
			Undecomposed	Decomposed	Undecomposed	Decomposed	Undecomposed	Decomposed
Neem	2001	20.60	1.424	1.610	0.120	0.312	0.32	0.131
	2002	16.20		2.017		0.257		0.210
	2003	39.00		1.780		0.300		0.240
	Mean	25.27		1.802		0.290		0.194

Hiwale (2004)

Potassium content of decomposed leaf litter was lowest in Neem (0.194 %) (Table 21.8). It is concluded that due to addition of litter by tree species, there was improvement in soil health, which is due to enrichment of soil through the release of entrapped nutrients and formation of organic matter.

References

Deshmukh NY (1957) Chromosome number in Neem. Indian Oilseeds J 3(1):45

Hiwale SS (2004) Technical bulletin on "Develop sustainable agri-horti production systems under rainfed conditions on marginal lands", CHES, Vejalpur, India pp 18–20

MacMillan HF (1935) Tropical planting and gardening, 4th edn. MacMillan & Co, London, 234 pp

Mitra CR (1963) Neem. Indian Central Oilseed Committee, Hyderabad, 190 pp

Nair NC (1959) Studies on Meliaceae. J Indian Bot Soc 38:367–378

Nair NC, Kanta K (1961) Studies on Meliaceae IV. Floral morphology and embryology of A. indica A. Juss., A reinvestigation. J Indian Bot Soc 40:382–396

Nair PKK, Yunus D (1980) Study of petal surface in the two taxa of Meliaceae. Curr Sci 49:822–824

Randhawa MS (1975) Flowering trees of India. Indian Council of Agricultural Research, New Delhi

Shahen S, Harode SM (1987) Assessment of certain heat values and their relationship in selected firewood and agricultural wastes. J Maharashtra Agric Univ 12:237–239

Watt G (1891) Dictionary of economic products of India, vol 5. W.H.Allen & Co., London, pp 211–221

Leucaena (*Leucaena leucocephala*) 22

Abstract

Subabool (*Leucaena leucocephala*) is a leguminous fodder-fuel-fertilizer adding tree/shrub which can grow even in poor soils under low and medium rainfall. It can be grown alone or intercropped with grass and other field crops to some extent. In semiarid climate, it is not possible to grow green fodder throughout the year. Subabool (*Leucaena leucocephala*) is one such tree which can meet the requirement of fodder along with fuel and in this area. It can be grown on field boundaries, roadsides, and boundaries of canals. It is fast growing, coppicing, and amenable to pruning. It is the most suitable species for agroforestry as an alley crop. The subabool was intercropped with fruit trees, viz., guava, ber, pomegranate, and kinnow.

22.1 Introduction

Subabool (*Leucaena leucocephala*) is a leguminous fodder-fuel-fertilizer adding tree/shrub which can grow even in poor soils under low and medium rainfall. It can be grown alone or intercropped with grass and other field crops to some extent. The majority of areas in India is semiarid or arid. Because of the climate, it is not possible to grow green fodder throughout the year. Subabool (*Leucaena leucocephala*) is one such tree which can meet the requirement of fodder along with fuel and in this area. *Leucaena leucocephala* (Lam.) De Wit. (Mimosoideae) has been introduced in many countries for its potentiality to produce firewood and nutritious green forage at a very short duration and rotation. The adaptability of giant varieties to diverse habitats and uses makes it further most important in the rural setup for augmenting the rural energy resources which are getting progressively scarce. It was introduced into Australia in the late nineteenth century, and it was naturalized in parts of northern Australia by 1920. At the Indian Grassland and Fodder Research Institute, Jhansi, the studies were initiated in early 1970, and by now various production systems have been evaluated (Pathak et al. 1981). Pathak et al. (1981) studied its biomass production under farm forestry at 3 and 4 years of age and found it to yield up to 11 t/row km of wood and 1 t/row km of dry forage. It is estimated to cover 2–5 million ha worldwide (Brewbaker and Sorensson 1990).

S. Hiwale, *Sustainable Horticulture in Semiarid Dry Lands*,
DOI 10.1007/978-81-322-2244-6_22, © Springer India 2015

22.2 Composition of Subabool Leaves

Nitrogen: 171.482 Kg/ha. Phosphorus: 10.472 Kg/ha. Potassium: 31.437 Kg/ha.

Subabool is dicotyledonous in nature and hence contain large amounts of nitrogen and protein. It also helps in soil and water conservation. It can be grown on marginal land, wasteland, and cattle grazing area without much care. If grown in alleys, the interspaces can be used for growing grasses like *Cenchrus* and leguminous species of *Stylosanthes*.

It can be grown on field boundaries, roadsides, and boundaries of canals. It is fast growing, coppicing, and amenable to pruning. The only problem is its seeds are dispersed all over and germinated; therefore, the same trees become nascent. It is mostly raised by seed.

22.3 Botanical Description and Genetic Variation

Leucaena leucocephala, formerly known as *L. glauca*, is a thornless long-lived shrub or tree which may grow to heights of 7–18 m. Leaves are bipinnate with 6–8 pairs of pinnae bearing 11–23 pairs of leaflets 8–16 mm long. The inflorescence is cream colored and globular in shape which produces a cluster of flat brown pods 13–18 mm long containing 15–30 seeds. Botanically, leucaena belongs to the family Mimosaceae; it is the best-known species of the *Leucaena* genus and has a variety of common names (Table 22.1). There are, however, at least 14 other species recognized in the genus.

The species may be distinguished on the basis of their tree size, flower color, leaflet size, and pod size (Anon 1990). These same authors consider the genus *Leucaena* to be an interbreeding complex capable of producing many interspecific hybrids. For instance, *L. leucocephala* crosses readily with *L. diversifolia* and *L. pallida* producing hybrids from which selection for improved growth form, psyllid resistance, and cold tolerance is possible. *Leucaena pallida*, in particular, has excellent seedling vigor, and hybridization of this species with *L. leucocephala* has the potential to produce a new highly productive and psyllid-resistant leucaena (Sorensson et al. 1994).

Table 22.1 Some common names of (*Leucaena leucocephala*.) de Wit (Brewbaker et al. 1985)

Common name	Countries
Leucaena	Australia, USA
Ipil ipil	Philippines
Lamtoro	Indonesia
Katin	Thailand
Yin ho huan	China
Kubabul or subabool	India
Koa haole	Hawaii
Tangantangan	Some Pacific islands
Cassis	Vanuatu
Guaje	Mexico
Huaxin	Central America (Maya)

22.4 Highlights

1. It contains about 24–33 % crude protein in its foliage.
2. It fixes air nitrogen in large quantity in soils.
3. The plant gives about 1 kg gum/plant/year during the first year itself.
4. It can attain a height of 5–6 m and basal diameter of 6–8 cm in a year if it is allowed to grow for fuel purposes; its top leaves are periodically lopped for animal feed.
5. It provides highly nutritive and palatable fodder during lean periods.

22.5 Uses

Leucaena leucocephala has a wide variety of uses, and it was this multiplicity of roles that led to the worldwide reputation of the species as a "miracle tree."

First and foremost, the leaves of leucaena are highly nutritious for ruminants, and many excellent animal production data have been published confirming the fodder value of leucaena. Leucaena poles are useful for posts, props, and frames for various climbing crops (Brewbaker et al. 1985). There is an opportunity to produce seedless triploid hybrids by crossing self-incompatible diploid species such as *L. diversifolia* (2x) with tetraploid species such as *L. leucocephala* (Brewbaker and Sorensson 1990).

22.6 Climate and Soil Adaptation

22.6.1 Temperature

Leucaena is a tropical species requiring warm temperatures (25–30 °C day temperatures) for optimum growth (Brewbaker et al. 1985). At higher latitudes and at elevated tropical latitudes, growth is reduced. Brewbaker et al. (1985) suggest that temperature limitations occur: above 1,000 m elevation within 10 °C latitude of the equator and above 500 m elevation within the 10–25 °C latitude zone. Leucaena is not tolerant of even light frosts which cause the leaf to be shed (Isarasenee et al. 1984).

22.6.2 Light

Shading reduces the growth of leucaena although this plant has a moderate tolerance of reduced light when compared with other tree legumes (Benjamin et al. 1991).

22.6.3 Rainfall Requirements and Drought Tolerance

Leucaena can be found performing well in a wide range of rainfall environments from 650 to 3,000 mm. However, yields are low in dry environments and are believed to increase linearly from 800 to 1,500 mm, other factors being equal (Brewbaker et al. 1985).

Leucaena is not tolerant of poorly drained soils, especially during seedling growth, and production can be substantially reduced during periods of water logging.

The work carried out at IGFRI, Jhansi, UP, indicated that growing plants at a closer spacing of 50 cm yield green fodder 118 t/ha/year. It is the most suitable species for agroforestry as an allay crop. Work was done at IGFRI with 12 different multipurpose tree species planted at three different spacing. After 1 year, the first cutting at 50 % height was done. Where maximum fodder production was recorded in subabool (8.73 q/ha), an average production of 4 years was recorded, which was highest among the 12 species.

22.7 Agroforestry Production System

The subabool was intercropped with fruit trees, viz., guava, ber, pomegranate, and kinnow. In the initial years, there was no fruit produced. However, in the initial year, fodder and fuel were obtained from subabool. 97.3 q/ha green fodder and 32.2 q/ha dry fodder were produced. In the sixth year, subabool was completely removed. Intercropping in mango yielded 32.3 q/ha green fodder and 11.0 q/ha dry fodder. In aonla-based cropping system also (250 plants), subabool yielded 15.4 qa/ha dry fodder.

22.8 Package of Practices

22.8.1 Seed Treatment

Because of the hard seed coat, the seeds should be treated either with hot water at 80 °C for a period of 2–3 min or with concentrated sulfuric acid for a period of about 5 min.

22.8.2 Nursery

Raise the seedlings by sowing treated seeds in polythene bags filled with 50 % soil and 50 % FYM during March. For better germination, either a bacterial culture or a pinch of soil from the kubabul growing plot should be added. In the absence of this, the seed could also be put directly on the site into the soil.

After clearing the area of bushes and rank growth vegetation, the land should be prepared thoroughly. After land preparation, pits of

$45 \times 45 \times 45$ cm^3 are dug at a distance of 4 m between the rows and 2 m between the plants.

22.8.3 Time of Planting/Direct Sowing

About two three-month-old seedlings grown in polythene bags should be transplanted in the pits after a good shower during July. In the available space in between the two rows of the trees, grass seeds may be sown in lines of 45 cm apart.

The treated seeds can also be sown directly after the onset of monsoon either in the fields, bunds, marginal lands, channels, tanks/river-banks, roadsides, boundary poles around wells, etc.

22.8.4 Fertilizer

Since the grass can also be grown in between the lines of subabool, apply a basal dose of 40 kg phosphorus and 20 kg nitrogen per hectare followed by 20 kg nitrogen when the grasses are about 1 month old. For intensive cultivation of kubabul, sow or plant at a spacing of 100 × 75 cm.

22.8.5 Weeding

Two weedlings, one in August and another in October, are recommended for better plant growth around the tree seedlings.

22.8.6 Green Fodder Yield

After 4 months of its growth, the plant can be harvested at a height of 1 m and successive cuts at an interval of 60–80 days. The total green fodder yield under an intensive cultivation of kubabul may be about 350 q/ha under rainfed conditions.

22.9 Subabool-Based Agri-Silvi Production System on Marginal Lands

Studies on 12–13-year-old subabool (*Leucaena leucocephala*) plants spaced 2 × 2 m intercropped with various arable crops indicated that it is not possible to grow intercrops in the subabool owing to close spacing and tall tree growth not allowing to pass sunlight and the dense root system of the tree crop. Growth parameters of the subabool increased at a slower space as plants have reached their last stage of growth. Mean plant height increased from 12.70 to 12.87 m. Mean DBH over the years increased from 114.63to 117.67 mm indicating an increase of 2.65 % (Hiwale 2004) (Table 22.2).

At the age of 12 years, mean dry biomass production of subabool was found to be 614.33 q/ha, which increased to 750.62 q/ha in the 13 years of age. The production from pasture component gets a boost by 5–7 times compared to the natural grazing lands. The pasture growth and production do not affect the tree growth. It can be concluded that at the spacing of 2 × 2 m *Leucaena* can be intercropped up to the age of 4–5 years

Table 22.2 Morphological parameters of subabool

Intercrops	Plant height (m)			DBH (mm)			Dry biomass produced by the main crop (q/ha)		
Year	2002	2003	Mean	2002	2003	Mean	2002	2003	Mean
Sunhemp	12.71	12.90	12.80	115.87	117.70	116.78	624.00	750.00	687.00
Fod. Jowar	12.70	12.92	12.81	114.60	116.40	115.50	580.00	800.00	690.00
Cenchrus	12.68	12.84	12.76	113.82	117.40	115.61	654.00	718.75	686.37
Stylosanthes	12.68	12.91	12.79	115.32	118.60	116.96	617.00	745.00	681.00
Green gram	12.73	12.87	12.80	113.27	117.30	115.28	598.00	727.50	662.75
Soya bean	12.72	12.80	12.76	114.90	118.20	116.55	613.00	762.50	687.75
Mean	12.70	12.87		114.63	117.67		614.33	750.62	
CD at 5 %	–	–	NS	–	–	NS	–	–	–

only; thereafter, there is reduction in yield of intercrops, and after 10 years there is complete failure of the crops. There is therefore a need to increase the spacing for raising intercrops or lopping should be practiced every year at least once (Hiwale 2004).

The economics of subabool-based cropping system indicated that maximum net return of Rs. 47,500/- can be obtained in subabool + maize and minimum in subabool + *Cenchrus* (Rs. 41,000/-). B:C ratio was maximum in subabool + maize (2.88) and minimum in subabool + *Cenchrus* (2.48) (Table 22.3). Economic returns from the silvipastoral system are less as compared to fruit tree-based cropping system. The economic returns are less because the maintenance costs over the years are higher compared to fruit-based cropping system where income from the system starts in the third year.

22.10 Leaf Litter Decomposition In Respect To Tree Species Involved in the System

Leaf litter fall and decomposition were studied in fruit tree species, viz., aonla (*Emblica officinalis*) and custard apple (*Annona squamosa*), and silvicultural tree species, viz., neem (*Azadirachta indica*) and subabool (*Leucaena leucocephala*). Organic recycling of nutrient through leaf litter fall and its decomposition plays an important role in sustainability of the agroforestry system in cultivated lands. Litter fall is the source of nutrient recycled to the soil from the plants. The decomposition of litter adds substantial quantity of nutrient to the soil through nutrient recycling. There is variation in the rate of decomposition and nutrient release pattern, which is governed by chemical composition of leaf and prevailing weather condition. The addition of leaf litter to the soil over the years helps in the improvement in soil fertility and aggregate stability (Hiwale 2004).

Table 22.3 Economics of subabool-based cropping system

Name of crop combination	Gross return of main crop (Rs.)	Net return (Rs.)	B:C ratio
Subabool + dolichos	60,000	43,500	2.64
Subabool + maize	64,000	47,500	2.88
Subabool + *Cenchrus*	57,500	41,000	2.48
Subabool + *Stylosanthes*	59,600	43,100	2.61
Subabool + moth bean	58,200	41,700	2.53
Subabool + soya bean	61,000	44,500	2.70
Mean	60,050	43,550	2.64

All the four tree species were found to be deciduous in nature, and a large amount of leaf litter is added by all the tree species to the soil. The studies on leaf litter fall and nutrient recycled carried out indicated that in all the tree species, leaf litter production increased over the years (on dry wt. basis). However, in the case of subabool, mean leaf litter fall/tree was just 3.805 kg/tree, and on per hectare basis it was highest, 9,513.33 kg/ha, which was due to the closer spacing adopted in the tree species (Table 22.4).

Recycling of N, P, and K varied with different tree species. However, on hectare basis, it was maximum in subabool leaf litter (171.482 kg N/ha, 10.472 kg P/ha, and 31.437 kg K/ha, respectively) due to more number of plants/ha.

22.11 Leaf Litter Decomposition and Its Nutrient Composition

Mean leaf litter decomposition was noted in subabool (44.15 %). The decomposition rates vary with substrate quality, climate, quality, and quantity of decomposer organism. Environmental

Table 22.4 Leaf litter fall (kg/tr. and kg/ha) and nutrient recycled (estimated g/tr. and kg/ha)

Name of tree species	Year	Dry leaf litter fall		Nitrogen		Phosphorus		Potassium	
		kg/tree	kg/ha	g/tree	kg/ha	g/tree	kg/ha	g/tree	kg/ha
Subabool	2001	3.46	8,650.00	62.21	155.52	3.80	9.50	11.42	28.54
	2002	3.44	8,615.00	62.48	156.19	3.80	9.51	11.42	28.56
	2003	4.51	11,275.00	81.09	202.72	3.80	12.40	14.88	37.20
	Mean	3.80	9,513.33	68.59	171.48	4.19	10.47	12.57	31.43

Table 22.5 Leaf litter decomposition rate and its nutrient content

Name of tree species	Year	% actual leaf litter decomposed after 2 months	Nitrogen (%)		Phosphorus (%)		Potassium (%)	
			Undecomposed	Decomposed	Undecomposed	Decomposed	Undecomposed	Decomposed
Subabool	2001	38.85	1.798	1.905	0.110	0.316	0.33	0.149
	2002	44.60		2.125		0.277		0.225
	2003	49.00		2.492		0.290		0.210
	Mean	44.15		2.174		0.294		0.195

factors which are most important in regulating the turnover rate of litter tend to be those that regulate the activity of microorganisms, i.e., soil moisture and nutrient availability (Hiwale 2004).

Subabool leaf litter (decomposed) had the highest N content (2.174 %). Higher N content in subabool can be attributed to atmospheric N fixation capacity of tree species (Table 22.5).

Phosphorus content of decomposed leaf litter was recorded in subabool (0.294 %). The phosphorus content of fresh leaf litter collected was low. The absolute amount of P in the residue showed biphasic pattern with initial low content followed by later slow buildup phase.

Potassium content of decomposed leaf litter was found in subabool (0.194 %). It is concluded that due to addition of litter by tree species, there was improvement in soil health, which is due to enrichment of the soil through the release of entrapped nutrients and formation of organic matter.

22.12 Second Rotation

Height and Diameter Growth. The plant growth of K 500 shows a great deal of variability on the roadside plantation in a moist site. Since the plants were harvested at a stump height of 20 cm, the growth in collar diameter shows variation in a moist site. Since the plants were harvested at a stump height of 20 cm, the growth in collar diameter shows variation due to the growth pattern and the number of coppice shoots. The average collar diameter ranged from 10.4 to 20.5 cm at 4 years and 14.6 to 43.9 cm at 5 years. The higher magnitude of variation at 5 years was due to the number of shoots allowed and the growth of the trees. The variability when grouped in diameter classes gave 3 and 4 classes in each cycle, respectively. Thus, at 4 years more than 60 % of plants were in the range of 9–13 cm, and at 5 years, more than 50 % plants were in the range of 13–19 cm. More than 22 % of plants were above 14 and 25 cm in diameter in 4 and 5 years. The average at 4 and 5 years were 15.42 and 26.48 cm, respectively. Plant height was also influenced by the diameter and varied according to the collar diameter with average of 9.58 and 11.23 m at the two respective durations. The mean annual increment of CD was thus 1.93 and 3.31 cm, respectively, and of height 2.39 and 2.25 m, respectively, at 4 and 5 years.

The diameter at breast height (DBH) gave corresponding variation as in case of CD and height with average of 14.87 and 23.4 cm at 4 and 5 years, respectively. In most of these cases, this high DBH was the product of 2 coppice shoots per stump. Annual average diameter growth was found to be 3.72 and 4.68 cm, respectively.

12 year old Subabool

Subabool forming dense cover on gully collar

22.13 Biomass Production

The aerial biomass was separated into bole, branch, leaf, and pod after harvest. The bole biomass ranged from 12.69 to 20.77 kg per tree in the three groups of plants with maximum variation in the lower diameter class plants. The variations were more pronounced at the branch level where the differences were 3- and 9-fold from low to high biomass. Leaf and pod biomass followed the similar proportion. At the level of total biomass, the differences were 1.5 and 3 times compared to 1.5 and 2 times of collar diameter. The non-photosynthetic/photosynthetic ratio increased with increasing diameter and production. Thus, on an average, 37.3 kg/tree of total aerial biomass was produced. The mean annual production of 9.32 kg/tree was obtained in this case.

In the fifth year, the plants were grouped into four from low to high diameter class. Fifty percent of plants in the lowest group gave the bole biomass of 25.6 kg. The production of bole biomass remained almost the same, up to 24 cm range of the collar diameter. But the bole biomass for the same diameter group at 5 years was higher than those at 4 years. Maximum bole biomass could go up to 56.8 kg/tree. The higher diameter groups and their quantum of branch biomass production increased with the increasing bole biomass production, and at various stages it equaled the bole biomass indicating the prominence of branches and their relative contribution in total plant biomass. This is quite common with this variety of leucaena where branch production has a greater prominence. Leaf biomass also increased steadily but the quantum of increase was not so much high. The production of pods did not show any definite trend. The total biomass production ranged between 46.2 to 102.2 kg/tree. The non-photosynthetic/photosynthetic ratio did not show any trend indicating the bole/branch relationship in this case. The ratio is lower compared to that of 4 years. The mean annual increment for height was 2.25 m, DBH 4.68 cm, and total biomass 14.39 kg. The high variability in the production of total biomass from 46.2 to 102.2 kg/tree could be ascribed to the plant type variability and the differential mobilization of nutrients. The proportion of such plant was only 11.1 %. Up to 77.8 % of plants averaged the production level of 46–61 kg/tree.

22.14 Forage and Fuel Production

When the production data were computed for fuel and forage production per row km, it was found to give 13.49 and 0.969 t/row km and 27.38 and 1.88 t/row km, respectively, at 4 and 5 years of age. Thus, if we see the current annual increment of these between the fourth and fifth years, more than double the production could be expected. Similar quantum of increment was found between the third and fourth years in the first cycle (Pathak et al. 1981). If the production values of 4 years in the first and second cycle are compared, we find an almost similar production with only 22.7 % increase. This was quite expected in a system of coppice growth which was faster to those from planted ones (Pathak et al. 1981).

22.15 Growth of Coppice Shoots

The leading shoots allowed after the thinning attained variable growth dimensions depending upon the diameter of their stumps. The variability was reflected for all the parameters, viz., height, collar diameter, and DBH. The mean height varied between 242 and 566 cm, collar diameter between 1.2 and 4.92 cm, and DBH between 0.93 and 3.97 cm. On an average the mean height up to 4.25 cm, CD up to 3.12 cm, and DBH of 2.52 cm could be attained in 1 year. This growth when compared to that after first cycle of harvest (Pathak et al. 1981) was found to be slightly lower which could be ascribed to reduction in the growth vigor after the second cycle of harvest or the environmental variability during these periods.

Forage Production from Coppice Shoots. The coppice shoots were harvested four times during the year in September, November, April, and June, leaving 2 shoots on each stump. The maximum number of shoots was removed in the first harvest which gradually decreased toward the last one. The very high basal diameter group produced up to 78 shoots/stump and its number gradually reduced toward low diameter stumps. The production of forage followed the pattern of the numbers, but at later dates this pattern was not even. Thus in a year, production from 0.71 to 3.26 kg/stump could be obtained which amounts to 0.7 t/row km in a year. Thus, in a year, the total forage from coppice and the felling go up to 2.58 t/row km. The production from coppice shoots was higher in this rotation, indicating a large number of shoots produced and the timely harvest. The thinning during the four harvests indicates the role of such a plantation in providing nutritious forage during the lean periods from such a system of farm forestry.

References

Anonymous (1990) Leucaena an important multipurpose tree, NFT highlights 90-01. NFTA, Waimanalo

Benjamin A, Shelton HM, Gutteridge RC (1991) Shade tolerance of some tree legumes. In: Shelton HM, Stür WW (eds) Forages for plantation crops, ACIAR proceedings no. 32. Canberra, pp 75–76

Brewbaker JL, Sorensson CT (1990) New tree crops from interspecific Leucaena hybrids. In: Janick J, Simon JE (eds) Advances in new crops. Timber Press, Portland, pp 283–289

Brewbaker JL, Hegde N, Hutton EM, Jones RJ, Lowry JB, Moog F, van den Beldt R (1985) Leucaena – forage production and use. NFTA, Waimanalo, 39 pp

Hiwale SS (2004) Develop sustainable agri-silvi-horti production system for marginal lands under arid conditions. Tech Bull 21:23–35

Isarasenee A, Shelton HM, Jones RM (1984) Accumulation of edible forage of *Leucaena leucocephala cv.* Peru over late summer and autumn for use as dry season feed. Leucaena Res Rep 5:3–4

Pathak PS et al (1981) Fuel and forage production from Leucaena Leucocephala under farm forestry

Sorensson CT, Shelton HM, Austin MT (1994) Seedling vigour of *Leucaena leucocephala, L. diversifolia, L. pallida* and their hybrids. Trop Grassl (in press)

23 Eucalyptus (*Eucalyptus sp.*)

Abstract

Eucalyptus is a diverse genus of flowering trees and shrubs in the family Myrtaceae. Members of this genus dominate the tree flora of Australia. Species of eucalyptus are cultivated widely in the tropical and temperate world, including the Americas, Europe, Africa, the Mediterranean Basin, the Middle East, China, and the Indian subcontinent. Some eucalyptus species have attracted attention from horticulturists, global development researchers, and environmentalists because of their desirable traits such as being fast-growing sources of wood, producing oil that can be used for cleaning, and functioning as a natural insecticide or having an ability to be used to drain swamps. Nowadays, regular plantations of eucalyptus are very common due to increasing demand for pole in construction work and also in paper industry. The biomass production on fresh wt. basis was 195 q/ha. in 10-year-old plant.

23.1 Introduction

Scientific classification – Kingdom: Plantae, Angiosperms, Eudicots, Rosids. Order: Myrtales. Family: Myrtaceae.

It is a diverse genus of flowering trees and shrubs (including a distinct group with a multiple-stem mallee growth habit) in the myrtle family, Myrtaceae. Members of the genus dominate the tree flora of Australia. There are more than 700 species of eucalyptus, mostly native to Australia, and a very small number are found in adjacent areas of New Guinea and Indonesia. One species, *Eucalyptus deglupta*, ranges as far north as the Philippines. Only fifteen species occur outside Australia, with just nine of these not occurring in Australia. Species of eucalyptus are cultivated widely in the tropical and temperate world, including the Americas, Europe, Africa, the Mediterranean Basin, the Middle East, China, and the Indian subcontinent, though most species do not tolerate frost.

Eucalyptus is one of three similar genera that are commonly referred to as "eucalypts," the others being *Corymbia* and *Angophora*. Many species, but far from all, are known as gum trees because they exude copious sap from any break in the bark. The generic name is derived from the Greek words ευ (*eu*) "well" and κάλυπτος (*kályptos*) "covered," referring to the operculum on the calyx that initially conceals the flower (Gledhill 2008).

S. Hiwale, *Sustainable Horticulture in Semiarid Dry Lands*,
DOI 10.1007/978-81-322-2244-6_23, © Springer India 2015

Some eucalyptus species have attracted attention from horticulturists, global development researchers, and environmentalists because of their desirable traits such as being fast-growing sources of wood, producing oil that can be used for cleaning, and functioning as a natural insecticide or having an ability to be used to drain swamps and thereby reduce the risk of malaria. Outside their natural ranges, eucalypts are both lauded for their beneficial economic impact on poor populations (Luzar 2007) and criticized for being "water guzzling" leading to controversy over their total impact (Santos 1997a, b).

On warm days, eucalyptus forests are sometimes shrouded in a smog-like mist of vaporized volatile organic compounds (terpenoids); the Australian Blue Mountains take their name from the haze.

23.2 Description

23.2.1 Size and Habit

A mature eucalyptus may take the form of a low shrub or a very large tree. There are three main habits and four size categories that these species can be divided into.

As a generalization, "forest trees" are single stemmed and have a crown forming a minor proportion of the whole tree height. "Woodland trees" are single stemmed although they may branch at a short distance above ground level.

"Mallees" are multistemmed from ground level, usually less than 10 m (33 ft) in height, often with the crown predominantly at the ends of the branchlets, and individual plants may combine to form either an open or closed formation. Many mallee trees may be so low growing as to be considered a shrub.

Two other tree forms are notable in Western Australia and are described using the native names "mallet" and "marlock." The "mallet" is a small- to medium-sized tree that does not produce lignotubers and has a relatively long trunk, a steeply branching habit, and often a conspicuously dense terminal crown. This is the normal habit of mature healthy specimens of *Eucalyptus occidentalis*, *E. astringens*, *E. spathulata*, *E. gardneri*, *E. dielsii*, *E. forrestiana*, *E. salubris*, *E. clivicola*, and *E. ornata*. The smooth bark of mallets often has a satiny sheen and may be white, cream, gray, green, or copper.

The term *marlock* has been variously used; in *Forest Trees of Australia*, it is defined as a small tree without lignotubers but with a shorter, lower-branching trunk than a mallet. They usually grow in more or less pure stands. Clearly recognizable examples are stands of *E. platypus*, *E. vesiculosa*, and the unrelated *E. stoatei*.

The term "morrell" is somewhat obscure in origin and appears to apply to trees of the Western Australian wheat belt and goldfields which have a long, straight trunk, completely rough barked. It is now used mainly for *E. longicornis* (red morrell) and *E. melanoxylon* (black morrell).

Tree sizes follow the convention of:

- Small – to 10 m (33 ft) in height
- Medium sized – 10–30 m (33–98 ft)
- Tall – 30–60 m (98–197 ft)
- Very tall – over 60 m (200 ft)

23.2.2 Leaves

Nearly all eucalyptus are evergreen but some tropical species lose their leaves at the end of the dry season. As in other members of the myrtle family, eucalyptus leaves are covered with oil glands. The copious oils produced are an important feature of the genus. Although mature eucalyptus trees are usually towering and fully leafed, their shade is characteristically patchy because the leaves usually hang downward.

The leaves on a mature eucalyptus plant are commonly lanceolate, petiolate, apparently alternate, and waxy or glossy green. In contrast, the leaves of seedlings are often opposite, sessile, and glaucous. But there are many exceptions to this pattern. Many species such as *E. melanophloia* and *E. setosa* retain the juvenile leaf form even when the plant is reproductively mature. Some species, such as *E. macrocarpa*, *E. rhodantha*, and *E. crucis*, are sought-after ornamentals due to this lifelong juvenile leaf form. A few species, such as *E. petraea*, *E. dundasii*, and *E. lansdowneana*, have shiny green leaves throughout their life cycle. *E. caesia* exhibits the opposite pattern of leaf development to most eucalyptus, with shiny green leaves in the seedling stage and dull, glaucous leaves in mature crowns. The contrast between juvenile and adult leaf phases is valuable in field identification.

Four leaf phases are recognized in the development of a eucalyptus plant: the "seedling," "juvenile," "intermediate," and "adult" phases. However, there is no definite transitional point between the phases. The intermediate phase, when the largest leaves are often formed, links the juvenile and adult phases.

In all except a few species, the leaves form in pairs on opposite sides of a square stem, consecutive pairs being at right angles to each other (decussate). In some narrow-leaved species, for example, *E. oleosa*, the seedling leaves after the second leaf pair are often clustered in a detectable spiral arrangement about a five-sided stem. After the spiral phase, which may last from several to many nodes, the arrangement reverts to decussate by the absorption of some of the leaf-bearing faces of the stem. In those species with opposite adult foliage, the leaf pairs, which have been formed opposite at the stem apex, become separated at their bases by unequal elongation of the stem to produce the apparently alternate adult leaves.

23.2.3 Flowers

The most readily recognizable characteristics of eucalyptus species are the distinctive flowers and fruit (capsules or "gumnuts"). Flowers have numerous fluffy stamens which may be white, cream, yellow, pink, or red; in bud, the stamens are enclosed in a cap known as an operculum which is composed of the fused sepals or petals or both. Thus, flowers have no petals, but instead decorate themselves with the many showy stamens. As the stamens expand, the operculum is forced off, splitting away from the cup-like base of the flower; this is one of the features that unites the genus. The name *Eucalyptus*, from the Greek words *eu-*, well, and *kaluptos*, cover, meaning "well covered," describes the operculum. The woody fruits or capsules are roughly cone shaped and have valves at the end which open to release the seeds, which are waxy, rod shaped, about 1 mm in length, and yellow brown in color. Most species do not flower until adult foliage starts to appear; *Eucalyptus cinerea* and *Eucalyptus perriniana* are notable exceptions.

23.2.4 Bark

Colored bark of *Eucalyptus deglupta* **native to S. E. Asia**

The appearance of eucalyptus bark varies with the age of the plant, the manner of bark shed, the length of the bark fibers, the degree of furrowing, the thickness, the hardness, and the color. All mature eucalypts put on an annual layer of bark, which contributes to the increasing diameter of the stems. In some species, the outermost layer dies and is annually deciduous, either in long strips (as in *Eucalyptus sheathiana*) or in variably sized flakes (*E. diversicolor*, *E. cosmophylla*, or *E. cladocalyx*). These are the gums or smooth-barked species. The gum bark may be dull, shiny, or satiny (as in *E. ornata*) or matte (*E. cosmophylla*). In many species, the dead bark is retained. Its outermost layer gradually fragments with weathering and sheds without altering the essentially rough-barked nature of the trunks or stems, for example, *E. marginata*, *E. jacksonii*, *E. obliqua*, and *E. porosa*.

Many species are "half-barks" or "blackbutts" in which the dead bark is retained in the lower half of the trunks or stems, for example, *E. brachycalyx*, *E. ochrophloia*, and *E. occidentalis*, or only in a thick, black accumulation at the base, as in *E. clelandii*. In some species in this category, for example, *E. youngiana* and *E. viminalis*, the rough basal bark is very ribbony at the top, where it gives way to the smooth upper stems. The smooth upper bark of the half-barks and that of the completely smooth-barked trees and mallees can produce remarkable color and interest, for example, *E. deglupta*.

23.3 Species and Hybridism

There are over 700 species of *Eucalyptus*. Some have diverged from the mainstream of the genus to the extent that they are quite isolated genetically and are able to be recognized by only a few relatively invariant characteristics. Most, however, may be regarded as belonging to large or small groups of related species, which are often in geographical contact with each other and between which gene exchange still occurs. In these situations, many species appear to grade into one another, and intermediate forms are common. In other words, some species are relatively fixed genetically, as expressed in their morphology, while others have not diverged completely from their nearest relatives.

Hybrid individuals have not always been recognized as such on the first collection, and some have been named as new species, such as *E. chrysantha* (*E. preissiana* × *E. sepulcralis*) and *E.* "rivalis" (*E. marginata* × *E. megacarpa*). Hybrid combinations are not particularly common in the field, but some other published species frequently seen in Australia have been suggested to be hybrid combinations. For example, *E. erythrandra* is believed to be *E. angulosa* × *E. teraptera* and due to its wide distribution is often referred to in texts.

Renantherin, a phenolic compound present in the leaves of some eucalyptus species, allows chemotaxonomic discrimination in the sections renantheroideae and renantherae (Hillis 1967), and the ratio of the amount of leucoanthocyanins varies considerably in certain species (Hillis 1966).

23.4 Related Genera

A small genus of similar trees, *Angophora*, has also been known since the eighteenth century. In 1995, new evidence, largely genetic, indicated that some prominent eucalyptus species were actually more closely related to *Angophora* than to the other eucalypts; they were split off into the new genus *Corymbia*. Although separate, the three groups are allied and it remains acceptable to refer to the members of all three genera, *Angophora*, *Corymbia*, and *Eucalyptus*, as "eucalypts."

Several eucalypts are among the tallest trees in the world. *Eucalyptus regnans*, the Australian mountain ash, is the tallest of all flowering plants (angiosperms); today, the tallest measured specimen named Centurion is 99.6 m (327 ft) tall. Coast Douglas-fir is about the same height; only coast redwood is taller, and they are conifers (gymnosperms). Six other eucalypt species exceed 80 m in height: *Eucalyptus obliqua*, *Eucalyptus delegatensis*, *Eucalyptus diversicolor*, *Eucalyptus nitens*, *Eucalyptus globulus*, and *Eucalyptus viminalis*.

23.5 Frost Intolerance

Most eucalypts are not tolerant of frost or only tolerate light frosts down to −5 °C (23 °F); the hardiest are the snow gums, such as *Eucalyptus pauciflora*, which is capable of withstanding cold and frost down to about −20 °C (−4 °F). Two subspecies, *E. pauciflora* subsp. *niphophila* and *E. pauciflora* subsp. *debeuzevillei* in particular, are even hardier and can tolerate even quite severe winters. Several other species, especially from the high plateau and mountains of central Tasmania such as *Eucalyptus coccifera*, *Eucalyptus subcrenulata*, and *Eucalyptus gunnii*, have also produced extreme cold-hardy forms, and it is seed procured from these genetically hardy strains that are planted for ornament in colder parts of the world.

An essential oil extracted from eucalyptus leaves contains compounds that are powerful natural disinfectants and can be toxic in large quantities. Several marsupial herbivores, notably koalas and some possums, are relatively tolerant of it. The close correlation of these oils with other more potent toxins called formylated phloroglucinol compounds (euglobals, macrocarpals, and sideroxylonals) (Eschler et al. 2000) allows koalas and other marsupial species to make food choices based on the smell of the leaves. For koalas, these compounds are the most important factor in leaf choice.

Eucalyptus flowers produce a great abundance of nectar, providing food for many pollinators including insects, birds, bats, and possums. Although eucalyptus trees are seemingly well defended from herbivores by the oils and phenolic compounds, they have insect pests. These include the eucalyptus longhorn borer *Phoracantha semipunctata* and the aphid-like psyllids known as "bell lerps," both of which have become established as pests throughout the world wherever eucalypts are cultivated. The eusocial beetle *Austroplatypus incompertus* makes and defends its galleries exclusively inside *Eucalyptus* plants.

23.6 Adaptation to Fire

Epicormic shoots sprouting vigorously occur from epicormic buds beneath the bushfire damaged bark on the trunk of a *Eucalyptus* tree.

23.6.1 Eucalyptus Forest in a State of Regeneration

Eucalypts originated between 35 and 50 million years ago, not long after Australia-New Guinea separated from Gondwana, their rise coinciding with an increase in fossil charcoal deposits (suggesting that fire was a factor even then), but they remained a minor component of the tertiary rainforest until about 20 million years ago, when the gradual drying of the continent and depletion of soil nutrients led to the development of a more open forest type, predominantly *Casuarina* and *Acacia* species.

The aridification of Australia during the mid-tertiary period (25–40 million years ago), combined with the annual penetration of tropical convection storms and associated lightning, deep into the continental interior stimulated the gradual evolution, diversification, and geographic expansion of the flammable biota. The absence of great rivers or mountain chains meant that there were no geographic barriers to check the spread of fires. From the monsoonal "cradle," fire-promoting species expanded into higher rainfall environments, where lightning was less frequent, gradually displacing the Gondwanan rainforest from all but the most fire-sheltered habitats.

The two valuable timber trees, alpine ash *E. delegatensis* and Australian mountain ash *E. regnans*, are killed by fire and only regenerate from seed. The same 2003 bushfire that had little impact on forests around Canberra resulted in thousands of hectares of dead ash forests. However, a small amount of ash survived and put out new ash trees as well. There has been some debate as to whether to leave the stands or attempt to harvest the mostly undamaged timber, which is increasingly recognized as a damaging practice.

23.7 Hazards

The two most common hazards of eucalyptus species to people are fire and falling branches.

23.8 Fire Hazard

Eucalyptus trees bent over due to the high winds and heat of the October 2007 California wildfires. They are located in the San Dieguito River Park of San Diego County and leaning west.

Eucalyptus oil is highly flammable (ignited trees have been known to explode) (Williams 2007); bushfires can travel easily through the oil-rich air of the tree crowns (Dold et al. 2005). Eucalypts obtain long-term fire survivability from their ability to regenerate from epicormic buds situated deep within their thick bark, or from lignotubers (Reid and Potts 2005), or by producing serotinous fruits.

In seasonally dry climates, oaks are often fire resistant, particularly in open grasslands, as a grass fire is insufficient to ignite the scattered trees. In contrast, a eucalyptus forest tends to promote fire because of the volatile and highly combustible oils produced by the leaves, as well as the production of large amounts of litter which is high in phenolics, preventing its breakdown by fungi and thus accumulates as large amounts of dry, combustible fuel (Reid and Potts 2005). Consequently, dense eucalypt plantings may be subject to catastrophic firestorms. In fact, almost 30 years before the Oakland firestorm of 1991, a study of eucalyptus in the area warned that the litter beneath the trees builds up very rapidly and should be regularly monitored and removed (Agee et al. 1973). It has been estimated that 70 % of the energy released through the combustion of vegetation in the Oakland fire was due to eucalyptus (U.S. Printing Office 2006). In a National Park Service study, it was found that the fuel load (in tons per acre) of nonnative eucalyptus woods is almost three times as great as native oak woodland (U.S. Printing Office 2006).

Some species of gum trees drop branches unexpectedly. In Australia, Parks Victoria warns campers not to camp under River Red Gums. Some Council in Australia such as Gosnells, Western Australia, have removed eucalypts after reports of damage from dropped branches, even in the face of lengthy, well-publicized protests to protect those particular trees (Thomson 2011). A former Australian National Botanic Gardens director and consulting arborist, Robert Boden, has been quoted referring to "summer branch drop" (Thistleton, John, 2008–2009). Dropping of branches is recognized in Australian literature through the fictional death of Judy in Seven Little Australians. Although all large trees can drop branches, the weight of eucalyptus wood is high because of its density and high resin content.

23.9 Cultivation and Uses

Eucalyptus was first introduced from Australia to the rest of the world by Sir Joseph Banks, botanist, on the Cook expedition in 1770. It was subsequently introduced to many parts of the world, notably California, Brazil, Ecuador, Colombia, Ethiopia, Morocco, Portugal, South Africa, Uganda, Israel, Galicia, and Chile. On the order, 250 species are under cultivation in California (Ritter 2012). In Portugal and also Spain, *eucalypts* have been planted in pulpwood plantations. *Eucalyptus* are the basis for several industries, such as sawmilling, pulp, charcoal, and others. Several species have become invasive and are causing major problems for local ecosystems, mainly due to the absence of wildlife corridors and rotation management. Eucalypts have many uses which have made them economically important trees and have become a cash crop in poor areas such as Timbuktu, Africa, and the Peruvian Andes, despite concerns that the trees are invasive in some countries like South Africa. Best known are perhaps the varieties karri and yellow box. Due to their fast growth, the foremost benefit of these trees is their wood. They can be chopped off at the root and grow back again. They provide many desirable characteristics for use as ornament, timber, firewood, and pulpwood. It is also used in a number of industries, from fence posts and charcoal to cellulose extraction for biofuels. Fast growth also makes eucalypts suitable as windbreaks and to reduce erosion.

Eucalypts draw a tremendous amount of water from the soil through the process of transpiration. They have been planted (or replanted) in some places to lower the water table and reduce soil salination. Eucalypts have also been used as a way of reducing malaria by draining the soil in

Algeria, Lebanon, Sicily (Global Eucalyptus Map 2009), elsewhere in Europe, in Caucasus (Western Georgia), and California (Santos 1997a, b). Drainage removes swamps which provide a habitat for mosquito larvae, but can also destroy ecologically productive areas. This drainage is not limited to the soil surface, because the eucalyptus roots are up to 2.5 m (8.2 ft) in length and can, depending on the location, even reach the phreatic zone.

23.9.1 Pulpwood

Eucalyptus is the most common short fiber source for pulpwood to make pulp (Nanko et al. 2005). *Eucalyptus globulus* (in temperate climates) and the hybrid of *Eucalyptus urophylla* × *Eucalyptus grandis* (in tropical climates) are the most used varieties in papermaking. The fiber length of eucalyptus is relatively short and uniform with low coarseness compared with other hardwoods commonly used as pulpwood. The fibers are slender, yet relatively thick walled. This gives uniform paper formation and high opacity that are important for all types of fine papers. The low coarseness is important for high-quality coated papers (Nanko et al. 2005). Eucalyptus is suitable for many tissue papers as the short and slender fibers give a high number of fibers per gram and low coarseness contributes to softness (Nanko et al. 2005).

23.9.2 Eucalyptus Oil

Eucalyptus oil is readily steam distilled from the leaves and can be used for cleaning and as an industrial solvent, as an antiseptic, for deodorizing, and in very small quantities in food supplements, especially sweets, cough drops, toothpaste, and decongestants. It has insect repellent properties and is an active ingredient in some commercial mosquito repellents. *Eucalyptus globulus* is the principal source of eucalyptus oil worldwide.

23.9.3 Dyes

All parts of *Eucalyptus* may be used to make dyes that are substantive on protein fibers (such as silk and wool), simply by processing the plant part with water. Colors to be achieved range from yellow and orange through green, tan, chocolate, and deep rust red (Bennet 2010). The remaining material after processing can be safely used as mulch or fertilizer.

23.10 Eucalyptus as Plantation Species

In the twentieth century, scientists around the world experimented with eucalyptus species. They hoped to grow them in the tropics, but most experimental results failed until breakthroughs in

Growth of Eucalyptus after first cut

the 1960s–1980s in species selection, silviculture, and breeding programs "unlocked" the potential of eucalypts in the tropics. Prior to then, eucalypts were something of the "El Dorado" of forestry. Today, eucalyptus is the most widely planted type of tree in plantations around the world (Bennet 2010) in Brazil, Paraguay, Uruguay, Australia, India, Galicia, and many more (Global Eucalyptus Map 2009).

Eucalyptus trees were planted in Israel by Jewish settlers in the beginning of the twentieth century in order to dry out marshy lands that caused diseases. These eucalyptus trees still exist today in many parts of Israel.

Eucalyptus trees were introduced to Sri Lanka in the late nineteenth century by tea and coffee planters, for wind protection, shade, and fuel. Forestry replanting of eucalyptus began in the 1930s in deforested mountain areas, and currently there are about 10 species present in the island. They account for 20 % of major reforestation plantings. They provide railway sleepers, utility poles, sawn timber, and fuel wood, but are controversial because of their adverse effect on biodiversity, hydrology, and soil fertility. They are associated with another invasive species, the eucalyptus gall wasp, *Leptocybe invasa* (Bandaratillake 2010; Rodrigo 2010).

23.11 Eucalyptus as an Invasive Species

Due to similar favorable climatic conditions, *Eucalyptus* plantations have often replaced oak woodlands, for example, in California, Spain, and Portugal. The resulting monocultures have raised concerns about loss of biological diversity, through loss of acorns that mammals and birds feed on, absence of hollows that in oak trees provide shelter and nesting sites for birds and small mammals and for bee colonies, as well as lack of downed trees in managed plantations. A study of the relationship between birds and eucalyptus in the San Francisco Bay Area found that bird diversity was similar in native forest vs. eucalyptus forest but the species were different. One way in which the avifauna changes is that cavity nesting birds including woodpeckers, owls, chickadees, wood ducks, etc. are depauperate in eucalyptus groves because the decay-resistant wood of these trees prevents cavity formation by decay or excavation. Also those bird species that glean insects from foliage, such as warblers and vireos, have population declines when eucalyptus replaces oak forest. Birds that do well in eucalyptus groves in California like tall vertical habitat like herons and egrets (possibly because redwood trees are less available) or have longer bills, which may

High density plantation of clonal Eucalyptus

Table 23.1 Growth and fresh biomass of *Eucalyptus*

			Main	Lateral
Age (year)	Plant height (m)	DBH (mm)	Stem fresh weight (kg/plant)	Stem fresh weight (kg/plant)
2nd	1.57	5.30	–	–
3rd	2.70	32.00	6.50	0.70
4th	4.69	51.00	14.70	1.30
5th	6.24	82.00	26.00	2.10
6th	8.05	95.00	37.60	2.80
7th	9.46	114.00	49.10	3.90
8th	10.38	121.00	58.80	4.80
9th	11.49	130.00	63.40	5.70
10th	11.84	139.00	71.50	6.40

play a role in preventing their nostrils from being clogged by eucalyptus resin/pitch (Suddjian 2004) (Table 23.1).

The growth and fresh biomass production of two principal forest species, i.e., subabul and eucalyptus, planted for rehabilitation of degraded lands are given in the above table. After 5 years, as per estimate, the value of timber from *Eucalyptus* was estimated to be Rs. 90,000/ha. The selected felling of subabul was also started for fuel wood production (Hiwale et al. 2007). It is obvious that at the 5th year, about 35 kg/plant/year additional dry biomass can be obtained in the form of main stem, lateral stem, and leaves. The main stem can be used as timber wood, lateral stem as fuel wood, and leaves as organic manure.

References

Agee JK, Wakimoto RH, Darly EF, Biswell HH (1973) Eucalyptus: fuel dynamics and fire hazard in the Oakland hills. Calif Agric 27:13–15. Retrieved 2011

Bandaratillake HM (2010) Eucalyptus plantations in Sri Lanka: environmental, social, economic and policy issues

Dold JW, Weber RO, Gill M et al (2005) Unusual phenomena in an extreme bushfire. In: Proceedings of the 5th Asia Pacific Conference on Combustion, Adelaide

Eschler BM, Pass DM, Willis R, Foley WJ (2000) Distribution of foliar formylated phloroglucinol derivatives amongst eucalyptus species. Biochem Syst Ecol 28(9):813–824

Gledhill D (2008) The names of plants. Cambridge University Press, p 158. Retrieved 2013, England

Hillis WE (1966) Variation in polyphenol composition within species of Eucalyptus l'herit. Phytochemistry

Hillis WE (1967) (Polyphenols in the leaves of eucalyptus: a chemotaxonomic survey-II.: the sections renantheroideae and renantherae. Phytochemistry

Hiwale SS, Raturi GB, Bagle BG, More TA (2007) Technical bulletin on Fruit tree based farming systems for the dry lands. CHES, Vejalpur, Gujarat

Luzar J (2007) The political ecology of a "Forest Transition": eucalyptus forestry in the Southern Peruvian. Ethno Bot Res Appl

Nanko H, Button A, Hillman D (2005) The world of market pulp. WOMP, LLC, Appleton, pp 107–109

Reid JB, Potts BM (2005) Eucalypt biology. In: Reid JB et al (eds) Vegetation of Tasmania. Australian Government, Canberra, pp 198–223

Ritter M (2012) The Jepson manual, 2nd edn. University of California Press, Berkeley, pp 913–914

Rodrigo M (2010) Another pest, another tree. Sunday Times, 27 June 2010

Santos RL (1997a) Section three: problems, cares, economics, and species. In: The eucalyptus of California. California State University, California

Santos RL (1997b) Section two: physical properties and uses. In: The eucalyptus of California. California State University

Suddjian DL (2004) Birds and eucalyptus on the central California coast: a love-hate relationship. David Sudan Biological Consulting. Archived from the original. Retrieved 2013

Thomson C (2011) Tree man's gum an eastern states 'widow maker'. WA Today. Retrieved 29 Jan 2011

U.S. Government Printing Office (2006) A transcontinental legacy: fire management, resource protection, and the challenges of Tasmanian Blue Gum. Retrieved 2011

Williams C (2007) Ignition impossible: when wildfires set the air alight. New Scientist 2615

24 Bamboo (*Bambusa vulgaris sp.*)

Abstract

Bamboo is one of the fastest-growing plants on the planet. Demand is drastically increasing due to industrial revolution like pulp for paper and rayon, laminated bamboo, parquet flooring, ply bamboo, bamboo composites, and charcoal coupled with mechanization of traditional sector like shoot processing, chopsticks, agarbatti, toothpick production, and bamboo handicrafts. At present the demand for bamboo is 26.9 million t as against the supply of 13.47 million tons. Bamboo could actually play an important role in forest and landscape restoration. Bamboo-based agroforestry has a very wide scope in India as it provides higher economic returns to the farmers, improves the soil, and provides raw material to the industry as well as for domestic use of the rural community. Its performance in semi-arid rainfed areas on gully collars is encouraging. It not only helped in gully collar stabilization but produced 40–50 bamboo every third year cycle.

24.1 Introduction

Bamboo is one of the fastest-growing plants on the planet. At present the demand for bamboo is 26.9 million t as against the supply of 13.47 million tons. Bamboo could actually play an important role in forest and landscape restoration. With adequate attention, investment, and the right standards in place, it could become a major renewable and sustainable crop – if we can update our outmoded view of it. Demand is drastically increasing due to industrial revolution like pulp for paper and rayon, laminated bamboo, parquet flooring, ply bamboo, bamboo composites, and charcoal coupled with mechanization of traditional sector like shoot processing, chopsticks, agarbatti, toothpick production, and bamboo handicrafts. According to estimates, bamboo-based activities could easily generate 8.6 million additional jobs in India and thus enable 5.01 families to cross the poverty line. Therefore, there is a good scope of bamboo-based agroforestry systems in the country. Distribution, ecological preference, and common uses of the bamboo species suitable for agroforestry in India are given in Table 24.1.

Kleinhenz and Midmore (2002) showed that bamboo leaf N level was more responsive to N

S. Hiwale, *Sustainable Horticulture in Semiarid Dry Lands*,
DOI 10.1007/978-81-322-2244-6_24, © Springer India 2015

Table 24.1 Bamboo species suitable for agroforestry in different climatic zones of India

Species	Ecological preference, altitude, and distribution in Indian states	Common uses
Bambusa affinis	Moist tropical region, Tripura	Making javelins, angling rod, furniture, etc. (solid culm)
Bambusa balcooa incense	Moist tropical, 600 m, West Bengal, Bihar, NE India	Building construction, ladder, sticks, edible shoots, wood chip industry
B. bambos syn. B. arundinacea	Moist tropical and semiarid zone, up to 1,200 m, northeast and south India, Maharashtra	Building construction, ladder, tent pole, baskets, mats, agricultural implements
B. nutans	Moist tropical, 600–1,500 m, Himalayan tracts, NE India, W. Bengal	Construction, basketry, handicrafts, pulp and paper
B. pallida	Moist tropical, 700–2,000 m MSL, NE India, North Bengal	Construction, mats, basketry, edible shoots, incense sticks, pulp and paper
B. polymorpha	Tropical zone, Madhya Pradesh, Kerala, Tamil Nadu, NE India	Building construction, incense sticks, fiberboards, roofing, edible shoots, pulp and paper
B. vulgaris	Moist tropical zone, up to 1,200 m, Madhya Pradesh, Chattisgarh, Bihar, Uttar Pradesh	Construction, basketry, handicrafts, hats, food grain containers, pulp and paper, saline and sandy area reclamation
Bambusa tulda	Moist tropical zone, NE India	Construction, basket making, mats, toys, handicrafts, pulp and paper
Dendrocalamus asper	Tropical zone (exotic species)	Edible shoots, construction, basketry, ladder, pulp and paper
D. giganteus	Moist tropical zone, 1,200 m, Arunachal Pradesh, Assam, Manipur, Nagaland, W Bengal and NW Himalayas	Construction, ladder, pulp and paper
D. strictus	Semidry areas, up to 1,000 m, throughout India except in NE	Construction, pulp and paper, land reclamation, etc.
Melocanna baccifera	Moist tropical, 600 m, NE India, Assam, Manipur, Meghalaya, Mizoram, Tripura, West Bengal	Construction, mats, toys, wall plates, screens, hangers, hats, baskets, food grain containers, pulp and paper, edible fruit
Pseudo-oxytenanthera stocksii	Tropical zone, 800 m, Goa, Karnataka, Kerala, Western Ghats	Construction, baskets, pulp and paper

fertilizer application when the leaf N concentration was below 3 %. To promote bamboo shoot and culm (pole) production, irrigation is required when rainfall is insufficient (Lin 1996).

24.2 Benefit of Bamboo-Based Agroforestry

Bamboo-based agroforestry has a very wide scope in India as it provides higher economic returns to the farmers, improves the soil, and provides raw material to the industry as well as for domestic use of the rural community. Intercrop cultivation provides various crop products from the early stages of plantations and generates much higher income than any other system. Further, bamboo can be harvested every year from the fourth year onward and hence regular income starts much earlier than expected from any other woody component. Agroforestry practices also benefit the bamboo plants by sharing the inputs of irrigation, fertilizers, weeding, etc. applied to agriculture crops, and hence bamboo yield and quality would be better than unmanaged plantations. By planting bamboo in parts of landscapes, degraded lands could be restored to productive use, thereby alleviating some of the development pressures on forests. So, total returns from bamboo-based agroforestry can be much higher than other systems.

24.3 Bamboo-Based Agroforestry Models

Bamboo under agroforestry systems is suitable for intercropping, soil conservation, windbreak, riparian filter, and permaculture. Bamboos are planted by farmers in homesteads or small-scale farmlands and managed as groves or following some agroforestry models. In bamboo + agriculture crop model, bamboo can be planted at varied spacing of 4 m×4 m to 8 m×8 m, depending on the clump size of the species. Among different crops, finger millet, cowpea, bottle gourd, turmeric, sesame, and sweet potato were found more suitable. Intercrops can be taken up to 5–6 years after bamboo planting. Shade-tolerant crops like ginger, turmeric, large cardamom, and orchard grasses can be taken in later stage. Intercrops can be taken for longer period by keeping large spacing (10 m) between lines and low spacing between plants and within lines of bamboo. Intercropping did not exhibit adverse effect on survival, growth, and yield of bamboo. In bamboo + horticulture tree model, fruit trees, plantation crops, and rubber can be grown as the main crop and bamboo as hedgerows or windbreak on the boundaries of orchard. Besides bamboo can be grown with multipurpose tree species on wastelands or degraded lands. Bamboo and tree species can be chosen depending on the climatic condition and soil types of the area. Such models can also be established either by converting seminaturally mixed stands of forests or planting new ones. The ratio of bamboo to trees may be 7:3 or 8:2 and 6:1:3 or 7:1:2 for bamboo, conifer, and broadleaf trees. Crops like soybean, black gram, sweet potato, turmeric, zinger, and vegetables can be intercropped up to 3 years after planting. The planting time for bamboo and trees should be determined based on the growth rate of the tree species involved.

In bamboo + medicinal plant model, medicinal plants that can tolerate shade and suit the soil and climatic conditions of the site are grown with bamboo. Species like *Curcuma domestica, Withania somnifera, Catharanthus roseus, Cassia angustifolia, Aloe vera, Plantago ovata, Chlorophytum* sp. *(safed musli)*, etc. can tolerate shade and are suitable to grow as intercrop due to their wide adaptability. Bamboo + crops + fish ponds are very common in plain or wetlands of Assam, West Bengal, Orissa, Kerala, and Western Ghats. In this model, one to three rows of culm-/shoot-producing sympodial (clump form) bamboos are planted on the banks of the pond. Soybean, rye, and other crops can be intercropped between bamboo clumps to form a complete food chain. Crops can be used as food of fish feed. Bottom mud from the ponds may be dug out in winter and used as fertilizer for bamboo clumps.

24.4 Establishment and Management of Bamboo Clumps

Irrigation during dry months of initial 2–3 years is essential for survival and establishment of bamboo plants. Inappropriate culm management jeopardizes bamboo resources worldwide. Even simple management practices such as the removal of dead and dying culms, intensive cultural operations like tillage during non-shoot season, and the use of cover and mulches to protect rhizome and new shoots can significantly increase the productivity of congested clumps and bamboo yield. The use of fertilizer doses 100:50:50 kg/ha N: P: K can significantly increase the culm growth and yield. Fertilization at the beginning of the shoot season and during growing season can result in threefold increase in bamboo productivity. Generally, 6–7 years may be required in the proper establishment of a stand of tall bamboo on marginal lands.

24.5 Bamboo Growth and Yield

Bamboo can be grown quickly and easily. The shoots of *Bambusa tulda* elongate at an average rate of 70 cm per day. *Dendrocalamus giganteus* also grow at the same rate and can reach 25 m in 3 months. It grows three times as fast and can be harvested four times as often as eucalyptus. It can be harvested annually without depletion of the parent plant and without causing harvesting damage or deterioration of the soil. Annual productivity of bamboo species in pure plantations varies between 10 and 20 t/ha/year and with different species a range of 28–298 ton per hectare. Bamboo yield from the agroforestry system will be low due to less number of clumps per hectare. Further, there is great variation in bamboo yield between species, site quality, and climate and due to management within same species. In India, bamboo yield per hectare is dismally low mainly due to poor management, unscientific and over-exploitation, and low investment in bamboo cultivation. Agroforestry-based cultivation can solve this problem.

An alternative however is to harvest in bundles. The culms are cut, collected, and bound with chords or strings into bundles. These bundles can be trimmed in length to any desired size. In bundles material deterioration is much more limited, and bundles can be kept in dry conditions. Unlike chips, they can be stored for several months before they are transported to industry, according to Prof. Liese, advisor of this part of the project. This aspect makes bamboo very interesting for wood industry. For both chipping and bundling, existing harvesting machines can be used without adaptations (CRA 2000).

Bamboo clump developed

Bamboo on gully collar

Cut bamboo after 10 years

24.6 Wood Industry

Today several bamboo products are on the market in Europe. Bamboo poles and sticks are used for a variety of applications, such as in tree nurseries and orchard as supports for trees, as sticks for young plants of tomato and begonia, and in niche markets like broom and exotic furniture (Liese 1985).

Technically bamboo can be used for the production of plywood, boards, and panels. This has been done in various countries in Asia like India, Malaysia, and China and also in America (Lee et al. 1996; Dagilis and Turcke 1996).

24.7 Income and Returns

Most of the bamboos are cultivated for culms and/or edible shoots. Bamboo-based agroforestry gives high profitability and better economic returns in comparison with sole cropping. Economically its cultivation ensures IRR of more than 25 % almost equal to eucalyptus, but returns are recurrent on annual basis after 5–6 years up to 30 years or more without recurring investments on plantations. Bamboo plantations for production of bamboo culms and edible shoots can provide Rs. 72,000 from the fourth year onward by giving Rs. 32,000 from bamboo culms and Rs. 50,000 from edible shoots. *Bambusa bambos-based* agroforestry in Tamil Nadu gave annual net returns of Rs. 13,300 when pigeon pea was intercropped in 1:1 rows at 3×3 m^2 spacing. In nonarable land of the middle Gangetic plains, bamboo with grass gave a B:C ratio of 1.76 with 55 % IRR. Bamboo and anjan grass-based silvipasture system for gullied lands of Mahi ravines in Gujarat gave a benefit-cost ratio of 1:1.85 for a 20-year period.

24.8 Improving Soil Properties

Bamboo-based agroforestry has positive effects on soil profile. Organic matter content, available nitrogen and phosphorus content increase under bamboo-based agroforestry systems. Soil pH and electrical conductivity also improve under its system. Experiments show that bamboo-based agroforestry system had higher organic carbon in the soil in comparison to other land-use systems. The exchangeable K, Ca, and Mg decrease with increasing distance from the bamboo clumps.

24.9 Controlling Soil Erosion

Bamboo plantations are known to conserve topsoil because of extensive rhizome and adventitious root system lying primarily in the top layers of soil. It develops a profuse root mat in the uppermost soil layer. The root and rhizome of different species of bamboo are mostly confined in 0–40 cm soil depths, which also make them important in securing the hydrological function of catchments and rivers. It stabilizes soils on slopes and riverbanks, preventing erosion and landslips. *D. strictus*-based agroforestry system with crops, i.e., wheat, soybean, mustard, niger, green gram, black gram, and pigeon pea, played a significant role in control of soil erosion and improvement of fertility of degraded agricultural lands in Madhya Pradesh.

24.10 Bamboo as Carbon Sink

Bamboo sequesters atmospheric carbon faster than many fast-growing trees. Its fast-growing canopy generates up to 35 % more oxygen than an equivalent stand of trees. There are reports that bamboos sequester as much as 17 metric tons/ha of CO_2 from the air, which makes it an extremely efficient replenisher of fresh air. India has about 122 million tons of bamboo growing stock. On an average, the bamboo biomass equals 22 t/ha. Bamboo is the most potent plant to tackle the challenge of climate change.

24.11 Large- and Mass-Scale Propagation of Bamboo Using Tissue Culture

Bamboos can be propagated by seeds; rhizome, culm cuttings, or branch cuttings; clump division or offset planting; and marcotting or layering. Some bamboos can be propagated only through one of these methods, while other species can be propagated through different methods. But the efficiency of these classical propagation methods varies greatly.

In the laboratory of Oprins Plant, a very efficient technology has been developed for the mass-scale propagation of different temperate and tropical bamboos based on axillary branching (Gielis and Oprins 1998).

24.12 Summary

Bamboo under agroforestry systems is suitable for intercropping, soil conservation, windbreak, riparian filter, and permaculture. They can be planted by farmers in homesteads and small-scale farmlands or managed as groves. Besides, bamboo can be grown with multipurpose tree species on wastelands or degraded lands. Bamboo-based agroforestry gives much higher returns than the other systems.

References

CRA (2000) Bamboo for Europe, Final report by B. Michot and Y. Schenkel. CRA, Gembloux, 92 pp

Dagilis TD, Turcke DJ (1996) An economical bamboo particle board. In: Proceedings of the Vth international bamboo workshop, Bali, Indonesia, Volume 3: engineering and utilization. INBAR, New Delhi, pp 156–173

Gielis J, Oprins J (1998) The strategic role of biotechnology in mass scale production of woody bamboos. In: El Bassam N, Behl RK, Prochnow B (eds) Sustainable agriculture for food, energy and industry. James and James (Science Publishers) Ltd., London, pp 167–171

Kleinhenz V, Midmore DJ (2002) Improved management practices for culinary bamboo shoots, RIRDC publication no. 02/035. Rural Industries Research and Development Corporation, Canberra

Lee AWC, Bai X, Peralta PN (1996) Physical and chemical properties of strandboard made from Moso bamboo. For Prod J 46(11/12):84–88

Liese W (1985) Bamboos-biology, sylvics, properties, utilization. Schriftenreihe der GTZ, no 180, 132 pp

Lin QY (1996) Cultivation techniques for *Dendrocalamopsis oldhamii*. In: Rao IVR, Sastry CB (eds) Bamboo, people and the environment, pp 50–55. Proceedings of the Vth International Bamboo Workshop and the IVth International Bamboo Congress, Ubud, Bali, Indonesia, 19–22 June 1995. International Network for Bamboo and Rattan, New Delhi

Pasture Species-Cenchrus

25

Abstract

Cenchrus ciliaris is one of the important grass species which survives under semiarid rainfed conditions. India has a live stock population of about 343 million, besides poultry, yet the production of milk and other milk products is the lowest in the world, the main reason for such is the lack of availability of sufficient quantity of quality fodder to the milking animals particularly in the rainfed areas. *Cenchrus* in one of the grass species which is found to provide palatable fodder to the tune of 80 q/ha on fresh weight basis in two to three cuttings in an intercropping system with fruit and forest trees.

25.1 Introduction

India has a live stock population of about 343 million, besides poultry, yet the production of milk and other milk products is the lowest in the world, the main reason for such is the lack of availability of sufficient quantity of quality fodder to the milking animals (Mann et al. 1989). The per capita availability of milk is 100 g/head/day against the minimum requirement of 201 g/head/day (Anon 1997). Intensive dairy farming has been the order of the day. Therefore, the fodder demand is throughout the year. The system of fodder production varies from region to region, place to place, and farmer to farmer. The availability of fodder in the summer season is the major problem in semiarid areas. Nowadays, the increasing cattle population is putting lot of pressure on the existing Gochar land; also the Gochar land is reducing owing to increased industrial and infrastructure land requirements. The balance sheet of animal green fodder, crop residue, and concentrates shows their availability to be 224.08 m tons, 231.05 m tons, and 31.6 m tons, respectively, against the requirement of 611.99 m tons, 609.79 m tons, and 95.4 m tons, respectively, with a deficit of 387.91 m tons of green fodder, 378.74 m tons of crop residue, and 63.8 m tons of concentrates. Thus, to meet the gap between the availability and requirment, there is a need to increase productivity of fodder per unit area. The only way to do it is to go for cropping system with components of fodder grasses, trees, or shrubs. The concept of agri-silvi, horti-silvi, agri-horti-silvi production systems will meet the requirement of food, fodder, and fuel for humans as well as the cattle.

S. Hiwale, *Sustainable Horticulture in Semiarid Dry Lands*,
DOI 10.1007/978-81-322-2244-6_25, © Springer India 2015

In semiarid dry lands, the various fodder crops like cowpea, jowar, guar, and moth are grown which are ready to harvest in 50–60 days and yield about 3–6 q/ha of green fodder. *Cenchrus ciliaris* is one of the important grass species which survives under semiarid rain fed conditions. With the identification of high-yielding varieties, biomass production can be increased up to 200–250 q/ha in semiarid rainfed conditions.

25.2 Varieties

Pusa giant, IGFRI-S- 3108, IGFRI-S- 3133, C-357, and C-358 are some of the varieties identified which are high yielding.

25.3 Package of Practices

1. Rainfall
   Rainfall in semiarid condition is 750 mm; however, there is lot of variability in respect to the number of rainy days in which the rainfall occurs.
2. Soil
   The grass can be grown on light to medium soils.
3. Sowing time
   The seed should be sown on the onset of monsoon preferably in the months of June–July.
4. Seed rate
   The seeds are very light in weight and hence seed rate of 5–6 kg/ha is sufficient. There is possibility of raising the crop by transplanting clumps from the already established plantation.
5. Spacing
   As the crop is raised under rainfed condition, a spacing of 50 × 30 cm is the best.
6. Manuring
   FYM at the rate of 5 t/ha along with N-45 kg, P-20 kg, K-20 kg/ha.
7. Weeding
   One or two weeding in the rainy season will help in the establishment of good stand.
8. Yield
   In the semiarid rainfed condition, 300–350 q/ha grass can be harvested per ha. There can be two to three cuttings at 50–60 days interval at 6–10 cm aboveground level. It is a perennial grass so once planted it can survive without much care.

25.4 Cropping System

With the decreasing land availability and increasing population to feed, raising the pure crop of grasses is not possible. Also in order to augment the meager income of the poor farmers under semiarid rainfed conditions, it is advisable to grow it as intercrop in a horti-silvi or agri-silvi or silvi- pasture cropping systems. Work carried out at CHES, Vejalpur, indicated that the grass is of perennial type and 2–3 cuttings can be taken up; apart from this the grass helped in the stabilization of gully collars and earthen check dams prepared to control soil and water losses in a horti-silvi pastoral system (Hiwale 2004).

The pasture species was incorporated in various fruit trees and silviculture based cropping system. The yield of *Cenchrus* in the interspaced started from 38.87 in the second year and increased to 94.1 q/ha on fresh weight basis which will help in meeting the fodder requirement of the animals reared by the farmer.

At CSWRI, Avikanagar, Rajasthan, India, the *Cenchrus* grass production has more than doubled due to theadoption of improved cultivation practices (23.6 q/ha of dry fodder from 10.6 q/ha). They also advised mixed sowing of legumes along with grasses to maintain soil fertility.

In another experiment on jujube-buffel grass cropping system in arid environment, Sharma and Vashistha (1985) reported that the system produced both dry biomass from the jujube and *Cenchrus* to the tune of 7.52–12.62 q/ha.

References

Anonymous (1997) Hand book of agriculture. ICAR, New Delhi

Hiwale SS (2004) Studies on agri-silvi pastoral system. In: Technical bulletin on Develop Agri –Silvi- Horti production system for marginal lands, CHES, Vejalpur, Gujarat, pp 18–21

Mann JS, Hasan B, Beniwal RK (1989) Adopt silviculture system for more fodder in arid and semiarid regions of Rajasthan. Indian Farm 39:33–35

Sharma SK, Vashistha BB (1985) Evaluation of jujube–buffel grass horti–pastoral system under arid environment. Ann Arid Zone 24(4):303–309

26 Pasture Species-Stylosanthes

Abstract

It is a fodder species rich in protein and therefore introduced in India from Australia. It is found to perform better in semiarid rainfed conditions. The production started in the second year seed sawing and one cutting was done with a fresh biomass production of 31.41 q/ha. increasing to 62.70 q/ha in the fourth year when intercropped with guava. Apart from this it helped in improvement of soil health by reducing the pH from 7.22 to 7.12 and organic matter content from 0.196 to 0.343 % over the years.

26.1 Intercropping

Intercropping of Stylosanthes in guava Hiwale (2004)

It is a legume which is rich in protein content. It can be mixed with green grass fodder in a balanced manner to improve protein uptake of the cattle's. It is originated from Australia.

S. Hiwale, *Sustainable Horticulture in Semiarid Dry Lands*,
DOI 10.1007/978-81-322-2244-6_26, © Springer India 2015

26.2 Varieties

There are two species of stylo, viz. (1) *Stylosanthes hamata* (2) *Stylosanthes scabra*.

26.3 Climate

Hot climate favors the growth of Stylosanthes. Areas receiving rainfall from 250 to 900 mm.

26.4 Soil

Grey well drained medium textured. The plantation of stylo is raised by direct seed sowing in the month of June–July (after the break of the monsoon).

26.5 Land Preparation

After clearing the area of bushes and rank vegetation and with the advent of monsoon, the first land preparation is carried out by mould board plough followed by further two operations using desi-plough. One light ploughing followed by one or two harrowing may be enough. As the size of seeds is small, a fine tilth of soil is necessary. Therefore, the clods should be broken up.

26.6 Fertilizer

Basal application of 5 t of farm yard manure. This is followed by application of 10 kg N/ha and 30 kg P_2O_5/ha. Fifty kilograms of urea and 300 kg of super phosphate per hectare should be mixed with soil before the last operation of land preparation.

26.7 Time of Sowing

July after first heavy shower of rain.

26.8 Seed Rate

10 kg/ha.

26.9 Spacing

The legume seeds are sown in lines spaced 50 cm apart. The seeds are lightly covered with soil. Depth of sowing can be about 1.25 cm. The seeds are dropped in lines by hand. The seeds are mixed with sand and farmyard manure before sowing. Covering of seed is not necessary. In case the seed drills are used, the depth of sowing should not be more than 1 cm.

26.10 Weeding

Stylo species cannot stand competition from weeds during the emergence of seedlings and their establishment. Two weedlings will be necessary.

26.11 Irrigation

Stylosanthes species normally do not need irrigation, but, if the sowing is done late or when the area concerned is dry, one or two irrigations in the season may enhance the flowering, pollination, seed setting and seed outturn.

26.12 Mixed Cropping

Using a seed rate of 6 kg/ha, legume seeds are sown at a spacing of 1.5 m between every two rows of grass. The legume seeds are sown after first heavy shower of rain at 1.5 m interval. Only after germination of the legume seeds, the grass seedlings are planted in between the legumes at a spacing of 50 cm.

Table 26.1 Fodder production in guava based intercropping

Crop	Second year		Third year		Fourth year		Mean	
	Fresh wt. kg	Dry wt. kg	Fresh wt. kg	Dry wt. kg	Fresh wt. kg	Dry wt. kg	Fresh wt. kg	Dry wt. kg
Stylosanthes hamata	31.41	9.67	51.02	12.66	62.70	14.70	48.26	12.34

Raturi and Hiwale (1993)

26.13 Inoculation with Bacterial Culture

Legume

culture of rhizobium bacteria is essential for assured establishment of legume. This is not easily available in India. However, even if one small tin (e.g. cigarette tins) of soil where the legume has already been grown earlier, as in this institutes, and used with legume seed before sowing it would be beneficial.

26.14 Time of Cutting

In the first year cutting should not be taken. Seed collection should be done. In subsequent years only one cut can be taken as stylo is an annual. Therefore it is advisable to maintain two plots for stylo, one for seed collection and the other for cutting only in September–October. Light grazing or cutting at the height of 15 cm, when the plants have attained a height of 30 cm, helps in profuse branching and thus enhanced seed yield (Table 26.1).

26.15 Green Fodder Yield

200 q/ha

References

Hiwale SS (2004) Technical bulletin on "Develop sustainable Agri-silvi-Horti Production system for Marginal lands under arid conditions". Central Horticultural Experiment Station, Vejalpur, 389340

Raturi GB, Hiwale SS (1993) Horti-Silvi-Pastoral system for increased productivity of marginal and degraded lands under rainfed conditions. Adv Hortic For 3:1993

Part IV

Alternate Land Use Systems Semiarid Rain Fed Areas (Horti–Agri, Horti–Silvi–Pastoral, Horti–Silvi)

Alternate Land-Use Systems or Sustainable Development

27

Abstract

The work was carried out to develop sustainable agri-silvi-horti production system of marginal lands. Aonla-based cropping system was found to be best suited for the rainfed areas of western India with maximum income, production, income generation, and improvement in soil fertility status.

Aonla with arable crops like *Sesamum* and maize are found to be the highest in production, LER (1.71), net return (Rs. 68,715/-), and B: C ratio of 7.92, respectively. Aonla + maize was found to be the most suitable for the semiarid area as maize is a traditional crop of the farmers of the region and also provides fodder for their cattle. Therefore, though the income is low, the technology is well adopted by the farmers. Adaptability of aonla is clearly established as despite the prevalent drought-like situation yield up to 55 q/ha. Was obtained followed by custard apple-based cropping system. There was overall improvement in soil health.

Increased growth and biomass production recorded in Neem + fodder jowar was the best. The system also resulted in improvement in soil fertility status. Agri-horti production system was found to be highly remunerative and the income started from 2 to 3 years of plantation.

Bioassay studies carried out with four tree species and nine test crops revealed that generally Neem was found to have a suppressing effect on most of the growth parameters, whereas fruit tree species of custard apple and aonla were found to have beneficial effect.

Studies on leaf litter fall and decomposition revealed that aonla produced the highest leaf litter than all the fruit tree species, whereas in silvicultural species, subabul was found to produce the highest leaf litter. Large amounts of nutrients were recycled by subabul followed by aonla.

Studies on soil moisture content carried out with the help of TDR meter at CHES, Godhra, indicated that during crop growth period, the moisture content in subabul soils was least and was maximum in custard apple moisture content in different system (5–17 %).

Staggered contour trench planting was found to be the best for reducing runoff and soil loss effectively. Runoff and soil loss was higher in custard

S. Hiwale, *Sustainable Horticulture in Semiarid Dry Lands*,
DOI 10.1007/978-81-322-2244-6_27, © Springer India 2015

apple and Neem; comparatively less runoff and soil loss was recorded in aonla.

Incidental PAR and higher LAI of the cropping system resulted in higher dry matter production and conversion coefficient. Maximum LAI, dry matter production, and conversion coefficient were recorded in subabul indicating that silvicultural species have more carbon accumulation than fruit tree species, where fruits act as vigorous sink for carbohydrates.

Root distribution studies carried out by excavation method revealed that maximum root biomass and aboveground biomass was recorded in subabul. Aonla produced maximum root depth and in Neem maximum horizontal root spread. Maximum root to crown spread ratio was recorded in subabul. Custard apple has very small but efficient root system both by weight and by spread. All the tree species have root system, which is within the canopy of the tree and hence will not interfere with the intercrops, and thus the competition between the species for water and nutrients reduced.

Studies on quality of charcoal produced by tree species revealed that the wood specific gravity was maximum in Neem. Percent charcoal production was highest in custard apple. Charcoal produced from aonla took maximum time to burn. Maximum ash was produced by 1 kg aonla charcoal. Though maximum percent charcoal was produced by custard apple, the amount of charcoal produced is very less compared to other tree species.

Sustainable resource management through horticulture-based farming systems attempted to demonstrate an ideal model for production utilization of resource-poor situations in semiarid areas of western India. Rainwater stored and recharged in down profile of the gullies is being utilized for life-saving irrigation at critical crop growth stages and also for establishment of new plantations. The tree component adds leaf litter to the soil, helping in recycling of the nutrients back to the soil. Apart from this, trees helped in reducing runoff losses and thereby soil erosion. The system has given higher economic return as compared to traditional system of maize + pigeon pea.

27.1 Introduction

Exponential increase in human and livestock population in India has put tremendous pressure on natural resources, owing to not only degradation of principal resources like land, water, and vegetation but also reduced per capita availability of land. In some places, the extent of land degradation has crossed such critical limits that it is difficult to bring it back to its original state. The unscientific cultivation of annual crops in rainfed areas, as usually is the case, accelerates runoff, soil loss, as well as other degradation processes causing fragile ecosystem. In India about 53.34 million tons of top soil (16.4 t/ha) containing 5.4 million tons of plant nutrients is lost annually due to agriculture and associated activities (Dhruva Narayan and Ram Babu 1983). The prime agricultural lands continue to be diverted to nonagricultural uses like industrialization and urbanization with the result that marginal rainfed lands are brought under cultivation with least attention on resource conservation. Out of 143 million hectares of arable land in India, nearly 74 % is rainfed, which contributes to about 42 % of the total food production. However, soil moisture stress due to vagaries of weather, moderate to severe soil erosion, and poor soil fertility are the major limiting factors of low productivity in rainfed areas.

Though India has made significant stride in food production in the recent past, major emphasis was given on irrigated and resource-rich situations. However, in rainfed areas, the marginal farmers are still practicing traditional way of farming and struggling below the poverty line. Therefore, this trend needs to be discouraged. Virtually, a sizable area (171 m ha) in our country is categorized as degraded lands. Such lands are subjected to problem of water and wind erosion, salinity, alkalinity, acidity, waterlogging, gullied and ravine lands, shifting cultivation, etc. and are either lying vacant or underutilized. Although the lands have their own biophysical farming constraints, they can be utilized efficiently on sustained basis for fulfilling the basic needs of food, fruit, fuel, fodder, fiber, fertilizer, etc. through horticulture-based farming systems and by adopting practically feasible, economically viable, and eco-friendly technologies developed in recent past. Considering the above facts, horticulture-based composite farming systems were evaluated from 1985 to 2000 to establish a viable model for rehabilitation and utilization of semiarid rainfed areas at Vejalpur, Godhra, Gujarat.

Fruit trees play a vital role in crop diversification. Their inclusion in the system reduces the risk of crop failures. Apart from this, the fruit trees are known to increase the income of the farmer substantially over traditional monocropping. Fruit cultivation has so far been associated with fertile soil with assured irrigation, but presently the irrigated fertile land is already being used for meeting the food requirement of an ever-increasing population and industries. The marginal and degraded lands in arid and semiarid region of the country with practically no irrigation facilities are to be used for increasing fruit production with simultaneous need to meet the basic demand for food, fodder, and fuel. Hence, no time should be lost in developing highly productive, self-sustainable farming systems fully adapted to the precarious ecology of arid regions.

In the broadest sense, the term agri-horti-silvipastoral system encompasses any and all techniques that attempt to establish or maintain forests, horticultural crops, forage trees, and pasture grasses on the same piece of land. The science and practice of agri-horti-silvipastoral system aim at systematically developing land-use system and practices where the positive interaction between trees and crops is maximized. This seeks to achieve more productive, sustainable, and diversified output from the land than is possible with conventional monocropping system. In this system, the fruits and vegetable crops provide seasonal revenue, while forest trees are managed at 5–10 years' rotation giving extra returns of timber, fuelwood, and fodder. The approach aims at optimizing the use of land, water, and vegetation in an integrated way and thus helps to alleviate drought, moderate floods, prevent soil erosion, improve water availability, and increase production of fruits, vegetables, fuels, and fodder on sustained basis.

27.2 Alternate Land-Use Systems or Sustainable Development

Nearly 60 % of the land is under rainfed condition, which contributes only 40 % to the national food basket. Crop production in dry areas is merely dependent on rainfall which is seasonal, erratic, and highly variable in space and time. The evaporation in dry areas is high, leading to drought. The occurrence of drought is inevitable in dry land areas. Because of it, crop production under the traditional land-use system in dry land areas suffers a lot. The traditional pattern of land use is the basic cause for poor economic condition of the farmers in dry land regions.

The disastrous effect of drought cannot be eliminated but the losses can be minimized to a reasonable extent by application of scientific methods of management, but these management options are short lived and call for permanent solution; the only solution lies in judicious management of limited resources in an integrated manner to ensure adequate supply of food, fodder, and fuel. Therefore, alternate land-use systems suit to such situation as a drought-proofing measure and offers great scope to stabilize and increase the productivity of farm as a whole.

27.2.1 Advantages of Alternate Land-Use System

Some of the advantages of alternate land-use system are as follows:

These systems utilize the off-season rainfall for productive purpose which otherwise go waste. They help in reducing the pressure on forest cover by supplying timber, fruits, fuel, fodder, etc. and thus help in conservation and development. The top feed tree species grown as component of the system will provide much-needed fodder and also serve as fodder bank in times of drought and during the lean months. When the fodder shortage looms large.

These systems are helpful in sustaining and conserving land productivity. Many leguminous tree species fix free atmospheric nitrogen and the same will be returned back to the soil in the form of fallen leaves or green leaf manure, thus helping the farmer in improving the soil chemical as well as its physical properties.

These systems can be adopted in extreme soil conditions of soil acidity and alkalinity to reduce the effect on the performance of the crop and restore the soil productivity. These systems are also helpful in conserving the soil as well as nutrients by reducing erosion and runoff to a great extent.

The trees and shrubs can withstand drought better, because of their capability for tapping water and nutrients present in the deeper layers of the soil which are beyond the reach of arable crops. These systems are capable of meeting the demands of several raw materials and forest-based industries like paper mills, wood industries and sport goods, etc. They also increase employment opportunities for the rural poor and the landless during the off-season as most of the activities are labor intensive.

27.2.2 Resource Management in Alternate Land-Use System

In the broadest sense, the term horti-silvipastoral encompass any and all the techniques that attempt to establish or maintain forest, horticultural crops, forage trees, and pasture grasses on the same piece of land. The science and practice of horti-silvipastoral system aim at systematically developing land-use system and practices. This seeks to achieve more productive, sustainable, and diversified output from the land than is possible with conventional cropping system. The approach aims at optimizing the use of land, water, and vegetation in an integrated manner and thus helps alleviate drought, moderate floods, prevent soil erosion, improve water availability, and increase production of fruits, vegetables, fuel, and fodder on sustained basis.

The high demand for fodder and fuelwood have stripped off much of the protective cover of the land in the catchment areas of rivers and hills, bringing down the forest cover to just 11 % from the required 33 %.

In India, out of the 148 million hectare land, about 50 % of the area has been classified as degraded land; the challenge is to make ecological resources more productive and their exploitation less hazardous. About 97.1 million hectare of drought-prone area is in semiarid tropics. Considering the fact that rainfed areas comprise more than half (73.7 %) of the scarce land resources of the country which produce about one third of the food, thus, rainfed agriculture should no longer remain unstable and low yielding; greater efforts are therefore needed to increase production in rainfed areas.

According to FAO (1978) world soil resource report, in dry lands, rainfall is less than half the potential evapotranspiration. In such areas, the prime concern is moisture conservation. Thus, the strategy for development needs to be directed toward more efficient use of natural finite resources like land and water.

27.3 Resource Constraints and Its Management

27.3.1 Constraints in Attaining Sustainability

Despite the desire to increase the production potential of the land, farmers are not able to attain full productivity/unit area owing to their socio-economic status, technology availability, transfer and acceptance, soil/land constraints, and climate change.

27.3.2 Criterion for Evaluation of Sustainability

Risk assessment, production assessment, maintenance of production over time, stability of type, impact of farming system, degradation, economics of the system profitability, and environmental soundness.

27.3.3 Soil

Major constraints are soil health, soil erosion, and soil salinity.

27.3.4 Overcoming Constraints

Soil health can be improved by addition of organic matter, atmospheric nitrogen fixation, and leaf litter fall in the cropping system. Soil erosion control can be achieved through planting methods (staggered contour trench planting, alley or hedgerow planting, shelterbelts, and soil cover with mulches and crop residue.). Soil salinity can be overcome to some extent by growing tolerant crops, soil amelioration, and breeding strains resistant to salinity.

27.3.5 Water

Major constraints are limited precipitation, erratic distribution, and high evaporative demand.

27.3.6 Overcoming Constraints

Water harvesting, choice of suitable crop, mulching, reduced deep percolation, use of anti-transpirants, and establishment of orchards through in situ budding/grafting.

27.3.7 Climate

Extremes of temperatures, high wind velocity, and high vapor pressure deficit.

27.3.8 Overcoming Constraints

Climate resilience agriculture, alternate cropping systems, raising shelterbelts, mulching and grass cover, raising cover crops, and for low temperature damage control.

27.4 Major Alternate Land-Use Systems

27.4.1 Agri-horticultural System

Agri-horticultural system essentially consists of a tree component, which should be a fruit tree and field crops. This system mainly focuses on higher income per unit area and suitable for better soil groups. The enterprise requires high initial investment and recurring expenditure. The tree crops give fruits which are perishable in nature and need quick marketing facility as well as grading and value addition for better returns. This system also provides employment opportunities during the off-season.

Hardy fruit trees like ber (*Zizyphus mauritiana*), guava (*Psidium guajava*), phalsa (*Grewia asiatica*), and pomegranate (*Punica granatum*) suit this system as tree component; as most of the fruit trees are woody perennials, they take minimum number of years to bear fruits. Judicious management of trees and understory crops like cereals and pulses during the initial stages could be profitable and helpful to the farmers to meet the working expenses when there is no income from the tree component.

27.4.2 Agri-silvicultural System

Alley cropping is a typical example of agri-silviculture in which crops are grown in alleys formed by the hedgerows of trees or shrubs; the main objectives of alley cropping are to get green and palatable fodder from hedgerows in a dry season and produce reasonable quantum of grain and stover from the alley during cropping season. A remarkable feature of the system is that lopping (cutting back the hedgerows to a convenient height of 90–120 cm) is done at the time of sowing of field crops to avoid shedding effect. *Leucaena* species is the best suitable hedgerow component, whereas field crops like sorghum, bajara, and red gram can be grown as field crops in the alley formed by the hedge.

27.4.3 Silvipastoral System

Silvipastoral system is suitable for marginal and degraded land and most suitable where fodder shortage looms large. The system invariably consists of a top feed tree species carrying grasses and legumes as understory crops.

Dry land farmers having larger holdings and keeping part of their farm land for a longer period for one reason or another can go for the system which could provide both fodder and fuel. The trees like subabul (*Leucaena leucocephala*), Shisham (*Dalbergia sissoo*), gum arabica (*Asiatica senegal*), Khejri (*Prosopis cineraria*), Babul (*Asiatica nilotica*), etc. can be grown as tree component, whereas grasses like buffel grass (*Cenchrus ciliaris*), Rhodes grass (*Cloris gayana*), Sudangrass (*Sorghum sudanese*), Guinea grass (*Panicum maximum*), deenanath grass (*Pennisetum pedicellatum*), and legumes like stylo (*Stylosanthes hamata*) and cowpea (*Vigna uguiculata*) can be grown as pasture component.

For better soils of the dry land region, the food-based agri-horticulture systems are suitable for increasing the production from the unit area and unit time, whereas for the poor soils, the fodder and fuel-based systems like silvipastoral are good for stabilizing production. The alternate land-use systems are efficient in utilizing the limited resources of the dry land and at the same time conserving the same. Since these systems are based on sustainable farm management practices, there is no doubt that they will go a long way to enhance and stabilize the productivity of the region as a whole. The sustainable development of dry land region could only be achieved by substituting with the alternate land-use systems that are technically sound, practically feasible, economically viable, and socially acceptable.

27.5 Agri-horti Production System on Marginal Lands

At CHES, Vejalpur, under semiarid rainfed conditions, various fruit crop-based cropping system, viz., aonla, ber, custard apple, guava, and mango, were developed. Similarly agri-silvipastoral systems with Neem and *Leucaena* were evaluated. The results are presented below

27.5.1 Aonla-Based Agri-horti Production System on Marginal Lands

Aonla cv. NA-7 (7–9 years old) under rainfed condition with 10×10 m spacing was intercropped with various arable crops successfully. All the growth parameters showed increase despite prevalent drought-like situation in the area. Evapotranspiration surpassed the rainfall, creating a shortage of available moisture required for plant growth. Aonla proved to be hardy and withstood vagaries of monsoon better than any other fruit tree as can be seen from growth and yield data (Tables 27.1 and 27.2).

Data on yield in respect to main crop, though found to be influenced by drought condition. It

Table 27.1 Morphological parameters in agri-horti production system (aonla)

Aonla with intercrop	Plant height (m)	Stock diameter (cm)	Scion diameter (cm)	Plant spread (m) (N–S)	Plant spread (m) (E–W)
Aonla + okra	4.84	21.88	18.31	4.80	4.50
Aonla + maize	5.11	21.14	17.69	4.73	4.67
Aonla + green gram	4.92	21.04	18.71	4.81	4.53
Aonla + *Sesamum*	5.02	23.32	18.82	4.88	4.77
Aonla + pigeon pea	5.04	33.36	18.65	4.52	4.67
Aonla + moth bean	4.83	21.34	19.68	4.69	4.56

Hiwale (2004)

Table 27.2 Yield in agri-horti production system (aonla)

Aonla with intercrop	Main crop	Intercrop	Pure crop
Aonla	69.39		
Aonla + okra	69.75	20.28	10.80
Aonla + maize	69.72	12.43	16.84
Aonla + green gram	68.69	2.30	6.55
Aonla + *Sesamum*	68.82	3.01	4.02
Aonla + pigeon pea	68.90	0.83	1.57
Aonla + moth bean	68.89	1.15	1.83

Hiwale (2004)

Table 27.3 Biomass productions of intercrops and land equivalent ratio (LER)

Aonla with inter crops	Land equivalent ratio (LER)	Fresh biomass production (q/ha)	
		Intercrops	Pure crops
Aonla + okra	1.71	18.18	74.74
Aonla + maize	1.66	173.47	229.72
Aonla + green gram	1.65	48.10	63.14
Aonla + *Sesamum*	1.58	175.78	180.66
Aonla + pigeon pea	1.47	26.19	54.72
Aonla + moth bean	1.55	29.14	58.06

improved in the third year owing to proper distribution of rainfall. Yield reduction was to the tune of 27.49–29.77 % in the second year; however, in the third year, the yield increased by more than 49 % over the previous year, thus discounting the yield reduction.

Intercropping aonla with okra was found to be more productive than all other intercrops. The reduction in yield was compensated by higher yield of main crops, giving higher net returns compared to sole cropping.

The system resulted in maximum utilization of land area, making maximum use of rainwater. The land equivalent ratio was least when aonla was intercropped with pigeon pea and highest with aonla + maize system. Biomass production, though increased in most of the cases, showed variation over the years. Among all the intercrops, maximum biomass was produced in maize (Table 27.3), thus providing valuable fodder for cattle.

Economic analysis of aonla-based cropping system for the past 3 years revealed that in the second year maximum net return of Rs. 75,000/- and Rs. 79,050/- was obtained in aonla + okra

Table 27.4 Economic analysis of cropping systems

Aonla with intercrops	Gross return of main crop (Rs.)	Gross return of Inter crop (Rs.)	Net return (Rs.)	B:C ratio
Aonla + okra	69,421	11,580	68,715	6.55
Aonla + maize	69,720	6,631	66,955	7.58
Aonla + green gram	68,687	3,578	63,120	7.12
Aonla + *Sesamum*	68,823	7,781	67,618	7.92
Aonla + pigeon pea	68,900	1,333	61,798	7.32
Aonla + moth bean	68,890	1,808	62,275	7.17

Hiwale (2004)

cropping system, proving it to be superior to all followed by aonla + maize (Table 27.4). Maize is the main cereal crop of the region and maize straw is an important fodder crop for the cattle. Hence, the system is recommended for adoption as it fulfills all the three requirements of food, fodder, and fuel. Benefit-cost ratio was maximum in all the 3 years in aonla + *Sesamum*-based cropping system, which was due to low cost of cultivation required for *Sesamum* cultivation.

27.5.2 Custard Apple-Based Agri-horti Production System on Marginal Lands

Custard apple (5–7 years old) planted at 5 × 5 m spacing was intercropped with various arable crops. The results indicated that it is possible to increase the sustainability of the fruit tree-based land-use system even under rainfed condition. Vegetative growth parameters showed an increasing trend despite prevalent drought conditions. Plant height increased from 2.5 to 3.17 m. Similarly stem diameter increased from 57.80 to 72.45 mm and plant spread from 1.91 to 2.55 m (Table 27.5).

Yield of main crop increased over the years from 18.93 q/ha to 34.50 q/ha, which indicated that monsoon failure does not have influence on growth and yield of custard apple. The yield increase was three times to the tune of 106.95 % over the years, thus proving it to be hardy tree

Table 27.5 Morphological parameters of custard apple in agri-horti production system

Custard apple with intercrops	Plant height (m)	Stock diameter (mm)	Plant spread in m (N–S)	Plant spread in m (E–W)
C. apple + okra	2.89	66.35	2.39	2.33
C. apple + maize	2.77	67.81	2.26	2.35
C. apple + green gram	2.71	66.36	2.20	2.23
C. apple + *Sesamum*	2.69	64.30	2.15	2.19
C. apple + pigeon pea	2.75	65.04	2.35	2.40
C. apple + moth bean	2.73	66.88	2.12	2.29

Table 27.6 Yield/ha

Custard apple with intercrops	Main crop	Intercrop	Pure crop
C. apple + okra	26.07	22.26	31.67
C. apple + maize	23.98	9.05	15.22
C. apple + green gram	22.85	1.84	3.06
C. apple + *Sesamum*	23.69	2.68	3.79
C. apple + pigeon pea	23.55	1.37	2.01
C. apple + moth bean	24.51	2.14	2.99

Hiwale (2004)

Table 27.7 Economic analysis of cropping systems – custard apple

Custard apple with intercrops	Gross return of main crop (Rs.)	Gross return of intercrop (Rs.)	Net return (Rs.)	B:C ratio
C. apple + okra	12,154	10,839	13,964	1.54
C. apple + maize	11,671	4,192	9,101	1.23
C. apple + green gram	12,098	2,645	7,276	0.97
C. apple + *Sesamum*	12,032	6,265	11,134	1.53
C. apple + pigeon pea	12,515	2,047	7,186	1.02
C. apple + moth bean	11,719	3,800	8,245	1.12

Hiwale (2004)

species. Yield increase in the second year in respect to *Sesamum*, moth bean, and green gram might be due to their drought hardy nature and owing to the absence of rainfall during flowering. But the yield of *Sesamum*, green gram, and moth bean decreased in the third year. Maximum yield per hectare in pure crops and intercrops was recorded in okra, proving its superiority to other intercrops. Thus, intercropping custard apple orchard with okra was found to be more productive than all the other intercrops tried (Table 27.6).

Economic analysis of custard apple-based cropping system indicated that maximum net return of Rs. 13,964/- can be obtained in custard apple + okra with a benefit-cost ratio of 1:1.54 averaged over the past 3 years followed by custard apple + *Sesamum* (1:1.53) (Table 27.7).

Intercropping in Custard apple

Table 27.8 Biomass production in agri-horti production system (custard apple)

	Fresh biomass of intercrops (q/ha)				Dry biomass of intercrops (q/ha)			
Name of crop	2001	2002	2003	Mean	2001	2002	2003	Mean
C. apple + okra	50.55	60.90	49.80	53.75	21.67	27.81	15.50	21.66
C. apple + maize	158.8	140.39	169.7	156.3	33.98	28.32	40.34	34.33
C. apple + green gram	45.85	53.88	48.51	49.41	24.00	30.24	18.7	24.47
C. apple + *Sesamum*	136.25	216.88	137.32	163.48	30.00	14.68	43.81	29.24
C. apple + pigeon pea	58.10	58.88	39.43	52.14	13.05	11.44	13.6	12.52
C. apple + moth bean	46.10	49.12	29.73	41.65	10.05	10.33	90.65	9.99

Hiwale (2004)

Table 27.9 Economics of mango-based agri-horti production systems

	Yield q/ha				
Crop combinations	Main crop	Intercrop	Gross output (Rs./ha)	Gross input (Rs./ha)	Net return (Rs./ha)
Mean of three years					
Mango	15.40	–	15,400.00	1,666.67	13,733.00
Mango + cowpea	15.72	47.67	30,034.33	5,000.00	25,034.00
Mango + okra	16.06	67.46	36,303.67	5,000.00	31,303.00
Mango + *Dolichos*	15.93	23.74	23,055.33	4,666.67	18,388.67
Mango + cluster bean	16.36	23.33	19,221.33	4,666.67	14,554.67
Mango + chilli	15.73	21.07	22,054.33	5,000.00	17,054.33
Mango + brinjal	16.00	60.64	23,836.00	4,666.67	19,369.58
Maize + pigeon pea[a]	05.96	2.57	06,833.33	1,500.00	05,333.33

[a]Traditional system (Hiwale et al. 2007)

Maximum mean dry biomass was produced by custard apple + maize-based cropping system (34.33 q/ha) and minimum by custard apple + moth bean-based cropping system (9.99 q/ha). Biomass production, though increased in most cases, showed variation over the years (Table 27.8).

Growth and biomass of various intercrops grown in custard apple orchard were carried out at different crop growth stages. The results indicated that plant height was highest in *Sesamum* and least in cowpea. Root length of different intercrops measured denoted that growth of the roots under rainfed conditions was very less in all the species (0–20 cm depth only). As such there was no competition between the intercrops and tree crops for moisture or nutrients. Fresh and dry weight of different intercrops recorded revealed that the initial buildup was slow. The growth improved with the onset of monsoon resulting in increased conversion coefficient. Moisture content on percent basis worked out at different stages of growth of intercrops was between 80 and 90 % except in pigeon pea where it dropped to 50 % and slowly went up to 70 %.

27.5.3 Mango-Based Cropping System

In the present investigation, mango was planted at 10 × 10 m spacing and six vegetables, i.e., cowpea, okra, *Dolichos*, cluster bean, chilli, and brinjal, and two grasses, i.e., *Cenchrus* and *Stylosanthes*, were grown in the interspaces of mango (Table 27.9). These combinations were compared with the traditional system of the farmers, i.e., maize + pigeon pea. Data presented in Table 27.9 indicate that the yield of mango increased with increasing age of trees but there was no definite trends in the yield of intercrops. The yield of mango was better with intercropping (15.72–16.6 q/ha) than sole mango orcharding (15.4 q/ha). Among intercrops, okra had given the highest yield (67.46 q/ha) followed by brinjal and cowpea. The grass yield was as high as 81.48 q/ha in *Stylosanthes* and 72.50 q/ha in *Cenchrus*. These combinations are compared with traditional system, i.e., maize + pigeon pea, the yield of traditional crops was very low even after including the yield of both crops (8.53 q/ha). However, the yield is not the only criterion of evaluating a system; therefore, economics of the system was also calculated based on existing local market prices. The 3-year average data revealed that mango + okra has given maximum return (Rs. 31,303/ha) followed by mango + cowpea (Rs. 25,034/ha). Growing of grasses as intercrop has given good yield but economically they were far below than these two combinations. On the other hand, the traditional system has given the lowest return (Rs. 5,333/ha), which was markedly less than even the sole mango orcharding.

27.5.4 Guava-Based Cropping System

The guava cultivar Allahabad Safeda was planted at closer spacing (5 × 5 m). The plant has less gestation period and an average 28.93 q/ha fruit yield was recorded (Table 27.10). The mean fruit yield after three consecutive harvests under sole plantation was 28.93 q/ha which did not show much variation with intercropping. This indicates that the intercrops do not have any adverse effect on the yield of the main crop. In general, the yield of intercrops reduced during the 4th year due to poor rainfall distribution during the crop season. Among different vegetables, brinjal ranked first (63.57 q/ha) followed by okra (60.80 q/ha). Though, on the yield basis, the guava-based agri-horti production system was better than mango-based systems due of accommodation of more number of plants per ha and also less gestation period, the economics was better in mango-based systems due to high selling price of mango fruits. Overall guava + okra had given highest return (Rs. 27,179/ha) followed by guava + cowpea (Rs. 23,493/ha). The traditional system again gave the lowest return.

27.5.5 Ber-Based Cropping System

The ber is another promising crop of the region. The 3-year average data (Table 27.11) revealed that sole plantation of ber has given 79.13 q/ha fruit yield and there was no much variation in yield of ber when intercropped with either vegetables or grasses. Among intercropping of vegetables, brinjal ranked first (63.04 q/ha) followed by okra (54.91 q/ha) and cowpea (41.18 q/ha). The yield of grasses was reduced markedly in the 5th year of planting and the overall performance of grass yield with ber was poor than other perennial components. The annual return of ber + okra ranked first (Rs. 49,708.33 q/ha) followed by ber + cowpea (Rs. 45,722.66 q/ha) and ber + brinjal (Rs. 41,479.00 q/ha). The traditional systems of farming again rank lowest and given about nine times less than return as compared to orcharding of ber and about 12 times less return than the best combination of ber + okra. Thereafter, the yield of intercrops decreased due to shading effect of ber trees.

Table 27.10 Economics of guava-based agri-horti production system

Crop combinations	Yield q/ha Main crop	Yield q/ha Intercrop	Gross output (Rs./ha)	Gross input (Rs./ha)	Net return (Rs./ha)
Guava	28.93	–	14,466.66	2,333.33	12,133.33
Guava + cowpea	28.7	47.81	28,826.33	5,333.33	23,493.00
Guava + okra	28.56	60.80	32,524.33	5,333.33	27,179.00
Guava + *Dolichos*	28.76	24.23	21,654.33	5,333.33	16,321.00
Guava + cluster bean	29.16	22.96	21,473.33	5,266.66	16,306.66
Guava + chilli	28.50	18.45	19,786.00	5,333.33	14,452.66
Guava + brinjal	28.73	63.57	22,312.91	5,333.33	16,979.58
Maize + pigeon pea[a]	6.16	2.43	6,733.33	1,500.00	5,233.33

[a]Traditional system (Hiwale et al. 2007)

Table 27.11 Economics of ber-based agri-horti production system

Crop combinations	Yield q/ha Main crop	Intercrop	Gross output (Rs./ha)	Gross input (Rs./ha)	Net return (Rs./ha)
Ber	79.13	–	39,566.66	3,000.00	36,566.66
Ber + cowpea	78.73	41.18	51,722.66	6,000.00	45,722.66
Ber + okra	78.46	54.91	55,708.33	6,000.00	49,708.33
Ber + *Dolichos*	79.00	23.98	46,695.00	6,000.00	40,695.00
Ber + cluster bean	79.43	28.08	48,141.66	6,000.00	42,141.66
Ber + *Cenchrus*	79.20	58.42	42,521.00	4,000.00	38,521.00
Ber + *Stylosanthes*	78.50	67.71	42,605.00	4,000.00	38,635.00
Ber + chilli	79.06	22.44	46,266.33	6,000.00	40,266.33
Ber + brinjal	79.20	63.04	47,480.00	6,000.00	41,479.58
Maize + pigeon pea[a]	5.80	2.20	6,183.33	1,500.00	4,683.33

[a]Traditional system (Hiwale et al. 2007)

27.6 Agri-silvi Production System on Marginal Lands

27.6.1 Neem Based Agri-silvi Production System on Marginal Lands

Five- to six-year-old Neem (10 × 5 m) were intercropped with various intercrops to maximize production potential of marginal lands. The plant height increased from 5.70 m (5th year) to 6.38 m (6th year) indicating an increase of 11.38 %. The effect of different intercrops on the growth parameters of the main crop showed it varied from 5.66 m in the case of *Stylosanthes* to 6.31 m in the case of fodder jowar. However, differences were nonsignificant. The stem diameter increased from 17.29 cm in the fourth year to 20.98 cm in the fifth year indicating an increase of 21.34 %. The effect of different intercrops on the stem diameter of the main crop revealed that it varied from 17.92 cm (*Stylosanthes*) to 20.71 cm (fodder jowar). The N–S plant spread ranged from 3.80 m in the fourth year to 4.53 m in the fifth year showing an increase of 21.34 %. It was maximum (4.55 m) in moth bean and minimum (3.93 m) in *Stylosanthes*. The E–W plant spread increased from 3.67 m (5th year) to 4.67 m (6th year) indicating an increase of 26.43 %. The effect of different intercrops on the effect of plant spread (E–W) of the main crop revealed that it varied from 4.03 m in the case of Anjan to 4.52 m in the case of fodder jowar (Table 27.12).

Table 27.12 Morphological parameters and biomass production of Neem in agri-silvi production system

Intercrops	Plant height (m)	Stem diameter (cm)	Plant spread (m) (N–S)	Plant spread (m) (E–W)
Sunhemp	6.25	19.32	4.21	4.20
Fodder jowar	6.31	20.71	4.14	4.52
Cenchrus	6.09	19.12	4.15	4.03
Stylosanthes	5.66	17.92	3.93	4.27
Moth bean	6.15	19.12	4.55	4.21
Soybean	5.88	19.50	4.49	4.38

Hiwale (2004)

The mean Neem dry biomass production was found to increase from 54.48 q/ha in fifth year to 56.23 q/ha indicating the biomass addition of 1.75 q/ha. Biomass production in different intercrops was maximum 57.47 q/ha in Neem + fodder jowar. Raising of various intercrops in the interspaces revealed that substantial quantity of carbon in the form of biomass could be harvested making maximum utilization of precipitation and sunlight. The mean maximum biomass production was in the Neem + fod. jowar (98.63 q/ha) (Table 27.13).

Economic analysis of Neem-based cropping system indicated that maximum mean net return of Rs. 14,543/- can be obtained in Neem + fodder jowar-based cropping system with a benefit-cost ratio of 1:3.32 averaged over the past 2 years followed by Neem + *Cenchrus* (1:3.01) and Neem + sunhemp (1:2.72) (Table 27.14).

Intercropping in Neem

Table 27.13 Biomass production (q/ha) of Neem-based cropping system

Intercrops	Dry biomass produced by main crops (q/ha)	Fresh biomass produced by intercrops (q/ha)	Dry biomass produced by intercrops (q/ha)
Sunhemp	54.32	79.77	24.92
Fodder jowar	57.47	98.63	29.93
Cenchrus	55.65	89.75	27.52
Stylosanthes	56.52	56.96	13.46
Moth bean	53.10	29.55	16.31
Soybean	55.05	63.45	14.86

Table 27.14 Gross return, net return, and B.C. ratio of Neem-based cropping system

Intercrops	Gross return (Rs.)	Net return (Rs.)	B:C ratio
Sunhemp	16,379	11,812	2.720
Fodder jowar	19,107	14,543	3.315
Cenchrus	17,992	13,415	3.010
Stylosanthes	9,013	4,436	0.990
Moth bean	10,970	8,642	1.935
Soybean	13,999	9,421	2.155

Hiwale (2004)

27.6.2 Subabul-Based Agri-silvi Production System on Marginal Lands

Twelve-year-old subabul (*Leucaena leucocephala*) plants spaced 2×2m which were intercropped with various arable crops indicated that it is not possible to grow intercrops in the subabul owing to their close spacing and tall tree growth which do not allow sunlight to pass and dense root system of tree crop. Growth parameters of subabul showed increase at a slower pace as plants have reached their last stage of growth. Mean plant height increased from 12.70 to 12.87 m. Mean DBH over the years increased from 114.63 to 117.67 mm indicating an increase of 2.65 % (Table 27.15).

At the age of twelve, mean dry biomass production of subabul was found to be 614.33 q/ha, which increased to 750.62 q/ha at the age of thirteen. The production from pasture component gets a boost by 5–7 times compared to the natural grazing lands. Pasture growth and production does not affect the tree growth. It can be concluded that at the spacing of 2×2 m,

Table 27.15 Morphological parameters of subabul

	Plant height (m)			DBH (mm)			Dry biomass produced main crops (q/ha)		
Intercrops	2002	2003	Mean	2002	2003	Mean	2002	2003	Mean
Sunhemp	12.71	12.90	12.80	115.87	117.70	116.78	624.00	750.00	687.00
Fod. jowar	12.70	12.92	12.81	114.60	116.40	115.50	580.00	800.00	690.00
Cenchrus	12.68	12.84	12.76	113.82	117.40	115.61	654.00	718.75	686.37
Stylosanthes	12.68	12.91	12.79	115.32	118.60	116.96	617.00	745.00	681.00
Green gram	12.73	12.87	12.80	113.27	117.30	115.28	598.00	727.50	662.75
Soybean	12.72	12.80	12.76	114.90	118.20	116.55	613.00	762.50	687.75

Hiwale (2004)

Table 27.16 Economics of subabul-based cropping system

Name of crop combination	Gross return of main crop (Rs.)	Net return (Rs.)	B:C ratio
Subabul + *Dolichos*	60,000	43,500	2.64
Subabul + maize	64,000	47,500	2.88
Subabul + *Cenchrus*	57,500	41,000	2.48
Subabul + *Stylosanthes*	59,600	43,100	2.61
Subabul + moth bean	58,200	41,700	2.53
Subabul + soya bean	61,000	44,500	2.70

Hiwale (2004)
Cost of cultivation: Rs. 16,500/-(1,500/-x 11 year); sale rate – Fuelwood at Rs. 80/q

Leucaena can be intercropped up to the agc of 4–5 only; thereafter, there is reduction in yield of intercrops, and after 10 years, there is a complete failure of the crops. There is therefore need to increase the spacing for raising intercrops or lopping should be practiced every year at least once.

Economic analysis of subabul-based cropping system indicated that maximum net return of Rs. 47,500/- can be obtained in subabul + maize and minimum in subabul + *Cenchrus* (Rs. 41,000/-). B:C ratio showed maximum in subabul + maize (2.88) and minimum in subabul + *Cenchrus* (2.48) (Table 27.16). Economic returns from silvipastoral system are less as compared to fruit tree-based cropping system. The economic returns are less because the maintenance cost over the year is higher compared to fruit-based cropping system where income from the system starts in the 3rd year.

27.7 Allelopathic Influence of Horti-silvi Tree Species on Arable Crops Under Semiarid Conditions

Allelochemicals produced by one plant are known to influence the germination, growth, development, and metabolism of other plants, which may be beneficial (allelopathy) or harmful due to continuous leaf shedding of such trees and addition of allelochemicals in the soil over a long period of time. Therefore, an experiment was initiated to know the allelopathic effect of leaf leachate of four tree species, viz., custard apple (*Annona squamosa*), aonla (*Emblica officials*), Neem (*Azadirachta indica)*, and subabul (*Leucaena leucocephala*), on germination, seedling growth, and seedling vigor index of nine arable crops, viz., soybean (*Glycine max*), maize (*Zea mays*), okra (*Abelmoschus esculentus*), sunhemp (*Crotalaria juncea*), green gram (*Phaseolus aureus*), pigeon pea (*Cajanus cajan*), fodder jowar (*Sorghum vulgar*), *Sesamum* (*Sesamum indicum*), and moth bean (*Phaseolus aconitifolius*) under ambient condition at Central Horticultural Experiment Station, Vejalpur (22° 41′38″ North and 73° 33′ 22″ East), Gujarat, with the annual rainfall of 750 mm. The leaf leachate of various tree species has suppressing as well as beneficial effect on germination and seedling vigor index in some arable species. Neem was found to suppress most of the growth parameters, whereas subabul, custard apple, and aonla promoted them. Custard apple, aonla, and subabul were found to have a beneficial effect on soya bean, green gram, pigeon pea, and *Sesamum* and suppressing effect on okra, fodder jowar, sunhemp, maize, and moth bean.

27.7.1 Introduction

Agroforestry is a collective name for land-use systems in which woody perennials are deliberately grown on same piece of land as agricultural crops. It is an established fact that one plant may influence the germination, growth, development, and metabolism of other plants through the release of some chemicals (allelochemicals), which may be harmful or beneficial (allelopathy) due to continuous leaf shedding of such trees and addition of allelochemicals in the soil over a long period of time (Suresh and Rai 1988). The effect of these chemicals on other plants is known to be dependent on the concentration as well as the combinations in which one or more substances are released into the environment (Putnam and Tang 1986). Tree leaves are the potent source of allelochemicals. However, toxic metabolites are distributed in all other parts of plants, and the primary effects of allelopathy on crop production are the result of toxins released from tree litter on/or in the soil (Gill 2001).

Considerable progress has been made in understanding the nature of allelochemicals and the extent of negative effects and potential benefits in annual cropping systems. Most allelopathy literature is based on laboratory studies or bioassays conducted under controlled condition using unrealistic concentrations of aqueous leachates of extracts of tree plants (Nandal et al. 1994). Some research workers emphasized on the need for investigation of the allelopathy of various tree species used in agroforestry (Rao et al. 1994: Kaur et al. 1999). While identifying suitable tree species for agroforestry, efforts should be made to select the species with the least allelopathic activity (Kaur and Rao 2000; Gaba 1987). The present investigation was conducted to examine the allelopathic response of four horti-silvi tree species on seed germination, seedling growth, and vigor index of different arable crops cultivated in the agri-horti-silvi production system under rainfed conditions of semiarid regions of Gujarat state.

The allelopathic effects of custard apple (*Annona squamosa*), aonla (*Emblica officinalis),* Neem (*Azadirachta indica),* and subabul (*Leucaena leucocephala*) were studied on various agricultural crops, viz., soybean (*Glycine max*), maize (*Zea mays*), okra (*Abelmoschus esculentus*), sunhemp (*Crotalaria juncea*), green gram (*Phaseolus aureus*), pigeon pea (Cajanus *cajan* L.), fodder jowar (*Sorghum vulgar*), sesame (*Sesamum indicum* L.), and moth bean (*Phaseolus aconitifolius*) at Central Horticultural Experiment Station, Vejalpur (22° 41′38″ North and 73° 33′ 22″ East), Gujarat, India. It lies on 110 m above mean sea level with the annual rainfall of 750 mm.

27.7.2 Germination Percentage

Out of the nine crops tried, germination was significantly inhibited in three crops, viz., okra, moth bean, and *Sesamum* significantly. Germination was suppressed in all the other crops but non-significantly. Maximum suppression was observed in Neem leaf extract except in soybean and green gram. Germination was enhanced in soybean, maize, and fodder jowar, whereas in the case of okra and sunhemp, there was suppression in some and promotion in some (Table 27.17). The phytotoxins and tannins present in the extract of aonla and Neem might have suppressed germination in the agronomical crops. In case of *Leucaena*, it may be due to mimosine content in the leaves. The suppressing effect of *Albizia*, *Azadirachta, Eucalyptus*, and *Syzygium* on wheat seed germination was reported by Kaur and Rao (2000). The findings revealed that allelopathic interactions are crop specific and are not suppressed in all but are also promoted in same species. Banwarilal (2000) reported promotion of seed germination when *Syzygium* extract was used at low concentration. However, at higher concentration it has suppressing effect. The ratio used (20 g/1,000 ml water) in preparing the left leachate eliminates the chances of osmotic

Table 27.17 Effect of tree leaf leachates on germination percent of different intercrops

Crops	*Glycine max*	*Zea mays*	*Abelmoschus*	*Crotalaria*	Phaseolus mungo	*Cajanus cajan*	*Sorghum*	*P. aconitifolius*	*Sesamum*
Annona	68.88	93.33	84.44	97.78	97.77	88.89	88.88	92.50	91.37
Emblica	62.22	93.33	71.11	95.55	100.00	84.44	88.88	88.37	90.37
Azadirachta	64.44	88.89	64.44	95.55	100.00	77.77	77.77	83.37	80.25
Leucaena	66.66	88.89	86.66	100.00	97.77	79.99	88.88	92.62	91.75
Control	62.22	82.22	75.55	97.78	100.00	88.89	77.77	94.94	94.00
CD at 5 %	NS	NS	9.85	NS	NS	NS	NS	3.699	2.107

Table 27.18 Effect of tree leaf leachates on shoot length (cm) of different intercrops

Crops	*Glycine max*	*Zea mays*	*Abelmoschus*	*Crotalaria*	*P. aureus*	*Cajanus cajan*	*Sorghum*	*P. aconitifolius*	*Sesamum*
Annona	4.55	7.74	2.39	4.52	9.57	5.37	4.45	6.93	4.02
	55.82	**−17.92**	**−43.76**	**−29.48**	**3.23**	**6.34**	**22.93**	**−0.13**	**85.25**
Emblica	5.49	7.12	3.67	4.34	10.11	5.23	3.25	6.54	3.05
	63.39	**−24.49**	**−13.65**	**−32.29**	**9.06**	**3.56**	**−10.22**	**−17.84**	**40.55**
Azadirachta	2.70	4.28	2.23	3.74	8.77	4.28	3.35	5.45	1.97
	34.33	**−54.61**	**−47.53**	**−41.65**	**−5.39**	**−15.25**	**−7.46**	**−31.53**	**−9.22**
Leucaena	2.95	8.38	2.63	3.55	9.17	6.19	3.39	6.98	4.03
	31.86	**−11.13**	**−38.11**	**−44.62**	**−1.08**	**22.57**	**−6.35**	**−12.31**	**85.71**
Control	2.01	9.43	4.25	6.41	9.27	5.05	3.62	7.96	2.17
CD at 5 %	0.89	2.13	1.32	0..49	1.07	1.38	1.66	0.578	0.236

Figures in bold are percent values

inhibition of seed germination (Richardson and Williamson 1988), which might have resulted in promotion of germination in some and suppression in some sensitive species.

27.7.3 Shoot Length and Root Length

The effect of different tree leaf leachates on length of shoot of different test crops revealed that there was suppression of shoot length in maize, okra, sunhemp, and moth bean. Maximum suppression of shoot length was due to Neem. However, in soybean, pigeon pea, fodder jowar, and *Sesamum* and green gram, there was promotion of shoot length. Maximum promotion was observed with custard apple followed by aonla leaf leachate (Table 27.18). A similar trend was observed in respect to root length except in maize, okra, sunhemp, green gram, pigeon pea, fodder jowar, moth bean, and *Sesamum* and soybean. Maximum suppression was observed due to Neem leaf leachate. Maximum promotion of root length over control was maximum in maize with subabul extract; soybean, okra, and green gram with aonla extract; pigeon pea with subabul; and fodder jowar and *Sesamum* with custard apple extract. There was total suppression in moth bean and sunhemp in all the extracts (Table 27.19). Bhatt and Todaria (1990) observed that root development was significantly suppressed by *Eucalyptus terectonics* extract when applied to kidney bean. Dalal et al. (1992) reported an inhibitory effect of *Zizyphus mauritiana* on the growth of chickpea and mustard. Narwal (1994) reported no allelopathic effect or stimulatory effect in the case of *Zizyphus jujuba*. Bisla and Nandal (1992) also reported stimulatory effect of aqueous extract of perennials on growth and biomass production of arable crops.

27.7.4 Fresh Shoot Weight

Shoot weight was promoted in soybean, maize, and green gram. There was promotion in *Sesamum* and reduction in some like green gram, pigeon pea, fodder jowar, and *Sesamum*, whereas in the case of okra, sunhemp, and moth bean,

Table 27.19 Effect of tree leaf leachates on root length (cm) of different intercrops

Crops	*Glycine max*	*Zea mays*	*Abelmoschus*	*Crotalaria*	*P. aureus*	*Cajanus cajan*	*Sorghum*	*P. aconitifolius*	*Sesamum*
Annona	1.52	4.97	7.53	1.11	4.97	2.95	4.53	2.60	1.04
	4.19	**10.69**	**–4.44**	**–21.27**	**2.89**	**–31.11**	**15.27**	**–13.33**	**26.83**
Emblica	1.72	4.23	10.29	1.31	5.27	2.38	2.61	1.65	0.87
	17.81	**–5.79**	**30.58**	**–7.09**	**9.11**	**–5.78**	**–35.59**	**–45.00**	**6.09**
Azadirachta	0.44	3.63	6.02	0.72	3.70	1.50	2.51	1.47	0.62
	–69.86	**–19.15**	**–23.60**	**–48.94**	**–23.39**	**–33.33**	**–36.13**	**–51.00**	**–24.39**
Leucaena	0.65	5.65	7.43	0.97	4.81	2.99	3.08	2.42	0.92
	–55.48	**25.83**	**–5.71**	**–31.20**	**–0.41**	**32.89**	**–21.63**	**–19.33**	**12.13**
Control	1.46	4.49	7.88	1.41	4.83	2.25	3.93	3.00	0.82
CD at 5 %	0.45	0.94	2.25	0.42	0.62	1.08	1.80	0.305	0.125

Figures in bold are percent values

Table 27.20 Effect of tree leaf leachates on fresh shoot weight (mg) of different intercrops

Crops	*Glycine max*	*Zea mays*	*Abelmoschus*	*Crotalaria*	Phaseolus mungo	*Cajanus cajan*	*Sorghum*	*P. aconitifolius*	*Sesamum*
Annona	122.00	387.67	207.67	213.33	265.27	93.00	40.73	686.20	221.52
	108.6	**8.59**	**–33.58**	**–18.05**	**5.55**	**20.67**	**35.77**	**–1.55**	**20.96**
Emblica	124.00	356.33	245.33	214.13	251.33	82.00	29.33	594.60	188.43
	112.07	**–0.19**	**–21.53**	**–17.75**	**–0.13**	**6.39**	**–2.23**	**–14.69**	**2.89**
Azadirachta	94.00	235.33	204.67	192.33	215.67	74.33	24.23	555.47	157.76
	60.77	**–34.08**	**–34.54**	**–26.12**	**–14.19**	**–3.55**	**–19.23**	**–20.30**	**–13.85**
Leucaena	82.00	460.67	272.67	194.13	230.33	116.60	25.13	671.57	191.91
	40.24	**29.04**	**–12.79**	**–25.43**	**–8.35**	**51.29**	**–16.23**	**–3.65**	**4.80**
Control	58.47	357.00	312.67	260.33	251.00	77.07	30.00	697.00	183.12
CD at 5 %	40.39	150.67	69.54	26.43	31.14	19.93	9.99	35.251	18.583

Figures in bold are percent values

there was maximum reduction in shoot weight in the leaf leachate of Neem. Maximum promotion was recorded in soybean with aonla; maize with subabul, green gram, and fodder jowar; *Sesamum* with custard apple; and pigeon pea with subabul (Table 27.20). The results are in line with that of Dalal et al. (1992) who reported an inhibitory effect on the growth of chickpea and mustard and Bisla and Nandal (1992) on stimulatory effect of aqueous extract of perennials. Narwal (1994) reported no allelopathic effect or stimulatory effect in the case of *Zizyphus jujuba* on growth and biomass production of arable crops.

27.7.5 Fresh Root Weight

Root weight was suppressed under all the treatments in soybean, okra, sunhemp, green gram, and moth bean, whereas there was promotion with *Sesamum* and reduction in maize, green gram, red gram, fodder jowar, and *Sesamum.* Neem suppressed the root weight maximum in all the test crops except in soybean, green gram, and red gram (Table 27.21). Dalal et al. (1992) reported an inhibitory effect of *Zizyphus mauritiana* on the growth of chickpea and mustard.

27.7.6 Dry weight of Shoot and Root

Shoot and root dry weight was promoted in soybean, maize, green gram, and *Sesamum.* There was reduction in some like green gram, pigeon pea, and fodder jowar, whereas maximum reduction was in Neem leaf leachate in okra, sunhemp, and moth bean. Maximum promotion was recorded in soybean with aonla; maize with subabul; green gram, *Sesamum*, and fodder jowar with custard apple; and pigeon pea with subabul

Table 27.21 Effect of tree leaf leachates on fresh root weight (mg) of different intercrops

Crops	*Glycine max*	*Zea mays*	*Abelmoschus*	*Crotalaria*	Phaseolus mungo	*Cajanus cajan*	*Sorghum*	*P. aconitifolius*	*Sesamum*
Annona	8.66	183.67	30.00	23.27	72.67	38.67	4.60	93.25	39.93
	−55.97	**9.33**	**−22.42**	**−19.39**	**18.87**	**8.41**	**4.54**	**−10.03**	**65.55**
Emblica	16.13	171.33	13.20	17.33	72.67	36.80	3.73	78.70	28.32
	−17.99	**1.98**	**−65.86**	**−39.97**	**18.87**	**3.17**	**−15.23**	**−23.93**	**17.41**
Azadirachta	12.00	119.67	5.13	14.00	46.53	35.13	3.33	70.60	20.64
	−38.99	**−28.77**	**−86.73**	**−51.51**	**−3.81**	**−1.51**	**−24.32**	**−31.89**	**−14.43**
Leucaena	11.47	231.33	36.33	19.33	58.80	42.60	4.33	89.57	31.93
	−41.68	**37.69**	**−6.05**	**−33.04**	**−24.21**	**19.43**	**−1.59**	**−13.58**	**32.38**
Control	19.67	168.00	38.67	28.87	61.13	35.67	4.40	103.65	24.12
CD at 5 %	7.66	59.69	18.58	5.59	20.34	14.47	1.29	7.634	3.103

Figures in bold are percent values

Table 27.22 Effect of tree leaf leachates on dry wt. of shoot and root (mg) of different intercrops

Crops	*Glycine max*	*Zea mays*	*Abelmoschus*	*Crotalaria*	Phaseolus mungo	*Cajanus cajan*	*Sorghum*	*P. aconitifolius*	*Sesamum*
Annona	34.55	256.75	53.25	144.50	277.25	73.50	23.25	137.50	15.75
	111.18	**29.34**	**−55.90**	**−0.68**	**10.35**	**22.50**	**22.37**	**−3.17**	**47.19**
Emblica	34.75	146.75	50.00	139.25	265.50	68.75	14.25	124.75	13.75
	112.41	**−26.07**	**−58.59**	**−4.29**	**5.67**	**14.58**	**−25.00**	**−12.15**	**28.50**
Azadirachta	21.81	139.25	40.75	104.5	187.50	55.75	16.25	115.7	9.25
	−33.31	**−29.84**	**−66.24**	**−28.17**	**−25.37**	**−7.08**	**−14.47**	**−18.52**	**−13.55**
Leucaena	20.46	268.0	58.00	139.50	241.00	68.50	19.00	131.50	14.25
	25.06	**35.12**	**51.96**	**−4.12**	**4.08**	**14.17**	**0.00**	**−7.39**	**33.18**
Control	16.36	198.50	120.75	145.50	251.25	60.00	19.00	142.00	10.7
CD at 5 %	2.55	53.05	15.07	28.30	40.21	10.39	5.77	NS	3.38

Figures in bold are percent values

(Table 27.22). The results are in line with that of Dalal et al. (1992) who reported an inhibitory effect on the growth of chickpea and mustard, whereas Bisla and Nandal (1992) reported stimulatory effect of aqueous extract of perennials. Narwal also (1994) reported no allelopathic effect or stimulatory effect in the case of *Zizyphus jujuba* on growth and biomass production of arable crops.

27.7.7 Seedling Vigor Index

The seedling vigor index was estimated as per the procedure of Abdulbaki and Anderson (1973). Leaf leachate of aonla, custard apple, Neem, and subabul when applied to different intercrops grown in the system under ambient condition revealed that there was suppressing as well as beneficial effect on germination percentage and seedling growth. Leaf leachates of Neem when applied had suppressing effect on germination and seedling growth. However, custard apple leaf leachate improved germination and seedling growth. Seedling vigor index was promoted in all the leaf leachate in soybean. Maximum seedling vigor index was recorded with custard apple, aonla, and subabul, whereas Neem leaf leachate suppressed in most of the test crops. There was suppression in some and promotion in some (Table 27.23).

27.8 Leaf Litter Decomposition in Respect to Tree Species Involved in the System

Organic recycling of nutrient through leaf litter fall and its decomposition plays an important role in sustainability of agroforestry system in culti-

Table 27.23 Effect of tree leaf leachate on seedling vigor index of different intercrops

Crops	*Glycine max*	*Zea mays*	*Abelmoschus*	*Crotalaria*	Phaseolus mungo	*Cajanus cajan*	*Sorghum*	*P. aconitifolius*	*Sesamum*
Annona	418.10	1186.92	837.64	550.50	14 21.57	739.56	798.14	880.97	461.98
Emblica	448.61	1059.29	992.69	539.86	1538.00	642.59	520.84	723.94	350.03
Azadirachta	202.34	0703.12	530.34	426.15	1247.00	449.51	455.73	577.26	207.73
Leucaena	223.99	1247.13	871.79	452.00	1366.82	734.31	663.93	869.85	449.83
Control	215.90	1144.50	916.42	761.64	1410.00	648.89	587.16	1041.35	286.22

Table 27.24 Leaf litter fall (kg/plant) and estimated nutrient recycled (g/plant)

Name of tree species	Leaf litter fall (dry wt. basis) (kg/plant)	Nutrient recycled (g/pl)		
		Nitrogen	Phosphorus	Potassium
Aonla	9.94	79.26	10.98	35.95
Custard apple	1.58	20.19	1.60	3.96
Neem	4.89	69.71	5.90	15.69
Subabul	3.44	62.48	3.80	11.42
CD at 5 %	0.542	7.060	0.608	1.83

Hiwale (2004)

vated lands (Agrawal 1980). Litter fall is the source of nutrient recycled to the soil from the plants. The decomposition of litter adds substantial quantity of nutrient to the soil through nutrient recycling. There is variation in the rate of decomposition and nutrient release pattern, which is governed by chemical composition of leaf and prevailing weather condition. Addition of leaf litter to the soil over the years helps in improving soil fertility and aggregate stability (Young 1997).

Tree species like *Emblica officinalis*, *Annona squamosa*, *Azadirachta indica*, and *Leucaena leucocephala* are found to be deciduous in nature. Thereby, they add a lot of leaf litter to the soils resulting in addition of nutrients, improving soil health. The present investigation revealed that Indian gooseberry produced maximum leaf litter adding higher quantity of N, P, and K to the soil. The decomposition rate was however faster in *Leucaena*. The analysis of leaf litter revealed that custard apple litter had maximum nitrogen and phosphorus content.

All the four tree species (aonla, Neem, custard apple, and subabul) were found to be deciduous in nature and large amount of leaf litter is added by all the tree species to the soil. The data on leaf litter fall (on dry weight basis) of different tree species was recorded. The results revealed that the maximum leaf litter was produced by aonla (9.94 kg/plant) followed by Neem (4.89 kg/plant) and it was lowest in custard apple (1.58 kg/plant) (Table 27.24).

There were significant differences in the case of nitrogen, phosphorus, and potassium recycled by different tree species. Among these, aonla leaf litter recycled maximum nitrogen, phosphorus, and potassium to the soil, i.e., 79.26 g, 10.98 g and 35.95 g, respectively, whereas minimum nutrient were recycled (20.19 g N/pl.) 1.60 g P/pl. and 3.96 g K/pl.) by custard apple leaf litter (Table 27.24).

27.8.1 Leaf Litter Decomposition and Its Nutrient Composition

The data revealed that there were significant differences among different tree species in respect to leaf litter decomposition and its nutrient composition. After the period of two months, maximum actual leaf litter decomposition percentage was noted in subabul (44.60 %) and minimum in Neem (16.20 %). The decomposition rates vary with substrate quality, climate, and quality and quantity of decomposer organism (Swift et al. 1979).

The role of climate in litter decomposition has been reported by several workers (Moore 1986). Environmental factors which are most important in regulating the turnover rate of litter tend to be

Table 27.25 Leaf litter decomposition rate and its nutrient content

Name of tree species	% actual leaf litter decomposed after two months	Available nutrient (%)					
		Nitrogen		Phosphorus		Potassium	
		Undecomposed	Decomposed	Undecomposed	Decomposed	Undecomposed	Decomposed
Aonla	24.60	0.796	2.637	0.11	0.325	0.36	0.227
Custard apple	26.00	1.270	3.217	0.10	0.477	0.25	0.250
Neem	16.20	1.424	3.017	0.12	0.257	0.32	0.310
Subabul	44.60	1.798	2.732	0.11	0.277	0.33	0.325
CD at 5 %	7.420	0.098	0.647	0.02	0.029	0.09	0.074

Hiwale (2004)

those that regulate the activity of microorganisms, i.e., soil moisture and nutrient availability (Berg and Ekbohm 1991).

The increase in nitrogen content in the litter may be attributed to microbial immobilization. Among the tree species, custard apple leaf litter (decomposed) had highest nitrogen content (3.217 %) and it was on par with Neem (3.017 %), whereas it was found lowest in aonla decomposed leaf litter (2.637 %), which was on par with subabul leaf litter (2.732 %). Increase in nitrogen content may be due to conversion of carbon into Co_2 due to faster oxidation of soluble carbon compound and subsequent weight loss (Kumar and Deepu 1992). The accumulation of N could be from herbivorous fractional precipitation as reported by Swift et al. (1979) (Table 27.25).

Phosphorus content of decomposed leaf litter was recorded maximum in decomposed custard apple (0.477 %) followed by aonla (0.325 %) and subabul (0.277 %) and it was minimum in decomposed Neem leaf litter (0.257 %). The phosphorus content of fresh leaf litter collected was low which might be due to efficient translocation of nutrients from aging foliage during the dry season (Das and Ramakrishna 1985; Pande and Sharma 1988). The absolute amount of P in the residue showed biphasic pattern with initial low content followed by latter slow buildup phase (Table 27.25).

Potassium content of decomposed leaf litter was found highest in subabul (0.325 %), which was at par with Neem (0.310 %), and was noted the lowest in aonla (0.227 %) (Table 27.25). As per the finding of Bocock (1964) and Kunhama (1994), potassium content was reduced because potassium is not structurally bound in organic compound and also quick leachability in the water could be a major reason for decline phase.

27.8.2 Nutrient Composition of Soil

Analysis of soil samples of the experimental site showed improvement in soil nutrient composition in respect to major nutrients. The results showed that there was reduction in pH and significant increase in organic matter content. Soil pH was nonsignificantly influenced by different tree crops at 0–15 and 15–30 cm depth. However, data revealed that there was reduction in soil pH compared to control (Table 27.26).

Organic carbon percent of the soil was maximum under aonla (0.51 %) which was on par with custard apple and it was least in control (0.32 %) (Table 27.26). Soil nitrogen and potassium content were significantly influenced by leaf litter of different tree species, whereas phosphorus content showed nonsignificant differences among different tree species. Maximum nitrogen content was observed in aonla plot (169.97 kg/ha), whereas phosphorus and potassium content were highest in custard apple plot, i.e., 105.57 kg P_2O_5/ha and 11.77 kg K_2O/ha, respectively.

It can be concluded that overall improvement in soil physicochemical properties mainly because of addition of large quantity of leaf litter by different tree species because of sufficient soil moisture triggering activity of soil micro *fauna*. It is concluded that due to addition of litter by tree species, there was improvement in soil health which is due to enrichment of soil through the release of entrapped nutrients and formation of organic matter (Young 1997).

Leaf litter fall in Aonla

Table 27.26 Soil analysis

Name of tree species	pH 0–15 cm depth	pH 15–30 cm depth	Organic carbon (%)	N (kg/ha)	P (kg/ha)	K (kg/ha)
Aonla	6.29	6.60	0.51	169.97	8.25	101.05
Custard apple	6.64	6.39	0.50	162.04	11.77	105.57
Neem	6.70	7.05	0.39	145.14	10.94	76.07
Subabul	6.43	7.10	0.34	149.30	9.46	87.72
Control	6.81	7.35	0.32	130.32	8.08	74.47
CD at 5 %	NS	NS	0.13	22.04	NS	21.32

27.9 Root Distribution, Production of Above- and Belowground Biomass of Horti-silvi Tree Species

Root distribution studies were carried out by excavation method. The data on fresh and dry weight of below- and aboveground biomass production for the past two years was recorded. Maximum mean belowground biomass (50.32 kg/tree) and mean aboveground biomass production (267.07 kg/tree) were recorded in subabul. It was least in custard apple (2.96 kg/tree and 13.45-kg/tree, respectively). However, mean root shoot ratio was maximum in aonla (0.59) and minimum in subabul (0.18). Custard apple produced the least above as well as belowground biomass. Root distribution in aonla revealed that most of the roots were distributed in 0–60 cm depth and 0–120 cm radial distance. Maximum in 0–30 cm depth. In the case of custard apple, the roots are shallow root system. In subabul and Neem however, there was better distribution of roots (depth wise) up to 60 cm depth. Radial distribution in all the four tree species was up to 120 cm. Maximum plant production can be obtained with relatively small root system. (Van Noorwijik 1989).

Silvipastoral tree species were found to have more deep vigorous root system compared to fruit trees probably because of diversion of synthesized food to the major sink, i.e., fruits. Per cent root weight ranged from 15.85 to 37.37 % of the total biomass produced by the tree, which is found to be low due to soil type, and growing of crops under rainfed conditions (Table 27.27).

Table 27.27 Shoot and root biomass (on fresh wt. basis) (kg/tree) of different horti-silvi species

Name of tree species	Year	Main root weight	Lateral root weight	Total weight	Aboveground biomass	Root wt. to total biomass (%)
Aonla (8 years old)	2002	6.50	28.80	35.30	60.50	
	2003	7.80	30.95	38.75	63.55	
	Mean	7.15	29.87	37.02	62.02	37.37
Custard apple (4 years old)	2002	0.90	1.60	2.50	12.00	
	2003	1.20	2.22	3.42	14.90	
	Mean	1.05	1.91	2.96	13.45	18.03
Subabul (12 years old)	2002	22.00	26.90	48.90	264.00	
	2003	23.50	28.25	51.75	270.15	
	Mean	22.75	27.57	50.32	267.07	15.85
Neem (5 years old)	2002	12.00	26.20	38.20	104.00	
	2003	14.80	27.95	42.75	114.85	
	Mean	13.40	27.07	40.47	109.42	26.99
CD at 5 %	2002	1.193	1.864	2.920	3.786	
	2003	1.784	2.965	3.939	5.100	

Hiwale (2004)

Table 27.28 Root spread and crown characteristics and root: shoot ratio of different horti-silvi species

Name of tree species	Year	Root per shoot ratio	Rooting depth (cm)	Horizontal root spread (m)	DBH (cm)	Aerial crown spread (m^2)	Root spread crown spread ratio
Aonla (8 years old)	2002	0.58	117.30	3.30	16.00	4.52	0.73
	2003	0.61	119.00	3.50	18.00	4.79	0.73
	Mean	0.59	118.15	3.40	17.00	4.65	0.73
Custard apple (4 years old)	2002	0.21	84.50	1.16	10.00	1.30	0.89
	2003	0.23	86.00	1.25	12.50	2.49	0.50
	Mean	0.22	85.25	1.20	11.25	1.89	0.69
Subabul (12 years old)	2002	0.18	110.80	2.17	23.00	1.50	1.450
	2003	0.19	113.00	2.29	24.80	1.60	1.438
	Mean	0.18	111.90	2.23	23.90	1.55	1.44
Neem (5 years old)	2002	0.37	129.40	4.12	17.00	3.90	1.06
	2003	0.37	131.00	4.25	20.95	4.50	0.95
	Mean	0.37	130.20	4.18	18.97	4.20	1.01
CD at 5 %	2002	0.045	4.479	0.209	2.168	0.311	0.224
	2003	0.033	5.136	0.427	2.533	0.534	0.134

Hiwale (2004)

27.9.1 Root Spread and Crown Characteristics

Neem recorded maximum mean root depth (130.20 cm) and mean horizontal root spread (4.18 m) indicating faster belowground root growth compared to other tree species whereas custard apple recorded minimum mean root depth (85.25 cm) and mean horizontal root spread (1.20 m) indicating slow belowground growth. Mean aerial crown spread was found highest (4.65 m^2) in the case of aonla and lowest (1.55 m^2) in subabul indicating that aonla is having more spreading growth and subabul upright. Data on mean root/crown spread ratio revealed that it was maximum in subabul (1.44) and was least in aonla (0.73) (Table 27.28).

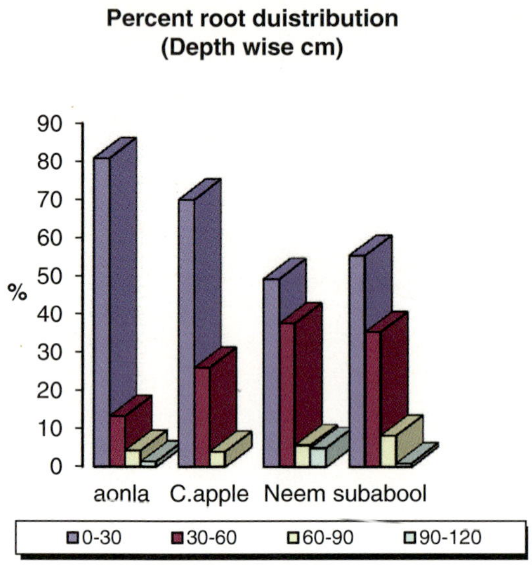

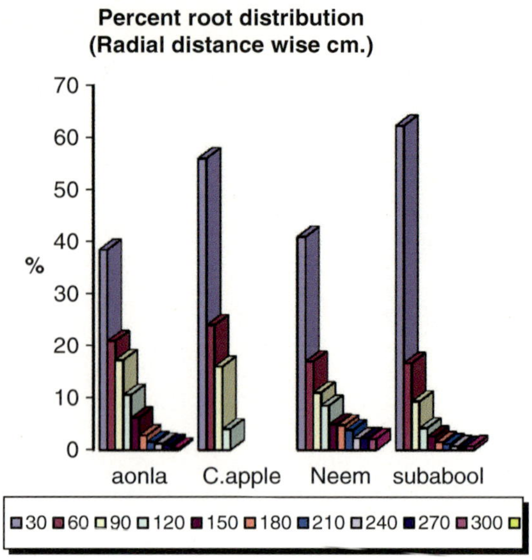

There is a relation between lateral spread of shoots and crown and the internal spread of roots with few exceptions. However, in most cases, root characteristics across plant species vary independently of shoot characteristics. Root shoot ratio was maximum in aonla and least in custard apple, which means that aboveground growth was faster in custard apple than belowground growth, whereas in aonla it is almost equal (Table 27.28). There is fraction equilibrium between root and shoot growth (Brower 1983).

Subabool root system

Aonla root system

27.10 Specific Gravity and Charcoal Production and Its Quality from Different Horti-silvi Tree Species

Specific gravity and charcoal production was studied from the wood of different tree species, viz, aonla (*Emblica officinalis*), custard apple (*Annona squamosa*), Neem (*Azadirachta indica*), and subabul (*Leucaena leucocephala*). Among all the experimental tree species, the maximum specific gravity (0.94) was noted in Neem wood followed by subabul (0.90) and aonla wood (0.885), whereas it was minimum (0.80) in custard apple wood.

The highest percent charcoal of 41.77 % was recorded in the case of custard apple wood, whereas, it was lowest in the case of Neem (13.02 %). One kg of charcoal of each tree species was burnt in *Sagadi*. The visual observations were made while burning the charcoals. It was seen that there was sparking in aonla and subabul charcoal but it was only at the beginning of the burning of the charcoal, whereas there was no sparking in Neem and custard apple.

It took 1.42–2.12 h for complete burning of 1 kg charcoal. Short duration was recorded in the case of Neem (1.42 h) and custard apple (1.47 h), whereas the duration of burning of charcoal was more in aonla (2.12 h) and subabul (2.02 h). While burning, the smoke intensity was not felt by all tree species. The color of the flame of fire during the burning of charcoal was red to pure red, in case of Neem and subabul, and was yellowish red in custard apple, whereas it was golden yellow in aonla. The charcoal obtained from the different tree species differed in respect to ash content per kg of charcoal. The highest ash content per kg of charcoal was recorded from aonla (88.25 g) and lowest in subabul (57.55 g) (Table 27.29).

Table 27.29 Specific gravity and percent charcoal production of different horti-silvi species

Name of tree crops	Year	Specific gravity	Wood burnt (kg)	Charcoal produced	% charcoal produced	Ash produced (g/kg)	Burning of charcoal (hours)
Aonla	2002	0.90	28.70	5.25	18.29	90.00	2.15
	2003	0.87	20.00	3.44	17.18	86.50	2.10
	Mean	0.885			17.735	88.25	2.125
C. apple	2002	0.79	5.20	2.20	42.30	63.00	1.55
	2003	0.81	20.00	8.25	41.25	60.25	1.40
	Mean	0.80			41.775	61.625	1.475
Neem	2002	0.97	71.80	9.70	13.50	70.00	1.50
	2003	0.91	20.00	2.51	12.55	65.20	1.35
	Mean	0.94			13.025	67.60	1.425
Subabul	2002	0.91	186.50	32.58	17.61	60.00	2.08
	2003	0.89	20.00	3.19	15.95	55.10	1.55
	Mean	0.90			16.78	57.55	2.025
CD at 5 %	2002	0.037	3.868	1.075	1.301	2.307	0.168
	2003	0.042	–	0.298	1.491	2.543	0.099

Hiwale (2004)

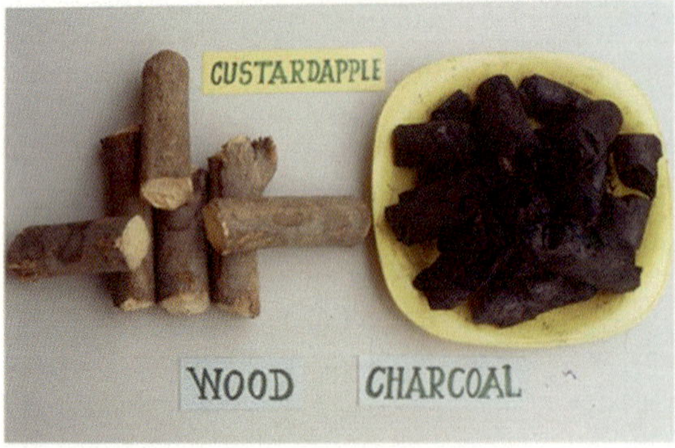

27.11 Fertility Status of Agri-horti Production System

27.11.1 Aonla-Based Cropping System

At CHES, Godhra, soil analysis of the system showed that there was improvement in respect to all the parameters recorded. The pH and electrical conductivity of the soil has decreased. Organic carbon content of the soil improved over the years probably due to addition of leaf litter by tree species and residue by intercrops. Similarly NPK content of the soil increased. Mean nitrogen and phosphorus content was maximum in green gram (186.60 kg N/ha and 8.25 kg P_2O_5/ha) being a leguminous crop, whereas mean potassium content was maximum (270.974 kg K_2O/ha) in soil intercropped with maize (Fig. 27.1). Soil organic carbon content had improved considerably in all the intercrops; it was maximum in moth bean (0.699 %) The pH and EC decreased marginally (Table 27.30).

27.11.2 Custard Apple-Based Cropping System

Soil analysis showed that there was improvement in respect to all the parameters. The pH and electrical conductivity of the soil was decreased. Organic carbon content of the soil improved due to addition of large quantity of leaf litter by tree species. NPK content of the soil increased in the system over control. Mean nitrogen content was maximum in pigeon pea (182.187 kg N/ha), and mean P and K content was maximum in soil intercropped with okra i.e., 13.47 kg P_2O_5/ha and 190.30 kg K_2O/ha, respectively. The pH and EC decreased marginally. Maximum soil organic carbon content was found in maize (0.616 %) (Table 27.30, Fig. 27.2).

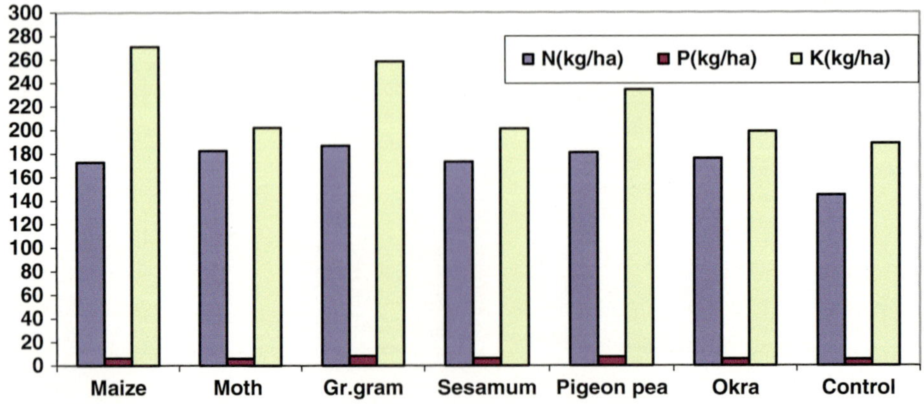

Fig. 27.1 Soil NPK content in aonla

Table 27.30 Mean physicochemical characteristics of soil

Intercrops	pH	EC (mmhos/cm^2)	O.C. (%)
Aonla + maize	6.79	0.149	0.693
Aonla + moth bean	6.37	0.137	0.699
Aonla + green gram	6.37	0.160	0.625
Aonla + *Sesamum*	6.49	0.140	0.630
Aonla + pigeon pea	6.28	0.152	0.606
Aonla + okra	6.65	0.163	0.652
Control	7.03	0.194	0.376
C. apple + maize	6.49	0.161	0.616
C. apple + moth bean	6.44	0.180	0.596
C. apple + green gram	6.50	0.125	0.592
C. apple + *Sesamum*	6.21	0.179	0.485
C. apple + pigeon pea	7.00	0.173	0.539
C. apple + okra	6.37	0.166	0.400
Control	6.91	0.194	0.298
Neem + F. jowar	7.58	0.156	0.398
Neem + moth bean	6.59	0.157	0.514
Neem + soybean	6.91	0.166	0.533
Neem + stylo	7.12	0.166	0.343
Neem + *Cenchrus*	7.56	0.158	0.461
Neem + sunhemp	7.24	0.151	0.257
Control	8.17	0.174	0.287
Subabul + maize	6.95	0.035	0.267
Subabul + *Dolichos* bean	6.04	0.046	0.359
Subabul + sunhemp	6.96	0.045	0.381
Subabul + *Cenchrus*	6.70	0.042	0.369
Subabul + stylo	6.42	0.043	0.278
Subabul + F. jowar	6.70	0.031	0.377
Control	7.22	0.034	0.196

Hiwale (2004)

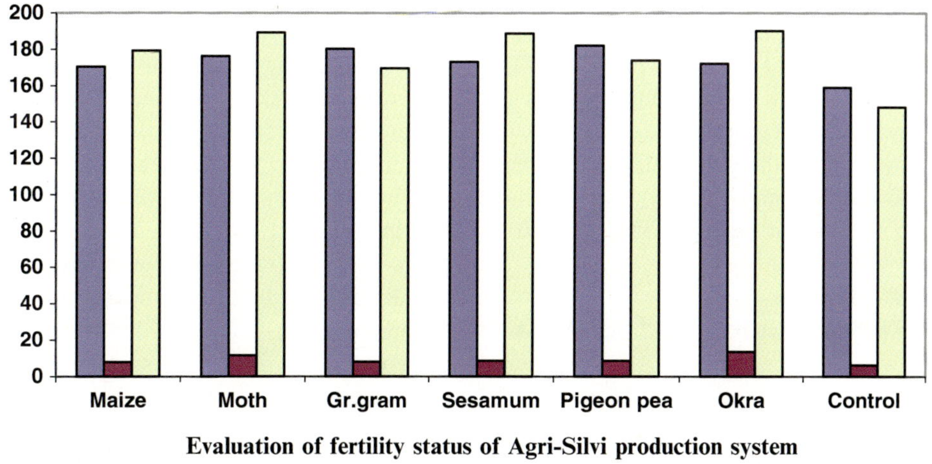

Fig. 27.2 Soil NPK content in custard apple

27.11.3 Neem-Based Cropping System

Soil analysis of the system for the past three years in respect to pH, EC, organic carbon (%), and NPK content showed that there was improvement in respect to all the parameters recorded. The pH of the soil decreased marginally. Electrical conductivity of the soil was decreased. Soil mean organic carbon content was found maximum in soya bean (0.533 %).

Organic carbon content of the soil improved probably due to addition of leaf litter by tree species and residue by intercrops. Similarly NPK content of the soil increased in all the intercrops over control. Mean nitrogen and phosphorus content was recorded maximum in sunhemp (184.500 kg N_2O/ha and 24.46 kg P_2O_5/ha, respectively), whereas mean potassium content was noted maximum in soil intercropped with *Cenchrus*, i.e., 270.10 kg K_2O/ha. (Table 27.30) (Fig. 27.3).

27.11.4 Subabul-Based Cropping System

Soil analysis showed improvement in respect to all the parameters recorded. The pH and EC decreased marginally. Maximum soil organic carbon content was found in sunhemp (0.381 %). Mean N content was maximum in *Dolichos* bean (159.70 kg N_2O/ha), and mean P content was maximum in *Stylosanthes* (9.46 kg P_2O_5/ha), whereas maximum K content was noted in soil intercropped with maize (112.11 kg K_2O/ha) (Fig. 27.4). The soil of subabul though showed improvement are poorest among all the cropping system because the plantation is done at the gully collars and due to steep slope added biomass carried away by rainwater, also soil erosion are higher (Table 27.30).

27.12 Rainfall, Runoff, and Soil Loss

The annual rainfall index varied from 21.47 to 1,099.4 with a mean of 601.12 (Table 27.31). About 98.1 % of the annual erosion index was concentrated during four months June to September. July and August contributed the highest EI_{30} values of 81.21 % of the annual value. The erosion index distribution curve showed that August is the most erosive month with EI_{30} value of 276.84, whereas September is the least erosive month with EI_{30} value of 31.69 (Fig. 27.5).

Maximum runoff and soil loss was recorded in cultivated uncropped soil (122.14 mm and 6.36 t/ha, respectively). It was minimum in no-tillage (36.90 mm and 0.49 t/ha, respectively). Wood perennials and tree crops characteristically produce relatively large amount of above- and belowground biomass. Because of their perennial nature, there is a continuous addition of organic matter and biomass to the soil, thereby reducing runoff and ultimately soil erosion.

Different fruit tree plantations behaved differently to the imposed treatments. Runoff and soil loss was higher in custard apple plantation due to total shading of leaves and new growth took 20–30 bays to cover the soil. Comparatively less runoff and soil loss was recorded in aonla as the canopy of the plant was well developed before the onset of monsoon. The tree canopy unless low and dense does not provide an effective soil cover. The canopy alone only reduces rainfall erosivity by the order of 10 % and may sometimes increase it (Wiersum 1985). Soil cover can have a considerable influence on the rate of erosion. The USLE cover factor for annual crops and 0.4–0.8 and those for natural vegetation are in the range of 0.01–0.001 (Roose 1977).

Studies on rainfall, runoff, and soil loss under different agroforestry systems carried out for three years in the entire four tree species (aonla, custard apple, Neem, and subabul) showed that in custard apple plot, the runoff and soil loss was higher compared to other tree species. Soil loss was highest in cultivated soil uncropped. It was less in cultivated but cropped soil with different intercrops, which provided cover to the soil resulting in less soil loss as compared to open soil. Minimum soil loss was recorded in subabul due to close plantation at 2×2 m spacing and also the complete canopy coverage of the soil and higher binding of soil by the root system (Table 27.32).

The total rainfall causing runoff ranged over the years from 29.28 to 33.55 % (Table 27.31).

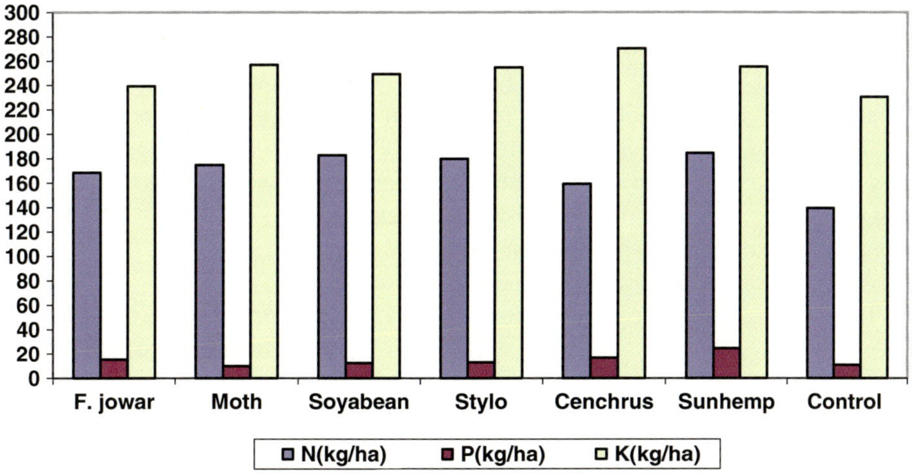

Fig. 27.3 Soil NPK content in Neem

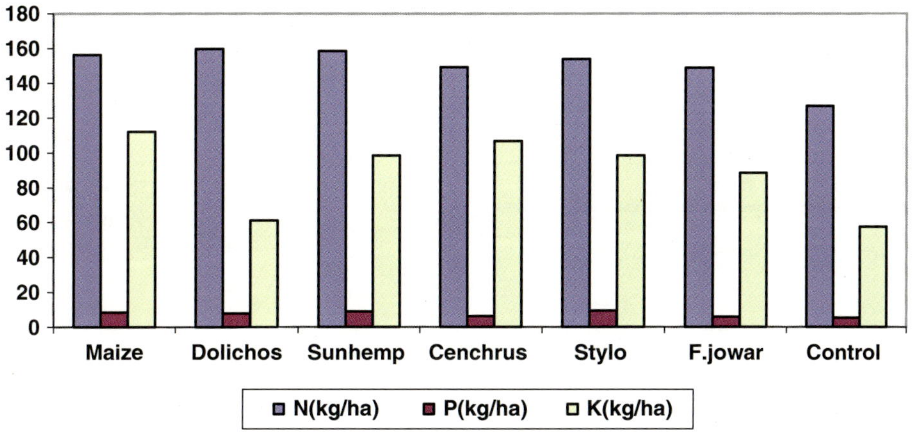

Fig. 27.4 Soil NPK content in subabul

Table 27.31 Monthly and annual rainfall

Year/month	June	July	August	September	October	Annual
1998	69.74	219.08	84.12	500.83	61.21	934.98
1999	10.59	78.82	89.68	54.69	27.69	261.47
2000	88.36	199.71	90.14	14.92	0.00	363.13
2001	59.33	168.71	53.40	0.00	0.00	281.13
2002	172.48	143.61	47.23	26.83	0.00	390.15
2003	67.4	433.81	321.21	30.22	0.00	852.64
2004	27.53	151.12	497.23	0.00	0.00	675.91
2005	277.30	347.10	92.8	302.4	0.00	1019.6
Mean	96.59	217.74	159.45	116.23	11.11	601.12
% of annual	16.06	36.22	26.52	19.33	1.84	–
Std deviation	83.92	117.25	162.32	201.02	23.70	304.12
Coefficient of variation	88.28	53.74	102.08	173.56	213.29	50.76

Hiwale (2014)

Though precipitation is low in dry lands, rainfall to runoff ratio of 25 % occurs even in light soil. The results indicated that least soil loss was caused in aonla tree crop probably due to low slope (0.5 %) and faster canopy development of tree species. Comparative performance of control measures taken to control runoff and erosion losses revealed that no-tillage was the best in reducing soil loss followed by staggered contour trenching. The system is better because it allows percolation of water deeper near the root zone of plants and helps in plant establishment.

The soil loss was least in no-tillage and maximum in cultivated uncropped land followed by cultivated cropped and staggered trench planting. A similar trend was recorded in all the cropping systems.

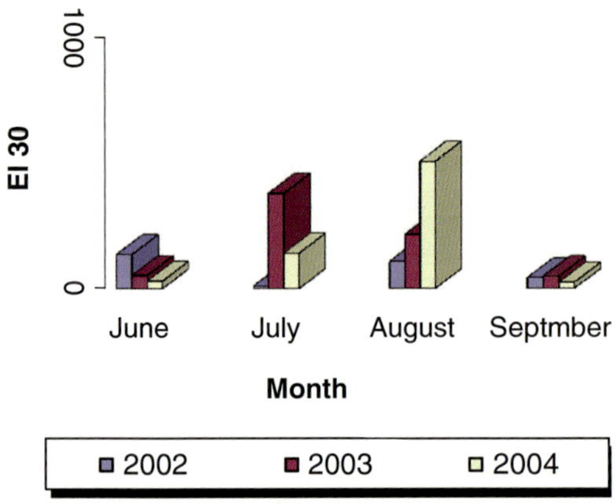

Fig 27.5 Monthly EI 30 index 2002–2004

27.13 Soil Moisture Content in Different Cropping Systems

The results of the studies carried out at CHES, Vejalpur, showed that the water content of upper layer at 100 cm depth was low and then it increased at 200 mm, again showing a depression at 300 mm, and then increased from 400 mm depth to 1,000 mm depth. The moisture content over the time at different depths increased after the rainfall up to 15 days and then dropped only at 100 mm depth, whereas it increased at other depths, indicating soil moisture evaporation and use by plant in the upper 100 mm layer. Most of the active roots are also observed in this layer.

Table 27.32 Rainfall, runoff, and soil loss in different cropping systems

		Runoff (mm)			% runoff to rainfall			Soil loss (q/ha)		
Sr no.	Treatment	2002	2003	2004	2002	2003	2004	2002	2003	2004
A	Custard apple									
1	Cultivated uncropped	115	147	104.7	32.63	29.28	33.55	5.3	7.6	6.9
2	Cultivated cropped	83.54	98.38	66.24	23.87	19.59	21.28	1.6	2.15	4.08
3	Uncultivated uncropped	64.94	76.10	61.51	18.55	15.16	19.74	0.6	0.98	1.23
4	Staggered contour trench planting	46.44	27.46	40.27	13.26	5.47	12.90	0.8	1.24	2.12
B	Aonla									
1	Cultivated uncropped	94.82	76.81	72.3	27.03	15.83	23.17	3.8	4.9	3.2
2	Cultivated cropped	66.43	46.09	19.52	18.98	9.18	6.25	1.8	1.79	1.65
3	Uncultivated uncropped	53.81	36.88	21.64	15.37	7.34	6.93	0.5	0.65	0.32
4	Staggered contour trench planting	29.72	16.53	11.58	8.49	3.25	3.71	1.03	1.20	0.72
C	Neem									
1	Cultivated uncropped	107.6	119.8	110.6	30.74	23.86	35.44	6.8	8.48	7.1
2	Cultivated cropped	45.18	43.06	43.28	12.91	8.47	13.61	2.1	2.79	3.15
3	Uncultivated uncropped	34.37	29.04	45.71	9.82	6.81	14.66	0.97	0.72	1.36
4	Staggered contour trench planting	18.42	14.69	33.85	5.26	2.92	10.71	1.46	1.92	1.82

Hiwale (2004)

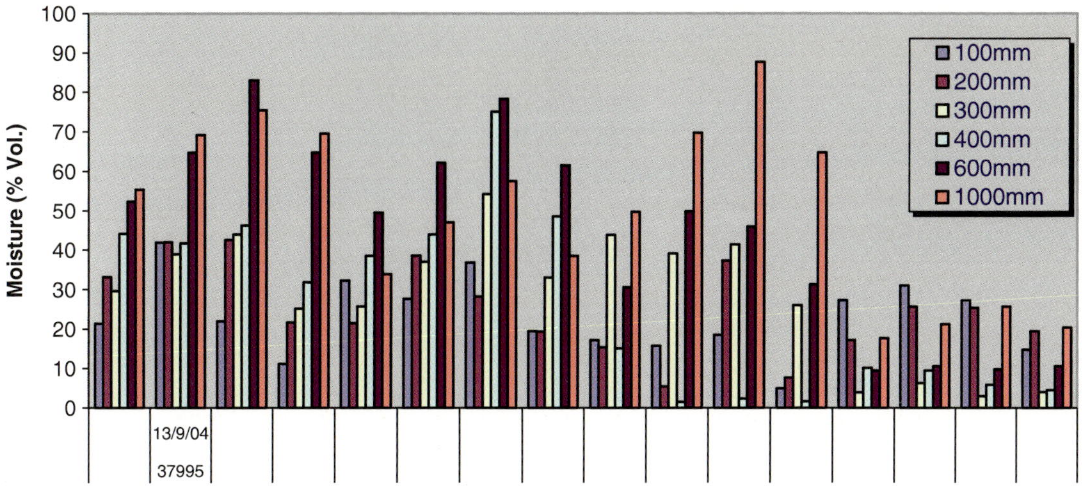

Fig. 27.6 Moisture content (% vol.) in different depth of nearer to various trees

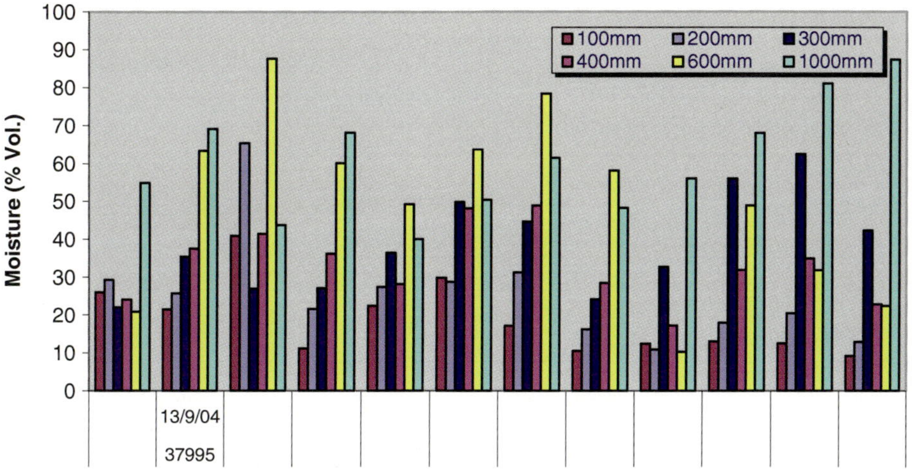

Fig. 27.7 Moisture content (% vol.) in different depth of between various trees

The rainfall intensity reduced after 16/8 to 24/8 which was the least;15 days after cessation of rainfall moisture content dropped by almost 30–50 % at all the depths except at 1,000 mm depth, indicating faster soil evaporation and loss by transpiration up to 300 mm and deep percolation of water to 600 and 1,000 mm depth; on 27/9 the intercrop plants started showing wilting symptoms as the water content was reduced to the level of permanent wilting point, whereas tree species still continue their growth without apparent sign of water deficit probably the root system was able to extract water stored in deeper layers. Thus, even after rainy season effective use of soil water can be achieved by mixing tree crops. It is evident from the data that 300 mm is the point of zero balance (ZFP), i.e., above which the water is lost by evapotranspiration and below which there is drainage (Figs. 27.6 and 27.7).

In semiarid areas 30–60 % of rainfall is lost to the atmosphere through soil evaporation. Shading of ground by trees and/or crop canopy should reduce the first-phase soil evaporation rates; second-phase soil evaporation rates are determined by soil hydraulic properties and should be independent of shade (Black et al. 1969). Soil moisture

Table 27.33 Tree volume, leaf area index, biomass production, and conversion coefficient in tree species

Tree species	Tree volume (m^3)	Leaf area index	Biomass production (kg/plant) Aboveground	Biomass production (kg/plant) Belowground	Total biomass (kg/plant)	Conversion coefficient (e) kg/µmol m^2/s
Neem (5 years old)	25.25	25.38	71.0 (109.42)	26.99 (40.47)	97.99	2.81
Leucaena	30.34	34.70	195.0 (267.07)	15.85 (50.32)	210.85	6.43
Aonla (8 years old)	25.71	20.50	36.0 (57.02)	37.37 (62.02)	73.37	1.40
Custard apple	8.89	16.55	2.96 (3.45)	18.03 (7.0)	20.99	0.78

Figures in bracket are fresh weight values

content changes with time. The rate of change in water content increases abruptly. This point is intercepted as thc time when the roots of the annual plant have reached this depth. Decrease in soil water content before time is taken as drainage in soil water and after this time as root abstraction.

27.14 Tree Volume, Leaf Area Index, Biomass Production, and Conversion Coefficient in Tree Species

Silvicultural species have higher biomass production compared to fruit trees and biomass production is associated with leaf area index, i.e., the higher the leaf area index, the higher the biomass production (Table 27.33). The highest biomass was produced by subabul which may be due to its faster growth and adaptability to the climate followed by Neem. Similar trend in respect to conversion coefficient was recorded.

27.15 Photosynthetic Efficiency Conversion Coefficient, Stomatal Conduction, and Transpiration Rate in Different Cropping Systems

Agroforestry systems are more productive than the sole cropping primarily because of its greater light interception, better moisture utilization, and increased soil sustainability. The yield advantage in intercropping system in terms of dry matter is quite high in all the systems. In trees species light interception was higher in subabul and was lowest in custard apple, which clearly influence on dry matter production and leaf area indcx and conversion coefficient of the tree species (Fig. 27.8).

The intercrops grown between the tree species light interception obtained in pigeon pea as it is a slow-growing plant. The ultimate result is higher leaf area index and higher dry matter production with highest conversion coefficient. Rapidly transpiring vegetation will cause vapor pressure to increase faster than the temperature of soil. Relative increase in temperature and vapor pressure will in agroforestry if there is dry soil between the trees or decrease if there is crop between the trees. As the age of the plant increases, the leaf area index (LAI) also increases resulting in higher dry matter production. However, as LAI increases, mutual shading of leaves also increases and shaded leaves contribute less to the dry matter accumulation. There is therefore optimum diameter for maximum dry matter accumulation, which is reached when the largest number of leaves which receive sunlight for photosynthesis is greater than the rate of translocation and there is no accumulation in leaves. The rate of photosynthesis is influenced by solar radiation and CO_2 concentration (Fig. 27.9).

27.16 Technology Developed

Agri-horti production system work was carried out under the NAT Project "Develop sustainable Agri-Silvi-Horti production system" under rainfed condition on marginal lands at CHES,

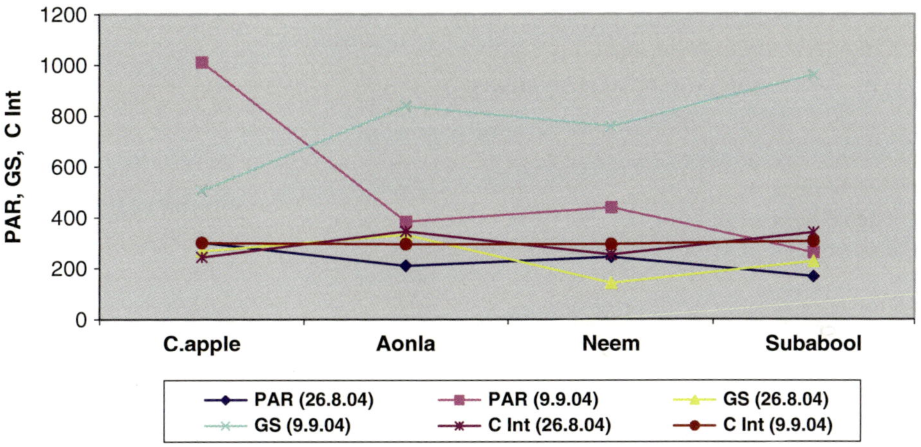

Fig. 27.8 Photosynthetic parameters of agri-horti cropping systems

Fig. 27.9 PAR, DM (Kg), LAI, volume (m^3) and conversion coefficient (*e*) of tree species

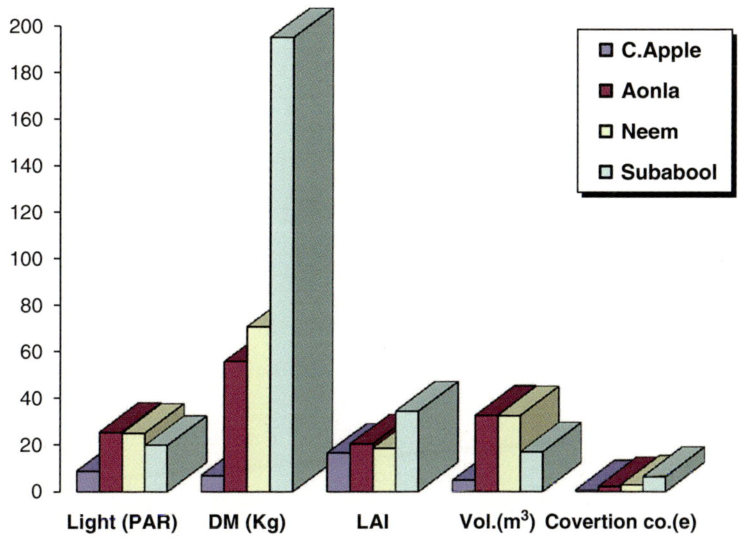

Vejalpur, Gujarat. Since the last two years clearly demonstrated that despite the drought conditions prevalent in the area during 2001 and 2002, aonla-based cropping system withstood vagaries of monsoon and though in 2001 all the agriculture crop species failed. Aonla produced normal crop to the extent of 60 q/ha in the seventh year of age. In the year 2002 also the system resulted in increased productivity of aonla crop to the tune of 75 q/ha. Though the rainfall was just sufficient for the crop to survive, agronomical crop produces some yield.

Economic analysis of the system carried out revealed that the highest net income per 1 ha of Rs. 69,720/- was obtained from aonla + maize cropping system compared to other cropping systems. The Highest B.C. ratio of 1:7.92 was however obtained in aonla + *Sesamum* cropping system. Custard apple-based cropping system though withstood drought resulted in less revenue generation of Rs. 13,964/- per ha in custard apple + okra cropping system, with a B.C. ratio of 1:1.54. Aonla crop can be grown profitably under purely rainfed conditions successfully; thus, it can be concluded that aonla-based cropping system was found to be drought hardy and is highly remunerative under purely rainfed conditions compared to all the other cropping systems.

27.17 Impact Assessment of Aonla-Based Cropping System

At CHES,Vejalpur, aonla-based intercropping system was found to be highly remunerative and paying proposition under rainfed conditions and the system proved to be one that mitigate drought effect better than the existing crop which failed during the past 2 years. The technology has been adapted for large-scale plantation in rainfed areas of western India in coarse cereal-based cropping system of Maharashtra and oil seed-based cropping system of Gujarat, M. P., and Rajasthan. Apart from this, area under cotton-based cropping system of Gujarat and Maharashtra was found to be the most suitable.

According to the latest statistics, percent area of aonla to net cropped area on national bases was highest in cotton-based cropping system (0.87 %) closely followed by coarse cereal-based (0.47 %) and oil seed-based (0.35 %) cropping system. The existing agri-horti production technology for aonla was refined with specific refinement like identification of suitable variety, grafting/budding technology for orchard establishment. Intercropping with various arable crops to get additional income from the unit area and rejuvenation of existing old plants by heading back and budding with superior variety.

Expected adoption of the technology was at the initial level, percentage of area to total cultivated area was 0.056 %, and minimum adoption level 3.404 %. The number of years required to achieve maximum adoption level is 5 years. The impact of refined technology resulted in reduction of adoption lag by 5 years and the number of neighboring farms that have taken up the technology has increased to 8 and the number of other villages who have taken up the technology has gone up to 24.

Soil improvement was distinct under different cropping systems. Because of regular cultivation and addition of biomass to the soil, it improves organic matter content and NPK content of the soil. Maximum leaf litter was added to the soil by aonla, thereby recycling maximum amount of NPK back to the soil. Large amount of leaf litter added to the soil by the tree crops and also covering of soil surface helped in reduced soil losses, thereby maintaining soil sustainability (Table 27.34).

Table 27.34 Impact of TAR/IVLP technology assessed/refined/feedbacks

Technology assessed	Refinements made over the existing technology	Feedbacks to researchers
1. Agri-horti	Variety NA-7	1. NA-7 compared with other variety
	In situ budding	2. Water harvesting should be studied in depth
	Intercropping	3. Time of in situ budding and rejuvenation standard and yield studies can be taken up
	Rejuvenation	4. Improvement in soil health over the years to be worked out
		5. Comparative yield studies should be taken up

1. Target domain (technology is targeted to)

Name of technology	Region/state/ districts	Name of production systems	Name of the crops
Aonla based	Semiarid region – Maharashtra	Coarse cereal based	Aonla + maize
Agri-horti	Gujarat	Oil seed based	
Production system	M.P. Rajasthan	Cotton based	

2. Performance of the new technology assessed/refined at farmer's field in comparison to the existing technology

Economic advantage	Assessed	Refined
Name of the crop	Maize	Aonla + maize
Per hectare yield increase (%)	See Annexure-I	See Annexure-I
Per hectare cost reduction (%)	See Annexure-I	See Annexure-I

3. Expected adoption

Parameters	Assessed	Refined
Name of the crop	Maize	Aonla + maize
Year of initial adoption	–	1999
Initial level of adoption (as percentage of area/livestock population in target domain)	–	0.056
Maximum adoption level (%)	–	3.404
Number of years required to achieve maximum adoption level	–	5 years
Number of years the new technology will be replaced by a more superior technology	–	20. rs.

4. Area (cropping system wise) distributed under aonla as on date

	Net cropped area on national basis (ha)	Aonla area under different cropping systems (ha)	Percentage area of aonla to net cropped area
Coarse cereal based	1,568,000	7,404.60	0.47
Oil seed based	1,792,000	6,329.60	0.35
Cotton based	896,000	7,763.65	0.87

5. Impact of technology dissemination through large-scale on-farm demonstration

Particular	Assessed	Refined
Name of the crop	Maize + *Cajanus*	Aonla + maize
Reducing adoption lags (years)		5 years
Total number of farmers covered under demonstration		–
Total area (ha) covered under demonstration		2 ha
Number of farmers who had taken a demonstration earlier and are continuing		–
Number of neighboring farmers have taken up the technology		8
Number of other villages who have taken up the technology from these demonstrations		24

6. Yield advantage – aonla per ha

Year	Yield (q/ha)	Gross return (Rs.)	Net return (Rs.)	Cost of cultivation (Rs.)
1994–1995	–	–		10,970[a]
1995–1996	–	–		1,500
1996–1997	07.80	07,800	05,800	2,000
1997–1998	25.80	25,800	23,800	2,500
1998–1999	39.30	39,300	36,800	2,500
1999–2000	61.70	61,300	58,700	3,000
2000–2001	75.40	75,400	71,900	3,500
2001–2002	53.30	53,300	49,800	3,500

[a]Includes cost of orchard establishment, aonla sale rate Rs. 1,000/- per q

7. Yield per ha, return, and cost of production of maize

Year	Yield (q/ha)	Gross return (Rs.)	Net return (Rs.)	Cost of cultivation (Rs.)
1994–1995	10.40	2,080	580	1,500
1995–1996	9.70	1,940	440	1,500
1996–1997	9.40	2,350	850	1,500
1997–1998	8.90	2,225	725	1,500
1998–1999	8.30	2,075	575	1,500
1999–2000	7.40	2,960	1,460	1,500
2000–2001	5.60	2,240	744	1,500
2001–2002	6.30	3,150	1,650	1,500

Maize sale rate: 1994–1996 Rs. 200/q, 1996–1999 Rs. 250/q, 2000–2001 Rs. 400/q, 2002 Rs. 500/q

8. Yield/ha, return, cost of production, and yield advantage in aonla-based agri-horti production system

	Yield (q/ha)			Net return (Rs.)			Yield increase				
Year	Main crop	Intercrop	Pure crop	Main crop	Intercrop	Pure crop	Main crop	Intercrop	Pure crop	Per ha	%
1994–1995	–	10.40	16.42	–	580	1,284	1,0970[a]	1,500	2,000	–	–
1995–96	–	9.70	17.51	–	440	1,502	1,500	1,500	2,000	–	–
1996–1997	7.80	4.40	17.94	5,300	850	2,485	2,000	1,500	2,000	–	–
1997–1998	25.80	8.90	15.23	23,330	725	1,807	2,500	1,500	2,000	2.28	228
1998–1999	39.30	8.30	14.19	36,800	575	1,047	2,500	1,500	2,500	3.35	335
1999–2000	61.70	7.40	13.36	58,700	1,460	2,844	3,000	1,500	2,500	5.17	517
2000–2001	75.40	5.60	15.32	71,900	744	3,592	3,500	1,500	2,500	5.29	529
2001–2002	53.30	6.30	16.48	49,800	1,650	5,740	3,500	1,500	2,500	3.62	362

Sale rate: *Aonla* Rs. 1,000/q, *maize* Rs. 200/q (1994–1996), Rs. 250/q (1996–1999), Rs. 400/q (2000–2001), Rs. 500/q (2002–2003)

[a]Includes cost of orchard establishment

References

Abdulbaki A, Anderson JD (1973) Vigor determination in soybean seeds by multiple criteria. Crop Sci 13:630–663

Agrawal RK (1980) Physicochemical status of soil under khejri (*Prosopis cineraria*). In: Mann HS, Saxena SK (eds) Khejri (*Prosopis cineraria*) in the Indian deserts. CAZRI, Jodhpur, pp 32–36

Anonymous (1984) Detail soil survey of CHES, Godhra, Gujarat by NBSS & LUP, Center Baroda

BanwariLal (2000) Beneficial allelopathic effect of *Syzygium cumini* litter fall biomass on arable crops. Indian J Agro For 2:53–57

Berg B, Ekbohm G (1991) Litter mass loss rates and decay pattern in some needles and leaf litter types. Long term decomposition in a Scots pine forest VII. Can J Bot 69:1449–1456

Bhatt BP, Todaria NP (1990) Studies on allopathic effect of some agroforestry tree crops of Gaharwal Himalayas. Agrofor Syst 12:251–258

Bisla SS, Nandal DS (1992) Influence of aqueous leaf extract of eucalyptus and poplar on germination and seedling growth in winter crops. In: Tauro P, Narwal SS (eds) Proceedings of first national symposium on Allelopathy in Agro eco system. HAU, Hissr

Black TA, Garduer WR, Thartel GW (1969) The prediction o evaporation, drainage of soil water storage foe a bare soil. Soil Sci Soc Am J 33:655–660

Bocock KI (1964) Changes in the amount of dry matter, nitrogen, carbon and energy in decomposing woodland and leaf litter in relation to activities of the soil fauna. J Ecol 2:273–284

Bocock KL, Gilbert G (1957) The disappearance of leaf litter under different wood land conditions. Plant Soil 9:179–185

Brower R (1983) Functional equilibrium sense or non sense Netherland. J Agric Sci 31:335–348

Challa P, Ravindra V (1999) Allelopathic potential of mango leaves for weed management in Rose (*Rosa hybrida* cv. Happiness) basins. Allelopath J 6(1):75–80

CSWCRTI (1980) 25 Years of research on soil and water conservation in semi arid deep black soils. Monograph No. 1. Central Soil and Water Conservation Research and Training Institute, Research Center, Bellary, Karnataka, pp 189

Dalal MR, Dahiya DS, Sarmah MK, Narwal SS (1992) Suppression effect of arid zone trees on plant stand and growth of crops. In: Tauro P, Narwal SS (eds) Proceedings of first national symposium on allelopathy in agro-ecosystem. HAU, Hissar, pp 127–128

Das AK, Ramakrishna PS (1985) Litter dynamic in khasi pine of north East India. For Ecol Manage 10:135–153

Dhruva Narayan VV, Ram Babu (1983) Estimation of soil erosion in India. Irrig Drain Eng 109(4):419–434

FAO (1978) Report on the FAO/UNEP expert consultation on methodology for assessing soil degradation. FAO, Rome, 70 pp

Gaba RK (1987) Role of allelopathic in social forestry. In: Khosla PK, Kohli RK (eds) Social forestry for rural development. ISTS, Solan

Gill AS (2001) Tips for the success of agro forestry. Indian Farm 3:21–24

Hiwale SS (2000) Annual report of national agricultural technology project on develop sustainable agri-horti-silvi production system on marginal land, pp 25–54

Hiwale SS (2001) Ber (*Zizyphus mauritiana* Lamk.) – a suitable agro-forestry species for semi-arid conditions for western India. Paper presented in national workshop on reorientation of diversified agro-forestry system for profitable rural industries, held at G.B. Pant University of Agriculture and Technology, 26–27 March 2001

Hiwale SS (2004) Technical bulletin on develop sustainable agri–horti on marginal lands under rain fed condi-

tions. Central Horticultural Experiment Station (C.I.A.H.), Vejalpur-389340
Hiwale SS (2014) Paper entitled "Increasing productivity of marginal lands through rainfed fruit cultivation in semiarid tropics". In: ISTS –IUFRO conference on sustainable resource management for climate change mitigation and social security, Ludhiana, 13–5 March 2014
Hiwale SS, Raturi GB (1996) Propagation of Ber by *In – Situ* budding. Indian Hortic 41(3):25–26
Hiwale SS, Raturi GB, Bagle B.G, More TA (2007) Technical bulletin fruit tree base cropping systems for dry lands
Hussain FI, Bong–Seop J (1991) Allelopathic effect of walnut plants (*Juglans regia*) on four crop species. Korean J Bot 34:93–100
Info@treesftf.org. How to calculate the amount of CO^2 sequestered per tree
Kaur A, Rao PB (2000) Allelopathic effect of four agro forestry tree species on seed germination and seedling growth of certain varieties of wheat. Acta Bot Ind 28:9–14
Kaur A, Shankdhar SC, Rao PB (1999) Allopathic studies of Eucalyptus, Leucaena and popular tree species on seed germination and seedling growth of certain varieties of wheat. Proc Acad Environ Biol 8:205–215
Kumar BM, Deepu JK (1992) Litter production and decomposition dynamics in deciduous forests of Western Ghats in peninsular India. For Ecol Manage 50(2):181–201
Kunhama TK (1994) Nutrient content of decomposition of leaf biomass of selected woody tree Species. M. Sc. thesis submitted to Kerala Agricultural University, Trichur, India
Lal R (1988) Soil erosion control with alley cropping. In: 5th international soil conservation conference, Bangkok
Lal B, Singh HK, Khosla OPS (1997) Effect of litter biomass of MPTs and nitrogen sources on growth and yield of forage crops. Indian J Soil Conserv 25(1):46–50
Lewis J, Vosti S, Ericson PJ, Guevera R, Tomich T (2002) Alternative to slash and burn in brazil, Summary report, pp 32–33
Moore AM (1986) Temperature and moisture dependence of decomposition rates of hard wood and coniferous leaf litter. Siol Biol Biochem 18:427–435
Mutanal SM, Nadagoudar BS, Patil SJ (2001) Economic evaluation of an agro–forestry system in hill zone of Karnataka. Indian J Agric Sci 71(3):163–165
Nandal DPS, Bisla SS, Narwal SS, Kaushik JC (1994) Allelopathic interactions in agro-forestry systems. In: Narwal SS, Tauro P (eds) Allelopathy in agriculture and forestry. Scientific Publishers, Jodhpur, pp 92–130
Narwal SS (1994) Allelopathy in crop production. Scientific Publisher, Jodhpur, 288 p
Pande PK, Sharma SC (1988) Litter nutrient dynamics of some plantation at New Forest, Dehra Dun. J Trop For 4:339–349
Patnaik US, James VV, Chittaranjan S, Ramnath B, Subbayan R (1982) Conservation ditching for soil and water conservation in deep black soils. Annual report, Central Soil and Water Conservation Research and Training Institute, Dehra Dun, pp 180
Putnam AR, Tang CS (1986) The science of allelopathy. Wiley, New York
Ramakrishna A, Singh B (1974) Wasteland News 3(1): 79–87
Rao PB (1988) Effect of environmental factors on seed germination and seedling growth in *Quercus floribunda* and *Cupressus torulosa* tree species of central Himalaya. Ann Bot 61:531–540
Rao OP, Saxena AK, Singh BP (1994) Allelopathic effects of certain agro forestry tree species on the germination of wheat, paddy and gram. Ann For 2:60–64
Raturi GB, Hiwale SS (1993a) Horti-silvi-pastoral system for increased productivity of marginal and degraded lands under rainfed conditions. Adv Hortic For 3:179–186
Raturi GB, Hiwale SS (1993b) Horti-silvi-pastoral system for increased productivity of marginal and degraded lands under rainfed conditions. In: Trivedi AP, Gupta GN (eds) Chapter in book on "Afforestation of Arid Lands". Scientific Publisher, Jodhpur, pp 364–369
Raturi GB, Hiwale SS (1999) Horti-Silvi-Pastoral system – a boon to the tribal farmers. In: Lead paper presented in national seminar on sustainable horticultural production in tribal areas, held at C.H.E.S., Ranchi on 25–26 July 1999
Richardson DR, Williamson GB (1988) Allopathic effect of Shrubs of the Sandpine shrubs on Pines and grasses of the sand hills. For sci 34:592–605
Roose EJ (1977) Erosion et ruissellment en Afrique de l'ouest. Vingt annees de measures en Petites parcelies experimentales. Travaux et Documents de L'ORSTOM, Office dela Research Scientifique et Technique Outre-Mer, France No. 78, 108 pp
Sharma KMS, Dhillon MS, Dhingra KK (1967) Presence of germination inhibitors in the leaf leachates of some farm grown trees. Indian Forester 113:816–820
Suresh KK, Rai RSV (1988) Allelopathic exclusion of understory by a few multipurpose trees. Int Tree Crop J 5:143–151
Swift MJ, Heal OW, Anderson JM (1979) Decomposition in terrestrial ecosystems. University of California Press, Berkeley, pp 46–87
Tondon HLS (1984) Methods of analysis of soils, plants, water and fertilizer. Fertilizer Development and Cultivation Organization, New Delhi
Van Noorwijik RJ (1989) Functional interpretation of root densities in the field for nutrient and water uptake edits Kustecher warzeslogic and Nutzenwadung, Gumpenstin Iraning, pp 262–279
Wiersum KF (1985) Effects of various vegetative layers in an *Acacia auriculiformis* forest plantation on surface erosion in Java, Indonesia. In: EI-Swaify SA, Molenhauer WC (eds) Soil erosion and conservation
Young A (1988) Agro forestry in the control of soil erosion by water. Agro For Abstr 1(2/3):39–48
Young A (1997) Agroforestry for soil management, 2nd edn. CAB International/ICRAF, New York, pp 47–109

Part V

Post Harvest Studies

Prolonging Shelf Life of Some Semiarid Fruits

28

Abstract

Arid zone fruits being highly perishable have to be marketed immediately after harvest. The postharvest losses range from 25 to 30 %. Chudawat et al. reported that most of the fruits could be stored for only up to 3–4 days at room temperature. Enzyme evolution of fruits on the tree is not triggered due to the presence of natural inhibitors. Various chemicals are known to play a significant role in suppression of enzyme activity (Desai and Deshpande; J Maharashtra Agric Univ, 9:24–26, 1978). Considering its short shelf life, work in arid zone was initiated to study the effect of ripening retardants on enzyme activity in arid zone fruits.

The fruits showed increased loss of weight and a fall in percentage of marketable fruits. The total soluble solids, acidity, ascorbic acid, and starch content declined, while activities of amylase, peroxidase, catalase, and pectin methyl esterase increased along with the evolution of ethylene. The content of sugar increased first and then registered a fall. Treatment with $CaCl_2$, GA_3, and bavistin delayed the ripening processes and the effect was pronounced in fruits with bavistin 1,000 ppm or gibberellic acid 200 ppm was most effective in altering most of the physical, physiological, and biochemical changes, giving maximum shelf life in terms of marketable fruits.

28.1 Introduction

The fruits are cherished bounties of nature and important in human diet as a major source of vitamins and minerals. The total production of fruits to date is 32.9 million tons. India occupies the first position (8.92 % share) in world fruit production. However, the capita per consumption is just 30 g/head.

Arid zone fruits being highly perishable have to be marketed immediately after harvest. Arid zone fruits such as ber, custard apple, guava, aonla, Jamun, etc. are highly perishable in nature as compared to some tropical and subtropical fruits; also, they are grown in remote areas and, hence, need special attention to extend their shelf life. Nowadays, they are extensively cultivated as they fetch remunerative prices in the market.

S. Hiwale, *Sustainable Horticulture in Semiarid Dry Lands*,
DOI 10.1007/978-81-322-2244-6_28, © Springer India 2015

Postharvest handling can play a major role in reducing fruit losses by following definite methodology, viz., timing of harvest, precooling, treatment with chemicals, packing, transportation, and marketing. Postharvest losses can be prevented by adopting various methods such as pre- and postharvest treatments with certain chemicals, ripening retardants, ethylene absorbents, ethylene antagonists, fungicidal dips, etc. These play a vital role in retarding the rate of ripening, thereby delaying the physicochemical and enzymatic activity in treated fruits which ultimately lead to increased shelf life of fruit. The experiment described below was, therefore, carried out to increase the shelf life of all the fruits.

With this view in mind and to develop a postharvest method which is simple, cost effective, and can be conveniently used on farm, which can help in extending the shelf life of these fragile fruits.

28.2 Treatment Details

T1 – Calcium chloride 2 % plus packing in polythene bags

T2 – Gibberellic acid 200 ppm plus packing in polyethylene bags

T3 – Bavistin 1,000 ppm plus packing in polyethylene bags

T4 – Calcium chloride 2 % plus packing in polythene bags with $KMnO_4$-coated silica gel

T5 – Gibberellic acid 200 ppm plus packing in polyethylene bags with $KMnO_4$-coated silica gel

T6 – Bavistin 1,000 ppm plus packing in polyethylene bags with $KMnO_4$-coated silica gel

T7 – Packing in polyethylene bags with $KMnO_4$-coated silica gel

T8 – Packing in polyethylene bags (control)

Upon observation on physiological, physical, chemical, and biochemical characteristics, oxidative and cell wall softening enzymes along with ethylene evolution were recorded at the 2nd, 4th, and 6th day of storage.

The results of the studies revealed that weight loss and ripening increased with increase in storage period (Tables 28.1, 28.2, and 28.3). Treating the fruits with various chemicals and storing them in polyethylene bags along with $KMnO_4$-coated silica gel were most effective in reducing the rate of increase in percent weight loss and ripening as compared to control. The maximum physiological loss in weight was observed in ber followed by aonla. The maximum ripening

Table 28.1 Physiological loss in weight (%)

Treatments	Ber	Custard apple	Aonla	Jamun	Guava
T1	3.44	5.48	1.57	2.82	3.05
T2	3.20	5.83	1.60	2.61	3.05
T3	3.44	6.19	1.56	2.57	2.88
T4	0.91	3.48	1.00	2.40	2.27
T5	0.76	3.28	0.86	2.29	2.50
T6	0.70	3.20	0.84	2.33	2.08
T7	4.24	6.60	2.24	3.17	3.30
T8	4.86	8.28	2.46	4.49	4.65
Mean	2.65	5.29	1.52	2.83	2.99
CD at 5 %	0.51	0.94	0.34	0.20	0.89

Table 28.2 Ripening (%)

Treatments	Ber	Custard apple	Guava
T1	67.35	41.66	14.25
T2	69.05	38.89	12.03
T3	67.35	40.74	11.48
T4	63.14	31.48	9.26
T5	63.14	24.07	7.59
T6	66.41	25.41	6.85
T7	72.93	42.59	18.96
T8	75.14	51.85	27.40
Mean	68.22	37.03	13.45
CD at 5 %	2.53	4.13	3.62

Table 28.3 Marketable fruits (%)

Treatments	Ber	Custard apple	Aonla	Jamun	Guava
T1	98.15	93.52	97.41	60.94	97.77
T2	98.52	95.37	98.14	68.23	98.52
T3	99.63	100.00	98.52	64.39	97.78
T4	99.63	97.22	98.89	66.04	99.26
T5	99.26	98.15	99.63	73.29	99.26
T6	99.63	100.00	100.00	73.89	99.26
T7	97.04	88.89	97.03	57.89	95.92
T8	94.29	80.93	96.29	44.80	93.33
Mean	98.32	94.26	98.24	63.68	97.64
CD at 5 %	3.32	5.42	3.70	2.67	5.37

Table 28.4 Spoilage (%)

Treatments	Ber	Custard apple	Aonla	Jamun	Guava
T1	1.48	11.11	2.59	34.09	2.59
T2	1.48	6.94	1.85	36.77	2.59
T3	0.37	1.39	1.48	38.95	2.59
T4	0.37	4.16	1.11	35.63	0.74
T5	0.74	2.78	0.37	28.38	0.74
T6	0.00	0.00	0.00	27.78	0.74
T7	2.96	18.05	2.96	45.44	4.07
T8	5.92	29.16	3.70	54.70	5.55
Mean	1.67	9.20	1.76	37.72	2.36
CD at 5 %	1.90	6.59	2.50	3.84	3.60

percent is observed in ber as compared to other fruits. The reduction in weight loss is due to packing in polyethylene bags, which had a significant effect on reducing the respiration. Similar results were also reported by Dhoot et al. (1984) in guava and Jain et al. (1979) in ber.

The highest percentage of marketable fruits were observed in fruits treated with bavistin (1,000 ppm) and packed in polyethylene bags with $KMNO_4$-coated silica gel (100 %) (T6). The maximum percentage of marketable fruits was observed in custard apple and aonla at T6 (100 %). The higher percentage of marketable fruits could be obtained because there had been reduced percent loss in weight and also spoilage. Waval and Atale (1989) reported increased marketability of fruits in mango when the fruits were treated with different growth regulators.

Percent spoilage increased with increase in storage period (Table 28.4). Maximum spoilage was observed in custard apple and ber. Treating the fruits with bavistin (1,000 ppm) and then packing them in polythene bags containing in $KMnO_4$-impregnated silica gel was effective in reducing the percent spoilage of fruits (Table 28.5).

Organoleptic rating was also highest in response to this treatment. It was maximum in aonla followed by guava. This indicates that bavistin treatment prevented microbial damage, while packing in $KMnO_4$-coated silica gel slowed down ripening process. Gupta and Srivastava (1979) observed that bavistin was effective in controlling the storage rot in mango.

Table 28.5 Organoleptic rating

Treatments	Ber	Custard apple	Aonla	Guava
T1	50.33	51.66	68.88	62.66
T2	53.33	55.00	68.77	63.33
T3	53.33	51.66	69.44	65.00
T4	66.66	56.66	71.68	66.66
T5	75.33	58.33	81.66	71.66
T6	73.33	58.33	77.93	73.33
T7	46.66	48.33	46.10	49.66
T8	43.33	45.00	45.22	48.33

Table 28.6 Acidity %

Treatments	Ber	Custard apple	Aonla	Jamun	Guava
T1	0.21	0.28	1.85	0.79	0.43
T2	0.20	0.28	1.79	0.79	0.45
T3	0.20	0.28	1.80	0.78	0.44
T4	0.22	0.30	1.94	0.76	0.47
T5	0.22	0.32	1.99	0.74	0.47
T6	0.21	0.31	1.96	0.79	0.46
T7	0.20	0.29	1.59	0.79	0.42
T8	0.19	0.26	1.55	0.81	0.42
Mean	0.21	0.29	1.81	0.78	0.45
CD at 5 %	0.07	0.13	0.32	NS	0.10

Acidity and ascorbic acid content decreased at a slower rate, when the fruits were treated with calcium chloride (2 %), gibberellic acid (200 ppm), and bavistin (1,000 ppm) as a postharvest dip for 5 min and packed in polyethylene bags with $KMnO_4$-coated silica gel in T4 and T6 as compared to control (Table 28.6). The slow rate of degradation of acidity and ascorbic acid content in fruits packed in polyethylene bags with chemical treatments may be due to reduced loss in weight and reduced enzymatic activity, thereby reducing the rate of ripening. Gupta et al. (1987) reported slow rate of ripening in ber cv. Umran. As the storage period increased, the content of acidity and ascorbic acid declined.

The total soluble solids decreased with increase in storage period. The maximum TSS (19.66., 22.78., 12.33., 14.57., and 11.51° Brix in ber, custard apple, aonla, Jamun, and guava, respectively) was observed in untreated control (T8). It was minimum in calcium chloride 2 %

Table 28.7 Ascorbic acid content (mg/100 g)

Treatments	Ber	Custard apple	Aonla	Jamun	Guava
T1	78.20	14.00	375.71	13.69	181.01
T2	83.26	15.53	358.04	14.65	188.24
T3	81.38	15.25	367.47	15.90	193.95
T4	86.33	16.71	376.89	18.78	191.71
T5	87.98	16.22	389.84	18.55	189.72
T6	88.22	16.26	389.84	18.37	191.94
T7	75.97	14.61	348.36	13.24	176.56
T8	70.35	14.62	329.07	11.10	171.09
Mean	81.46	15.40	366.90	15.53	185.55
CD at 5 %	3.58	2.00	23.35	2.02	10.40

Table 28.8 Total soluble solid content

Treatments	Ber	Custard apple	Aonla	Jamun	Guava
T1	18.20	21.94	12.00	13.20	10.67
T2	18.00	22.28	11.94	13.07	10.84
T3	18.37	21.44	11.83	13.10	10.69
T4	18.30	21.11	12.28	12.30	10.29
T5	18.42	21.77	11.61	12.10	10.28
T6	17.68	21.61	11.67	11.93	10.18
T7	18.60	22.28	12.06	13.67	11.27
T8	19.66	22.78	12.33	14.57	11.51
Mean	17.86	21.90	11.97	12.99	10.71
CD at 5 %	1.01	0.97	0.48	0.77	0.92

plus packing in polythene bags with KMnO4-coated silica gel (T4) in custard apple. The increased content of TSS might be due to reduced starch hydrolysis during the early period. In the later stage of storage, decreases in TSS could be due to rapid utilization of sugar. Singh et al. (1981) in guava reported similar results (Tables 28.7 and 28.8).

Reducing sugar and total sugar showed an increasing trend with storage period (Tables 28.9 and 28.10) in all the fruits. The rate of increase was slow and steady in fruits treated with calcium chloride (2 %), gibberellic acid (200 ppm), and bavistin (1,000 ppm) and packed in polyethylene bags with $KMnO_4$-coated silica gel. The increased retention might be due to suppression of activity of hydrolytic enzymes like alpha-amylase. Similar results were reported by Dhoot et al. (1984).

Increased alpha-amylase activity was observed with decrease in starch content of the fruits. Starch content decreased as the storage period increased (Table 28.11). The rate of increase in

Table 28.9 Reducing sugar content

Treatments	Ber	Custard apple	Aonla	Jamun	Guava
T1	4.24	14.23	4.19	8.44	4.14
T2	4.06	14.34	4.10	8.41	4.09
T3	3.97	14.48	4.10	8.33	4.29
T4	3.87	12.96	4.19	8.02	3.98
T5	3.81	13.91	4.14	7.68	3.37
T6	3.71	13.55	3.97	7.72	4.79
T7	4.48	14.60	4.10	9.19	5.05
T8	4.97	15.15	4.27	9.60	5.26
Mean	4.14	14.15	4.13	8.42	4.37
CD at 5 %	0.31	0.71	0.16	0.38	0.39

Table 28.10 Total sugar content (%)

Treatment	Ber	Custard apple	Aonla	Jamun	Guava
T1	8.19	14.63	4.91	8.26	7.91
T2	8.33	14.16	5.00	8.14	7.91
T3	7.94	13.98	4.98	8.26	7.95
T4	7.93	13.59	4.89	6.90	7.11
T5	8.04	13.37	4.56	6.79	7.01
T6	8.54	13.28	4.48	6.79	7.80
T7	9.50	14.23	4.86	9.06	8.50
T8	9.90	15.30	4.89	9.76	9.07
Mean	8.55	14.07	4.82	7.99	7.87

Table 28.11 Starch content

Treatments	Ber	Custard apple	Aonla	Jamun	Guava
T1	1.47	3.33	2.64	1.12	1.92
T2	1.65	3.14	2.69	1.10	2.07
T3	1.62	3.33	2.60	1.09	1.92
T4	1.70	3.16	2.80	1.16	2.22
T5	1.83	3.24	2.86	1.13	2.27
T6	1.77	3.37	2.78	1.19	2.20
T7	1.31	2.72	2.58	1.08	1.73
T8	1.19	2.62	2.44	0.97	1.56
Mean	1.57	3.11	2.67	1.11	1.99
CD at 5 %	0.20	0.21	0.18	0.09	0.14

alpha-amylase activity was slower in fruits treated with calcium chloride (2 %), gibberellic acid (200 ppm), and bavistin (1,000 ppm) and packed in polyethylene bags with $KMnO_4$-coated silica gel. Desai and Deshpande (1978) reported decreased amylase activity in banana treated with gibberellic acid and kinetin.

28.3 Summary

It is very clear from the above observations that there was retardation of ripening by application of various chemicals followed by packing them in polyethylene bags with $KMnO_4$-coated silica gel. These treatments also reduced the rate of degradation of various chemicals by retarding the activity of various oxidizing enzymes and ethylene evolution up to 4 days of storage. Thus, it is possible to increase the shelf life of semiarid fruits up to 4–9 days after harvesting as compared to control.

References

Desai BB, Deshpande PB (1978) Hydrolytic and oxidative enzymes during banana ripening. J Maharashtra Agric Univ 9:24–26

Dhoot LR, Desai UT, Rane DA (1984) Studies on shelf life of guava fruits with polythene packing and chemical treatments. J Maharashtra Agric Univ 9(2):185–188

Gupta JH, Srivastava VP (1979) Control of post harvest diseases of mango. Mango workers meeting tech.doc, Panaji, Goa, pp 206–206

Gupta OP, Siddique S, Chauhan KS (1987) Evaluation of various calcium compounds for increasing shelf life of ber fruits. Indian J Agric Res 21(2):65–67

Jain RK, Chitkara SD, Singh K (1979) Effect of various post harvest treatments on storage of Umran (*Zizyphus mauritiana* Lamk.). Haryana J Hortic Sci 8(3/4):181–185

Singh BP, Singh HK, Chauhan KS (1981) Effect of post harvest calcium treatments on storage life of guava fruits. Ind J Agric Sci 51:44–47

Waval KN, Atale PW (1989) Studies to prolong shelf life of mango fruits. Acta Hortic 231:771–775

29 Postharvest Enzymatic Activity of Some Arid Zone Fruits as Influenced by Chemical Treatments and Storage Period

Abstract

Arid zone fruits being highly perishable have to be marketed immediately after harvest. The postharvest losses range for 25–30 %. Chundawat et al. (Haryana J Hortic Sci 5:130–136, 1976) reported that most of the fruits could be stored for only up to 3–4 days at room temperature. Enzyme evolution of fruits on the tree is not triggered due to the presence of natural inhibitors. Various chemicals are known to play a significant role in suppression of enzyme activity (Desai and Deshpande; Physiol Plant 44:238–240, 1978). Considering its short shelf life, work in arid zone was initiated to study the effect of ripening retardants on enzyme activity in arid zone fruits.

29.1 Introduction

Arid zone fruits being highly perishable have to be marketed immediately after harvest. The postharvest losses range from 25 to 30 %. Chundawat et al. (1976) reported that most of the fruits could be stored for only up to 3–4 days at room temperature. Enzyme evolution of fruits on the tree is not triggered due to the presence of natural inhibitors. Various chemicals are known to play a significant role in suppression of enzyme activity (Desai and Deshpande, 1978). Considering its short shelf life, work in arid zone was initiated to study the effect of ripening retardants on enzyme activity in arid zone fruits.

The experiment was carried out by applying various treatments to the freshly harvested fruits with ripening retarding chemicals; fungicidal dip along with packing in polyethylene bags with and without $KMnO_4$-coated silica gel (an ethylene absorbent) and enzyme activity was recorded at 0, 3, 6, and 9 day's interval. Mature, fully developed fresh fruits of uniform size and free from blemishes were collected from all sides of the tree of guava. cv. Allahabad Safeda, ber cv. Gola, aonla cv. NA-7, and custard apple cv. Sindhan were selected for applying the treatments.

29.2 Treatment Details

T1 – Calcium chloride 2 % plus packing in polythene bags

T2 – Gibberellic acid 200 ppm plus packing in polyethylene bags

T3 – Bavistin 1,000 ppm plus packing in polyethylene bags

S. Hiwale, *Sustainable Horticulture in Semiarid Dry Lands*,
DOI 10.1007/978-81-322-2244-6_29, © Springer India 2015

T4 – Calcium chloride 2 % plus packing in polythene bags with $KMnO_4$-coated silica gel
T5 – Gibberellic acid 200 ppm plus packing in polyethylene bags with $KMnO_4$-coated silica gel
T6 – Bavistin 1,000 ppm plus packing in polyethylene bags with $KMnO_4$-coated silica gel
T7 – $KMnO_4$-coated silica gel and packing in polyethylene bags
T8 – Packing in polyethylene bags (control)

29.3 Enzymatic Changes

29.3.1 Peroxides Activity (Change in OD/min/ml)

There were significant differences in respect to chemical treatment as well as storage period on peroxide activity in all the fruits. Chemical treatments with calcium chloride (2 %), gibberellic acid (200 ppm), and bavistin (1,000 ppm) plus packing in polyethylene bags with $KMnO_4$-coated silica gel reduced the activity of the enzymes. Maximum peroxide activity was reduced in guava, and whereas it was least in ber, no peroxidase activity could be recorded in aonla probably due to the presence of alkaloids in the fruit (Table 29.1). Though the peroxide activity was low in ber and custard apple compared to guava, they ripened faster due to the fewer amount of enzyme that might be essential to trigger ripening.

The senescence of the tissue is controlled due to lower peroxide activity, which suppressed IAA oxidation and in turn suppressing the biosynthesis of ethylene, thus preventing degradation of metabolites. It is reported that enzyme activity involved plant growth regulators and control of enzyme activity is linked with indigenous plant growth regulators. Dilley (1969) reported that plant growth regulators had antagonistic effect on the biogenesis of endogenous ethylene. The results are in agreement with that of Rao and Chundawat (1990) in Basrai banana and Khader et al. (1988) in mango cv. Mallika.

29.3.2 α-Amylase Activity

The analysis of the data revealed that there were highly significant differences among the various chemical treatments. Minimum alpha-amylase activity was observed in gibberellic acid (200 ppm). Maximum alpha-amylase activity was recorded in untreated control (T8) (Table 29.1).

In aonla, guava, and custard apple, minimum alpha-amylase activity was observed in fruits treated with bavistin 1,000 ppm plus packing in polyethylene bags with $KMnO_4$-coated silica gel (T6), whereas in ber, minimum alpha-amylase activity was observed in gibberellic acid 200 ppm plus packing in polythene bags with $KMnO_4$-coated silica gel (T5). It was maximum in untreated fruits (T8) (Table 29.3). Dhoot et al. (1984)

Table 29.1 Peroxides and amylase activity in semiarid zone fruits

	Peroxides			Alpha amylase mg of maltose/15 min			
Treatment	Ber	Guava	C. apple	Ber	Aonla	Guava	C. apple
T1	0.05	0.12	0.09	5.16	5.86	6.47	5.54
T2	0.04	0.09	0.065	5.21	5.81	6.29	5.61
T3	0.055	0.105	0.085	5.38	5.80	6.28	5.22
T4	0.05	0.115	0.085	4.55	4.96	6.23	4.31
T5	0.03	0.08	0.065	4.43	4.51	5.90	4.43
T6	0.045	0.1	0.070	4.55	4.46	6.18	4.25
T7	0.06	0.135	0.11	5.43	6.49	6.88	6.76
T8	0.07	0.165	0.135	6.17	6.32	7.20	7.86
CD	0.01	0.05	0.01	0.39	0.47	0.325	0.54

reported decreased amylase activity in guava treated with gibberellic acid and kinetin.

Increase in storage period increased the alpha-amylase activity, and maximum alpha-amylase activity was recorded after nine days of storage. Chemical treatments of fruits with calcium chloride (2 %), gibberellic acid (200 ppm), and bavistin (1,000 ppm) plus packing in polyethylene bags with $KMnO_4$-coated silica gel lowered and slowed down the breakdown of starch, which was mainly due to suppression of alpha-amylase activity in these treatments as compared to control.

29.3.3 Pectin Methyl Esterase (PME) (ml of 0.01 NaOH)

All the chemically treated fruits stored in polythene bags along with $KMnO_4$-coated silica gel showed reduced enzyme activity. Pectin methyl esterase activity was significantly reduced in T6. Maximum PME activity was recorded in aonla and minimum in guava. Among the chemicals, calcium and bavistin reduced PME activity to the minimum (Table 29.2).

The PME activity increased as the storage period increased in guava and custard apple, whereas it reduces after 6 days in ber and aonla. Softening occurs in fresh fruits after maturity, and peak of ripening is generally associated with fairly narrow range of firmness. This change is brought about by the coordinated action of hydrolytic enzymes on pectin and other carbohydrates (Sinclair and Joliffs 1961). The results are in conformity with the studies carried out by Pal and Selvaraj (1987) in guava and Singh (1989) in mango.

29.3.4 Catalase (ml H_2O_2 Oxidized/min/ml)

In ber, guava, aonla, and custard apple, chemical treatment of fruits and storage in polythene bags with $KMnO_4$-coated silica gel were found to be the best for reducing enzyme activity and increasing shelf life up to 9 days in the case of guava and aonla and to 6 days in the case of ber and custard apple. The highest activity of enzymes was recorded in custard apple and least in ber (Table 29.2). The activity of enzymes was influenced significantly by the storage period in all fruits except guava, where it reduced down on the 9th day of storage. The senescence of the tissue is controlled due to lowered catalase activity, which suppressed IAA oxidation and in turn suppressing the biosynthesis of ethylene, thus preventing degradation of metabolites. It is reported that enzyme activity involved plant growth regulators and control of enzyme activity is linked with indigenous plant growth regulators. Dilley (1969) reported that plant growth regulators had antagonistic effect on the biogenesis of endogenous ethylene. The results are in agreement with that of Khader (1989).

Table 29.2 PME and catalase activity in arid zone fruits

	PME (ml of 0.1 N NaOH)				Catalase (μl of H_2O_2 oxidized/ml)			
Treatment	Ber	Aonla	Guava	C. apple	Ber	Aonla	Guava	C. apple
T1	1.61	2.59	1.60	1.88	376.49	619.32	536.37	657.63
T2	1.58	2.52	1.01	1.90	386.75	618.86	532.89	639.68
T3	1.59	2.72	1.47	1.86	417.06	629.40	531.58	622.24
T4	1.31	2.82	1.23	1.74	367.78	638.45	501.73	535.06
T5	1.29	2.80	1.19	1.67	359.76	635.77	488.36	545.37
T6	1.34	3.06	1.24	1.72	372.12	640.01	492.82	546.97
T7	1.85	3.31	1.71	2.10	424.80	625.66	548.81	654.66
T8	2.06	3.16	2.20	2.27	411.71	622.50	587.77	714.36
CD	0.165	0.175	0.165	0.17	24.07	NS	53.51	97.13

Table 29.3 Ethylene activity in arid zone fruits

	Ethylene mg/kg			
Treatment	Ber	Aonla	Guava	C. apple
T1	0.0065	0.014	0.0155	0.0104
T2	0.0065	0.015	0.0149	0.0105
T3	0.00635	0.0152	0.0147	0.0106
T4	0.00545	0.0135	0.0125	0.0095
T5	0.0052	0.0129	0.0124	0.0082
T6	0.00605	0.0133	0.0119	0.00915
T7	0.00715	0.0159	0.0159	0.01165
T8	0.0072	0.0173	0.0176	0.0132
CD	0.0032	0.0012	0.0032	0.0012

29.3.5 Ethylene Activity (mg/kg/h)

Minimum ethylene activity was observed in bavistin 1,000 ppm plus packing in polyethylene bag with $KMnO_4$-coated silica gel (T6) in aonla, guava, and custard apple. In ber, minimum ethylene activity was observed in gibberellic acid 1,000 ppm plus packing in polythene bags with $KMnO_4$-coated silica gel (T5). Maximum ethylene activity was recorded in untreated control (T8) (Table 29.3). Dostal and Leopard (1967) suggested that gibberellic acid would retard the senescence of fruit by opposing the action of ethylene produced during ripening. Dilley (1969) reported that plant growth regulators had antagonistic effect on the biogenesis of endogenous ethylene. Bavistin in combination with ethylene absorbent prevented the enzyme, thereby reducing the senescence process. The results are in line with those of Rao and Chundawat (1986) in banana. In respect to storage period, the minimum ethylene activity was recorded in three days of storage and the maximum in nine days of storage in pooled mean data.

References

Chundawat BS, Singh JP, Kainsa R, Gupta OP (1976) Post harvest studies on guava fruits. I. Effect of packing and storage period on quality of fruits. Haryana J Hortic Sci 5:130–136

Desai BB, Deshpande PB (1978) Chemical control of ripening in banana. Physiol Plant 44:238–240

Dhoot LR, Desai UT, Rane DA (1984) Studies on shelf life of guava fruits with polythene packing and chemical treatments. J Maharashtra Agric Univ 9(22):185–188

Dilley DR (1969) Hormonal control of fruit ripening. Hortic Sci 4:11–14

Dostal HC, Leopard AC (1967) Gibberellins delays ripening of tomatoes. Science 158:1979–1980

Khader SESA (1989) Delaying ripening by post harvest treatment of gibberelic acid in mango. Indian J Hortic 46(4):444–447

Khader SESA, Singh BP, Khan SA (1988) Effect of gibberelic acid as post harvest treatment of mango fruits on ripening, amylase and peroxides activity and quality during storage. Sci Hortic 36:261–266

Pal DK, Selvaraj Y (1987) Biochemistry of papaya (*Carica papaya* L.) fruit, ripening changes in DNA, RNA proteins and enzymes of mitochondrial carbohydrates, respiration and phosphate metabolism. J Hortic Sci 62:117–124

Rao DVR, Chundawat BS (1986) Effect of certain chemical retardants on ripening changes of banana cv. Lacatan at ambient temperature. Prog Hortic 18(3/4):189–195

Rao DVR, Chundawat BS (1990) Effect of certain ripening retardant on post harvest behaviour of Basrai banana at non-refrigerated temperature. Adv Hortic For 1:33–48

Sinclair WB, Joliffs VH (1961) Pectic substances of 'Valencia' oranges at different stages of maturity. J Food Sci 26:125

Singh DB (1989) Effect of post harvest treatments on ripening changes and quality of mango fruits (*Mangifera indica* L.) Cv.Kesar and Amrapali. Ph.D thesis submitted to the Gujarat Agriculture University, Anand

30 Value Addition in Underutilized Fruits

Abstract

Underutilized fruits are known for their wider adaptability to soil and climatic conditions in semiarid rainfed areas. Apart from this, they are a rich source of vitamins and minerals. Many of them have medicinal values par excellence. The fruits however have short fruiting season and are highly perishable in nature. There is a glut during the fruiting season and there is no market value. Value addition in these fruits with simple technologies which the farmers can adopt at their farm is the need of the hour. Efforts made at the station were successful and many products, viz., dried and dehydrated fruits, RTS, squash, fruit bars, candies, powders, wines, and condensed fruit juices through solar drying, were prepared. All the products were stored for a year under ambient conditions under central Gujarat conditions except condensed fruit juices. Fruits like ber, aonla, pomegranate, bael, date palm, sapota, and wood apple which are available locally were used. Chemical analysis of these fruits indicated that though there was a decrease in vitamins and mineral composition, they retain sufficient amount of them to meet the requirement of human body like vitamin C content in aonla squash or candies or mineral content in wood apple powder and date palm fruit bar. Use of these products will definitely help in arresting availability low nutritive food among the adivasi women and children.

30.1 Introduction

There is 30 % postharvest loss of fresh fruit, due to lack of postharvest technology for its value addition. There are large numbers of plants (though there are no regular plantations) of the highly nutritive fruits scattered all over the semiarid tract, and there is potential for developing cottage processing industry in the backward areas. Large quantity of these fruits is available in the season, and value addition will increase the meager income but will also create employment opportunity in the nearby places.

1. To develop protocols for value addition in semiarid fruits at cottage level
2. To standardize the quality of the product

S. Hiwale, *Sustainable Horticulture in Semiarid Dry Lands*,
DOI 10.1007/978-81-322-2244-6_30, © Springer India 2015

3. To develop technologies that can be adopted at farmer's level with the least infrastructure facilities
4. To improve value addition of underutilized semiarid fruit
5. To provide employment to rural poor by establishing self-help groups raising their income by sale of value-added products

As per the expert consultation on postharvest technologies for ensuring food security and value addition for enhanced income held in Dec. 2004 in Bangkok, postharvest losses in fruits are to the tune of 30 % of the total fruit harvest (Anon 2004). Control of postharvest losses and adding value to the fruit crops has therefore been recognized as an effective means of increasing the availability of fruits to the ever-increasing population at affordable prices. The losses can be avoided by two ways, either by developing the infrastructure facilities so that the fruits can be transported or stored and then transported to the markets or by developing the technology for value addition of the fruits. However, data indicate that hardly 1 % of the total production is utilized for value addition. With the increase in area under the fruits due to efforts made by the government, the production is already on increase, but postharvest losses cannot be reduced. Therefore, to sustain the growth, both in area and production, and at the same time maintain the profitability, there is need to develop simple, low-cost technologies resulting in reduction in postharvest losses, and avoid glut in the market, so that the fruit cultivation remains a profitable preposition to the farmers. This will not only help in developing the rural infrastructure but result in employment generation and augmentation of meager income.

The semiarid fruits are mostly consumed fresh in the local market and have to be disposed off as they are perishable in nature; hence, there is great scope for developing technology for value addition of these fruits. Apart from this, the consumer and fruit processing industry discard large quantity of inedible parts of the fruits. Amount of these wastes is as high as 70 % in apple. In mango, it ranges from 27 to 50 %, grapes 5–10 %, citrus 50 %, and banana 24–46 (Murti et al. 1976). The waste material (seed and pomace) arising out of value addition can be used for the preparation of cattle feed, extraction of pectin and oil, and preparation of paper. Canning of fruit slices, dehydration and dying, and preparing powder in ready-to-use form can open new vistas in this field (Singh and Singh 2005). Relevant work done on various aspects is being reviewed to get ideas, which can be applied to value addition of semiarid fruit.

India has done a good progress on the horticulture map of the world with a total annual production of fruits touching 45 million tons during 2003–2004 as mentioned in the Indian horticultural database 2005 published by the National Horticultural Board. Our share in the world production is about 10.1 % in fruits and 14.4 % in vegetables. The horticultural crops cover about 8 % of the total area contributing about 20 % of the gross agricultural output in the country. Though we are the second largest producer of fruits, our per capita consumption is very low. Apart from this, 30 % of the fruits is wasted due to the perishable nature of this commodity and inadequate facilities for transport, storage, and processing. Although the R&D efforts on the development of postharvest handling has helped in reducing the spoilage, considerable losses continue to occur. Even if 10 % of these losses could be prevented during the glut season at the production centers by converting them into new categories of value added products will result in the saving of Rs. 6,750 million (Potty 1988) and addition to it the revenue realized by sale of the product prepared will be to the tune of Rs. 10,000 million. This is the case with the major fruits grown in our country. In the case of underutilized fruits, there is practically no work on value addition, though they are available a plenty locally and are known since ages to contain fair amount of minerals and vitamins and were used by our ancestors to live a healthy life. Apart from this, they have medicinal properties. Thus, their consumption can help in overcoming the problem of malnutrition prevalent among the tribes in the backward areas. Nowadays, the area under these crops is increasing as peoples are now realizing the potential of these crops and their ability to grow under rainfed conditions naturally.

There are two approaches which will help in achieving the goal. One is by the creation of cold storage chains for storage of these fruits in the region where there is large-scale production; this will help in reducing the on-farm losses and will make the fruit available to the market and industry for a longer period helping in avoiding the glut situation during the season. Another approach is to process the fruits into various products with minimum capital and instrumentation. This will add to the value of the product. With increasing urbanization and population, the food habits of people are changing resulting in increased demand for such products.

In spite of all the efforts, only 2 % of the total production is used for the preparation of various products compared to 40–60 % in developed countries. Thus, there is scope for preservation and processing of the fruits which will help in extending the consumption period. Diversification of product range can definitely add to the marketability of the products. Establishment of processing unit in the area where availability of one or the other produce throughout the year is assured can reduce post harvest losses, the products prepered can be sold in distant markets and can provide employment to the locals. There is tremendous scope for developing technologies for value addition in these underutilized crops.

30.2 Various Factors That Should Be Considered While Setting Up a Processing Unit

- They are available locally in plenty but for a limited period.
- Reduce the postharvest losses.
- Generate additional income by value addition and marketing the produce.
- Improve the nutrition of the local population by consumption of their own processed product throughout the year.
- Generate employment.
- Develop processed product protocols which are not capital intensive or energy intensive and are useful to small-scale cottage industry.
- The product should have export value.
- Alleviating poverty in rural areas through additional income generation and employment creation requires identification and development of opportunities for diversification and adding values to on- and off-farm activities and integrating them into supply chain.
- There is need to create market-oriented rural production and value addition enterprise that are well developed with required service providers.
- There is need to encourage the development of low-cost, practically feasible, and simple technologies that are relatively easy to use and locally implementable that will help farmers to safely store their products for a longer time.

30.2.1 Problem Identification

Underutilized fruit crops such as wood apple, khirni, karonda, ber, aonla, etc. growing in the local areas can play a major role not only in partially solving the problem of malnutrition but will also help in augmenting the meager farm income in backward areas. Some work with chemical preservative treatment and storage under cold condition has been done; however, no work on value addition without cold storage and use of chemical preservatives has been reported. Underutilized fruits are perishable in nature and if not harvested timely are lost. Apart from this, being low value fruits, its marketability is also difficult leading to higher losses as compared to major fruits. Processing of these fruits is practically negligible. It is therefore of utmost importance to process these naturally occurring fruits to meet the nutritional requirement of the people of the areas so that they get supplemental nutrition in their daily diet throughout the year and earn additional income (Table 30.1).

30.3 Products

30.3.1 Juice

Pilot-scale studies on production of ready-to-serve drink from red-pulped guava revealed that

Table 30.1 Important underutilized fruits and value-added products that can be prepared from them

Name of fruit	Scientific name	Value-added products
Ber	*Zizyphus mauritiana* Lamk.	Candy, preserve, RTS, chips (dried) dehydrated ber, squash, fruit bar
Aonla	*Emblica officinalis* Gaertn.	Preserve, candy, syrup, squash, RTS, pickle (oilless), chutney, dried shreds, triphala, chyawanprash, fruit bar (pure and mix), seed oil, cider
Bael	*Aegle marmelos* Corr.	Preserve, nectar, squash, cider, canned bael slices
Karonda	*Carissa carandas* L.	Pickle, candy, jelly, jam, preserve, wine
Wood apple	*Feronia limonia* Corr.	Squash, powder, fruit bar, chutney, jelly
Date palm	*Phoenix dactylifera* L.	Biscuits, fruit bar, RTS, squash, syrup, and wine
Ker	*Capparis decidua*	Pickle, culinary preparation, powder, chutney
Lasora	*Cordia myxa*	Pickle, vegetable
Khejri	*Prosopis cineraria* L.	Vegetable, biscuits
Cactus pear	*Opuntia ficus-indica* L.	Vegetable, fruit bar, salad
Custard apple	*Annona squamosa* L.	Ice cream, basundi
Fig	*Ficus carica*	Dried figs, fruit bar
Mulberry		Juice, squash, syrup
Khirni	*Mimusops hexandra*	Dehydrated fruits, fruit bar
Mahua	*Madhuca indica*	Dried flowers, seed oil, wine
Chironji	*Buchanania lanzan*	Dried kernels, fruit bar
Phalsa	*Grewia asiatica*	Juice, squash, syrup
Palmyra palm	*Borassus flabellifer*	Kernels
Pomegranate	*Punica granatum* L.	Juice, squash, syrup, carbonated drink, powder, anardana, wine
Jamun	*Eugenia jambolana*	Juice, RTS, squash, syrup, carbonated drink

in lye peeling, pulp yield was more (73.68 %) than in hand peeling (58.68 %). From 100 kg guava, 247 L of RTS could be prepared with a benefit-cost ratio of 1:1.79 and Rs 5.45/kg (Bhuvaneshwari and Tiwari 2007). Fruit preparation of clarified juice includes pulping, treating the pulp with enzymatic preparation, filtrating, and packing. The yield of juice ranges from 60 to 80 %. At room temperature, a reaction time of 16 to 18 h. is recommended. The Brix acid ratio of juice is adjusted, filtered, and heated to 74 °C for a few seconds to inactivate the enzyme (Sethi et al. 2005).

Extracting, clarifying, and concentrating the juice up to 70° Brix using forced-circulation evaporation, centrifugation, and addition of SO_2 and ascorbic acid are found to reduce the nonenzymatic browning. Pure and clear fruit juice is prepared by enzyme treatment (Kadam 2001). Yeast fermentation in phalsa juice was effectively controlled with 500 ppm sodium benzoate (Anand 1980).

In grapes, Khalil (1990) observed that the juice is preserved using chemical preservative and stored at room temperature. An enzymatic treatment has been developed to hasten the process of argol precipitation to obtain clear juice, which is treated with 0.04 % ultrazyme 100 (peak enzymes) at 45 °C, and cleared juice is filtered, which can be either bottled after adjusting Brix acid ratio or used for concentration. Rokhade et al. (2006) reported that RTS beverage from Jamun was prepared with 14 % juice and 0.15 % citric acid and by adjusting the TSS to 14° Brix with sugar. The mixture was found to be nutritious and had least microbial load. They also standardized the protocol for preparation of ber juice and found that treatment of pulp with pectinase II enzyme at 6 g/kg pulp increased juice recovery to 56.37 % compared to 30.85 in control resulting in very good scores.

30.3.2 Preserves

They form in about 4.5 % of the total fruit and vegetable products. Commonly used fruits for making murabba are apple, amla, harhad, bael, bihi, karonda, and pineapple. Sethi and Anand (1993) made various products from jackfruit. To retain better texture in the preserve, steep apple overnight in firming agents like 3 % calcium nitrate or 1 % calcium chloride solution containing 0.1 % KMS before blanching and steep in sugar syrup.

Aonla preserve prepared by using the optimum maturity fruits can be kept longer and has better organoleptic qualities (Jain et al. 1984). Karonda fruits are dyed red before it is made into preserve (Sethi and Anand 1977). The industry treats fruits directly into sugar than in syrup to reduce labor costs. Sethi et al. (1980) standardized a method to prepare semidry amla preserve segments at 56° Brix by dipping them in a solution containing glucose and sucrose with preservatives like potassium sorbet (0.45 %) and KMS (0.1 %).

30.3.3 Fruit Bar

A simple method was standardized by Singh (2009); the pulp of the fruit was prepared after mixing with strong and suitable preservative sprayed in trays in cross-flow air dryer set at 70 °C. Roy and Singh (1979) dried the pulp in the form of sheet, keeping 10 % moisture after adding 500–2000 ppm SO_2. The sheet was cut into pieces and further dried to 4 % for grinding to powder. Guava fruit slab was prepared by adding 10 % sugar and 1,500 ppm SO_2 to the pulp and drying to 14.5 % moisture (Roy and Singh 1979). Toffee is prepared by mixing 100 parts pulp with 40 parts sugar and 4.5 parts glucose, 10 parts skim milk powder, and 6 parts hydrogenated fat and cooling the mixture to 8.5 % moisture (Table 30.2).

Wood apple fruit bar

30.3.3.1 Flow Chart for the Preparation of Aonla, Wood Apple, Mango, and Green Date Fruit Bar

500 g pulp of ripened fruit → Put in mixer grinder → Take milk powder 50 g dissolved in hot water and add to the pulp → Add 50 g ghee waste (material remaining at the bottom after preparation of ghee) → Add 300 g sugar → Add cinnamon powder 5 g. → Heat for 15 mins → Spread in tray precoated with ghee → Dry in oven for 2–3 days → Cut into pieces → Pack in aluminum foil → Store in ambient condition.

30.3.3.2 Dehydration and Drying

In India, dehydration of fruits is an important method to considerably reduce the spoilage. In the composition of fruit, stage of maturity influences the final dehydrated product. Before dehydration, pre-drying by lye peeling, sulfur fumigation, and chemical treatment have to be done to improve the quality of dehydrated product. The dehydration of fruit can then be done by sun drying, solar dehydration, dehydration in oven, freeze drying, osmotic dehydration, reverse osmosis, foam drying specifically the micro-flake

Table 30.2 Table fruit bar

Product	TSS° Brix	Acidity (%)	pH	Vit. C	Reducing sugar (%)	Total sugar (%)
Sapota bar	88.83	0.70	1.75	11.45	3.87	9.65
Wood apple bar	93.06	1.77	2.86	22.67	19.81	22.43
Date palm bar	101.32	0.19	1.92	12.53	1.56	5.84
Mixed wood apple aonla bar	83.50	2.32	3.13	40.34	9.87	12.05

Hiwale (2012)

process, improved drum drying, etc. (Sethi et al. 2005). Drying has been the most successful method for long-term storage of fruits. Some of the work done in our country on this aspect will help in planning the experiments.

30.3.3.3 Solar Drying of Fruits

Various solar dried products were prepared. Analysis of the fruits indicated that percent dry weight of different products varied with the type of fruit. Maximum drying was recorded in aonla powder 88.38 % and least in khirni 57.5 %. The total soluble solid content was maximum in aonla candy 104.60 ° Brix; it was least in anardana 44.60 ° Brix. The percent total sugar content varied from 30.37 % in sapota to 7.29 % in aonla powder. The reducing sugar content showed similar trend. Data on vitamin C content indicated that it was highest in aonla powder 416.16 mg/100 g. and was least in khirni 11.67 mg/100 g. The pH of different products did not show much variation (5.85 in khirni to 3.04 in aonla powder). The temperature of the solar drier was higher throughout the day than in open atmosphere (Tables 30.3 and 30.4).

Table 30.3 Temperature of solar drier

Time	Open	Solar drier
10 am	23	34
12 pm	27	45
4 pm	27	30

30.3.4 Aonla Candy

Take 5 kg of fruits→Wash and keep in deep freeze for 2 days→Remove and bring to room temperature and dip in normal water→Fruit press to remove the seed and skin and separate the pieces→Add 175 g salt – keep it overnight→Wash pieces to remove extra salt and then dip in water→Add sugar, 50 % of fruit weight→Keep it in the shade and allow it to absorb→Sugar for 3 days, if required add more sugar→Allow it to dry on a newspaper for another 2 days. After 2 days, final aonla candies are developed.

30.3.5 Aonla Shreds

The analysis of data of aonla shreds treated with different salt concentrations ranging from 2 to 30 % and stored at ambient conditions up to 2 years indicated that solar drying of aonla shreds after treating them with salt resulted in increased percent dry weight as the salt concentration increased. It increased from 13.6 % in 2 % salt treated shreds to 32 % in 30 % salt treatment. Ascorbic acid content was found to be significantly influenced by the application of salt. It showed a decreasing trend with increased concentration of salt (3,072 to 510 mg/100 g). After 1 year of storage under ambient condition, ascorbic acid content decreased drastically from 288.17 mg/100 g in 2 % salt-treated shreds to

Table 30.4 Solar drying of fruits

Product	Fresh wt. g.	Dry wt. g.	Moisture (%)	TSS ° Brix	Acidity (%)	Reducing sugar (%)	Total sugar (%)	Vitamin C mg/100 g	pH
Wood apple powder	960.00	204.00	21.25	56.71	14.82	8.06	9.51	95.26	4.46
Aonla powder	1600.0	215.00	11.62	96.10	14.34	5.25	7.29	416.6	3.04
Aonla candy (salt treated)	800.00	315.00	39.37	104.6	0.99	8.00	13.76	93.34	3.25
Khirni	447.00	190.00	42.50	56.20	0.79	14.28	20.78	11.67	5.45
Sapota	600.00	108.00	18.00	52.80	0.48	16.24	30.37	12.56	5.85
Anardana	230.00	66.00	28.69	44.60	4.48	11.65	18.28	25.15	4.25

Table 30.5 Aonla shreds (two-year storage)

Treatment	Dry wt.%	Vitamin C mg/100 g			% Acidity			TSS ° Brix		
		1/11	3/12	1/13	1/11	3/12	1/13	1/11	3/12	1/13
2 % salt	13.6	3098	285.7	25.4	2.4	1.2	0.77	31.5	26.4	19.5
3 % salt	13.93	2,775	257.2	27.7	1.9	1.1	0.82	44.1	32.6	24.7
5 % salt	15.2	975	142.9	22.7	1.5	1.1	0.88	45.3	33.5	17.7
10 % salt	17.6	675	114.3	44.0	1.2	1.3	0.86	46.2	36.2	26.3
20 % salt	24.0	525	104.7	31.4	1.5	1.4	0.60	42.0	29.4	22.0
30 % salt	32.0	480	114.3	24.6	1.3	0.9	0.46	67.2	36.8	23.9
Control	25.0	1,275	85.7	18.5	0.9	0.7	0.51	12.6	12.2	11.4
CD 5 %	3.67	166.8	35.13	3.61	0.25	0.38	0.11	1.54	1.83	2.27

94.13 g/100 g in 30 % salt-treated shreds indicating that as the concentration of salt increased, ascorbic acid content of aonla shreds decreased. Aonla shreds treated with 2 % salt helped in higher retention of ascorbic acid and acidity. Whereas shreds treated with 10, 20, and 30 % salt retained better color and appearance, higher retention of vitamin C (44 mg/100 g) in treatment 10 % salt. Percent acidity and TSS showed similar results (0.86 % and 26.3 ° Brix). Thus, treating aonla shreds with salt prevents browning and the shreds can be stored up to 2 years with the highest vitamin C storage (44 mg/100 g pulp) (Tables 30.5 and 30.6).

Table 30.6 Aonla shreds (two-year storage)

Treat	Total sugar (%)		Reducing sugar (%)	
	3/12	1/ 13	3/12	1/13
2 % salt	21.05	16.78	5.0	4.7
3 % salt	24.60	17.82	5.5	4.9
5 % salt	25.64	19.54	6.3	5.2
10 % salt	35.86	21.12	7.8	5.4
20 % salt	33.20	20.53	7.4	5.3
30 % salt	38.41	27.29	8.6	6.1
Control	20.28	19.72	3.7	2.9

Aonla shreds

Aonla oil less pickle

Table 30.7 Aonla oilless pickle

Treatment	Aonla pieces (g)	Water extracted (g)	Wt. of pieces (g)	Final product (g)	Acidity (%) 10/12	Acidity (%) 2/13	TSS 10/12	TSS 2/13	Ascorbic acid mg/100 g 10/12	Ascorbic acid mg/100 g 2/13
10 % salt	500	43	541	621	1.33	0.22	21.0	12.12	137.25	40.95
20 % salt	500	117	537	617	1.31	0.26	24.25	12.75	136.5	35.45
30 % salt	500	147	497	577	1.17	0.34	29.50	16.50	129.0	29.86
40 % salt	500	153	481	561	1.11	0.36	35.25	24.25	122.87	23.07
50 % salt	500	163	475	555	0.97	0.38	38.75	26.62	121.25	21.77
CD 5 %	–	–	–	–	0.09	0.034	3.31	2.32	8.44	2.44

30.3.5.1 Flow Chart for the Preparation of Aonla Shreds

Take mature selected fruits → Wash the fruits in running water → Make shreds with shredder → Remove seeds and separate the shreds → Add the required salt concentration and mix thoroughly → Remove the water oozed out of the shreds → Dry the shreds in the sun on newspaper → Weigh the final produce till constant weight arrives → Store in polythene bags in ambient condition → The shreds can be used to make aonla powder.

30.3.6 Oilless Pickle

Oilless aonla pickle with 10–50 % salt was prepared and stored under ambient condition; as the salt concentration increased, the amount water extracted also increased. The final product decreased as the amount of salt increased. The pickle prepared in 10 and 20 % salt spoiled after 25 and 30 days of storage. However, pickle with 30, 40, and 50 % salt did not spoil under ambient condition without oil. Thus, oilless pickle with aonla pieces when treated with 30 % salt was found to be the best treatment as it has the lowest conc. of salt. Organoleptic tests also indicated that preparation of oilless pickle with 30 % salt was the tastiest of the treatments. Analysis of the pickle indicated that percent acidity decreased as the concentration of salt increased. However, in respect to the total soluble solids, it was reverse. The ascorbic acid content of the fruit decreased with increased concentration of salt, both initially and 6 months after storage. Acidity, TSS, and ascorbic acid content decreased considerably during the storage period. Aonla pieces with 10 % salt retained higher ascorbic acid (40.95 mg/100 g pulp) and kept their quality up to 1 year under ambient conditions (Table 30.7).

30.3.6.1 Flow Chart for the Preparation of Aonla Pickle

Take mature selected fruits → Wash the fruits with running water → Boil them in water for 15 min → Remove seeds and separate the species and dry pieces on paper → Add the required salt → Soak the fruits in salt for 12 h → Remove the water oozed out of fruits → The extra salt on fruits was removed by dipping them in hot water → Put groundnut oil in a pan, when become hot → Add methi to it → After a while, add the masala mixture prepared → Add pieces of fruit to it → Mix thoroughly till the masala sticks to the fruits → Cool and fill in the bottles → Keep the lid of the bottle open → Mix once a day → Store at room temperature.

30.3.7 Aonla Squash

The aonla fruit squash with different concentrations of juice from 25 % t, 50 %, and 75 % and its ambient storage without application of chemical preservative were studied. There was reduction in ascorbic acid content to the tune of 85 % after 1 year of storage, which further declined to 90 % after 2 years of storage. However, reduction in TSS was at slower pace, which was just 6.62 % in squash, where 25 % juice was used after 1 year of storage and 20.30 % after 2 years of storage. Maximum reduction in TSS was observed in squash where 75 % juice was used. The reduction

Table 30.8 Different juice sugar combinations and product composition

	Ascorbic acid mg/100 g			TSS °Brix						Sugars (%)	
Treatment	12/11	9/12	1/13	12/11	9/12	1/13	Ca (%)	Mg (%)	K (%)	Reducing	Total
25 % juice +75 % sugar syrup	366.8	53.0 (85.5)	31.2 (91.5)	78.8	72.0 (8.62)	62.8 (20.3)	0.05	0.15	0.064	30.2	53.0
50 % juice +50 % sugar syrup	445.4	64.0 (85.6)	40.2 (90.9)	63.4	37.4 (41.0)	34.0 (46.3)	0.14	0.04	0.14	28.6	48.6
75 % juice +25 % sugar syrup	510.2	81.4 (84.0)	43.6 (91.4)	53. 0	22.8 (56.9)	17.8 (66.4)	0.15	0.05	0.122	23.2	45.2
CD 5 %	36.53	2.90	3.79	3.78	3.63	3.89	0.02	0.02	0.017	4.37	2.93

Hiwale (2013)

Table 30.9 Fruit juice concentrates (solar drying)

									Vit. C mg/100 g	
Juice	Juice fresh (ml)	Solar dried juice (ml)	TSS fresh ° Brix	TSS after 4 months	Acidity % fresh	Acidity % after 4 months	Total sugar %	Red sugar %	Fresh	4 months
Aonla	250	35	56	30	18.5	6.16	94.29	29.2	2,725	94.29
Aonla	250	60	28	18	12.71	4.75	73.32	22.76	2,175	78.90
Aonla	250	100	15	12	8.39	3.39	52.38	18.8	1,125	52.38
Aonla	250	21.83	60	32.6	20.79	7.45	98.7	31.7	2,785	110.6
Pure juice			8	–	2.31	–	8.34	5.86	531	–

was 56.98 % in the first year and 66.41 % in the second year of storage. Mineral composition of squash indicated that Ca content after 1 year of storage was 0.05 % in 25 % juice to 0.15 % in 75 % juice. Mg content was reverse. Potassium content was 0.06 to 0.12 %. Organoleptic analysis revealed that still the squash is not spoiled and can be used for RTS preparation even after 2 years of storage (Table 30.8).

30.3.8 Fruit Juice Concentrates (Solar Drying)

Fruit juice concentrate was prepared by solar drying of fresh fruit juice for 7–18 days to 75 %, 50 %, and 25 % concentration, and its composition was recorded after 4 months of storage. There is reduction in TSS, acidity, and total and reducing sugars as well as vitamin C content. The juice could be stored up to 6 months under refrigerated storage; thereafter, it lost its organoleptic quality (Table 30.9).

Application of membrane technology for fruit juice concentration was studied by Tondon (2002). It was found that due to thermal evaporation, there is loss of aroma and flavor. Fruit juice could be clarified by ultrafiltration and reverse osmosis to maintain their natural color flavor and aroma. Orange juice can be powered by puff drying, freezing, or foam met drying (Singh 2009). The flavor is invariably added to make up the lost flavor. Similarly, in juice powdered by thin film evaporation method concentrating it to 41° Brix, ascorbic acid decreased from 2 to 1.75 mg in the final concentration indicating 15 % loss. Adding lime oil reinforced the flavor of the concentrate. Grape juice can be concentrated to 62° Brix. Enzyme-clarified juice however can be concentrated to 72° Brix under vacuum in a forced circulation evaporator and can be kept for 6 months at –18 °C (Khalil 1990). Sharma et al. (2001) observed decline in the overall acceptability of ready to serve drink prepared from galgal juice concentrate stored for 6 months at room temperature. Mango pulp is concentrated by two methods: in the first method, the pulp is concentrated in a forced circulation evaporator 30 ° Brix which facilitates easy removal of the concentrate from the evaporator (Ramteke 1987). In the second method (split process), mango pulp is treated with pectic enzyme. Serum is separated from fibrous pulp residue to obtain 40° Brix concentrate (Gupta and Girish 1988).

30.3.8.1 Flow Chart of Aonla Juice Concentrates

Select mature fruits → Wash them – make it to shreds → Extract juice by squeezing → Allow it to settle down → Decant the clear juice and remove the settlement → Filter the juice through double-layer muslin cloth → Concentrate the juice to different concentrations (75 %, 50 %, 25 %) and dry it in solar driers (11,9,7 days) → Keep in it the refrigerator.

30.4 Utilization of Fruit Waste

A large quantity of inedible parts of fruits is discarded by the consumer and fruit processing industry. The amount of these wastes is as high as 70 % in apple. In mango, it ranges from 27 to 50 %, grapes 5–10 %, citrus 50 %, and banana 24–46 % (Murti et al. 1976). Various produces like starch, fat, pectin, vinegar, bear, etc. can be prepared from these wastes. A number of products like starch, fat, pectin, vinegar, beer, and oil can be prepared from the waste (Maini and Anand 1993).

30.5 Export and Marketing of Agricultural Commodities

The export of fresh and preserved fruits from the country is very low, except for some fruits like mango, banana, cashew, grapes, and pomegranates. The total fresh fruit export stands at 2,646 thousand tons with a value of 365 thousand rupees in 2008–2009, whereas the processed fruit and vegetables export stands at 708 thousand tons which is worth Rs. 262 thousand in 2008–2009. And it was found to remain static in 2009–2010 and 2010–2011 (Table 30.10).

30.6 Extent of Value Addition

Data on the extent of value addition in different countries indicated that though India is the second largest producer of fruits in the world, hardly

Table 30.10 Export of horticulture produce from India

Products	2008–2009		2009–2010		2010–2011	
	Qty Mt	Value Rs.	Qty Mt.	Value Rs.	Qty Mt.	Value Rs.
Fresh fruits and vegetables						
Fresh onions	1,670,186.29	182,752.21	1,664,922.39	231,942.98	1,163,472.58	174,155.41
Other fresh vegetables	505,285.46	68,020.32	419,241.35	73,185.9	490,914.05	89,293.61
Walnuts	5,696.34	14,123.63	9,073.38	19,789.51	5,244.58	15,650.59
Fresh mangoes	83,703.18	17,071.25	74,460.61	20,053.98	59,220.77	16,292.13
Fresh grapes	124,627.97	40,861.28	131,153.61	54,533.89	99,311.83	41,206.32
Other fresh fruits	256,768.53	43,086.84	260,675.43	52,283.32	253,850.99	48,964.74
Total	264,6267.77	365,915.53	2,559,526.77	451,789.58	2,072,014.8	385,562.8
Processed fruits and vegetables						
Dried and preserved vegetables	147,861.22	49,641.51	124,613.5	53,207.48	110,173.91	51,697.09
Mango pulp	173,013.6	75,298.9	186,197.85	74,460.77	171,929.43	81,400.66
Other processed fruits and vegetables	387,126.42	137,179	397,978.17	143,550.63	340,067.97	131,635.53
Total	708,001.24	262,119.41	708,789.52	271,218.88	622,171.31	264,733.28

Bijay Kumar et al. 2011

2 % of the produce is being used for value addition, whereas small countries like Thailand, Israel, and Malaysia processed 50 % of the produce. There is therefore scope for the Indian fruit industry to increase value addition and earn foreign exchange (Tables 30.11 and 30.12).

30.6.1 New Postharvest Opportunities

Value addition, product development, and market development.

- Alleviate poverty in rural areas through additional income generation and employment creation.
- Identify and develop opportunities for diversification and add values to on- and off-farm activities and integrate them into supply chain.
- There is need to create market-oriented rural production and value addition enterprise that are well developed with required service providers.
- There is need to encourage the development of low-cost, practically feasible, and simple technologies that are relatively easy to use and locally implementable that will help farmers to safely store their products for a longer time.

30.6.2 Broad Fiscal Initiatives

- Zero excise duty on processed products based on fruits and vegetable, dairy, meat/fish/poultry. Food ingredients are pectin, yeast, etc.
- Excise duty reduces to 8 % on ready-to-eat food and instant food mixes.
- Custom duty on packaging machine reduces from 15 to 5 %.
- Income tax rebate on 100 % of profit for 5 years and 25 % for the next 5 years for new fruit and vegetable units.

30.6.3 Entrepreneurship Development

Twenty-nine food processing training centers (FPTCs) were set up to train rural entrepreneurs in different parts of the country. 470 entrepreneurial development programs (EDPs) were launched to train potential entrepreneurs. There is marketing assistance to farmers to ensure returns for their produce.

Table 30.11 Extent of value addition

Country	Extent of value addition (%)
USA	83
Brazil	70
Malaysia	60–70
France	50
Thailand	50
Israel	50
New Zealand	40
China	10
India	<2

Table 30.12 Strategies for increasing value addition

Stage of development	Gaps	Action to be taken
Farmer's enterprises are fully integrated into supply chains producing products that meet market demand in terms of quality and frequency of supply	Food safety implementation, testing, certification	Research to respond to private sector requirement
	Acceptance of private sector role	
	Developing processing varieties	
Developing commercially oriented enterprises adds value and diversifies the product	Market assessment	Contract farming
	Supply chain integration	Forecast of consumer demand future market Incentives for new product development
	Intellectual property regime	Monitoring and analysis impact of policies

30.6.4 Krushak Bazaar

The establishment of Krushak Bazaar has an aim to train farmers, create primary rural markets, and launch awareness campaigns.

30.6.5 Terminal Market Complexes (TMCs)

Set up three terminal market complexes under the PPP mode in order to provide a fair share of consumer price to the producer as well as ensure high quality and hygiene of the produce. The TMC would operate on a hub-and-spoke format, wherein the terminal market (the hub) would be linked to a number of collection centers. The TMCs would enable farmers get proper prices for their surplus produce sold at market yards. TMCs would also offer sufficient facilities for cleaning and drying, grading, weighing, and bagging, on all market yards/submarket yards/temporary procurement centers engaged in procurement.

30.6.6 Market Yards

District market yards have to be established to facilitate the selling and procurement of crop products and ensure farmers can get a reasonable price for their produce. The minimum support price (MSP) mechanism is to be implemented effectively across the country so as to ensure remunerative prices for the farm produce. Agri-export zones (AEZs) would be established in PPP mode for agricultural and horticultural produce having export potential.

30.6.7 Government Recent Efforts

1. Horticultural train which links production clusters in the country and provides integrated cold chain solution
2. Specially designed freight containers (with ISO certification by international agency) with food grade thermal insulation on walls and ceilings from the inside. Ventilation holes in floor and top rails and three-door loading and unloading system

References

Anand JC (1980) Efficacy sodium benzoate to control yeast fermentation in phalsa juice. Indian J Hortic 17:138–140

Anonymous (2004) Expert consultation on post-harvest technologies for ensuring food security and value addition for enhanced income, Thailand, 1–3 December,2004

Bhuvaneshwari S, Tiwari RS (2007) Pilot scale processing of red flesh guava RTS beverage. J Hort Sci 2(1):50–52

Gupta TR, Girish K (1988) Studies on mango pulp filtration for concentration by pectinolytic enzyme. Indian Food Pack 42(6):7–14

Hiwale SS (2012) Preparation and storage of fruit bar under ambient conditions, Annual report. CIAH, Bikaner

Hiwale SS (2013) Preparation and storage of aonla shreds, *oil* less pickle and squash under ambient conditions, Annual report. CIAH, Bikaner

Jain SP, Tripathi VK, Ram HB, Singh S (1984) Effect of storage conditions on the keeping quality of fruit squashes. Ind FD Packer 28:33

Kadam SS (2001) New products from arid and semiarid fruits. Indian J Hortic 58(1&2):170–177

Khalil A (1990) Studies on preparation of fruit juice concentrate. Ph.D (Food Sci.) thesis, CFTRI, University of Mysore

Kumar B, Mistry NC, Singh B, Gandhi CP (2011) Indian horticulture data base pp 212

Maini SB, Anand JC (1993) Utilization of fruit wastes. In: Chadha KL, Pareek OP (eds) Advances in horticulture, vol 4. Malhotra Publication, New Delhi, pp 1967–2023

Murti IAS, Malathi HN, Rajlaxmi D, Bhattacharya SC (1976) Industrial wastes and by products of food processing industries in India – A survey. Indian Food Pack 30: Indian Food Pack 31(2):18–21

Ramteke RS (1987) Characterization and storage behavior of annona constituents. Ph.D thesis submitted to Mysore University, Mysore

Rokhade AK, Samira N, Sardespande M (2006) Value addition in Jamun fruit. Karnataka J Agric Sci 19(4):1024–1025

Roy SK, Singh RN (1979) studies on utilization of bael fruit for processing. II. Extraction of bael fruit pulp. Indian Food Pack 33(1):5–7

Sethi V, Anand JC (1977) Drying Karonda for making their preserve. Indian Food Pack 31:20–23

Sethi V, Anand JC (1993) Retention of nutrients in aonla. Indian Food Pack 39(4):42–48
Sethi V, Anand JC, Saxena SK (1980) Kinnow orange in juice and beverage making. Indian Hortic 25:13–14
Sethi V, Sethi S, Deka BC, Meena YR (2005) Processing of fruits and vegetables for value addition. Indus Publishing, New Delhi, pp 35–48
Sharma PC, Sharma SK, Kaushal BB (2001) Preparation and evaluation of some value added products from lemon (Citrus pseudo lemon tan.) fruits. Indian J Agric Sci 71:691–694
Singh IS (2009) Post harvest handling and processing of fruits and vegetables. Westville Publishing House, New Delhi, pp 95–109
Singh DK, Singh SK (2005) Physiology and post harvest management of horticultural crops. In: Salunkhe DK (ed) Post harvest biotechnology of fruits, vol I. CRC Press, Boca Rattan
Tondon DK (2002) Application of membrane technology for fruit juice concentration. In: National seminar cum workshop on Hi tech Horticulture and Precision farming Souvenir, pp 75–77

Glossary

Acclimatized Adapted to a environment.

Actionomorphic Having radially symmetric shape usually refers to the petals of flower.

Acuminate The shape of a tip or base of a leaf or perianth segment where the part tapers gradually and usually in a concave manner.

Adaptation It is the process by which organisms become more suited to survive and function in a given environment. It also refers to the result of this process.

Adnate Joined to or attached to.

Agri-Horti A cropping system where agronomic and fruit species are grown together.

Agri Silvi A cropping system where agronomic and sivicultural species are grown together.

Agroforestry A cropping system where agriculture and forestry species are grown together.

Air layering A method of vegetative propagation where a small part of bark is removed and moist moss grass is tied o initiate rooting.

Albumen Starchy and other nutritive material in a seed, stored as endosperm inside the embryo sac, or as periderm in the surrounding nucellar cells; any deposit of nutritive material accompanying the embryo.

Androceium All the male reproductive organs of a flower; the stamens.

Angiosperm A plant producing seed enclosed in an ovary. A flowering plant.

Annual A plant that completes its life cycle from germination to death within one year.

Anterior Front; on the front side; away from the axis.

Anther The pollen: bearing (terminal) part of the male organs (Stamen), borne at the top of a stalk (filament).

Anthesis Flower bud opening; strictly, the time of expansion of a flower when pollination takes place, but often used to designate the flowering period; the act of flower bud opening.

Apex The tip of an organ, the growing point.

Apical Pertaining to the apex.

Apiculate Having a short point at the tip.

Aril (arillus) A fleshy or sometimes hairy appendage or outer covering of a seed.

Arillate Provided with an aril.

Auricle Small ear: like projections at the base of a leaf or leaf: blade or bract.

Axil The upper angle formed by the union of a leaf with the stem.

Axillary Pertaining to the organs in the axil, e.g. buds, flowers, inflorescence.

Axis The main or central stem of herbaceous plant or of an inflorescence.

Basal Borne on or near the base.

Berry Type of fruit.

Bilabiate Two lipped.

Biomass Biological material derived from living organisms.

Bipinnate A pinnate leaf with primary leaflets themselves divided in a pinnate manner; of pinnate.

Biseriate In two rows.

Bisexual Having both sexes present and functional in one flower.

Blade The flattened part of a leaf; the lamina.

Bract A much: reduced leaf, particularly the small or scale: like leaves in a flower cluster or associated with the flowers; morphologically a foliar organ.

Bracteole A secondary bract; a bractlet.

Caduceus Falling off early, or prematurely, as the sepals in some plants.

Calyx The outer whorl of floral envelopes composed of the sepals.

S. Hiwale, *Sustainable Horticulture in Semiarid Dry Lands*,
DOI 10.1007/978-81-322-2244-6, © Springer India 2015

Carbon sequestration Accumulation of carbon through growing of plants.

Carinate Keeled; provided with a reproductive organs, comprising an ovary and a stigma, and containing one or more ovules.

Carpel One of the flowers' female reproductive organs, comprising an ovary and a stigma, and containing one or more ovules.

Character A morphological, anatomical or physiological feature of an organism; usually a product of the actions of both genotype and environment.

Chromosome Structure of DNA, protein and RNA found in cells.

Clonal selection Selection of a superior type from the clones.

Clone A group of plants that have arisen by vegetative reproduction from a single parent, and which therefore all have identical genetic material.

Confluent Merging or blending together.

Connate United or joined as one body or organ.

Convex Arched outward.

Cordate Heart: shaped, often restricted to the basal portion rather to the outline of the entire organ.

Coriaceous Of leathery texture.

Cotyledon Seed leaf; the primary leaf or leaves in the embryo.

Cross pollination The transfer of pollen from the anther of the flower of one plant to the flowers of a different plant.

Crustaceous Of hard and brittle texture.

Cultivar A race or variety of a plant that has been created or selected intentionally and maintained through cultivation.

Cutting A piece of matured stem used for vegetative propagation.

Cyme A broad, more or less flat: topped, determinate flower cluster, with central flowers opening first.

Cymose Inflorescence showing the cyme arrangement.

Deciduous Falling at the end of one season of growth or life, as the le of non: evergreen trees.

Decoction Herbal preparation made by boiling a plant part in water.

Degree day A unit that represents one degree of difference from a given point (as 65°) in the mean daily outdoor temperature and that is used especially to measure heat requirements.

Dehiscence The method or process of opening a seedpod or anther.

Derived Originating from an earlier form or group.

Dichotomous Forked, in 1: or 2 pairs.

Diocious Having male (staminate) and female (pistillate) flowers on different plants.

Diploid Having two sets of chromosomes.

Distichously Two ranked, with leaves, leaflets or flowers on opposite sides of stem and in the same plane.

Double cross Crossing of two single crosses one male sterile and the other male fertile

Downy Covered with short and weak soft hairs.

Drupaceous A fruit showing the characteristics of a drupe.

Drupe Seeded indehiscent fruit with seed enclosed in a stony endocarp; stone fruit.

Elliptic Oval in outline.

Endocarp The inner layer of the pericarp or fruit wall.

Endosperm The starch and oil: containing tissue of many seeds.

Entomophilous Insect pollinated.

Environment Sum total of external conditions, which influence an organism.

Epigynous Borne on or arising from the ovary; used of floral parts when the ovary is inferior and flower not perigynous.

Evapotranspiration Sum of evaporation and plant transpiration from the earth and ocean surface to the atmosphere.

Exalbuminous Without albumen.

Exocarp The outer layer of the pericarp or fruit wall.

Falcate Scythe: shaped; curved and flat, tapering gradually.

Family A group of individuals sharing a common parent/ancestor.

Fascicle A condensed or close cluster.

Fertilization Union of male and female gametes in sexual reproduction.

Filament Thread; particularly the stalk of the stamen, terminated by the anther.

Filiform Thread: shaped, long, slender and tenet.

Fulvous Dull, brownish: yellow.

Genotype The genetic constitution of an organism, acquired from its parents and available for transmission to its offspring.

Genus A group of related species, the taxonomic category ranking above a species and below a family.

Germplasm The sum total of hereditary material or genes present in a species.

Glabrous Not hairy.

Globose Globe: shaped.

Gynoecium All the female parts of a flower.

Heterogeneous A mixture of different types, usually different genotypes.

Heterozygous Having dissimilar pairs of genes foe any hereditary characteristic.

Homonym A scientific name given two or more times to plants of the same taxonomic rank but which are quite distinct from each other.

Hormone A substance occurring naturally in plants and modifies plant processes.

Hybrid Progeny from hybridization between two or more strains.

Hypocotyl The axis of an embryo below the cotyledons which on seed germination develops into the radical.

Indehiscent Not regularly opening, as a seedpod or anther.

Indigenous Native and original to the region.

Inflorescence The flowering part of a plant and especially the mode of its arrangement.

Insect Pest An insect species, which causes damage to a crop species; it may be a tissue feeder or sucking pest.

Isolation Separation of two or more plants, strains or populations to prevent mating among them. Usually achieved by distance and/or border rows.

IU (international unit) A unit used to measure the mass of certain vitamins and drugs based on their expected effects. For each substance to which this unit applies, there is an international agreement specifying the biological effect expected with a dose of 1 IU. Other quantities of the substance are then expressed as multiples of this standard example: 1 IU represents 45.5 micrograms of a standard preparation of insulin or 0.6 microgram of a standard preparation of penicillin. Consumers most often see IU's on the labels of vitamin packages: the equivalent of 1 IU is 0.3 microgram (0.0003 mg) for vitamin A, 50 micrograms (.05 mg) for vitamin C, 25 nanograms (0.000025 mg) for vitamin D 2/3 mg for vitamin E.

Lamellae A thin, flat plate or laterally flattened ridge.

Lanceolate Shaped like a lance head, several times longer than wide, broadcast above the base and narrowed toward the apex.

Lateral Side shoot, bud etc.

Locular Having a cavity or chamber inside the ovary, anther or fruit.

Mean It is the arithmetic average of a set of observations

Membranous Thin in texture, soft and pliable.

Mesocarp The fleshy middle portion of the wall of a succulent fruit between the skin and stony layer.

Mutation One of the methods of plant breeding by use of chemicals or irradiation.

Naturalized To cause a plant to become established and grow undistributed as if native.

Necrotic Death of cells or tissues through injury or disease.

Nectar Sweet secretion of glands in many kinds of flower.

Nectiferous Producing nectar.

Obconical Inversely conical, having the attachment at the apex.

Oblique Slanting, unequal sided.

Obovate Inverted ovate; egg: shaped, with the broadest part above.

Obtuse Blunt or rounded at the end.

Octaploids Having 8 times the basic number of chromosomes.

Orbicular Circular.

Ovary inferior With the flower: parts growing from above the ovary.

Ovary superior With the flower: parts growing from below the ovary.

Ovate Egg: shaped, with the broader end at the base.

Ovule The body which after fertilization becomes the seed.

Panicle A loose irregularly compound inflorescence with pedicle late flowers.

Paniculate Borne in a panicle.

Parameter A numerical quantity, which describes some characteristic of a population.

Paripinnate A pinnate (compound) leaf with all leaflets in pairs.

Pedicel A tiny stalk; the support of a single flower.

Pendulous More or less hanging or declined.

Perianth The floral envelope consisting of the calyx and corolla.

Pericycle The tissue of stele lying just inside the endodermis.

Perigynous Adnate to the perianth, and therefore around the ovary and not at its base.

Petal A division of the corolla; one of a circle of modifies leave immediately outside the reproductive organs, usually brightly colored.

Petiole The stalk of a leaf that attaches it to the stem.

Phenotype The morphological, physiological, behavioral, and other outwardly recognizable forms of an organism that develop through the interaction of genes and environment.

Pilose Hairy, especially with soft hairs.

Pinnate A compound leaf consisting of several leaflets arranged on each side of a common petiole.

Pollination It consists of pollen grains reaching the stigma of a flower.

Polygamous Bearing male and female flowers on the same plant.

Polyphyletic Having members that originated, independently, from more than one evolutionary line.

Polyploidy Having more than two sets of chromosome.

Prolate Having flattened sides due to lengthwise elongation.

Propagate To produce new plants, either by vegetative means involving the rooting or grafting of pieces of a plant, or sexually by sowing seeds.

Protandrous Refers to a flower where the shedding of the pollen occurs before the stigma is receptive.

Psilate Referring to a pollen grain having a smooth surface.

Pubescent Covered with hairs, especially short, sort and down like.

Pulvinus A swelling at the base of a leaf or leaflet.

Pyriform Pear shaped.

Qualitative Character Character showing distinct classes and little or no effect of the environment; governed by oligone(s).

Quantitative Character Character showing continuous variation and considerable effect of the environment.

Raceme A simple inflorescence of pediceled flowers upon a common more or less elongated.

Rachis The main stalk of a flower cluster or the main leafstalk of a compound leaf.

Ramification Branching.

Reticulate In the form of a network, net veined.

Retuse With a shallow notch at a rounded apex.

Rootstock The root system and lower portion of a woody plant to which a graft of a more desirable plant is attached.

Rotundate Nearly circular; orbicular to oblong.

Rufous Reddish: brown.

Rugose Wrinkled.

Rugulose Covered with minute wrinkles.

Sample A set of random observations taken from a population.

Scandent Climbing but not self: supporting.

Scarify To scar or nick the seed coat to enhance germination.

Sclerenchymatous Tissue composed of cells with thickened and hardened walls.

Scurfy Covered with tiny, broad scales.

Seed Part of a plant used for raising a seed or commercial crop.

Self pollination The transfer of pollen from the anther of a flower to the stigma of the same flower, or difference flower on the same plant.

Sepal A division of calyx; one of the outermost circles of modified leaves surrounding the reproductive organs of the flower.

Sericeous Bearing fine, usually straight, appressed hairs.

Serrate Having sharp teeth pointing forward.

Serrulate Finely serrate.

Sessile Without a stalk.

Sheath A tubular envelope.

Spathulate Gradually narrowing downward to a summit; spoon shaped.

Stamen One of the male pollen: bearing organs of the flower.

Staminode A sterile stamen, or any structure without anther corresponding to a stamen.

Stigma That part of a pistil through which fertilization by the pollen is effected.

Stipule An appendage at the base of the petiole, often appearing in pairs, one on each side, as in roses.

Style The usually attenuated portion of the pistil connecting the stigma and ovary.

Testa The outer seed coat.

Tetraploid Having four sets of chromosomes (twice the normal number chromosomes).

Tomentose Covered with a thick felt of radicals; densely pubescent with matted wool.

Tomentulose Rather tomentose.

Tomentum Closely matted, wooly hairs.

Transverse Crosswise in position.

Tricoporate Having three apertures in the pollen grain wall.

Tropism The movement of an organism in response to an external source of stimulus, usually toward or away from it.

Truncate Ending abruptly, as if cut off transversely.

Tuberculate Bearing tubercles, covered with warty lumps.

Unguiculate Narrowed, clawed.

Valvate Open by valves.

Vegetative propagation Propagation of plants by asexual means

Vivipery Germination of seed in the fruit itself

Zygomorphic Capable of division by only one plane symmetry.

Printed by Printforce, the Netherlands